HUMAN BIOLOGY

HUMAN BIOLOGY
AN EVOLUTIONARY AND BIOCULTURAL PERSPECTIVE

Edited by

Sara Stinson
Queens College, City University of New York

Barry Bogin
University of Michigan at Dearborn

Rebecca Huss-Ashmore
University of Pennsylvania

Dennis O'Rourke
University of Utah

 **WILEY-LISS**

A JOHN WILEY & SONS, INC. PUBLICATION

New York • Chichester • Weinheim • Brisbane • Singapore • Toronto

For ordering and customer service, call 1-800-CALL-WILEY.

Library of Congress Cataloging-in-Publication Data:

Human biology: an evolutionary and biocultural approach / edited by Sara Stinson . . . [et al.].
    p. cm.
  Includes bibliographical references and index.
  ISBN 0-471-13746-4
  1. Physical anthropology.   2. Human biology.   I. Stinson, Sara.

GN60.H82 2000
599.9—dc21
                                                                99-052180

Printed in the United States of America.

10   9   8   7   6

Contents

Preface

This book is a collaborative effort by members of the Human Biology Association to provide an introduction to the field of human biology. No other book presents the mix of topics covered here, especially with our focus on the evolutionary and biocultural nature of human adaptation and variation. The editors and the chapter authors designed this book to appeal to advanced undergraduate students, graduate students, their professors, and all others with an interest in the human sciences—including anthropology, biology, health sciences, and physiology.

Human biology deals with understanding the extent of human biological variability; explaining the mechanisms that create and pattern this variability; and relating biocultural variability to health, disease, aging, growth and development, demography, genetics, and the social issues derived from these areas of research. Human biology relies heavily on an evolutionary perspective to explain variation through space and time but also considers to be crucial the effect that human cultures have on our biology—a biocultural perspective.

In this book, we cover the major areas of human biology: genetic variation, variation related to climate, infectious and noninfectious diseases, nutrition, growth, aging, and demography. Each chapter is written by an authority or authorities in that field who provide expert coverage of each topic. Boxed text in the chapters explains the methods that human biologists use. Important terms are defined in the glossary, and each glossary term appears in bold type the first time it is used in a chapter. A list of recommended readings at the end of each chapter directs students to sources that will provide a good introduction to the topics covered in the book.

Acknowledgments

We thank all of the reviewers who so generously gave of their time to review the chapters in this volume and the members of the Human Biology Association for their continuing enthusiastic support of this project. Pamela Angulo did the monumental task of copyediting the manuscript and Amie Jackowski Tibble cheerfully saw the book through its production. Special thanks are due to our editors at Wiley-Liss—William Curtis, Robert Harington, and Luna Han—for their patience and valuable assistance.

Contributors

Cynthia M. Beall, Department of Anthropology, Case Western Reserve University, Cleveland, OH 44106

Barry Bogin, Department of Behavioral Science, University of Michigan-Dearborn, Dearborn, MI 48128

Douglas E. Crews, Department of Anthropology, The Ohio State University, Columbus, OH 43210

Peter T. Ellison, Department of Anthropology, Harvard University, Cambridge, MA 02138

Timothy B. Gage, Department of Anthropology, University at Albany, Albany, NY 12222

Gillian J. Harper, Department of Social Medicine, College of Osteopathic Medicine, Ohio University, Athens, OH 45701

Rebecca Huss-Ashmore, Department of Anthropology, University of Pennsylvania, Philadelphia, PA 19104

Fatimah L.C. Jackson, Department of Anthropology, University of Maryland, College Park, MD 20742

Francis E. Johnston, Department of Anthropology, University of Pennsylvania, Philadelphia, PA 19104-6398

Lyle W. Konigsberg, Department of Anthropology, University of Tennessee, Knoxville, TN 37996

William R. Leonard, Department of Anthropology, Northwestern University, Evanston, IL 60208

Michael A. Little, Department of Anthropology, Binghamton University, Binghamton, NY 13902

Dennis O'Rourke, Department of Anthropology, University of Utah, Salt Lake City, UT 84112

Mary T. O'Rourke, Department of Anthropology, Harvard University, Cambridge, MA 02138

Lisa Sattenspiel, Department of Anthropology, University of Missouri, Columbia, MO 65211

B. Holly Smith, Museum of Anthropology, University of Michigan, Ann Arbor, MI 48109

A. Theodore Steegmann, Jr., Department of Anthropology, SUNY at Buffalo, Amherst, NY 14261

Sara Stinson, Department of Anthropology, Queens College, CUNY, Flushing, NY 11367

Stanley J. Ulijaszek, Institute of Biological Anthropology, University of Oxford, Oxford, OX2 6QS, United Kingdom

Mark L. Weiss, Physical Anthropology Program, National Science Foundation, Arlington, VA 22230

CHAPTER 1

Theory in Human Biology: Evolution, Ecology, Adaptability, and Variation

REBECCA HUSS-ASHMORE

INTRODUCTION

In this chapter, I deal with the theoretical perspectives that human biologists have used in their attempts to document and explain biological diversity and variability in the human species. The very fact that human biologists deal with humans as a species, as a worldwide set of related and potentially interbreeding populations, is part of our theoretical stance. The similarities across the species and the differences within and among populations are equally important sets of data. To make sense of similarities and differences, we need a theoretical framework that explains how variation arises and why characteristics do or do not persist in given situations. We need to account for biological change over time as well as the distribution of traits in space. As a result, human population biologists—like most other biological scientists—use evolutionary theory as their primary explanatory framework.

Humans are a peculiar species of mammal: bipedal, omnivorous, relatively hairless, massively **encephalized**, intensely social, and reliant on complex learned behavior for survival. Genetically and morphologically, the species is diverse, with a high degree of both **genotypic** and **phenotypic** variation. Behaviorally, it is even more diverse. Individuals communicate by using thousands of different languages, are organized into societies with widely varying structures, and solve environmental problems with myriad technological solutions. Thus, humans are a species with a highly developed capacity for symbolic thought and representation; environmental manipulation; and invention, learning, and appreciation of social facts.

The expressed variety in human behavior is clearly not reducible to biology. However, characteristics such as language, kinship networks, and technological tinkering are found throughout the species. It is therefore reasonable to assume

Human Biology: An Evolutionary and Biocultural Perspective, Edited by Sara Stinson, Barry Bogin, Rebecca Huss-Ashmore, and Dennis O'Rourke
ISBN 0-471-13746-4   Copyright © 2000 Wiley-Liss, Inc.

that these characteristics are biological; that is, the propensity to engage in such behavior has a biological basis, even though the specific expression is modeled by culture. These human peculiarities have ramifications for theory. Although much of the biology of such a species can be dealt with in the same theoretical terms as the biology of any other mammal, some aspects, such as the pervasiveness (and survival value) of symbols and meaning, require perspectives that would not be needed for other species.

Whereas theory in human biology is almost synonymous with the synthetic theory of evolution, the need to deal with humans and the complexity of their experience has meant that some parts of general biological theory have been more and less useful. For example, the parts of evolutionary theory that deal with ecology and environmental relationships, adaptability, and behavior have been especially needed in human biology, but (because we deal with a single species) parts of the theory dealing with **macroevolution** and **speciation** have received less attention.

In addition, much of recent research on the biological functioning of humans has had medical implications. Some biomedical research is informed by evolutionary theory, but much is not—instead, it is concentrated on aspects of normal physiological function and the causes and consequences of pathology. This means that within one discipline we have multiple ways of explaining variation. The evolutionary perspective can be thought of as dealing with *ultimate* causes and measures (reproductive **fitness**) and the physiological perspective as dealing with *proximate* causes and measures (health and well-being). These are not opposing viewpoints ; they are complementary, and both are necessary. (See Box 1.1 for a discussion of complementary levels of explanation.)

Many excellent books and essays on evolution are available, and some are listed as recommended reading at the end of this chapter. Rather than duplicating these

---

BOX 1.1   COMPLEMENTARY LEVELS OF EXPLANATION IN HUMAN BIOLOGY

Daly and Wilson (1977), in their discussion of the evolution of sexual behavior, point out that an evolutionary question such as, "Why do birds sing in the spring?" can be answered on four different levels. They argue that any behavior typical of a given animal can be understood in terms of its (1) physiological control, (2) adaptive significance, (3) developmental history, and (4) evolutionary history.

Although Daly and Wilson refer specifically to behavior, these levels of explanation apply to other biological traits as well. Take as an example the rewarming response of human extremities exposed to cold. This biological response is found in many human groups, in which the skin temperature of extremities (e.g., hands, feet, noses, ears) exposed to cold first drops to near ambient temperature, then rises, and finally cycles between the two processes. This response can be explained in terms of its physiology, that is, the neurological response to temperature stimuli and the resulting changes in **capillary** size and blood flow to the extremities. The initial drop in skin temperature is caused by the constriction of capillaries under the skin (**vasoconstriction**), and the subsequent temperature rise by the dilation of the capillaries (**vasodilation**) The adaptive significance lies in the ability of this response to both conserve body heat in the cold (vasoconstriction) and to prevent tissue damage to exposed body parts (vasodilation).

We can assume a selective advantage for this trait in cold climates if individuals who show this response are more likely to survive and reproduce successfully under these circumstances. This response also has a developmental aspect: Adults who grow up in cold climates show a more efficient warming response than their children do. A developmental explanation, then, would talk about stages of neurological maturation and the influence of environmental exposure on neural control of capillary size. Finally, the evolutionary history of the trait can be inferred from looking at its distribution across populations (and in some cases, across related species).

Not all humans show the rewarming response, and not all who do are equally efficient. In general, populations with a long history of residence in cold climates show the most efficient rewarming, whereas those with tropical ancestry may suffer frostbite in the cold because the response is simply not present. In this case, it would probably not be very useful to look at other primates in tracing the evolutionary history of this feature, because few of them are exposed to tissue-damaging cold.

Daly and Wilson (1977) emphasize the need to use all these levels in explaining animal behavior. They apply equally well to other biological phenomena. The approaches are not contradictory; together they provide a more complete explanation of *why* we see a particular distribution of features across the species or within populations.

In other chapters in this book, many authors give examples of how these explanatory levels have been used to address problems in human biology. The approaches are not usually identified in the terms used here, but if you read carefully, you can pick out the type of explanation being used.

discussions, I give a brief outline of the basic tenets of Darwinian theory and its current critiques. Then, I discuss the theoretical areas that are most important for the science of human biology: ecology, human adaptability, and what we might call the "biomedical perspective." In addition, I introduce some theoretical areas that have not yet been widely applied within human biology but might be useful in formulating future research. I include the area of behavioral biology and approaches drawn from the social sciences, such as economics, cultural anthropology, and psychology.

EVOLUTIONARY BASICS

What do we mean by the "synthetic" theory of evolution? This term refers simply to the fact that the modern theory of evolution is a synthesis of Darwinian theory and the science of genetics. Darwinian theory revolves around the theory of **natural selection**. At its simplest, this theory has four basic tenets: (1) More organisms are produced than can survive; (2) organisms within a species vary in their traits; (3) some of this variation is heritable; and (4) variants best suited to the environment survive to be represented in the next generation.

Mendelian genetics provided a plausible explanation of how variation is inherited, and molecular and **cytogenetics** have clarified how variation arises at the level of the deoxyribonucleic acid (**DNA**) molecule. (Genetics is explained in more detail in Chapter 3.) Another way to say this would be (1) changes in DNA can produce phenotypic changes that are subject to natural selection; (2) phenotypes best suited to the environment are most likely to survive and reproduce; (3) phenotypes with greater reproductive success leave more of their **genes** to the next generation; and (4) a change in gene frequencies from one generation to the next is defined as evolution.

Darwinian Thinking

We speak of Darwinian theory because Charles Darwin was the first to publish a coherent theory of how biological change might take place over time such that one species might come to supplant another. Evolution itself was not Darwin's idea; in his time, the idea already had been considered and debated for at least 100 years. However, with the publication of Darwin's *Origin of Species* in 1859, evolution seemed to be an idea whose time had come. One year earlier, Alfred Russell Wallace had conceived of a similar mechanism to explain evolutionary change, and Darwin had been pushed reluctantly into print.

To understand the importance of Darwin's contribution, we must understand what was so different about the theory he proposed. The simple answer is that Darwin's primary contribution was the concept of natural selection, which provided a mechanism for evolution. But this idea was not really simple, because it required several radical changes in thinking about the natural world. Conventional wisdom held that species were individually created, each according to its own basic plan, and that Earth was only about 6000 years old. To accept this view meant that there was (1) no biological connection among neighboring species, (2) no historical connection from fossils to living species, and (3) no time for the gradual change from one form to another that a theory of evolution seemed to demand. It also meant that (4) individuals of any species were viewed as being essentially alike; the similarities that made them recognizable products of the basic species plan counted more than any differences that might be observed.

Darwin rejected the first three assumptions early in his career as a result of his reading and observations while serving as a naturalist aboard the *Beagle*. Rejecting the fourth assumption took him much longer. Mayr (1972) has argued that this change was essential to formulating a theory of natural selection and that it was the most radical of the shifts in Darwin's thinking. The idea that individuals vary and that this variation is linked to survival and reproductive success is at the heart of the whole theory.

Mayr sees Darwin's reasoning process as a series of five facts and three inferences. The first three facts were based on a reading of Malthus's *Essay on Population* (1798):

1. There is a potential geometric increase in the size of any given population.
2. Real populations are held at a relatively stable size.
3. Resources (especially food) are limited.

The inference he drew from these facts was that there is a struggle for survival among individuals in any population. If this were true, then what could account for which individuals survived? The second set of facts addressed this question:

4. Individuals vary; each is unique.
5. Variation is heritable; traits can be passed on to the next generation.

Darwin's association with animal breeders convinced him that a natural struggle for survival could select variants with favorable traits, just as breeders selected for the traits that they wished to propagate. Therefore, his second inference was that some individuals in a population survived because they were better suited to acquire resources or otherwise possessed favorable traits. In other words, they were better adapted to the environment. Darwin called this differential survival based on adaptive traits *fitness*, which is both the outcome of natural selection and the measure of adaptive success.

Darwin's third inference drew on the second: If organisms with favored traits survived in greater numbers and passed on those traits, then eventually, individuals with those traits would come to characterize the population. In other words, the population would become different from its ancestors and in time might even become a separate species. Thus, over many generations, cumulative changes would result in evolution.

I dwell on the process of Darwin's thinking because it is important for understanding how human biologists approach their subject matter. The concept of variability is central to our research, as it was to Darwin's theory. The large variation present in the human species means that natural selection has a lot of raw material to operate on. The fact that some traits are found more frequently in one population than in others also suggests that that some populations may have adapted to particular environmental conditions (although the discussion later shows that other explanations are possible).

Changing environments provide opportunities for selection; humans are increasingly exposed to either unintended or self-inflicted environmental change. Much of our current research in human biology is concerned with how different populations of humans, with their varying biological traits, respond to stressful or changing environmental conditions.

The synthetic theory of evolution has been one of the transforming ideas of the 20th century. For modern-day scholars, it is difficult to imagine an intellectual world without the concept of biological evolution. Even people who misunderstand the theory—or who don't believe it—can use the vocabulary and the concepts that arose from this one body of thought. It is a tribute to the power and pervasiveness of the evolutionary concept that terms such as adaptation, fitness, and natural selection are presently part of the everyday speech of any literate society. Of course, this was not the case in the middle of the 19th century. It has taken 150 years of debate to reach this point, and still we have no absolute consensus as to how evolution really works. Most of the current debate concerns the differences between mechanisms of macroevolution (speciation) and **microevolution** (changes in gene frequencies within a species) and the relative importance of the two processes.

Evolutionary Dynamics

Darwin was concerned with the process of speciation, which is precisely the part of evolutionary theory that is most widely debated today. Darwin's basic view was that evolutionary change was a gradual process of cumulative changes. Some present-day biologists have challenged this idea, arguing that gradual changes in the gene pool of a population are not enough to produce the reproductive isolation necessary for a new species to be defined. Instead, they feel that new species emerge through relatively rapid changes that have a major effect on phenotype or behavior.

Gould and Eldridge are two of the most vocal proponents of this theory, which they call "punctuated equilibria." Gould and Eldridge (1977) argue that small populations in marginal habitats are most likely to be under selective pressure and are most likely to be rapidly transformed if major genetic changes occur. They feel that these rapid speciation events are probably followed by long periods of stasis in which little further change occurs. These ideas are difficult to test, and the time scale used probably affects the conclusions you draw. Work with very short-lived species such as bacteria suggests that punctuation can be seen if you have enough data and sample frequently enough over enough generations (Elena et al. 1996). However, less-frequent sampling of the same population would make the change look gradual.

Human biologists have spent little time dealing with macroevolution. To be sure, we compare other species with humans to determine whether such functions as the tempo of growth are similar among the higher primates or in our **hominid** ancestors. To do this, we have to assume that we are **phylogenetically** related to the species we study. However, unlike paleontologists, we are less concerned with the actual mechanisms that produce new species. Some of the macroevolutionary trends that human biologists have addressed include changes in the pattern and timing of human growth (see Chapter 11), the relationship of brain size and metabolism to diet (see Chapter 9); (Ulijaszek 1995), and the selective role of secondary compounds in human food plants (Jackson 1991). In each of these cases, researchers have been concerned with ways in which humans differ from other primate species, including our own ancestors, and how these differences might have arisen.

However, human biology is very much concerned with microevolution, or changes in the genetic makeup of human populations. (The mechanisms of microevolution are discussed in detail in Chapter 4.) This brief introduction serves as background to the other theoretical topics discussed.

Travisano et al. (1995) attributed the diversity of life to three processes: adaptation, chance, and history. In a sense, our concern with microevolution is an attempt to figure out the relative contributions of these three. We are especially concerned with the trade-off between adaptation and chance. Adaptation is often overemphasized such that some biologists have invoked natural selection to explain almost any phenotypic difference, but many biological traits are not adaptive.

Beneficial **mutations** arise by chance, and the traits that they code for may be retained or lost due to **stochastic** processes such as **genetic drift**. Chance may be especially important in retaining biological variation at the molecular level.

The third process, history, refers to the fact that different populations start with slightly different gene pools, and that they experience different environmental conditions over time. Gould (1989) has argued for the importance of history, pointing out that any set of potential adaptations is constrained by what you have inherited, that is, the set of genetic and phenotypic tools with which you come equipped. This means that the course of evolution for any population at any time is contingent on prior historical events.

It is generally accepted that four basic mechanisms can change the frequency of genes and genotypes within a population: mutation, selection, **gene flow**, and genetic drift. Of these, mutation and gene flow are generally agreed to increase variation within a population, whereas selection and genetic drift can decrease intrapopulation variation. Mutation is the ultimate source of genetic variation, through the alteration of **bases** in the DNA molecule or through the rearrangement of genes during **meiosis**. Mutation provides the raw material on which selection can operate.

Gene flow is the exchange of genetic material between populations through the processes of migration and mating. In human populations, mobility and intermarriage have probably always been important means of maintaining genetic diversity. Historical forces such as droughts, wars, economic alliances, international trade, and colonialism have influenced the rate and location of gene exchange. Since the 19th century, global travel and population contact have undoubtedly increased rates of gene flow and thus are important mechanisms of microevolution.

Although Darwin believed that natural selection was the primary force driving evolutionary change, this assumption is now being questioned. Theorists such as Wright (1982) have argued that genetic drift, or random changes in gene frequencies, are much more important than has been recognized. Random change is likely to have larger effects in small isolated populations, where a given allele may be introduced and retained (or eliminated) by chance. In small populations, the loss of an individual and his or her genes could significantly reduce the overall genetic variability for the next generation.

Several of these small populations have been studied by human biologists, including the inhabitants of Tristan de Cunha, the Pitcairn islanders, the Yanomamo of Venezuela and Brazil, and the Amish of eastern North America. In each case, the population has gone through several bottlenecks and expansions where the genetic material available has been reduced or increased by chance. One of the notable results, especially well documented in the Amish, is the phenotypic expression of **recessive** traits. In this population, the increased number of people homozygous for recessive traits has led to a relatively high prevalence of genetic diseases. In the absence of modern medicine, we assume that natural selection would act on these phenotypes and might reduce the number of people carrying the harmful recessive alleles.

Kimura (1979) argued that many mutations are neutral; that is, they have no selective advantage or disadvantage. In many cases, the phenotypic changes that they code for are small or not apparent. In Kimura's view, neutral mutations account for much of the genetic variability in the human species. This concept does not mean that selection is not operating on different phenotypes but that it does not account for the diversity seen at the level of the DNA molecule. This idea makes sense if we realize that natural selection operates on phenotypes and affects genes indirectly through the fate of the phenotype. Because these neutral mutations have little phenotypic effect, they could be retained primarily by chance, contributing to increased human **heterozygosity**.

In any case, it has been difficult to demonstrate unequivocally the operation of natural selection in humans. Part of this difficulty is due to the fact that humans adapt to their environments through both genetic and nongenetic means. It also is due to the ways natural selection is measured, that is, through differential survival and reproduction (or differential contribution of genes to the next generation). As Kimura and others have pointed out, genes interact with each other and with the environment, so the correspondence between genotype and phenotype is not one-to-one. In addition, often no clear relationship exists between either genotype or phenotype and reproductive success. The long generation time of humans makes us a difficult species in which to document long-term outcomes of biological variation. Consequently, much of human biological research uses other, more proximate indicators of probable adaptive success (health, growth, work capacity, etc.). The thorny problem of measuring adaptive success is discussed in greater detail in the following sections.

Adaptation, chance, and history have all played a role in producing the biological diversity that we see in human populations. Adaptation is a central concept in Darwinian theory and remains important in human biology. However, the role of other mechanisms cannot be ignored if we are to understand the human species. Chance may be especially important in producing the genetic differences between populations. History has been less appreciated within biology in general, outside of its role in reconstructing phylogeny.

However, history may limit the ability to adapt, through the array of responses that humans bring to their environmental problems. Clearly, not all responses—either at the genetic or phenotypic level—are adaptive; sometimes, no truly adaptive response is available. The repertoire of available responses is partly the product of a population's history, that is, the challenges they have faced and the responses they have mounted in the past. Constraints on adaptation and the result of nonadaptive responses are likely to be important future theoretical directions in the field.

OTHER THEORETICAL DIRECTIONS

Whereas evolutionary theory underlies much of the work in human biology, other theoretical frameworks contribute to research and writing in this field. Some of these, such as ecology and behavioral biology, have been more widely applied to nonhuman species than to humans. Others, such as adaptability and biomedicine, have a distinctly human slant.

Ecology

At its simplest, ecology can be thought of as relationships between species and their environments. These relationships include the way any species makes its living (its niche) and the flows of resources, including energy, nutrients, and information, through that environment. Environment is an important aspect of Darwinian explanation, as it is the driving force in natural selection. Organisms have to meet the challenges of the environments in which they find themselves in order to survive, and those that do this best and most efficiently should be those with the greatest potential for Darwinian or reproductive fitness. Biological ecologists have developed elaborate models for predicting the behavior of various animal species in different environments and for evaluating the reproductive success of species by using different mating and feeding strategies.

Although environmental relationships are important in human biology, few researchers have tried to use the types of mathematical models found elsewhere in ecology. Instead, they have looked at the environment in terms of the challenges that it presents and tried to see how these challenges might be related to human biological traits. Baker has suggested that the environment should be divided into physical, biotic, and social aspects (Thomas 1997). All of these aspects present challenges that could affect human biological function (see Figure 1.1). They present conditions to which humans, as individuals or as groups, must adapt in order to survive and reproduce.

Environmental challenges can be divided further into stressors and resources. Stressors are conditions that threaten to disrupt normal biological function or **homeostasis**. Heat, cold, **hypoxia**, disease, and malnutrition are some of the stressors that have been investigated for humans. Critical resources are those environmental products necessary for survival and reproduction. They include such things as food, water, access to mates, and the raw materials for technology to acquire these things. For humans, they may include social relationships as well. *Any living organism has two basic adaptive tasks: to avoid stressors and to procure critical resources in sufficient quantities for survival and reproduction.*

One of the areas in which human ecology has followed the example of biological ecology is in the measurement of energy flow in different human groups. Several studies have looked at the ways in which particular localized human populations capture, transform, and use energy. The assumption is that energy is a critical resource for any living organism and that strategies for energy capture are major determinants of survival. Thus, part of the overall adaptive equation for human groups is the effectiveness and efficiency with which they tap into the flows of energy in their particular habitat.

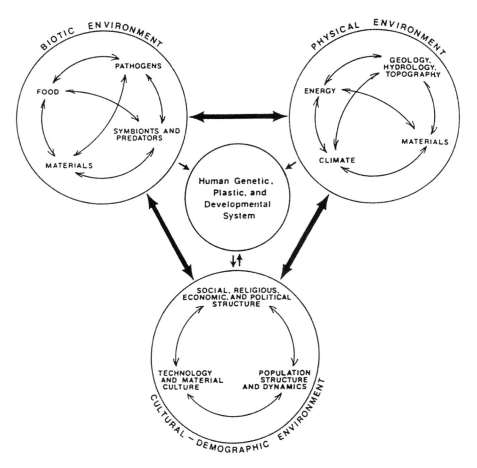

Figure 1.1 Aspects of the environment that affect human biological function. (*Source:* M.A. Little and J.D. Haas, eds. Human Population Biology, 1989, Oxford University Press, 298.)

Some of the factors that influence how successful a given human group may be in capturing sufficient energy are illustrated in Chapter 10 (see Figure 10.1). Clearly, environmental resources, technology, and division of labor are crucial variables in the successful quest for energy.

Past energy studies have investigated the role of ritual in regulating energy intake (Rappaport 1968), the ability of foragers to maintain energy balance in a harsh environment (Lee 1968), and the strategies used to cope with limited energy availability at high altitude (Thomas 1973). Rappaport (1968) argued that the cycle of warfare for the Tsembaga Maring of Papua New Guinea was a way of keeping the populations of both humans and domestic pigs in balance with environmental resources. Lee's (1968) work with the !Kung of Botswana overturned the notion that nonagricultural people were chronically on the brink of starvation. His classic study of these desert foragers showed that they could obtain a relatively balanced diet with 3–4 days per week of work for each adult. However, water was a critical resource in this environment; thus, time and energy spent on getting water, rather than food, determined the pattern of movement and settlement of the people.

In addition, Thomas's (1973) work in the Andes showed that child labor and trade with groups at lower altitudes were critical for survival in this energy-limited environment. Trade is one way of importing vital forms of energy into systems with a scarcity of resources. Animal traction and fossil fuels are substitutes for human energy that may raise the amount of food energy that can be captured from the local environment. Kemp's (1971) study of energy flow in the Canadian arctic demonstrated the importance of imported energy, in the form of guns and gasoline, for the hunting success of the Baffin Island Eskimo.

Other aspects of the environment that have interested human ecologists include its stability and predictability. How often and how regularly an environment changes affects how stressful it is and how available resources might be. It also affects the strategies that humans use to adapt. Some ecological theories address the question of the best strategies to use when resources are patchy, that is, clumped or irregularly distributed in time and space. For humans, these theories have been used to look at the foraging strategies of such groups as the boreal forest Cree and the Ache of Paraguay (Hill and Hurtado 1996; Winterhalder and Smith 1981). There also has been a recent interest in the effects of fluctuating or highly seasonal environments on human health and nutrition.

It is clear that humans not only use their environments but also change them. Populations in developing countries have seen major changes in their environments over only the past generation. These changes include climate and vegetation as well as social relationships and the nutritional resources available. These types of change are likely to accelerate during the next generation, such that few humans will have a lifestyle anything like that of their grandparents.

Because much of our species now lives in urban environments, we need more research on the ecology of the urban habitat (see Box 1.2). The attempt to define urban landscapes in terms of stressors and resources has just begun, so we have little information about the long-term effects of urban environments on human biology. How populations with different genetic makeups will respond to urban conditions is a major question for future research.

Human Adaptability

The role of the environment is also important to the study of human adaptability. We talk about not only the adaptations that currently characterize the species but also the ability of humans to respond appropriately to new environmental conditions. One of the most prominent features of our species is our ability to adjust to new conditions and the wide variety of strategies that we have for doing this. Our planet-wide distribution and our survival in diverse and stressful habitats is a result of our high degree of adaptability.

BOX 1.2   URBAN ECOLOGY AND HUMAN BIOLOGY

Numerous approaches have been taken to studying the ecology of urban areas. They reflect the disciplines that have most often been interested in the features of urban design and city life: sociology, psychology, geography, archaeology, and architecture. All of these areas are useful to the human biologist in trying to understand the adaptive challenges and biological outcomes of living in cities. For our purposes, we need to ask, What is unique about cities as human environments? What makes them different from the types of environments in which we evolved and have lived for millennia? How will these differences affect us as a biological species?

Early sociological writing on urbanism stressed the alienating aspects of city life. Wirth's famous essay (1938) typified cities as large, dense communities comprised of socially heterogeneous individuals. These features led to typically urban behaviors and attitudes, such as anonymity and impersonal relationships. The Chicago "school" of urban ecology took a more biological approach, in which competition for space led to cooperative relationships that facilitated survival. Park (1925), the best known proponent of this school, used ecological concepts of competition, dominance, invasion, and succession to explain human organization and behavior in the urban setting.

Moran (1979) and Odum (1971) argued that cities should be studied as ecosystems, with characteristic trophic organizations and flows of materials and energy. Using such an approach, cities can be seen as habitats with high **biomass**, low species diversity, and massive imports of energy from the surrounding countryside. Concern for the impact that cities have on other ecosystems and for the ability of the planet to absorb heat, chemicals, and waste generated by urban populations is increasing.

Anthropologists have looked at social organization in cities and at the strategies of different populations and different households for integration and survival in urban areas. Urban migrants are especially interesting, because they face new social, economic, and biological challenges when they move into the city. Studies of migrants show that they use a combination of individual and group strategies to find housing, jobs, marriage partners, and general social support. Successful migrants are often those who can act as individual decision makers when opportunities arise but retain social networks and group identity for times of crisis.

Bogin (1988) and Schell (1996) are two human biologists who have summarized what we know about the biological impacts of urban living. They agree that cities have both positive and negative effects on human health and functioning. Although pre-industrial cities had higher mortality than surrounding rural areas, industrial cities today usually have both lower fertility and lower mortality than their rural hinterlands. In general, urban children are taller and heavier for their age than rural children, but urban populations show higher rates of obesity, **hypertension**, coronary heart disease, cancer, and perhaps mental illness.

What accounts for the altered health profile of urban dwellers? Some of the difference is attributed to increased stress—increased noise, crowding, crime, and social isolation. These are the very factors that most interested early sociologists. However, we now recognize that differences in diet, exercise, sunlight exposure, and environmental pollution also can affect the health of urban populations. However, urban areas provide access to a wide array of resources, from health care to employment to potential mates. As human biologists, we have spent less time accounting for the resources than the hazards of urban areas.

Research on urban ecology might take any of many directions in the future. If we want to understand how humans cope with increased urbanization, we can look at the distribution of populations, pathogens, and insults across the urban landscape to see how they intersect. But we also need to look at *perceived* threats, and how urban dwellers map hazards and resources in space. Geographers have shown that "mental maps" are important for human use of space (see Figure 1.2).

How people define and perceive the threats and rewards in their environment can have a large biological impact and can affect the people, pathogens, and toxins to which they are exposed. Evolutionarily, cities are novel environments, built by humans, for humans. Our biological history as tropical primates and as socially dependent small bands of foragers can hardly have prepared us for these new habitats. Our responses as individuals, as populations, and as a species may form the core of human biological research in the 21st century.

Research in human adaptability has concentrated on the identification of salient environmental stressors and the human responses to them. The most thoroughly studied stressors have been environmental conditions such as heat, cold, and hypoxia; however, disease and nutritional deficiency also have been given attention in recent years. Some of the research carried out in these areas is discussed in Chapters 6, 7, and 12. The goal of much of this research has been to show the ability of such stressors to disrupt human biological function and to identify ways in which function is maintained or restored. Two of the most interesting findings are that different human populations have different responses to similar stressors and that similar responses vary in effectiveness across populations. The question then becomes, Why is this variation seen? Again, we are in the position of deciding how much is due to adaptation as opposed to chance or history.

Human biologists interested in adaptability have classified adaptive responses as genetic, physiological, developmental, and behavioral. Frisancho (1993) makes an important distinction between genetic adaptation and phenotypic adjustment or plasticity. Frisancho concentrates on *functional adaptation*, or the series of beneficial adjustments that humans make to the environment: "[A]djustment can be either temporary or permanent, acquired either through short-term or lifetime processes, and may involve physiological, structural, behavioral, or cultural changes aimed at improving the organism's functional performance in the face of environmental stresses" (Frisancho 1993, 4).

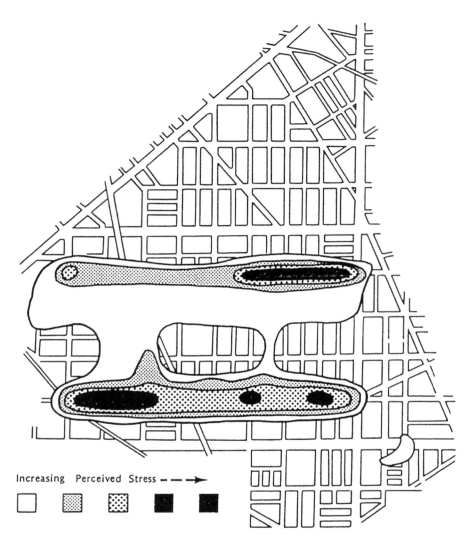

Increasing   Perceived   Stress – – →

Figure 1.2   A map of perceived environmental stress in a section of north Philadelphia. Areas of perceived danger and corridors of relative safety are shown. (*Source:* Gould and White 1972, 31)

Notice that this approach concentrates on changes possible within the lifetime of the individual and thus does not include genetic change. Although phenotypically different populations do have identifiable genetic differences, Frisancho argues that it has been difficult to show that these differences have arisen by adaptation rather than chance. In addition, much of the human biological variation in response to environmental stressors has been shown to result from either the process of growing up in a given environment or from being exposed to it for different lengths of time as an adult.

In explaining functional adaptation, Frisancho emphasizes the processes of **acclimatization** and **habituation**. Acclimatization refers to changes during the lifetime of an organism that reduce the harmful effects of naturally occurring environmental factors such as climate, nutrient imbalance, or disease. These changes may occur during the period of growth, in which case they are called **developmental adaptation** or **developmental acclimatization**. Examples of developmental adaptation are increased cold-induced vasodilation in people growing up in cold climates (see Box 1.1) and improved oxygen delivery in lowlanders raised at high altitude.

Changes in phenotype during growth may or may not improve survival and function. Schell (1995) made the point that a plastic response to the environment should be called an adaptation only if it can be shown to improve functional capacity. Slow growth in poorly nourished children is a plastic response to environmental circumstances but is not necessarily adaptive (see Chapter 12). Whereas phenotypic plasticity has been widely studied in human biology, habituation as a response has not. Habituation is the gradual reduction of response to repeated stimulation or the perception of stimulation. The ability to "tune out" urban noise after repeated or constant exposure is an example of habituation.

Although humans are capable of a wide variety of adaptive biological responses, much of what makes them such a flexible and adaptable species is the array of behavioral and cultural responses that they use. Some of these responses, such as language or the technology that permits clothing and heated dwellings in cold climates, seem obviously adaptive. However, other types of behavior, such as marriage patterns or cultural rules for food preparation, probably need to be investigated before we can say that they have functional significance or reduce environmental stress.

In general, human biologists have paid less attention to behavioral aspects of adaptability than to physiological and developmental aspects. However, behavior is obviously important in avoiding current stress and in setting up changed environmental conditions that may cause future stress. Cultural norms and traditions encode information for dealing with environmental challenges, but they also limit the available options for dealing with new environments. Your culture provides information about how to make or where to buy clothing, but it also tells you that not all possible ways of covering the body are acceptable (cardboard may keep off the rain, but it is not usually considered clothing in the context of Euro-American culture).

Culture can be considered as part of the human package of adaptive strategies, but it also can be a source of change that requires continued flexibility and adaptability to survive. Culture can produce stress and can selectively allocate impacts of stress to different portions of a population. Schell's (1992) work on lead toxicity is a good example. Schell showed that children of the urban poor have higher lead exposures and blood lead levels, which lead to cognitive impairment, behavioral difficulties, learning deficits, lower educational achievement, fewer employment opportunities, and continued poverty. Thus, cultural factors (lead paint, substandard urban housing, economic discrimination) have biological impacts that may affect the long-term survival and adaptive success of a portion of the human population.

Models have been devised to show how environmental conditions are linked to human adaptive responses. In general, the goal of such adaptive models is to

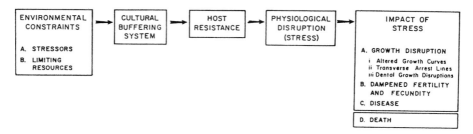

Figure 1.3 A stress model that shows features of the environment and human responses. (*Source:* Huss-Ashmore et al. 1982, 397)

predict what kinds of responses should be seen under what kinds of conditions. Models often are presented visually as flow charts, beginning with the environment, which is then appraised in some way by an organism, resulting in an array of responses. Figure 1.3 illustrates a simple model of this type and Figure 1.4 a very complex model.

Responses are classified according to type, level, onset time, and sustainability. The most successful organisms or populations should be those whose response choices are most appropriate for environmental conditions, that is, most effective and least costly.

Notice that outcomes can be measured not only in terms of reproductive fitness but also in terms of function. These functional or proximate indicators of adaptation, such as growth, work capacity, or health, have been widely used in studies of adaptability. Such indicators have the advantage of being observable within the lifetime of the individuals under study and thus within the lifetime of the investigator. The assumption made is that the ability to maintain growth or work capacity in a stressful environment is related to long-term fitness or the probability of persistence. In other words, these abilities are evidence of positive adjustments to environmental conditions.

For humans, responses to new environmental conditions are often a combination of physiological, developmental, and behavioral change. We would expect genetic change to be a response of last resort, for two theoretical reasons. First, genetic change takes a long time to implement, so it would be of little use for conditions that change rapidly. Relying solely on genetic change as an adaptive response would seriously reduce the flexibility of a species. Second, genetic change is costly in that it requires the deaths or reduced reproduction of less fit individuals. Well-adapted species should be able to buffer environmental stressors without resorting to genetic change.

Therefore, we would expect to see genetic adaptation in response to conditions that can't be handled by other means, for example, where the stressors affect most members of the population, where the stressors are present over multiple generations, and where no other effective means of coping are currently available. Hypoxia at high altitude is a stressor of this type, and we are just beginning to document **polymorphisms** that could confer survival benefits in high altitude populations (see Chapter 6; Beall et al. 1994).

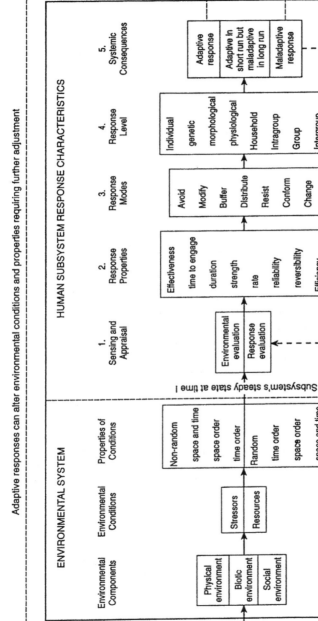

Figure 1.4 A complex flow chart that shows the relationship between features of the environment and the characteristics of human responses. (*Source:* Thomas et al. 1979, 6)

Research in human adaptability has a well-developed theoretical base, but two areas have been relatively neglected. One is the question of resource acquisition, and the other is the issue of cost. Although the ability to find and secure sufficient resources is half of the adaptive equation, human biologists have spent little time studying this area. In general, theories of how humans choose, distribute, and consume resources have been left to economists. Theytraditionally have had little interest in the biological outcomes of these processes. I noted earlier that rapidly changing environments, such as urban environments, will undoubtedly present humans with new and stressful conditions. However, they also will present new opportunities, new resources, and new questions:

What kinds of changes will humans have to make to take advantage of these new resources?

Are we biologically equipped as a species to make physiological changes that allow us to use new resources?

What kinds of new resources will be available in urban areas, in the industrialized or the developing world?

What kinds of strategies, in terms of growth, learning, household demography, or social organization, will give people the best access to the means for survival in new environments?

These are questions that human biologists of the next century will have to address if we are to understand the adaptive capacity of our species.

One final issue to consider is the question of cost. No adaptive strategy is free of costs, and part of the recipe for adaptive success is to use those response strategies that are most efficient, that is, that minimize cost in terms of critical resources or interference with other functions. As human biologists, we have spent little time researching the question of adaptive efficiency.

Cost can be figured in different currencies, such as time, energy, or adaptive flexibility. The whole question of how adaptive a response is depends not only on how effective it is in the short run but also how much it costs in the long run. Body size is one example. Large body size permits greater work capacity and greater energetic efficiency per unit of body mass but requires more energy overall to maintain. Conversely, small body size is more sustainable if energy is limited, but at the cost of lower endurance and greater risk of mortality. In the case of body size, human populations probably do not make conscious decisions; the relative benefits of large versus small size depend on the environment, available energy and technology, necessary work tasks, etc.

For behavioral strategies, however, costs may be consciously calculated. We know little about the way in which these cost–benefit decisions are made, especially when all available strategies are expensive, or when the trade-off is between costs in different currencies (time versus energy, for example). Good evidence indicates that humans may incur large costs to gain some desired goal, for example, sacrificing health or safety to acquire money. A sophisticated theory of human adaptability should consider how to account for these kinds of calculations.

Biomedical Theory

A biomedical approach is central to much of the research in human biology today. At its simplest, biomedicine can be characterized as interested in normal anatomy, normal physiological function, and the processes that cause pathology. More complex analyses stress that the biomedical model of disease is a cultural product—a specific way of looking at the human body and its disorders that reflects the cultural values and beliefs of industrialized societies (Rhodes 1990). Although some biomedical problems are phrased in terms of evolutionary theory, most are not. Even researchers who believe in evolution as a guiding theoretical principle often ignore it in practice. Therefore, much of the work in this area appears atheoretical because the problems of interest have been identified in very practical ways, as issues in public health or clinical medicine. These problems become of interest to human biology when they can be linked in some way to populations; that is, when their incidence, severity, or outcome varies in different identifiable human groups.

We should consider three different approaches when looking at the role of biomedicine in human biology. One is the "biomedical model" or philosophical principles that underlie medical investigation and practice. The second involves the principles and assumptions of **epidemiology**, and the third is the field of evolutionary medicine. Researchers interested in problems of human health and disease often do not distinguish among these approaches and may use all three in any given study. Articles in human biology journals may start with a statement of the problem based on a clinical concern, proceed to methods drawn from epidemiology, and conclude with a discussion of both public health outcomes and a possible evolutionary origin of the problem.

The "biomedical model" is the name given to the set of philosophical assumptions that underlie Western **allopathic medicine**, currently the dominant medical system in most industrialized countries of the world. Anthropologists and historians writing about medicine have emphasized the dualism inherent in biomedical theory. What they mean is that the body and mind are seen as separable entities; the body is part of the natural world, a bounded material entity that can be known and understood through scientific observation. Similarly, diseases are physical entities, "things" that occur or go wrong in identifiable locations in the body. This means that the body can be treated in isolation from the mind or the "spirit."

The body also can be reduced to its parts. Systems, organs, tissues, and cells are separable parts whose function and malfunction can be studied and treated. This approach gives a mechanistic quality to the ways that the body is envisioned and talked about. Medical anthropologists have termed the mechanistic approach "the machine metaphor," in which body systems are reduced to pumps and plumbing. Helman (1994) pointed out that these assumptions about the human body are an outgrowth of a science-oriented Western culture in which physical things are somehow more real than psychological things. He lists the main characteristics of biomedicine as (1) scientific rationality; (2) emphasis on objective, numerical measurement; (3) emphasis on physicochemical data; (4) reductionism, or the view that biological functions can be reduced to physical and chemical functions; and (5) emphasis on the individual patient as opposed to family or community (Helman 1994, 101).

Although human biologists have generally accepted this view of disease as a measurable set of physiological malfunctions, they have not confined themselves to individual patients. Their emphasis has been much more on the variation across populations in how normal bodies function and malfunction. Thus, they have emphasized that there is a range of variation in what is normal and that this range may be different from one population to another. Normal biochemical parameters for Euro-American males may be quite different than those for female African foragers, to use an extreme example, and the links between biochemical measures and the risk of disease also may vary. For human biologists, age, sex, ethnicity, body size, and physical fitness are necessary factors (among others) for interpreting the risk or likely outcomes of disease.

Epidemiologists also have been interested in the "who," "where," and "when" of disease. Epidemiology is a set of methods for determining the causes of disease from looking at who in the population is affected; where diseases occur in space and time; and the social, environmental, dietary, and lifestyle correlates of disease occurrence. Much effort is devoted to teasing out the links between observable disease and previous exposure to some pathogen, insult, or noxious agent. Epidemiologists have developed very rigorous research designs and sophisticated statistical techniques for assessing risk of disease for different groups of people. In general, human biologists share many of the research goals and have begun to borrow the methods.

One final area of biomedical research where human biologists are making visible contributions is the field of evolutionary or Darwinian medicine. The major premise of Darwinian medicine is that much of human disease and illness can be traced to our evolutionary background; our current biological design is the result of millions of years of evolutionary compromises. Some of these compromises (such as upright posture) enabled us to deal better with our environments (by seeing across savannas, or being able to carry lunch and the baby at the same time). However, this strategy may have associated costs, and the back pain of modern bipedal humans may be part of our payment.

In general, the goal of evolutionary medicine is to show how many of our current ills are related to evolutionary compromise or to the fact that we no longer live under conditions like those in which we evolved. Obesity, hypertension, diabetes, sudden infant death syndrome, and **osteoporosis** are conditions that may have evolutionary explanations and are currently being studied by human biologists.

Behavioral Issues

Throughout this chapter, I have emphasized that behavior is an important aspect of the human adaptive pattern. In fact, behavioral flexibility is arguably the most important trait that the human species has for dealing with ongoing environmental change. Such flexibility is undoubtedly part of our evolutionary heritage and as such contributes to the biological variation that we presently see in the species.

Despite the importance of behavior, few human biologists have integrated this trait into their research *at the theoretical level*. The discussion of biocultural aspects of disease in Chapter 8 and the discussion of behavior and the human life cycle in Chapter

11 are exceptions to this rule. In this section, I touch briefly on a several bodies of theory that human biologists might find useful or interesting for looking at behavior. They come from the fields of behavioral biology, economics, and psychology.

Behavioral biology, also known as evolutionary ecology, is an attempt to look at human social behavior within an evolutionary framework. Much of the emphasis is on trying to measure the degree to which different types of social behavior affect reproductive success. For humans, the types of behavior most closely related to reproductive success are mating and parenting. The trade-off between these two is that time and energy spent on mating reduces time and energy left for parenting, and vice versa. Many behavioral biologists are interested in understanding the circumstances under which more effort is put into either mating or parenting. Reproductive strategies are expected to vary by sex, with females competing for resources and males competing for females. Thus, males who control more resources should be able to compete more successfully for females, enhancing the probability of their reproductive success.

Behavioral biology has been widely criticized by both biologists and anthropologists. It makes the assumption that men and women should behave in ways that optimize their reproductive success, but there are good reasons why they might not. First is the possibility that optimal behaviors have not arisen in the species and thus could not be selected for. Current behaviors may be the results of evolutionary compromise, a "good enough" solution to environmental problems, rather than a best solution. Second, our history as a species has taken place in social groups and environmental conditions very different from those we face today. Modern reproductive strategies probably would differ from those that were adaptive for our ancestors. Both of these criticisms emphasize the importance of history in looking at human evolution. Despite the criticisms, some behavioral biologists are attempting to put together more rigorous studies that test for differences in reproductive success across cultures and within populations (Betzig 1997).

If reproductive strategies are constrained by history, so are other types of behavioral strategies. Economic behavior, or the procurement and control of resources, is important. If access to sufficient resources is part of the adaptive equation (and Darwin clearly thought it was), then behaviors designed to increase access or control can affect survival and reproductive success. Behavioral biologists have looked at economic behavior within an evolutionary framework, predicting under what circumstances foragers will stop searching for one type of food and be content with harvesting another. They also have looked at the reproductive advantages to controlling economic resources. Hill and Kaplan (1988) showed that among Ache foragers of Paraguay, men who are better hunters acquire more mates, and their children are more likely to survive. In addition, two other aspects of economics are now being used in biological research. One of these is a macroeconomic perspective drawn from political economy. The other is a microeconomic perspective drawn from household economics.

Thomas, Leatherman, and Goodman are three researchers who have looked at the effect of macro-level social and economic conditions on human biology and adaptability. Leatherman and Thomas (Leatherman et al. 1989) showed that increasing commercialization in highland Peru limited the ability of peasant households to produce sufficient food and to cope with the illness of their members. Goodman et al.

(1992) suggested that political control of economic resources may have affected the health of prehistoric populations as well. Research on health of archaeological populations at Dickson Mounds, Illinois, showed an increase in skeletal markers of ill health at the time when their more powerful neighbor, Cahokia, became a dominant political force in the region. The researchers concluded that local resources were siphoned off to feed the politically dominant community, reducing both availability and control at the local level.

Household economic models have also been used to look at the responses of peasant farmers to increasing commercialization in Africa. Work in Swaziland showed better levels of food consumption and better nutritional status for households growing hybrid maize than those growing cotton (Huss-Ashmore and Curry 1994). Diet also was better in households with more women employed off-farm. These conclusions make sense in both economic and adaptive terms. Specializing in nonfood cash crops can be a risky strategy, because household income (i.e., access to resources) depends on the world price of the commodity grown. You can't eat cotton, and if you can't sell it for an adequate price, you have few other options. Other studies have indicated that women's control of resources is an important factor in household health and nutrition (Frankenberger 1985). Women who work off-farm are more likely than men to invest their earnings in food and health care. From research like this, the adaptive significance of economic behavior is obvious.

Finally, recent work in psychology could be useful in examining the biological significance of human behavior. Work in perception and cognition is especially important. We know that humans, like other animals, have diverse modes of sensing and appraisal to make sense of the environments they live in. As human biologists, we have paid little attention to this aspect of behavior, despite its potential evolutionary importance. Perception and cognition are necessary steps to defining hazards and resources. They may take many forms, from visual and auditory cues about danger (e.g., the coughing of a leopard or a mugger, clouds signaling an oncoming storm) to the gustatory and olfactory appraisal of food (e.g., the scent of ripe peaches, the bitter taste of quinine). Humans make decisions about the resources to use or places to avoid based on both perception and memory (see Figure 1.2). A strong ability to map environmental features and to remember where things are is clearly adaptive for a mobile omnivore such as *Homo sapiens*. Cognitive and evolutionary psychologists are exploring these areas of human behavior to understand not only how they work but also how they might have evolved.

Social memory and meaning are also critical parts of the human adaptive repertoire that have not yet been explored within human biology. These areas often are studied by social and cultural anthropologists, usually without reference to biology. Yet a memory for social relationships, including kinship obligations and the hierarchy of power, is essential for living in social groups. Primatologists are looking at social memory in nonhuman primates and may provide information useful to human biologists.

The study of parenting strategies, economic behavior, and environmental perception may seem to be a long way from human variation and adaptation. However, I would argue that they are not so far apart. Human behavior is our most defining adaptive strategy and a critical factor limiting the other strategies available to us. Behavior and cognition affect how we define and map the hazards and resources in our social and natural environments. They affect with whom we mate and how many offspring we have, both factors in biological variation and success. Behavior and biology are linked in a complex web of interactions—at once cause and effect, dependent and independent variables.

CHAPTER SUMMARY

In this chapter, we have presented several theoretical positions that human biologists use or might use to explore and explain human biological variation. Of primary importance is biological evolution as an explanation of changes in the genetic makeup of populations. Human biology also utilizes ecological theory to analyze the environments to which humans adapt, human adaptability to characterize the numerous ways in which humans adapt to environmental stressors, and biomedicine to examine the health consequences of human variation. All are aimed at understanding the dynamics of humans as populations within a single, potentially interbreeding species. The complex interplay of biology and behavior makes humans a peculiar species, and also expands the range of theoretical perspectives needed. The following chapters demonstrate the use of evolutionary, ecological, adaptive, and biomedical perspectives.

RECOMMENDED READINGS

Bowlby J (1990) Charles Darwin: A New Life. Norton: New York.

Darwin CR (1898) On the Origin of Species by Means of Natural Selection, 6th ed., 1872. Appleton: New York.

Dennett DC (1995) Darwin's Dangerous Idea. Simon and Schuster: New York.

Mayr E (1991) One Long Argument: Charles Darwin and the Genesis of Modern Evolutionary Thought. Cambridge, Mass.: Harvard University Press.

REFERENCES CITED

Beall CM, Blangero J, Williams-Blangero S, and Goldstein M (1994) Major gene for percent of oxygen saturation of arterial hemoglobin in Tibetan highlanders. Am. J. Phys. Anthropol. 95: 271–276.

Betzig L (ed.) (1997) Human Nature: A Critical Reader. Oxford: Oxford University Press.

Bogin B (1988) Rural-to-urban migration. In CGN Mascie-Taylor and GW Lasker (eds.): Biological Aspects of Human Migration. Cambridge: Cambridge University Press, pp. 90–129.

Daly M and Wilson M (1977) Sex, Evolution, and Behavior. Boston: Willard Grant Press.

Darwin C (1859) The Origin of Species by Means of Natural Selection, or the Preservation of Favoured Races in the Struggle for Life. London: Murray.

Elena SF, Cooper VS, and Lenski RE (1996) Punctuated evolution caused by selection of rare beneficial mutations. Science 272: 1802–1804.

Frankenberger TR (1985) Adding a Food Consumption Perspective to Farming Systems Research. Washington, D.C.: Nutrition Economics Group, Technical Assistance Division, OICD/US Department of Agriculture.

Frisancho AR (1993) Human Adaptation and Accommodation. Ann Arbor: University of Michigan Press.

Goodman AH, Martin DL, and Armelagos GJ (1992) Health, economic change, and regional political economic relations: examples from prehistory. In R Huss-Ashmore, J Schall, and M Hediger (eds.): Health and Lifestyle Change. MASCA Research Papers in Science and Archaeology Vol. 9, Philadelphia: University Museum, University of Pennsylvania, pp. 51–60.

Gould P and White R (1972) Mental Maps. London: Penguin Books.

Gould SJ (1989) Wonderful Life: The Burgess Shale and the Nature of History. New York: Norton.

Gould SJ and Eldridge N (1977) Punctuated equilibria: the tempo and mode of evolution reconsidered. Paleobiology 3: 115–151.

Helman C (1994) Culture, Health and Illness, 3rd edition. Bristol, UK: Wright.

Hill K and Hurtado AM (1996) Ache Life History: The Ecology and Demography of a Foraging People. New York: Aldine de Gruyter.

Hill K and Kaplan H (1988) Tradeoffs in male and female reproductive strategies among the Ache: part 1. In L Betzig, M Borgerhoff Mulder, and P Turke (eds.): Human Reproductive Behavior: A Darwinian Perspective. Cambridge: Cambridge University Press, pp. 277–290.

Huss-Ashmore R and Curry JJ (1994) Diet, nutrition, and agricultural development in Swaziland. 3. Household economics and demography. Ecol. Food Nutr. 33: 107–121.

Huss-Ashmore R, Goodman AH, and Armelagos GJ (1982) Nutritional inference from paleopathology. In MH Schiffer (ed.): Advances in Archeological Method and Theory. New York: Academic Press, Vol. 5, pp. 395–474.

Jackson FLC (1991) Secondary compounds in plants (allelochemicals) as promoters of human biological variability. Annu. Rev. Anthropol. 20: 505–546.

Kemp W (1971) The flow of energy in a hunting society. Sci. Am. 225(3): 104–115.

Kimura M (1979) The neutral theory of molecular evolution. Sci. Am. 241: 98–126.

Leatherman TL, Thomas RB, and Luerssen S (1989) Challenges to seasonal strategies of rural producers: uncertainty and conflict in the adaptive process. In R Huss-Ashmore (ed.), with JJ Curry and RK Hitchcock: Coping with Seasonal Constraints (MASCA Research Papers in Science and Archaeology, Vol. 5). Philadelphia: University Museum, University of Pennsylvania, pp. 9–20.

Lee RB (1968) What hunters do for a living, or how to make out on scarce resources. In RB Lee and I de Vore (eds.): Man the Hunter. Chicago: Aldine de Gruyter, pp. 30–48.

Malthus TR (1798) An Essay on the Principle of Population, As It Affects the Future Improvement of Society with Remarks on the Speculations of Mr. Godwin, M. Condorcet, and Other Writers. London: J. Johnson.

Mayr E (1972) The nature of the Darwinian revolution. Science 176: 981–989.

Moran EF (1979) Human Adaptability. North Scituate, Mass.: Duxbury Press.

Odum HT (1971) Environment, Power, and Society. New York: Wiley-Interscience.

Park RE, Burgess EW, and McKenzie RD (1925) The City. Chicago: University of Chicago Press.

Rappaport RA (1968) Pigs for the Ancestors: Ritual in the Ecology of a New Guinea People. New Haven, Conn.: Yale University Press.

Rhodes LA (1990) Studying biomedicine as a cultural system. In TM Johnson and CF Sargent (eds.): Medical Anthropology: Contemporary Theory and Method. New York: Praeger, pp. 159–173.

Schell LM (1992) Risk focusing: an example of biocultural interaction. In R Huss-Ashmore, J Schall, and M Hediger (eds.): Health and Lifestyle Change (MASCA Research Papers in Science and Archaeology, Vol. 9). Philadelphia: University Museum, University of Pennsylvania, pp. 137–144.

Schell LM (1995) Human biological adaptability with special emphasis on plasticity: history, development and problems for future research. In CGN Mascie-Taylor and B Bogin (eds.): Human Variability and Plasticity. Cambridge: Cambridge University Press, pp. 213–237.

Schell LM (1996) Cities and human health. In G Gmelch and WP Zenner (eds.): Urban Life, 3rd edition. Prospect Heights, Ill.: Waveland Press, pp. 104–127.

Thomas RB (1973) Human Adaptation to a High Andean Energy-Flow System. Occasional Papers in Anthropology No. 7. University Park: Department of Anthropology, The Pennsylvania State University.

Thomas RB (1997) Wandering toward the edge of adaptability: adjustments of Andean people to change. In SJ Ulijaszek and R Huss-Ashmore (eds.): Human Adaptability: Past, Present, and Future. Oxford: Oxford University Press, pp. 183–232.

Thomas RB, Winterhalder B, and McRae S (1979) An anthropological approach to human ecology and adaptive dynamics. Yearbook of Physical Anthropology 22: 1–46.

Travisano M, Mongold JA, Bennet AF, and Lenski RE (1995) Experimental tests of the roles of adaptation, chance, and history in evolution. Science 267: 87–90.

Ulijaszek SJ (1995) Human Energetics in Biological Anthropology. Cambridge: Cambridge University Press.

Winterhalder B and EA Smith (1981) Hunter-Gatherer Foraging Strategies. Chicago: University of Chicago Press.

Wirth L (1938) Urbanism as a way of life. Am. J. Sociol. 44: 1–24.

Wright S (1982) Character change, speciation, and the higher taxa. Evolution 36: 427–443.

CHAPTER 2

History of Human Biology in the United States of America

FRANCIS E. JOHNSTON and MICHAEL A. LITTLE

INTRODUCTION

Why were human biologists among the first scientists to condemn the misuse of the term "race" in the mid-1950s, when others were pointing to physical characteristics as evidence of inferiority? Why were human biologists among the leaders in formulating an understanding of sickle-cell anemia—a dreaded, often fatal disease—as an outcome of the development of resistance to malaria? And why do so many human biologists focus their research on an understanding of the relationships between malnutrition and child development? The answers to these questions are at the root of unraveling exactly what human biology is, because what human biologists do is the key to understanding the common set of concepts and ideas that bind them together into a scientific discipline.

In seeking answers to the questions posed above, we should begin with an examination of the history of human biology. History is far more than a description of what happened in the past. The history of a science is the story of the unfolding of ideas about the nature of the things that concern its practitioners. And just as events that occur in the earlier life of an individual are important determinants of what happens in his or her later years, the current state of a science is best understood from the perspective of its history.

In this chapter, we discuss the history of human biology, largely in the United States. By restricting the coverage we do not mean to imply that human biology originated in the United States, that American human biologists have not had fruitful interactions with colleagues around the world, or that this country is the intellectual core of the discipline. In fact, the beginnings of human biology are found in the descriptions of human variation of the European Middle Ages when individuals concerned with the human body began to study its structure by dissecting

Human Biology: An Evolutionary and Biocultural Perspective, Edited by Sara Stinson, Barry Bogin, Rebecca Huss-Ashmore, and Dennis O'Rourke
ISBN 0-471-13746-4   Copyright © 2000 Wiley-Liss, Inc.

cadavers. Inevitably they began to encounter the many variations that exist and as their descriptions began to accumulate, so did an interest in the variability itself. Furthermore, in the course of their scholarly activities, human biologists of any country have always drawn on the work of colleagues in other parts of the world.

Our coverage in this chapter is restricted to the United States solely due to space limitations. For those interested in historical developments elsewhere, the International Association of Human Biologists has published a series detailing the development of the discipline in other countries. (Contact Professor G. F. De Stefano, Secretary General, Dipartimento di Biologia, University of Rome Tor Vergata, Via della Ricerca Scientifica, 00133 Rome, Italy.)

MAJOR DIVISIONS OF AMERICAN HUMAN BIOLOGY

No single scheme is used to divide human biology into its subdivisions. Because it is an integrated discipline, considerable overlap will exist, regardless of how it is organized. In terms of its history, the problem is even more difficult, because many of the scientists who helped shape human biology into its present state worked across much, if not all, of its breadth.

We decided to focus on four areas of concentration in this chapter: human variation, human adaptability, anthropological genetics, and growth and development. The major developments and the leading figures in each area are presented, albeit briefly, in an attempt to demonstrate the theoretical and methodological threads that have linked the early phases to their current orientation.

Human Variation

Interest in human variation and its causes has been fundamental to human biology from its earliest development in the 19th century. Much of the early history of human biology is tied to the history of physical anthropology in the United States and abroad. Before World War II, interests tended toward concerns with race, the significance of racial traits, and racial history. Since, the focus has turned to populations, relationships between biology and behavior, and population history and evolution.

Race and Early Typological Beliefs
Measurement and description of living humans and of skeletal remains constituted much of the activities of physical anthropologists during the latter half of the 19th and early 20th centuries. The organization of the data collected was guided by **typology**, a conceptual scheme dominant among natural historians and biological scientists of the time.

Typology is a method, a way of reducing a spectrum of variation to a smaller set of categories that can be managed more easily. Take, for example, a data set composed of the following five parameters taken on a sample of 100 adult females: height, weight, head circumference, shoulder breadth, and skin color. This data set would yield 500 pieces of information, challenging the investigator to develop an efficient way to describe the sample. A simple typological approach would be to calculate the averages of each measurement and combine them into the "type individual," that is, someone of average height, weight, skin color, etc.

But beyond the use of typology as a method is that it denotes an almost philosophical view of the nature of humankind. Rather than seeing human variation in the richness of its diversity, a typological view sees the human species as best represented by a set of types—stereotypes if you will.

The concept of typology presents two problems. First, it is almost impossible that anyone will match the type, which is nothing more that an idealized, statistical image that doesn't exist in the real world. Second, using a type to describe a population fails to capture the variation that exists. To the typologist, the set of averages becomes the reality, and variability is but an abstraction.

The ideas that dominated the field of human biology for nearly a century were based on typological beliefs, views of racial superiority and inferiority, some mysticism, and poorly conceived science (Brace 1982; Gould 1983). Most practitioners of physical anthropology were trained in medicine or anatomy and were concerned with characterizing the human body, particularly the male, as a set of types (i.e., races). Emphases were on fixed species and races and on attempts to identify and classify these "ideal types" of races. Such endeavors indicated that professionals concerned with physical anthropology were not in tune with the new ideas on evolution and variation developed by Charles Darwin in the mid-19th century.

By 1900, human populations were still viewed as races that were "fixed types"; that is, all important characteristics, thought to reflect heredity, were seen as relatively invariant. Races were viewed as "pure" (Negroid, Mongoloid, or Caucasoid) or a mixture of two or three of these primary races. Classification of all of these varieties of humankind was a primary academic activity. Such concepts as racial types, racial "purity," Caucasian superiority, and the dangers of racial mixing were part of scientific and popular beliefs alike.

One of the few anthropologists to challenge these beliefs was Franz Boas, whose late-19th century research (Boas 1897) illustrated how variable the growth of children and adolescents was. His study of eastern and southern European migrants to the United States at the end of the first decade of the 20th century laid to rest once and for all among human biologists the myth that human "types" were invariant and determined solely by heredity. By demonstrating differences in head form and height between migrants and their children raised in the United States, Boas (1911) showed that the environment had exerted a powerful force on human size and constitution. Races were not pure and invariant; their members were capable of dramatic changes arising from the environment within their own lifetimes.

Three biologically oriented anthropologists were most influential during the first half of the 20th century: Franz Boas (1858–1942), Ales Hrdlička (1869–1943), and Earnest A. Hooton (1887–1954). All were trained at least partly in Europe, and their ideas formed the framework for an American profession of physical anthropology.

Their roles were quite different. Boas, who worked for many years at Columbia University in New York City, conducted the most rigorous scientific studies by today's standards. His work on growth in children and adolescents was pioneering, and his studies of migrants served as models to be used by numerous younger physical anthropologists. Hrdlička, who spent most of his professional life at the Smithsonian Institution in Washington, D.C., was organizer and editor of the *American Journal of Physical Anthropology* (begun in 1918) and was instrumental in forming the American Association of Physical Anthropologists (in 1930). Hooton, from his base at Harvard University, wrote very successful books and texts (Hooton 1946) and trained most of the next generation of physical anthropologists.

Another influential figure from this period was Raymond Pearl (1879–1940). A population biologist and early demographer at Johns Hopkins University, Pearl was closely tied to anthropology and founded the journal *Human Biology*.

Plasticity, Variability, and Migration Models
Despite the earlier work of Boas, typological views of race and human variation persisted into the 1930s and 1940s. The prevailing perspective was hereditarian: Variability in any particular trait reflected differences that had been inherited from one's parents. The dynamic interplay between heredity and environment was little appreciated or understood. In particular, human biology was slow to accept the view of **plasticity**, the concept that sees the organism as responsive to the surrounding environment, especially during the formative years of growth and development.

Nonetheless, some human biologists carried out significant work on the effects of the environment on human biology. The work was conducted on migrants, following Boas' (1911) early study of European migrants to the United States. Harry Shapiro (1939) studied Japanese migrants to Hawaii; Marcus Goldstein (1943) studied Mexican immigrants to the United States; Gabriel Lasker (1946) began his work with studies of immigrant and American-born Chinese and later initiated studies of Mexican immigrants (Lasker 1952); and Fredrick Hulse (1957) compared Swiss immigrants to California with their nonmigrant kinsmen. The research designs were similar in that native residents were compared with first- and second-generation migrants.

During this period, human biology in the United States saw the first wave of researchers to reject strict racial typology and to pursue research that would uncover the influence of the environment on human biological characteristics. Models of migration to study human adaptation have continued to be used up to the present (Harrison 1966; Baker 1976; Malina et al. 1982; Mascie-Taylor 1984; Bogin 1995).

Post-World War II
The period following World War II was exciting for all biologists interested in population variability. Several developing threads of inquiry began to come together, and a new discipline began to emerge called population biology.

Interests in human adaptation to the environment arose during this time among scientists who, at that time, identified themselves as physical anthropologists. By 1950, two important publications demonstrated these interests. The first was the proceedings of a watershed conference held at Cold Spring Harbor organized by two distinguished scientists: Theodosius Dobzhansky, a geneticist, and Sherwood Washburn, a physical anthropologist (Warren 1950). Although several participants carried over some of the old emphasis on racial studies and classification, new ideas on human adaptation to the environment, human population genetics, **natural selection**, and evolutionary process dominated the symposium. This conference typified what Washburn (1951) later called the "new physical anthropology," in which use of the scientific method and hypothesis testing, within the context of evolutionary (and adaptation) theory, became the objectives of the refocused physical anthropology.

The second publication was *Races, a Study of the Problem of Race Formation in Man* (Coon et al. 1950). Although the title incorporated the outdated term "race," it nevertheless represented an entirely new way to interpret biological variation both within and among human populations. Previously, the basis of racial classification rested with nonadaptive (neutral) traits, that is, characteristics that did not change in response to the environment. Because a racial taxonomy was seen as a fixed thing, traits that could change with the environment would not provide the stability seen as essential to a classification.

In their book, Carleton Coon, Stanley Garn, and Joseph Birdsell rejected the emphasis on nonadaptive traits. Their idea—a revolutionary one at the time—was that racial categories were in fact adaptive categories, formed as populations responded through natural selection to the pressures of their environments. Human variation was significant, not a static phenomenon useful only for taxonomic purposes. Races were units that evolved as part of the dynamic interplay between a population's **gene pool** and its environment. The study of race was transformed from an exercise in taxonomy to the study of human adaptation.

These two works reflected the trends and changes in thinking that began to transform physical anthropology from a largely descriptive and anatomically based science with little theory to a more modern analytical science based on human population biology and evolutionary theory.

The rejection of a race as a fixed unit and taxonomy as the primary activity of human biology signaled the end of any real focus on race by human biologists. Among others, researchers such as Ashley Montagu and C. Loring Brace challenged the concept head-on. To Montagu, the evils of racism vitiated any legitimate use of the term and he suggested its replacement by "ethnic group." To Brace, human biological variation was not organized into neat packages that would permit the clear-cut delineation of groups. He argued that variables changed gradually across the geographical landscape, that they were distributed primarily as gradients (or **clines**).

Today, human biologists are not concerned with the study of race except to counteract its misuse in fostering stereotyping and discrimination. Human populations represent groups of individuals who interact with one another, perhaps sharing their **genes** and forming breeding units, or perhaps interacting as sociocultural and economic units pursuing a common set of goals. The nature of these interactions is complex and frequently particularistic with varying effects on biological patterns, but always too involved to be captured by any simplistic concept such as race. Human variation is viewed in dynamic terms, shaped by the interaction of genes and environments. Sometimes the result is adaptive; sometimes adaptive processes fail and disease results.

In the following section, we discuss "human adaptability" as the concept that has superseded "race" as the guiding concept in contemporary human biology.

Human Adaptability

Adaptation and Adaptability

According to Coon et al. (1950), the concept of adaptation to the environment was largely the adaptation of body morphology to climate, that is, extremes of cold and heat. These ideas were developed and expanded in the United Kingdom by Weiner (1954), Roberts (1952), and Barnicot (1959); in France by Schrieder (1951); and in the United States by M. Newman (1956), R. Newman (Newman and Munro 1955), and Baker (1958).

"Adaptation to the environment" as a paradigm for considerable human biology research had by this time begun to incorporate ecological theory as an expanded view of the human environment (Newman 1953; Baker 1962; Livingstone 1962). By the early 1960s, enough research had been done in these and other fields to provide a basis for publication of the first textbook in human biology (Harrison et al. 1964). The text was divided in five parts that were to reflect major areas in interest in the field over the next 30 years: Human evolution (Harrison and Weiner), Human genetics (Harrison), Biological variation in modern populations (Barnicot), Human growth and constitution (Tanner), and Human ecology (Weiner). Weiner's (1964) contribution to this text outlined a framework that was to be used over the next decade for the development of human biology research under the International Biological Programme (IBP) and its Human Adaptability (HA) Component.

Human Adaptability in the IBP

Human adaptability as a term of reference began to be used widely only during the years of the IBP (Baker 1991). At the beginning of the IBP, Joseph Weiner (1915–82), the HA Convener, conceptualized the field of human adaptability as a worldwide comparative study of the ecology of humankind, greatly needed because of the "vast changes [that] are affecting the distribution, population density, and ways of life of human communities all over the world" (1965, 7) "… [and that can be] measured in terms of health, fitness and genetic constitution" (1977, 3). Measures of health and fitness were expected to be derived from studies of human nutrition, child growth, physiology, work capacity, daily and seasonal activities, and disease status. Human adaptability research, then, focused efforts on the many ways in which human populations adapt biobehaviorally to their environments. Reviews of human adaptability perspectives from the 1960s are by Baker (1965), Lasker (1969), and Weiner (1966).

The Human Adaptability research program was organized at the IBP Paris Assembly in 1964. Four categories of research were identified: (1) survey of sample populations in conformity with a world scheme, (2) intensive multidisciplinary regional studies based on habitat contrasts, (3) special investigations on selected populations, and (4) investigations related to current World Health Organization (WHO) activities.

The first category included worldwide studies of child growth based on standardized procedures, as well as surveys of genetic **polymorphisms**. Under the second category, "The general aim was to elucidate physiological and genetic processes concerned in adaptation and selection in relation to climate and other environmental factors" (Weiner 1965, 8). Organization of the Human Adaptability research was provided by *Guide to the Human Adaptability Proposals* (Weiner 1965), a volume on field methods (Weiner and Lourie 1969), and an edited collection of human adaptability papers from the 1964 Wenner Gren symposium at Burg Wartenstein in Austria (Baker and Weiner 1966).

Multidisciplinary Research
As part of the human adaptability research, more than 230 projects were conducted worldwide under the umbrella of the IBP, and publications numbered in the thousands. In addition to promoting international research cooperation, the IBP also established a pattern of collaborative investigation called multidisciplinary research. Multidisciplinary research requires that scientists from diverse fields focus their efforts on a single problem or a single population. Although multidisciplinary research requires a great deal of organization and coordination, this kind of work offers many scientific advantages:

- Complex systems require extensive knowledge bases for their understanding, and few individuals have the background or capacity alone to understand a very complex system.
- Nearly all systems in today's world are affected by human action and involvement, and much of the earth is managed by humans. Accordingly, most management problems require scientific, technological, and human social expertise for their solution.
- Although humans are highly adaptable, their adaptability is limited. Limits to human adaptability that arise from rapid environmental change are best understood by multidisciplinary efforts.
- Multidisciplinarity allows large-scale problems to be approached from different perspectives and with less likelihood of major omissions and errors in judgment.
- Research in which scientists from many different backgrounds collaborate on a project broadens the base of experience and education of scientists and students alike and moves individuals out of the narrow confines of their own disciplines.

Figure 2.1 is a time scale for several multidisciplinary projects conducted over the past three and a half decades. Many were affiliated with the IBP, some were independent projects, and still others were affiliated with the Man and the Biosphere (MaB) Program of the United Nations Educational, Scientific, and Cultural Organization (UNESCO). Most of the projects centered on single populations of interest to human biologists and anthropologists. Underlying all of the projects was the theoretical concept of human adaptation to the environment as a reflection of human adaptability (Little et al. 1991).

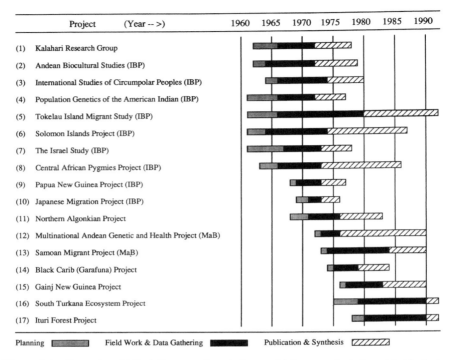

Figure 2.1   Several multidisciplinary projects in human biology conducted worldwide during the latter half of the 20th century. IBP, International Biological Programme; MaB, Man and the Biosphere Program of the United Nations Educational, Scientific, and Cultural Organization (UNESCO). Reprinted with permission from Little (1997).

Particular theoretical perspectives were applied to some of the projects listed in Figure 2.1. For example, *cultural and behavioral evolution* was an important perspective to the Kalahari Research Group; *microevolution* was a central issue to the Yanomama (Genetics of the American Indian) and Garifuna projects; *systems ecology* was a part of the framework of the South Turkana Ecosystem Project; and *health and environmental change* was the focus of the several projects that dealt with migrant populations.

Anthropological Genetics

Early Genetic Studies

The historical roots of genetics extend back to Gregor Mendel, whose experiments with peas in the 1860's first showed the particulate nature of the gene and the specificity of its information. His laws of **segregation** and **independent assortment** still form the basis for our understanding of the transmission of that information from one generation to the next. Also, his discovery of **dominant** and **recessive** alleles are essential in the study of the relationship between **genotype** and **phenotype**.

The development of a theoretical base for the analysis of genes in populations happened much later and owed much to the work of Hugo de Vries in the Netherlands and Thomas H. Morgan and Hermann Müller in the United States. Their research on primroses (*Primula*) and fruit flies (*Drosophila*) demonstrated the widespread occurrence and the spontaneous origins of genetic variation (now called mutations) and of the role of the environment as a mutagenic agent. Even though Darwin had described natural selection as the phenomenon that could shape the distribution of inherited variability across time more than 50 years before, he could not explain how new variants arose to provide the raw material for selection. De Vries, Morgan, and Müller were able to do so and, furthermore, to demonstrate that mutation was a regular, spontaneous, and natural event that did not require divine intervention.

The beginnings of a formal, quantitative theory began to take shape in the decades following World War I, with the studies by Haldane (1927) and Fisher (1930) in England and somewhat later by Wright in the United States. Their work was instrumental in the formulation of the "New Synthesis" theory, which expressed evolutionary change as the outcome of the interactions among **mutation**, **gene flow**, selection, and **genetic drift**. This theory led to the rise of two general approaches to the interpretation of patterns of gene frequency and of their course of change.

The first approach emphasized the role of natural selection in the course of evolution and was derived from the classic work of Fisher. Selection was seen as maintaining genetic variation both within and between gene pools. This interpretation was enunciated most forcefully by E. B. Ford, a British ecological geneticist who argued that essentially all genetic polymorphism was maintained by selection and, most importantly, that gene frequencies were likely to be kept in a steady state (equilibrium) by a balance between opposing selective forces (Ford 1965).

The second approach, advanced by Sewall Wright, emphasized the role of random statistical (**stochastic**) mechanisms in evolution, especially in small populations (Wright 1968–78). Wright's theories were among the first to link gene pools to the demographic structure of their populations. His work also signaled the introduction of **demography** into genetics. Demography is concerned with the dynamics of populations in terms of their size, growth, and structure and involves the analysis of such parameters as age structure, sex ratio, birth and mortality rates, and patterns of in- and out-migration. These aspects of populations can affect the patterns of their genetic variability by shaping the flow of genes within the boundaries of a population or by channeling the movement of genes into and out of its gene pool.

Human Genetics

In 1900, Karl Landsteiner provided the first description of the human ABO blood group system. Landsteiner (1868–1943) was an Austrian immunologist and pathologist who, as a research assistant at the Vienna Pathological Institute, showed that there were at least four major types of human blood that could be accounted for by the presence or absence of two **antigens**, labeled A and B. His discovery revolutionized **immunology** and human genetics. Not only did his work form the basis for modern immunology; the realization that a specific antigen is the product of a particular gene provided a method for detecting individual genes. Prior to World War

II, human biologists had based their research on measurements and descriptions of morphological variability. Landsteiner's discovery allowed human biologists to analyze variability at the level of the gene, to calculate the frequency of genes in populations, and ultimately to develop a quantitative theory of evolution. And as more antigen systems were discovered, the amount of genetic variation that could be detected and quantified increased enormously.

The movement of human biology into the genetic arena was sudden, headlong, and not without controversy and rancor. William Boyd, an immunologist, spearheaded the case the most forcefully by arguing for the abandonment of the traditional methods of **anthropometry** (body measurement), osteometry (measurement of bones), and anthroposcopy (description of traits) and their replacement with analysis of blood groups. In *Genetics and the Races of Man* (Boyd 1950), Boyd stressed that because morphology could be shaped by the environment, it was impossible to establish hereditary relationships from their analysis. The blood groups, on the other hand, were direct products of genes that did not change throughout the life of an individual.

It is of interest that even though Boyd introduced the concept of the gene into anthropology, he believed in the existence of races as real biological categories and advocated the use of gene frequencies as a superior method for racial taxonomies. He viewed the blood groups as selectively neutral, that is, nonadaptive and hence not subject to natural selection. It should be noted that later research has shown that, in fact, the frequencies of several human antigen systems (including the ABO blood groups) are shaped, at least to a degree, by natural selection and are part of the ways in which populations adapt to their environment. Nonetheless, even though Boyd's focus was on racial taxonomy, it heralded the beginning of human biology's entry into the study of human genetic variability.

The formulation of the synthetic theory of evolution and the publications of scientists such as Ford influenced the course of those human biologists in the 1960s and 1970s who studied human population genetics. Ford and other population biologists provided a theoretical basis for interpreting genetic **polymorphisms** as adaptations, and Boyd and other biochemical geneticists furnished the techniques for exploring the extent of polymorphism in the human gene pool. Human biologists launched their search for natural selection by a concerted effort to demonstrate the adaptive significance of specific genetic polymorphisms. Countless studies attempted to relate patterns of gene frequency variation to environmental factors, largely, disease (e.g., ulcers, cancer, plague, and cholera).

Overall, the results of this research were disappointing; however, in retrospect, they were not unexpected. Nonetheless, the course of the research, the theories that guided it, and the models that were developed have played a major role in the orientations of contemporary human biologists. In particular, the introduction of the concept of genetic adaptation and the need for finely contextualized ecological analyses have been as important a milestone in the history of human biology as any other event or phenomenon that can be recognized in its development. For example, Livingstone's (1958) classic analysis of the distribution and dynamics of hemoglobin S (sickle cell) in Africa was a defining moment in human biology, through the author's careful interweaving and integration of genetics, disease, culture, and population history within a carefully drawn ecological framework. It still stands as perhaps the leading example of the biocultural approach in human biology.

The acceptance by human biologists in the 1960s of genetic polymorphism as a measure of natural selection hindered the study of microevolutionary processes as stochastic phenomena, driven by probability, randomness, and demographic fluctuations. However, as researchers began to implement comprehensive research projects focused on small human isolated populations, the importance of all those phenomena that comprise genetic drift became increasingly evident. Distributional studies of gene frequencies in Oceania revealed none of the patterns suggested by the models of selectionists, whereas intensive investigations of populations in the rainforests of South America and Papua New Guinea revealed the importance in evolutionary change of population size, demographic structure, and unique historical events.

The implications of this realization were strengthened by the rise of neutralism among more mathematically oriented population geneticists, who argued that the degree of genetic polymorphism observable in the human species cannot be maintained by natural selection but instead represents stochastic processes acting on adaptively neutral alleles. Rather than population differences reflecting adaptations to different environments, the genetic separation reflected a longer historical separation between groups. In other words, whereas a selectionist interpretation would view genetic differences among populations as indicating different adaptations, a neutralist interpretation holds that the differences reflect separation over time. Although the proponents of these two interpretations are no longer as controversial as they were in the 1970s, the distinction still forms the basis for different understandings of the nature of evolution and human genetic variation.

Contemporary Trends in Anthropological Genetics

Even though this chapter deals with history, it is appropriate to note briefly how modern trends have followed on the work of earlier researchers.

In line with the rapid development of molecular biology, the focus of much of the study of human population genetic variation has been at the level of the **DNA** and the analysis of the molecular structure of specific genes. Inevitably, this study has become entwined with the Human Genome Project, a major initiative begun by the U.S. Department of Energy and the National Institutes of Health in 1987, described as a multidisciplinary effort to understand the basis of human heredity. The focus of the Human Genome Project is the characterization of the human genome—the complete collection of human genetic material, including the estimated 50,000 to 100,000 genes contained in human DNA. Similar efforts to study the human genome have been launched by Great Britain, France, Italy, Japan, and other countries.

Most of the impetus for the Human Genome Project has been clinical: determining the structure of alleles responsible for disease. However, population geneticists also are interested in the project for what they can learn about the molecular basis of differences among human populations and their gene pools. The Human Genome Project has not been without controversy. Some decry the enormous financial commitment to the project and the diversion of governmental research funds away from more traditional areas of interest.

Growth and Development

The description and analysis of human growth has occupied the research energies of human biologists almost from the time of the first systematic population-based studies in the latter third of the 19th century (Ulijaszek et al 1998). (For a comprehensive history of the study of growth, see Tanner 1981.) Four individuals stand out for the importance of their contributions to the development of the field as well as the breadth and the influence of their research: Franz Boas, T. Wingate Todd, Wilton Krogman, and James Tanner.

Boas is one of the most influential figures in the development of American cultural, linguistic, and physical anthropology. In addition to his analyses of population biological variability, Boas investigated and demonstrated the significance of differences in the timing of growth events among individual children. He argued effectively for the collection of **longitudinal** data, that is, regular examinations of the same children through the course of their development (see Tanner 1959 for an analysis of Boas' contributions to the study of growth).

Todd was a British anatomist whose contributions to growth were made after moving to the United States. His studies of **skeletal maturation** were influential in demonstrating the independence of biological age and chronological age and the importance of the rate of maturation as separate from the rate of growth. The various atlases of skeletal maturation—used to assess the maturation rates of children—produced by him and his team at Western Reserve University in Cleveland form the basis for contemporary assessments of the biological age of children (Todd 1937).

Krogman was a junior colleague of Todd's before coming to the University of Pennsylvania in 1948. Krogman's view of the "whole child" as the unit of study, and his voluminous writings have made him the "father" of growth studies in America. His abilities in his 60+ years as a teacher and the scores of students who studied with him during his career have made him one of the two most influential figures in the history of biological anthropology in the United States (Krogman 1941, 1948).

Tanner is important both historically and as the most widely known researcher internationally in the field of growth and development. As a human biologist and a clinician, his research has been basic to the synergism between basic research and studies of growth disorders. His work on the assessment of growth adequacy at the individual and population levels is unequalled to the present day.

Among human biologists, concern with the growth process has been driven by five themes:

- documentation of interpopulation variation in growth;
- relative contribution of heredity and the environment to the growth process;
- growth as an adaptive response to the environment;
- growth as an indicator of population health status; and
- evolution of the human growth curve.

Interpopulation Variation in Growth

The early naturalists traveled to locales outside of their European homelands, where they measured and observed the populations they encountered there. Subsequent research produced a considerable body of data on a range of anthropometric dimensions and indices. However, children were measured only rarely, and it was not until much later that we find any significant information about growth. In the Americas, almost all of this early information comes from studies of native North American children and was collected under the direction of Franz Boas and Clark Wissler (Wissler 1938, Boas 1940).

Heredity, Environment, and Growth

As noted earlier, to 18th and 19th century anthropologists, human biologists, and other natural scientists, anthropometric differences reflected genetic differences; the environment played little, if any, role. To most researchers, these differences provided the basis for racial classification, whereas to some, it provided a marker used for not only social and economic discrimination but also outright genocide.

Trained as a scientist (doctorate in physics), Boas moved to the United States in the late 1880s. As part of his broader research agendum, he began to conduct systematic studies on human growth, especially on the role of the environment in altering the course of growth. In a series of important and well-known publications (Boas 1940), he reported his measurements of European-born immigrants to the United States and their American-born children. This research, often ignored today, demonstrated clearly the significant alterations of growth in the children attributable to the different environments.

Growth as an Adaptive Response

The emergence of ecology—and, more specifically, of human ecology—stimulated a series of the comprehensive studies by human biologists carried out as part of the IBP. Major investigations of populations were launched in, for example, high-altitude, circumpolar, and arid ecosystems in an effort to uncover the pathways by which those individuals adapted to their often extreme environments.

The researchers refined the view of growth as the pathway to the adult, expressing the process as one in which the organism responds to environmental pressures by alterations of growth that result in more effective adaptations to those pressures. The scope of these studies was broad and a wide range of data was collected. In particular, Baker, who stimulated an impressive range of studies on high-altitude ecosystems and whose own research focused on Andean populations, saw the importance of growth as part of the process of adaptation. With Roberto Frisancho, Baker developed the concept of the **developmental adaptation**, which states that individuals who grow up in a particular environment are better adapted to that environment as adults than are others who move there as adults (Frisancho and Greksa 1989).

Growth as an Indicator of Health

As indicated earlier, the history of human biology shows a clear trend from an orientation that stressed genetic determinism to one that sees the environment as playing an important role in the human phenotype. This trend is seen also within growth studies, and the sensitivity of the growth process to the environment is one mechanism by which our species adapts. However, this sensitivity also makes the developing child vulnerable to the rigors of a harsh, constraining environment, and when the environment is one that impacts on all members of a community, the degree of **growth faltering** becomes a reliable measure of environmental quality. In his history of the study of human growth, Tanner (1981) called this effect "auxological epidemiology" and traced its beginnings to surveys of the growth of British factory children in the 19th century.

This use of growth data has had a significant effect on human biology, public health, and other disciplines that deal with the nutritional status and health of human groups. Although the formation of a strong base of theory and method is relatively recent (not more than 30 years old), human biologists such as Tanner, Robert Malina, and Reynaldo Martorell have played major roles in advising international bodies and in helping to design and implement nutritional surveillance programs.

Evolution of the Human Growth Curve

The final and most recent field of concentration in the history of growth studies in human biology is the evolution of the human growth curve. To be sure, human biologists have studied the growth of nonhuman primates for decades, and one of the first doctoral dissertations in American physical anthropology was Krogman's study of primate cranio-facial growth (Krogman 1929). Such research was based on a comparative anatomical approach and, as is characteristic of much paleoanthropology even today, was essentially an exercise in fossil taxonomy. In the 1960s, James Gavan began to study primate growth from the perspective of evolutionary biology and to examine the transformation of the growth process during the course of evolution (Gavan and Swindler 1966). However, the major impetus came from the work of Elizabeth Watts. Drawing on her studies of primates and humans, Watts (1986) was responsible for incorporating an evolutionary perspective into the study of human growth.

New Trends in Research

The study of child growth has, in the past several decades, widened its conceptual framework to a life-span approach. With this approach, the human organism has taken on a more dynamic aspect, such that variations in the process of growth have significant implications for adult status and health. Human biologists have increasingly directed their research at adults and at changes in morphology, body composition, and health associated with hereditary and environmental factors. The range of studies is broad and encompasses topics such as chronic disease (e.g., hypertension, osteoporosis), nutrition, and physiological status.

HUMAN BIOLOGY AND BIOMEDICINE

The interest of human biologists in the study of health and disease has been long-standing and can be traced to three related factors.

The first factor began to be apparent with the realization of the plasticity of much of human variability. Just as the impact of the environment on the growing child could result in a developmental adaptation, a harmful environment could also leave its mark. The record of growth thus became an indicator of the quality of child/environment interaction. Diversions of the growth patterns of children from a normal course became an important indicator of an underlying disease process, and evaluating growth became the basic approach for assessing the extent and severity of malnutrition in communities and other groups (Tanner 1986). With the development of **epidemiology** and the refinement of the concept of the risk factor, the association between variation and the potential for disease became even more important.

The second factor was that many human biologists have been physicians, dentists, and other health care professionals or were primarily affiliated with medical, dental, or similar institutions. Hrdlička, one of physical anthropology's most influential figures, held an M.D., and it has been remarked that he felt that an anthropologist should first be trained in medicine. Tanner earned an M.D. and a Ph.D., and Krogman's primary appointments at Western Reserve and the University of Pennsylvania were in schools of medicine. Albert Damon (1914–73), who worked on the health significance of variations in physique, was a physician who spent the last 10 years of his life in Harvard University's Department of Anthropology.

The third factor has been more recent in origin. An emerging new field is evolutionary medicine, which looks at evolutionary aspects of contemporary diseases of humans. In many ways, anthropological genetics has long been concerned with genes and disease, and Livingstone's work on malaria was an earlier precursor of the study of evolution and disease. But the field has expanded dramatically with the inclusion of "diseases of civilization," which result from the fact that adaptations to earlier, pre-industrial lifeways have become maladaptations in contemporary, urban society. Among the diseases that have been implicated are diabetes (Weiss 1993; Weiss et al. 1984), obesity, cardiovascular disease, and sudden infant death syndrome (McKenna and Mosko 1994). Although this focus is too recent to be discussed in historical terms, it is nonetheless today a major concern of human biologists and indicates the merging of the study of the normal and abnormal in the analysis of human biology (Johnston and Low 1984).

CHAPTER SUMMARY

Human biology deals with understanding the extent of human biological variability, with explaining the mechanisms that create and pattern the variability, and with relating it to health, disease, and the social issues that concern all individuals today. It stands at the junction of the biological and the social sciences and, along with anthropology, is frequently referred to as biocultural in nature. Contemporary human biology is the product of many branches of knowledge, but it is unique in integrating those branches and bringing them to focus on the essential and longstanding problem of human biological variation in space and across time.

BOX 2.1   PROFESSIONAL ORGANIZATIONS AND JOURNALS

By its nature, human biology is an eclectic field of study and as a result, human biologists are affiliated with numerous professional organizations and publish their research in a wide range of journals. However, many associations and journals are clearly identified with human biology and human biologists. Some of the most influential are listed here.

Scientific Association	Affiliated Journal
Human Biology Association	*American Journal of Human Biology*
Society for the Study of Human Biology	*Annals of Human Biology*
American Association of Anthropological Genetics	*Human Biology*
American Association of Physical Anthropologists	*American Journal of Physical Anthropology*
Society for the Study of Social Biology	*Social Biology*
The Galton Society	*Journal of Biosocial Science*

Human biology can be divided into the following subfields: human variation, human adaptability, anthropological genetics, and growth and development. The history of human biology and each of its subfields reflects the history of the way that humankind in general, and science in particular, conceptualizes human biological variability and its significance. Several developments have been crucial in the formation of contemporary human biology, including

- the rejection of typology;
- the rejection of the notion of races as fixed and unchanging categories and of racial taxonomy as the primary purpose of physical anthropology and human biology;
- the development of the science of genetics, including population and biochemical genetics;
- the discovery of the plasticity of human biology in response to the environment, especially during the course of child growth and development;
- the appreciation of the essentially adaptive nature of human variability and the importance of the concept of adaptability to the study of biological variation; and
- the application of the concepts and methods of human biology to problems of health and disease.

RECOMMENDED READINGS

Baker PT and Weiner JS (eds.) (1966) The Biology of Human Adaptability. Oxford: Clarendon.

Coon CS, Garn SM, and Birdsell JB (1950) Races: A Study of the Problems of Race Formation in Man. Springfield, Ill.: CC Thomas.

Gould SJ (1983) The Mismeasure of Man. New York: WW Norton.

Little MA and Haas JD (eds.) (1989) Human Population Biology: A Transdisciplinary Science. Oxford: Oxford University Press.

Livingstone FB (1958) Anthropological implications of sickle-cell gene distribution in West Africa. Am. Anthropol. 60: 533–562.

Spencer F (1981) The rise of academic physical anthropology in the United States, 1880–1980: A historical overview. Am. J. Phys. Anthropol. 56: 3534–364.

Spencer F (ed.) (1982) A History of American Physical Anthropology. New York: Academic Press.

Spencer F (ed.) (1997) History of Physical Anthropology, An Encyclopedia. New York: Garland.

Ulijaszek S, Johnston FE and Preece M (1998) Encyclopedia of Growth and Development. Cambridge: Cambridge Univ. Press.

Washburn SL (1951) The new physical anthropology. Trans. N.Y. Acad. Sci. Ser. 2, 13: 298–304.

Watts ES (1986) The evolution of the human growth curve. In F Falkner and JM Tanner (eds.): Human Growth: A Comprehensive Treatise (Vol. 2). New York: Plenum, pp. 153–166.

REFERENCES CITED

Baker PT (1958) The biological adaptation of man to hot deserts. Am. Natural 92: 337–357.

Baker PT (1962) The application of ecological theory to anthropology. Am. Anthropol. 64: 15–22.

Baker PT (1965) Multidisciplinary studies of human adaptability: theoretical justification and method. In JS Weiner (ed.): International Biological Programme: Guide to the Human Adaptability Proposals. London: Special Committee for the International Biological Programme, International Council of Scientific Unions, pp. 63–72.

Baker PT (1976) Research strategies in population biology and environmental stress. In E Giles, JS Friedlaender (eds.): The Measures of Man: Methodologies in Biological Anthropology. Cambridge, Mass.: Peabody Museum Press, pp. 230–259.

Baker PT (1991) Human adaptation theory: successes, failures and prospects. J. Hum. Ecol. Special Issue No. 1.

Baker PT and Weiner JS (1966) The Biology of Human Adaptability. Oxford: Clarendon Press.

Barnicot NA (1959) Climatic factors in the evolution of human populations. Proceedings of the Cold Spring Harbor Symposia on Quantitative Biology 24: 115–129.

Boas F (1897) The growth of children. Science 5: 570–573.

Boas F (1911) Changes in the Bodily Form of Descendants of Immigrants (Senate Document 208). Washington, D.C.: Government Printing Office, 61st Congress.

Boas F (1940) Race, Language, and Culture. New York: Free Press.

Bogin, B (1995) Plasticity in the growth of Mayan refugee children living in the United States. In CGN Mascie-Taylor and B Bogin (eds.): Human Variability and Plasticity. Cambridge: Cambridge Univ. Press, pp. 46–74.

Boyd WC (1950) Genetics and the Races of Man: An Introduction to Modern Physical Anthropology. Boston: Little-Brown.

Brace CL (1982) The roots of the race concept in American physical anthropology. In F Spencer (ed.): A History of American Physical Anthropology: 1930–1980. New York: Academic Press pp. 11–29.

Coon CS, Garn SM, and Birdsell JB (1950) Races: A Study of the Problems of Race Formation in Man. Springfield, Ill.: CC Thomas.

Fisher RA (1930) The Genetical Theory of Natural Selection. London: Oxford Univ. Press.

Frisancho AR and Greksa LP (1989) Developmental responses in the acquisition of functional adaptation to high altitude. In MA Little and JD Haas (eds) Human Population Biology,. New York: Oxford Press, pp. 203-221.

Ford EB (1965) Genetic Polymorphism. Cambridge, Mass.: MIT Press.

Gavan JA and Swindler DR (1966) Growth rates and phylogeny in primates. Am. J. Phys. Anthropol. 24: 181–192.

Goldstein MS (1943) Demographic and Bodily Changes in Descendants of Mexican Immigrants. Austin, Tx.: Univ. of Texas Institute of Latin American Studies.

Gould SJ (1983) The Mismeasure of Man. New York: WW Norton.

Haldane JBS (1927) A mathematical theory of natural and artificial selection. Part V. Selection and mutation. Proc. Cambridge Phil. Soc. 28: 838–844.

Harrison GA (1966) Human adaptability with reference to the IBP proposals for high altitude research. In PT Baker and JS Weiner (eds.): The Biology of Human Adaptability. Oxford: Clarendon Press, pp. 509–519.

Harrison GA, Weiner JS, Tanner JM, and Barnicot NA (1964) Human Biology: An Introduction to Human Evolution, Variation, Growth, and Ecology. London: Oxford Univ. Press.

Hooton EA (1946) Up From the Ape. New York: Macmillan.

Hulse FS (1957) Exogamie et hétérosis. Arch. Suisse d'Anthropol. Gen. 22: 103–125.

Johnston FE and Low SM (1984) Biomedical anthropology: an emerging synthesis. Yearbook Phys. Anthropol. 27: 215–227, 1984.

Krogman WM (1929) A study of growth changes in the skull and face of Anthropoids, with special reference to man. Thesis (Ph.D.)—University of Chicago, Dept. of Anthropology.

Krogman WM (1941) Growth of man. Tabulae Biologicae, Volume 20, The Hague: Junk.

Krogman WM (1948) A handbook of the measurement and interpretation of height and weight in the growing child. Monographs of the Society for Research in Child Development 13(3), 1-85.

Lasker GW (1946) Migration and physical differentiation. A comparison of immigrant and American-born Chinese. Am. J. Phys. Anthropol. 4: 273–300.

Lasker GW (1952) Environmental growth factors and selective migration. Hum. Biol. 24: 262–289.

Lasker GW (1969) Human biological adaptability. Science 166: 1480–1486.

Livingstone FB (1958) Anthropological implications of sickle-cell gene distribution in West Africa. Am. Anthropol. 60: 533–562.

Livingstone FB (1962) Population genetics and population ecology. Am. Anthropol. 64: 45–52.

Little MA, Leslie PW, and Baker PT (1991) Multidisciplinary studies of human adaptability: twenty-five years of research. J. Indian Anthropol. Soc. 26: 9–29.

Little MA, Leslie PW, and Baker PT (1997) Multidisciplinary research of human biology and behavior. In F Spencer (ed.): History of Physical Anthropology: An Encyclopedia, Volume 2, M–Z. New York, N.Y.: Garland Publishing, p. 196.

Malina RM, Buschang PH, Aronson WL, and Selby HA (1982) Childhood growth status of eventual migrants and sedentes in a rural Zapotec community in the Valley of Oaxaca, Mexico. Hum. Biol. 54: 709–716.

Mascie-Taylor CGN (1984) The interaction between geographical and social mobility. In AJ Boyce (ed.): Migration and Mobility: Biosocial Aspects of Human Movement. London: Taylor and Francis, pp. 161–178.

McKenna JJ and Mosko SS (1994) Sleep and arousal, synchrony and independence, among mothers and infants sleeping apart and together (same bed): an experiment in evolutionary medicine. Acta Paediatr. Suppl. 397: 94–102.

Newman MT (1953) The application of ecological rules to the racial anthropology of the aboriginal New World. Am. Anthropol. 55: 311–327.

Newman MT (1956) Adaptation of man to cold climates. Evolution 10: 101–105.

Newman RW and Munro EH (1955) The relation of climate and body size in U.S. males. Am. J. Phys. Anthropol. 13: 1–17.

Roberts DF (1952) Basal metabolism, race, and climate. J. R. Anthropol. Inst. 82: 169–183.

Schreider E (1951) Anatomical factors in body heat regulation. Nature 167: 823–824.

Shapiro H (1939) Migration and Environment. New York: Oxford Univ. Press.

Tanner JM (1959) Boas' contributions to knowledge of human growth and form. Am. Anthropol. 61: 76–111.

Tanner JM (1981) A History of the Study of Human Growth. Cambridge: Cambridge Univ. Press.

Tanner JM (1986) Growth as a mirror of the condition of society: secular trends and class distinctions. In Demirjian A (ed.): Human Growth: a Multidisciplinary Review. London: Taylor and Francis, pp. 3–34.

Todd TW (1937) Atlas of Skeletal Maturation. London: Kimpton.

Warren, KB (ed.) (1950) Origin and evolution of man. In: Proceedings of the Cold Spring Harbor Symposium on Quantitative Biololgy (Vol. 15). Cold Spring Harbor, NY: The Biological Laboratory.

Washburn SL (1951) The new physical anthropology. Trans. N.Y. Acad. Sci. Ser. 2, 13: 298–304.

Watts ES (1986) The evolution of the human growth curve. In F Falkner and JM Tanner (eds.): Human Growth: A Comprehensive Treatise (Vol. 2). New York: Plenum, pp. 153–166.

Weiner JS (1954) Nose shape and climate. Am. J. Phys. Anthropol. 12: 1–4.

Weiner JS (1964) Human ecology. In GA Harrison, JS Weiner, JM Tanner and NA Barnicot (eds.): Human Biology: An Introduction to Human Evolution, Variation, Growth, and Ecology (Part IV). London: Oxford Univ. Press, pp. 399–508.

Weiner JS (1965) International Biological Programme: Guide to the Human Adaptability Proposals. London: Special Committee for the International Biological Programme, International Council of Scientific Unions.

Weiner JS (1966) Major problems in human population biology. In PT Baker, JS Weiner (eds.): The Biology of Human Adaptability. Oxford: Clarendon Press, pp. 1–24.

Weiner JS (1977) The history of the Human Adaptability section. In KJ Collins and JS Weiner (eds.): Human Adaptability: A History and Compendium of Research. London: Taylor and Francis, pp. 1–23.

Weiner JS and Lourie JA (1969) Human Biology: A Guide to Field Methods (IBP Handbook No. 9). Philadelphia: F. A. Davis.

Weiss KM (1993) Genetic Variation and Human Disease. Principles and Evolutionary Approaches. Cambridge: Cambridge Univ. Press.

Weiss KM, Ferrell WE and Hanis CL (1984) A New World syndrome of metabolic diseases with a genetic and evolutionary basis. Yearbook Phys. Anthropol. 27: 153–178

Wissler C (1938) The American Indian, an Introduction to the Anthropology of The New World. Third edition. New York: Oxford Press.

Worthington EB (1975) The Evolution of IBP. Cambridge: Cambridge Univ. Press.

Wright S (1968–78) Evolution and the genetics of populations; a treatise. Volumes 1-4. Chicago: University of Chicago Press.

CHAPTER 3

An Introduction to Genetics

MARK L. WEISS

INTRODUCTION

Most simply, genetics is the study of heredity: the passage of information, in small units called genes, from one generation to the next. Yet, subsumed under this simple definition is a tremendous range of information and inquiry. Geneticists are concerned with the structure of the genetic material and how it is organized, how it operates, and how it is transferred across generations; this statement hints at a wide range of endeavors.

At the most basic level, genetics is the study of the chemical structure of the genetic materials and the chemical differences between the various forms of genetic materials. At a somewhat higher level, we examine the ways an individual gene is structured. We ask, for instance, do aspects of gene organization typify the beginning, middle, and end of a gene? Are all genes the same in this structuring? If not, what are the differences, and why do they exist?

Then, we observe that genes are organized into regions of chromosomes. Human chromosome 11, for instance, contains a region that includes genes that direct the production of a series of related, oxygen-carrying molecules that we use at different times in our lives. Why do we sometimes find that genes with related functions are linked together into units along a chromosome? Why is this not always the case?

At yet another level, we know that not all of the genetic material is organized into genes. In fact, something on the order of one-third of our genetic material is *not* part of any of the 100,000 or so genes in a cell. Rather, much of the genetic material occurs in units that we are only beginning to understand in terms of organization and function. We can go even further and ask, to what extent do observations regarding genetic structure hold true from one species to another? What causes variation?

Much excitement and media attention surrounds the analysis of molecular genetic structure, because many clues to the relationship among genes, health, and disease result from these efforts. However, understanding the molecular aspects of genetics is by no means the entire story.

Human Biology: An Evolutionary and Biocultural Perspective, Edited by Sara Stinson, Barry Bogin, Rebecca Huss-Ashmore, and Dennis O'Rourke
ISBN 0-471-13746-4   Copyright © 2000 Wiley-Liss, Inc.

Most of the earliest research in genetics revolved around understanding the means by which genes are passed from one generation to the next. Genetics was concerned with the formation of specialized cells that carry each parent's contribution to the next generation: How are these cells formed, and what are their special characteristics? What rules cover their contents? What are the processes that surround conception? Can we predict the passage of a trait from one generation to the next? These questions, too, direct geneticists' investigation.

Genetic investigation also entails understanding the behavior of genes within populations. Human populations differ in their adaptations and evolutionary histories. The field of population genetics is most focused on explaining how the history of groups can be deciphered from the analysis of their genes. Differences between groups in their adaptations to stresses also may be reflected in the genetic composition of different human groups.

Answering the questions of genetics is a huge task, and the complete answers (as far as we know them) would require explanation far beyond the scope of this chapter. However, this chapter is a brief sketch of some important aspects of molecular and cellular genetics as they apply to the general topic of human biological variation and adaptation. The relationship among population genetics, evolution, and the statistical analysis of genetic phenomena provides the basis for Chapters 4 and 5.

BASIC GENETICS

By now, you probably are familiar with the basic concepts of genetics that Gregor Mendel (1822–84) started to work out in the mid-1800s. Mendel was a monk who lived in Brno, a city in what is now the Czech Republic. His interest in science and heredity led him to conduct a series of experiments on the passage of traits across generations. As his experimental organism, he chose the sweet pea plant, partly because they had a number of easily scored, variable characteristics. A plant could have yellow seeds or green seeds, be tall or short, and so on. In following the passage of these traits from generation to generation, he noticed that sometimes progeny resembled one parent, sometimes the other—and sometimes neither.

Mendel used pea plants to deduce that each sexually reproducing organism has two copies of each **gene**, the basic unit of inheritance; each parent contributes one copy. The totality of a cell's genetic material is called its **genome**, and cells that contain two genes for each trait are in the **diploid** (Greek *diploos* = double) state. When an organism matures and prepares to produce its own offspring, it makes sex cells that contain one representative of each of the genes. Genes are "particulate factors" that do not blend together like paints in a pot but rather maintain their integrity as they pass from generation to generation; this concept forms the basis of Mendel's "Principle of **Segregation**" (or Mendel's first law). It states that sexually reproducing organisms possess two genes for each trait, but only one of each pair is passed on to each offspring from each parent.

Meiosis is the process by which sex cells, also called **gametes**, are produced. Thus, sex cells have only one-half the complement of genes of diploid body cells and thus are called **haploid** (Greek *haploos* = single). A gamete has one copy of a gene for each trait, instead of the two found in the body cells. During formation of gametes, the two genes for a trait separate—or segregate—from each other. A gamete does not receive two copies of the gene for one trait and zero copies of the genes for another. We also can say that each gamete has one copy of the gene for each trait. With the union of the haploid gametes at fertilization, the full diploid complement of genes is reestablished. Subsequent replication of this fertilized egg via a process called **mitosis** yields the many cells that make up the adult.

The particular versions of the gene that an individual has for a trait may or may not be identical. Some traits are so important that there may be little or no variation tolerated by **natural selection;** any individual with a variant dies. Some traits may tolerate mild variation, whereas yet others may go beyond tolerance and evolve to create significant variation.

Alternative versions of a gene are called **alleles** (Greek *allos* = other). In humans, we speak of the three alleles (*A*, *B*, and *O*) that result in different ABO blood types; one's blood type is a result of the alleles inherited from mother and father.

Because of a phenomenon known as **dominance**, the relationship between the genes one has for a trait and one's outward appearance is not always obvious. The genetic constitution is called a **genotype**, and the appearance is the **phenotype** (Greek *phainen* = to show). A person with two copies of the *O* allele for ABO blood groups (one inherited from the mother, one from the father) will have type O blood. However, because the *A* allele is dominant over *O* (i.e., *A* hides the presence of *O*), a person with one *A* allele and one *O* allele will have type A blood. Alleles that code for type O blood—or for any other trait that is not dominant—are called **recessive.**

Mendel, in his experiments on pea plants, had the brilliance to develop this idea of dominance and recessiveness. He went on to claim that the recessive allele, even though it is not apparent, is still present and can be passed on to subsequent generations. An individual with two identical versions of the gene (for instance, two A alleles for ABO blood type) is said to be **homozygous** (Greek *homo* = same; *zygos* = joined). An individual with two different alleles (e.g., an *A* and an *O*, or an *A* and a *B*) is genotypically a **heterozygote** (Greek *heteros* = different).

In referring to an organism's genotype, we talk about having genes for this trait and that trait. However, genes do not float as individual units within a cell. Rather, they are packaged into structures called **chromosomes** (Greek *khroma* = color; *soma* = body), which we can see quite clearly under the microscope. The normal human condition is to have 23 pairs of chromosomes, or 46 chromosomes. Members of a pair of chromosomes are said to be **homologous**; that is, they have genes for the same traits, but the exact form of the genes might not be identical. Different organisms have different numbers of chromosomes, and the number of chromosomes does not reflect the complexity of an organism. Some primates have more chromosome pairs than humans (e.g., chimpanzees have 48), and other primates have fewer.

The genes for a specific trait occupy a particular location or **locus** (plural = loci) on a specific chromosome; *ABO* genes are located on chromosome 9. Genes for many other traits also are found on chromosome 9, and genes for two traits that are

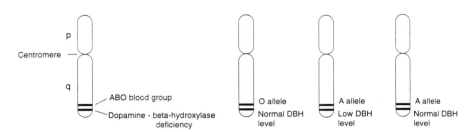

Figure 3.1 A simple rendering of human chromosome 9, showing two loci located on the long arm. Loci that control ABO blood groups and dopamine-β-hydroxylase (DBH) are linked (left-most chromosome). The other chromosomes illustrate three of six possible haplotypes, assuming that there are three alleles at the ABO locus and two at the DBH locus.

closely located on a chromosome are said to be **linked**. As genetic technology has expanded the number of linked variable traits that can be detected, scientists have found it useful to refer to **haplotypes**. Haplotypes are specific combinations of alleles for two or more linked loci (see Figure 3.1). For instance, two linked loci on human chromosome 9 are those for ABO blood groups and for an enzyme important in nerve transmission known as dopamine-β-hydroxylase (DBH) (Wilson et al. 1988). One haplotype might be the association of the *O* allele for the *ABO* locus and an allele for normal levels of DBH. Another haplotype for this region of chromosome 9 may associate the *A* allele for *ABO* and the allele for production of abnormally low levels of the DBH.

Depending on just how close any two loci are on a chromosome, one may find that the haplotypes are reconstituted through **recombination**. Recombination involves members of a chromosome pair lining up during meiosis, breaking, and crossing over so that genetic material is exchanged between members of the chromosome pair, and a new haplotype may be formed (see Figure 3.2).

For many years, anthropologists and human biologists concentrated their investigations of human genetic diversity on the frequencies of alternative forms of a single gene. Now, with the ability to easily study haplotypes, the frequencies of combinations of genes can be studied. For loci located close to each other on a chromosome, one often finds that a variant for one trait is often more commonly found in association with a particular variant for the second trait than is expected simply based on chance; this phenomenon is known as **linkage disequilibrium**. By detecting these deviations from expectations, linkage studies can uncover clues as to the operation of evolutionary forces in human populations as well as being medically useful.

Genes that control different traits may be linked; however, traits on different chromosomes (or even located far apart on the same chromosome) are **unlinked**. Mendel

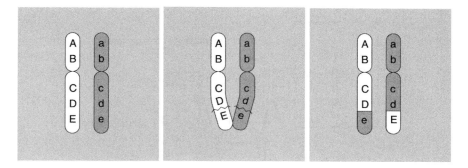

Figure 3.2 Schematic representation of crossing over. Two homologous chromosomes contain genes for the same traits but may have different versions, or alleles, for the traits. New linkage combinations can be generated when the chromosomes break and cross-unite.

noted this observation in his "Principle of **Independent Assortment**," which is sometimes referred to as his second law. Independent assortment means that un-linked genes are sorted into sex cells and passed on to the next generation independently of each other. In other words, the allele you inherit from a parent for ABO blood type has no relationship to the alleles you inherit from that parent for traits located on other chromosomes.

GENETIC MATERIAL

The basic knowledge of Mendelian heredity was worked out over the course of decades of investigation. Mendel described the basics of the behavior of the genes. However, knowledge of the genetic material's composition and the means by which a gene carries out 'its functions are much more recent, ongoing discoveries. In the following sections, I present some of the recently discovered complexities of defining a gene.

Although the definition of a gene as a unit of inheritance is still useable, more elaborate and refined statements now can be made about the hereditary material. Since the 1950s, we have known that genes are composed of substances called **nucleic acids**. Almost universally, the nucleic acid is deoxyribonucleic acid (**DNA**). However, some viruses lack DNA, and the hereditary material in its place is ribonucleic acid (**RNA**). These two nucleic acids share several features, and nucleic acids are capable of transmitting information in a coded form because of these shared features. In humans, DNA is the genetic material, but we also have many forms of RNA that help DNA carry out its tasks.

Nucleic acids are constructed from three elements. Both DNA and RNA have a backbone that is composed of two of the components: alternating sugar molecules and phosphate groups. In DNA, the sugar is 2-deoxyribose, and in RNA, it is ribose

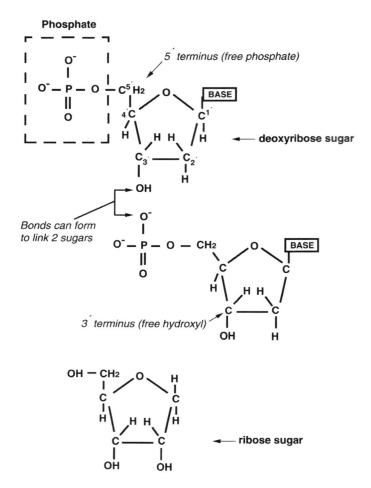

Figure 3.3 Structures of the nucleotides found in deoxyribonucleic acid (DNA). Carbon atoms of the sugar are numbered as indicated. The 3′ carbon of one sugar can bond to the 5′ carbon of another via the phosphate group. The sugar of ribonucleic acid (RNA), ribose, differs in that the 2′ carbon has a hydroxyl group.

(hence the names of the nucleic acids). As shown in Figure 3.3, the sugar molecules have a 5-carbon ring, and by convention, the carbon atoms are numbered as indicated. The 5′ carbon of one sugar is linked to the 3′ carbon of the next sugar. Through bonds involving the phosphate group, a long chain of alternating sugar and phosphate groups can be formed, and the chain has directionality. That is to say that at one terminus of the chain is a sugar that has a free phosphate on its 5′ carbon, and the other terminus has a free hydroxyl group on its 3′ carbon; these termini are called the 5′ and 3′ ends. Sometimes the 5′ end is called the upstream end and the 3′ is downstream, because, as you will see, DNA is replicated and read from the 5′ direction to the 3′ direction.

The third unit found in nucleic acids are nitrogen-containing **bases**. In DNA, four nitrogenous bases can be divided into two groups based on their chemical structure: **purines** and **pyrimidines**. Adenine and guanine are purines; thymine and cytosine are pyrimidines. Purines are a class of molecules with fused 5- and 6-carbon rings, whereas pyrimidines are molecules that have a 6-carbon ring (Figure 3.4). In RNA, the pyrimidine base uracil is substituted for thymine, rounding out the total of five bases used in building the nucleic acids.

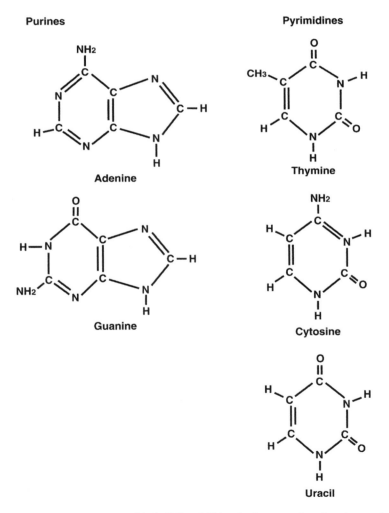

Figure 3.4 The four bases used in building DNA: adenine, guanine, thymine, and cytosine. RNA uses uracil in place of thymine.

A unit of one sugar, one phosphate group, and one base is called a **nucleotide**. When nucleotides are linked into long chains via 5′–3′ bonds, the structure is called a polynucleotide chain.

The sugars and phosphates form a long chain of repeating units, and the bases angle off this backbone at the sugar's first carbon position. If native DNA existed in this condition, these bases would be projecting into space. However, the bases form bonds with other bases in a very specific, predictable manner.

Because of this bonding ability, DNA occurs in a double-stranded configuration in most situations, so that a purine projecting from one of the backbones is bonded to

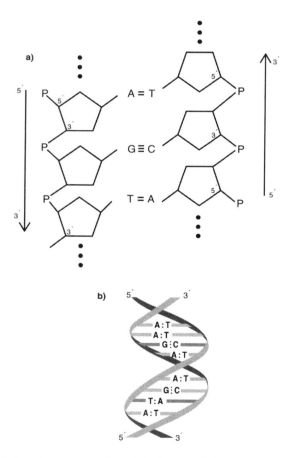

Figure 3.5 (a) Schematic representation of double-stranded DNA. The two backbones run in opposite directions. Adenine (A) always bonds to thymine (T) via two hydrogen bonds; guanine and cytosine (C and G) bond via three hydrogen bonds. (b) The sugar-phosphate backbones of double-stranded DNA twine around each other in a double helix.

a pyrimidine projecting from the other. In this way, the diameter of the DNA double strand is maintained at a regular diameter, ~20 angstroms (Å; 1 Å = one-ten millionth of a millimeter), over its length. Specifically, cytosine (pyrimidine) bonds to (and only to) guanine (purine), and the adenine (purine) bonds to thymine (pyrimidine) (see Figure 3.5). Because pyrimidines have only a single carbon ring, a pair of them would not be long enough to form a bridge; two purines would be too large. Thus, wherever cytosine projects from one strand, its opposing strand will contain guanine. The forces that hold these two bases together are three hydrogen bonds. In DNA, wherever a strand contains thymine, the opposing strand will contain adenine. These two bases are bonded by two hydrogen bonds. Adenine and thymine are said to be **complementary** to each other, as are guanine and cytosine. Although hydrogen bonds are rather weak chemical "glue," so many of these bonds unite the two strands that the resulting double strand is quite rigid.

DNA, then, is generally double-stranded. The chemistry of polynucleotide chains has two consequences of importance to us. The first is that it requires that the strands of DNA wrap around each other to form a double helix. You can picture the double helix as a ladder that has been twisted along its long axis into a spiral. The sugar-and-phosphate backbones form the sides of the ladder. Each base projects into the center space, where it bonds to its complementary base that extends from other backbone (Figure 3.5), forming the rungs of the ladder.

The second consequence is that the two backbones or polynucleotide strands run counter to one another. If we draw a double helix so that the 5′ end of one is at the top, then the 5′ end of the other strand will be at the bottom. The strands are antiparallel. As you will see, this directionality has great significance for DNA's function as well as our ability to manipulate and study the genetic material.

RNA, unlike DNA, exists as a single-stranded molecule, and as noted earlier, it serves as the genetic material in some viruses. In humans, RNA facilitates the completion of one of DNA's roles but is not the genetic material itself. For our purposes, it is most important for its involvement in the production of **proteins**.

DNA'S TWO ROLES

DNA needs to be able to do two things. First, by virtue of being the genetic material, it must be able to transmit information from one generation to the next. This task involves two related functions. The first function is that information must be carried across generations from parent to offspring, and after a new fertilized egg is formed, the information must be transmitted from the one initial cell to each of the descendent cells within the developing individual. In either case, the basic requirement is that the DNA must be able to replicate itself so that each descendent cell contains the required amount of DNA. The second function pertains to the translation of the information contained within the DNA, the reading of the genetic information, so to speak. In the following sections, I briefly describe the means by which each of these tasks is carried out.

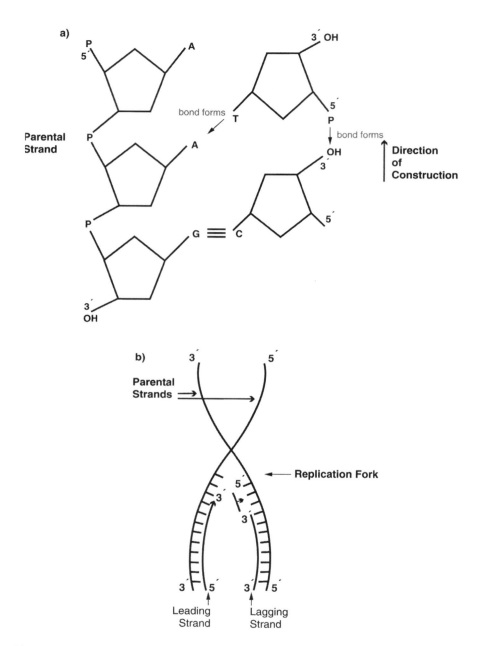

Figure 3.6 (a) Construction of DNA strands complementary to the parental strands. The newly constructed strand is bonded to its parental complement. (b) A larger scale view of DNA replication showing the leading and lagging strands and the replication fork.

DNA Replication

The ability of DNA to create copies of itself with great fidelity rests on the specificity with which bases bond to each other. At the time a cell is making new copies of the DNA, a series of events takes place which result in the production of two identical double helices from the original one.

DNA replication is **semiconservative**. Consider a segment of double-stranded DNA. We can call each of its two strands "parental." It has been experimentally demonstrated that when the parental DNA replicates, each of the two descendent double helices contains one of the "parental" strands and one newly synthesized strand. In other words, each of the parental strands has directed the production of its own complement, with which it then bonds into a new double helix (see Figure 3.6a). In very simple terms, this ability is, again, based on the complementarity of the bases; wherever one strand is constructed with the base adenine, the other strand will have the base thymine, and vice versa. Likewise, where one has guanine, the other will have cytosine.

How does this semiconservative replication proceed? A region of the double helix "melts"—that is, it opens up—as the weak hydrogen bonds between the opposing bases are broken. This opening leaves each strand with a segment where the bases are not bonded. These free sections serve as **templates** for the construction of new strands, and the area where the double strand has been opened is known as a replication fork. **Enzymes** (proteins that speed up reactions), specifically one group known as DNA polymerases, are involved in replication. Mammals have five polymerases, but the ones known as DNA polymerase-δ and polymerase-α are the two most important.

Consider again the directionality of DNA that results from the way sugar and phosphate are bonded in the backbone. DNA polymerases act by adding nucleotides in the 5′-to-3′ direction of a growing polynucleotide chain (see Figure 3.6b). Looking at the replication fork in Figure 3.6b, one can see that it is a relatively simple matter to keep exposing more and more of the 3′ end of one strand to serve as a template for replication. This strand is called the leading strand, and DNA polymerase-δ is involved in leading strand synthesis.

However, as the fork moves upward in the illustration, more and more of the 5′ end of the second strand is being exposed, and DNA polymerases cannot add nucleotides to the 5′ end. So, how is the second strand replicated? DNA polymerase-α builds short segments in the 5′-to-3′ direction that are complementary to the exposed part of the second strand. These segments then are added to the growing strand, which is called the lagging strand. The mechanics of constructing the lagging strand are rather complicated and involve several other steps.

One additional factor is necessary for DNA replication to commence: the presence of a primer molecule. Polymerases cannot start adding nucleotides simply anywhere along a segment of DNA. They need to have a primer. Several different sorts of DNA primers are known. They can be proteins or RNA, but the primer always provides the free 3′ end to which nucleotides can be added. The net result of all of the action of the primers and polymerases is the production of two double helices of DNA, each identical to the original double strand (see Figure 3.7).

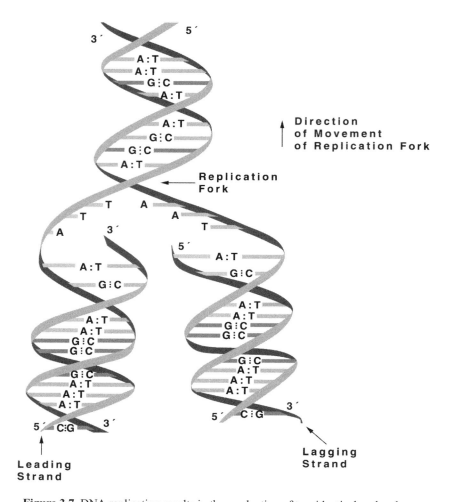

Figure 3.7 DNA replication results in the production of two identical molecules.

The fidelity of DNA replication is quite high; that is, DNA has the ability to reproduce itself with few errors. Generally, wherever the template strand of DNA contains a guanine, its complement will be constructed with a cytosine; where is the strand contains an adenine, the complement will be constructed with a thymine, and so forth. However, the system is not foolproof. Sometimes the wrong base is inserted into one of the growing chains. For instance, consider a position in the DNA that has guanine on one strand and cytosine on the other. Through a chemical reaction, the cytosine may be converted into a uracil.

Even though uracil is not usually found in DNA, it is a pyrimidine that can bond to adenine. When this segment of DNA that contains the mismatched uracil–guanine

(U–G) pair undergoes replication, one of the resulting molecules of DNA will be as it should: The double helix produced when the guanine-containing strand serves as the template will have a cytosine–guanine (C–G) pair at this position, as it originally did. However, the other new double helix will be produced using uracil as the template at this position. Because uracil is complementary to adenine, the original C–G pair could not result; rather, this double helix will have a uracil–adenine (U–A) pair. In subsequent rounds of replication, thymine would be substituted for the uracil. In effect, this substitution would change the bases from C–G to thymine–adenine (T–A). Organisms contain various means to recognize mismatched base pairs such as the U–G. DNA repair mechanisms often can rectify the error. Sometimes the error escapes detection and is the basis for some of the genetic variation we see in the world. An alteration in the sequence of the DNA, through this mechanism or others, is called a **mutation**.

DNA Makes Protein

We have an idea how the DNA can reproduce itself, but what does DNA actually do inside the cell? The short answer is that most genes direct the production of proteins or protein subunits, called **polypeptide chains**, which are then assembled into proteins (or another set of products, which I will discuss later).

What exactly are proteins? Structurally, they are molecules constructed out of a series of building blocks called **amino acids**. Twenty such amino acids are arranged into resulting protein molecules that are capable of carrying out different functions. Some proteins are rather small, composed of just a dozen or so amino acids, whereas others are many times larger. One protein involved in blood clot formation is called fibrinopeptide, and primate fibrinopeptides are generally 9–15 amino acids long. The most concentrated protein in the blood is albumin and it is ~600 amino acids long. CFTR (cystic fibrosis transmembrane conductance regulator) is a protein involved in transporting chloride ions across cell membranes and when defective can result in the disease cystic fibrosis. It is 1480 amino acids long.

The sequence of bases in the DNA tells the cell's machinery which amino acids to insert, one by one, into a chain of amino acids. This genetic message is conveyed from generation to generation as the DNA is replicated and passed along. DNA technology has made it a relatively simple matter to determine the sequence of a particular piece of DNA, so let's look at an actual example. Figure 3.8 lists the DNA sequence of part of the gene that produces one of the components of hemoglobin, the β-chain, in most human beings. Here is how the coded message in the DNA is converted into the final product.

The DNA resides in a special compartment of the cell called the **nucleus**. However, proteins are not produced in the nucleus. They are assembled in association with **ribosomes**, which are structures found outside the nucleus, in the **cytoplasm**. So, there must be a process by which the information contained in the DNA is transmitted from the nucleus to the site of protein manufacture. This process is diagrammed in Figure 3.9. Rather than having the DNA journey out of the nucleus, a

62041 gagccacacc ctagggttgg **ccaat**ctact cccaggagca gggagggcag gagccagggc

                                   Cap

62101 tgggc**ataa**a agtcagggca gagccatcta ttgctt**acat** ttg**cttctg**a cacaactgtg

                     EXON1 Met Val His Leu Thr Pro Glu Glu Lys Ser Ala Val

62161 ttcactagca acctcaaaca gacacc   ↑ atg gtg cac ctg act cct gag gag aag tct gcc g

       Thr Ala Leu Trp Gly Lys Val Asn Val Asp Glu Val Gly Gly Glu Ala Leu Gly ↓INTRON1

62221 tt act gcc ctg tgg ggc aag gtg aac gtg gat gaa gtt ggt ggt gag gcc ctg ggc ag   gt

62281 tggtatcaag gttacaagac aggtttaagg agaccaatag aaactgggca tgtggagaca

62341 gagaagactc ttgggtttct gataggcact gactctctct gcctattggt ctattttccc

62401 acccttaggc tgctggtggt ctacccttgg acccagaggt tctttgagtc ctttggggat

62461 ctgtccactc ctgatgctgt tatgggcaac cctaaggtga aggctcatgg caagaaagtg

62521 ctcggtgcct ttagtgatgg cctggctac ctggacaacc tcaagggcac ctttgccaca

Figure 3.8 DNA sequence of a 5′ segment of the human β-globin gene. This sequence is a fraction of a 73,308-base sequence of a β-globin cluster obtained from the National Center for Biotechnology Information (NCBI: www.ncbi.nlm.nih.gov). Numbers at left refer to positions within the NCBI sequence. Arrows show boundaries of the first exon. The Cap site is involved in the modification of mRNA so as to increase its longevity. Other sequences that regulate transcription are shown in boldface. The three-letter abbreviations above the bases of exon 1 indicate the amino acids coded in this protein.

coded copy of the information is constructed by using a version of the nucleic acid RNA; because this molecule will carry the DNA's message, it is called messenger RNA (**mRNA**). Earlier, I mentioned that RNA differs from DNA in two ways: The sugar in its backbone is slightly different from that of DNA, and it uses uracil instead of thymine. The process by which mRNA is constructed is called **transcription**.

During transcription, a short segment of the DNA double helix is separated, exposing the bases of each of the strands. We know that wherever unpaired guanine juts off a strand of DNA, the complementary base cytosine will be attracted and likewise for the other complementary pairs. As in replication, polymerase enzymes (in this case, RNA polymerase) are needed for the reaction to proceed, as are a number of other proteins called transcription factors. Three different RNA polymerases exist, but the one responsible for mRNA construction is called polymerase II. The transcription factors and RNA polymerase II are first attracted to a sequence of DNA bases that promote the initiation of the process. This sequence is called a promoter, and although it does not ultimately affect the structure of the protein product, it is considered part of the gene. The importance of the promoter and other 5′, or upstream, sequences is that whereas their detailed structures may vary between organisms and from gene to gene within any organism, the sequences are often broadly conserved from species to species, and from gene to gene. This conservation indicates significant evolutionary importance. Conserved upstream sequences for hemoglobin's β-chain are indicated in Figure 3.8.

RNA polymerase forms a small "bubble" where it locally unwinds the double helix of DNA. Using only one of the exposed DNA strands, it constructs a complement to that strand out of ribonucleotides. By convention, we call the DNA strand that is used as a template the "antisense" strand; the other DNA strand is the "sense" strand. The RNA molecule is a complement to the antisense strand and thus has the same sequence as the DNA's sense strand, except that it contains uracil where the sense strand has thymine. As RNA polymerase moves down the DNA toward the end of the gene, the mRNA molecule rolls off the DNA, and the two DNA strands reunite. When the polymerase reaches another important sequence at the end of the gene, the termination sequence, it detaches from the DNA.

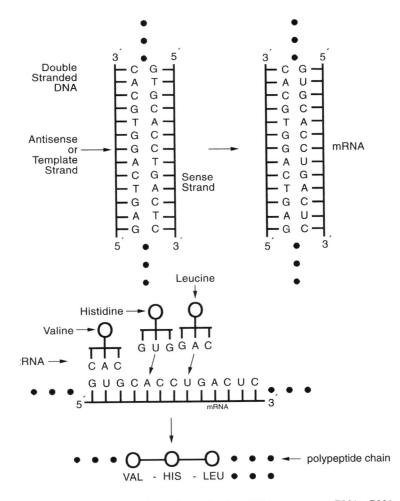

Figure 3.9 A simplified view of protein synthesis. mRNA, messenger RNA; tRNA, transfer RNA; Val, valine; His, histidine; Leu, leucine.

The completed mRNA then moves from the nucleus to the ribosomes, where the polypeptide chain is actually assembled during **translation**. A protein is constructed out of amino acids. Because the chemical bonds that unite one amino acid to another are peptide bonds, a string of amino acids is called a polypeptide chain. If this chain is capable of performing a function, then it is called a protein.

FIRST POSITION	SECOND POSITION				THIRD POSITION
	U	C	A	G	
U	UUU } Phe UUC } UUA } Leu UUG }	UCU UCC UCA Ser UCG	UAU } Tyr UAC } UAA } Stop UAG }	UGU } Cys UGC } UGA } Stop UGG } Trp	U C A G
C	CUU CUC CUA Leu CUG	CCU CCC CCA Pro CCG	CAU } His CAC } CAA } Gln CAG }	CGU CGC CGA Arg CGG	U C A G
A	AUU } AUC } Ile AUA } AUG } Met	ACU ACC ACA Thr ACG	AAU } Asn AAC } AAA } Lys AAG }	AGU } Ser AGC } AGA } Arg AGG }	U C A G
G	GUU GUC GUA Val GUG	GCU GCC GCA Ala GCG	GAU } Asp GAC } GAA } Glu GAG }	GGU GGC GGA Gly GGG	U C A G

The twenty amino acids commonly used in building proteins and their abbreviations

Ala	Alanine
Arg	Arginine
Asn	Asparagine
Asp	Aspartic acid
Cys	Cysteine
Gln	Glutamine
Glu	Glutamic acid
Gly	Glycine
His	Histidine
Ile	Isoleucine
Leu	Leucine
Lys	Lysine
Met	Methionine Methionine usually serves to intiate protein synthesis
Phe	Phenylalanine
Pro	Proline
Ser	Serine
Thr	Threonine
Trp	Tryptophan
Tyr	Tyrosine
Val	Valine

Figure 3.10 The genetic code. Three-letter codes in the mRNA indicate the amino acids they specify. Read down the left margin for the first letter, across the top for the second letter, and down the right margin for the third letter. The resultant codes and the abbreviations for the amino acids they represent are given in the body of the table; abbreviations are defined at the bottom. Thus, AAA is lysine (Lys), and AGU is serine (Ser).

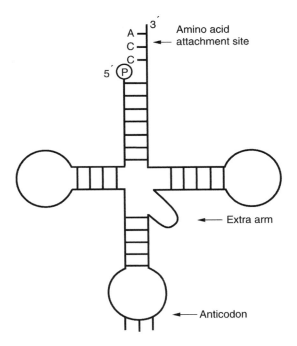

Figure 3.11 Representation of the structure of tRNA. The anticodon consists of three bases. tRNA with a particular anticodon will always carry a specific amino acid at the attachment site.

In the ribosomes, the bases making up the mRNA are read in codes of three letters. We know that 20 amino acids are used in making proteins. If single bases specified amino acids, there would only be four instructions—not enough to do the job. If a set of two bases constituted an instruction, 16 combinations would be possible (AA, AC, AG, AU, CA, CG, CU, CC, etc.)—still not enough to do the job. With three-letter codes, 64 different combinations are possible (Figure 3.10)—more than enough to specify the 20 amino acids.

In fact, the many combinations make redundancy possible. More than one "word" can mean "insert a particular amino acid." This reasoning seems sound, and experiments in the 1960s showed that it is correct. These three-letter codes or words are called **codons**. Furthermore, we know that codons are read in a nonoverlapping sequence. That is, bases 1, 2, and 3 are read as one unit, a codon; bases 4, 5, and 6 constitute the next codon. Bases 2, 3, and 4 do not operate as a codon. In Figure 3.10, all possible codons are shown, with the amino acids that they specify. With very rare exceptions, the genetic code is universal. The codon UUU, for instance, codes for phenylalanine in all organisms, from wheat plants to fish and humans.

Protein synthesis involves one more set of molecules at this stage: transfer RNA (**tRNA**). These single-stranded RNA molecules are ~75–100 base long (Figure 3.11). They all have a characteristic structure called a four-armed cloverleaf. In fact, some genes are not translated into protein but function directly as RNA; this class includes the DNA that directs the production of tRNA molecules.

In tRNA, the most important feature is the **anticodon**, which is a sequence of three bases in the middle of the anticodon arm. The three bases of the anticodon are capable of bonding with the three bases of a codon in the mRNA. Each tRNA with a particular anticodon on one arm will have a specific amino acid on the "acceptor arm" of the cloverleaf. The ribosomes facilitate the union of the mRNA and its complementary tRNA. After the two RNAs are linked, the amino acid on the acceptor arm is bonded to the growing polypeptide chain, which is then detached from the tRNA.

In Figure 3.10, three codons are labeled Stop. They are termination codons and are not represented by any tRNA. When one of these codons is reached, no amino acid is added to the chain, and the polypeptide chain is released to perform its function.

There are Stop codons and there is also a Start codon: AUG. Reference to the genetic code shows that AUG specifies the amino acid methionine. Two versions of tRNA complement this codon. One inserts a methionine wherever necessary in an elongating polypeptide chain, but the other version has several special properties. The second kind of tRNA is always used to initiate the construction of a polypeptide. The initial methionine may be removed from the final polypeptide chain.

The result of protein synthesis is a chain of amino acids. The sequence of amino acids is called the primary structure of the polypeptide. The polypeptide backbone can, however, fold into a secondary structure. Furthermore, the final three-dimensional placement of all the atoms in the polypeptide is called its tertiary structure. A protein composed of more than one polypeptide chain is said to be multimeric in organization. Other structures, such as sugar molecules, also can be incorporated into the final protein's structure. These additions to the protein can route it to its proper location within the cell. Proteins that must enter the nucleus to carry out their functions, for instance, have several short markers added that allow them to cross the nuclear membrane. In this way, proteins find their appropriate locations in the body.

CLASSIC MARKERS

Proteins carry out a multitude of tasks; they are produced in a tremendous variety of amino acid sequences and hence many shapes and sizes. Some proteins form the structural components of cells; others perform enzymatic functions, catalyzing metabolic reactions; yet others transport nutrients or gases through the blood or across membranes; and some, such as the ABO blood group molecules, help the immune system to identify which substances and cells belong in our bodies.

For many decades, anthropologists and human biologists were able to study molecular diversity only at the level of the proteins. In the early 1900s, Karl Landsteiner (1868–1943) started to unravel the molecular basis of the ABO blood groups, and some time later, the genetics of the system was described. Over the course of much of the 20th century, a sizable number of genetically variable proteins were described in humans. Anthropologists and human population biologists used the ABO blood groups and a host of other variable proteins as **markers**, or tags, for tracing human population histories and adaptations. If two populations showed similarities in the frequencies of a number of these classical markers, one might infer that they shared a recent common ancestor, or possibly similar genetic adaptations to environmental stresses. Several compendia of this information are now available (Cavalli-Sforza et al. 1994; Mourant et al. 1976).

We now know that the proteins are several steps removed from the genes. Because of this, comparisons of proteins tend to underestimate genetic differences; most of the techniques used to study proteins are blind to at least some underlying differences. Yet, until the techniques of the biotechnology revolution became available, allowing the study of the genes themselves, the classic markers provided a highly informative surrogate for assessing genetic variation.

STUDYING THE DNA

We all have heard about the biotechnology revolution—cloning and the like. The massive increase in knowledge of the hereditary material has largely come about through a few major developments during the past 20 years or so. One was the development of efficient and simple means to modify the DNA in a predictable fashion.

To a great extent, this capability relies on a class of enzymes called **restriction enzymes,** or restriction endonucleases. These enzymes can recognize specific DNA sequences, usually four or six bases long. The enzyme then cuts the DNA in a defined way. Figure 3.12 shows the enzyme EcoR1 cutting DNA at its target sequence of GAATTC. Whenever this target is encountered, the DNA is cut between the G and the A. Notice that a sequence of GAATTC on one strand will correspond to the sequence of CTTAAG on the other. Remembering that the two strands run antiparallel, we see that the DNA of both strands of the double helix will be cut. Restriction enzymes are a powerful tool, allowing the predictable cutting of DNA. Because of the predictability, one can then engineer ways of recombining the DNA in a predictable fashion to produce recombinant DNA.

A second development with major impacts on human genetics was the invention of a technique called the **polymerase chain reaction (PCR)**. Under ideal conditions, PCR allows a researcher to start with as little as one copy of a piece of DNA and produce almost limitless copies (Figure 3.13).

This ability to make many copies of DNA from a small number of original molecules is of great value in many anthropological settings. It allows detailed analysis of human variation from small quantities of starting material. DNA donated by research subjects can be conserved. This benefit is particularly important if it is unlikely that the donor can be contacted again for further assistance. Because anthropologists often conduct research that involves people from remote locations, this point is very important. Ancient DNA reclaimed from bone or other tissue samples—again, material in very short supply—may be the starting point, providing a time depth to evolutionary studies. Minute quantities of genetic material gathered at crime scenes can

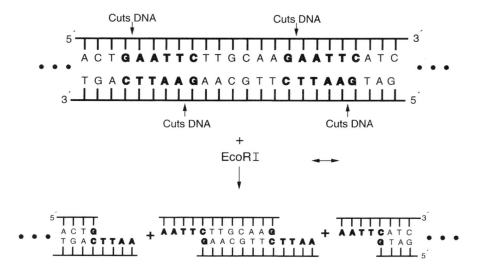

Fig. 3.12 The restriction enzyme EcoRI cuts double-stranded DNA at the sequence GAATTC. The overhanging "sticky ends" serve as a means to reunite DNA in new combinations.

be analyzed in fine detail for the purpose of criminal proceedings. Likewise, samples from accident scenes can aid in human identification.

Gene Structure

Living things can be divided into two types based on the structure of their cells: those with a nucleus to house their genetic material, and those without. Organisms that do have a nucleus are known as **eukaryotes**, and those that do not (e.g., bacteria) are called **prokaryotes**.

For many years, it was thought that the structure of genes in prokaryotes and eukaryotes was the same. The DNA had a base sequence that was colinear with the polypeptide chain that it produced; that is, every three bases in the DNA directed the insertion of an amino acid into a polypeptide chain. In the late 1970s, it was discovered that this is generally not the case in eukaryotes. Eukaryote genes are split into alternating DNA segments; those that code for amino acids are said to be expressed, but intervening sequences of DNA, although transcribed into mRNA, are not translated into an amino acid sequence. A DNA sequence that is expressed is called an **exon**, and the unexpressed sequences are called **introns**, or intervening sequences (IVSs).

The totality of a gene's DNA is transcribed into mRNA. Eukaryotic cells have evolved a mechanism that splices out the mRNA complement of introns and splices that of the exons together in the same order that they appear in the DNA (Figure 3.14). The mature mRNA is then transported to the ribosomes for protein synthesis. So that the splicing will be predictable, introns have sequences at each end that signal the splicing sites.

As one might guess, these signals are quite conserved in evolution. A mutation that alters a splice site would likely affect the ultimate structure of the protein under construction, which in turn could affect the organism's reproductive prospects. Other than this constraint, the sequence of bases in an intron appears free to vary within wide limits; it doesn't matter much what the intron's DNA sequence is, but it does appear important that the intron be present and of a certain size.

Exons, because they direct the construction of proteins, are much more highly conserved in evolution. A close look at DNA sequences indicates that exons tend to be shorter than introns. Some introns are in fact tens of thousands of bases long.

One interesting observation that may prove to be important in evolutionary studies is that some genes have alternative splice sites. Thus, an exon's transcript is spliced to that of a second exon under one set of conditions; under other circumstances, the first transcript is spliced to that of a different exon. Therefore, the proteins that are produced would be partly the same (the part specified by exon 1) and partly different.

BOX 3.1   THE POLYMERASE CHAIN REACTION

The polymerase chain reaction (PCR) is a technique that allows the production of millions of copies of a specific piece of DNA from as little as one precursor molecule. It is not yet an easy matter to replicate the entire DNA from a cell; it is much easier to do so for defined areas.

The technique, which is really quite simple in concept, involves repeated rounds of DNA replication in a test tube. It involves mixing the DNA one wants to replicate, a supply of bases, a DNA polymerase, and primers that can serve as a starting point for the action of polymerase. To replicate the particular segment of interest efficiently, the primers are designed to correspond to short sequences at either end of the target.

As illustrated in Figure 3.13, double-stranded DNA is heated so as to separate, or denature, the strands. As the mixture cools, the primers, which are short (10–20 bases long) pieces of single-stranded DNA bond to their corresponding sequences (shown as shaded boxes in the figure). The two primers then set the boundaries for the segment of DNA that will be amplified. With each primer as a starting point for DNA replication, the polymerase adds nucleotides and synthesizes a complementary strand. This is the end of cycle 1.

In cycle 2, the tube is again heated to denature the two molecules of DNA. Each of the double-stranded molecules forms the basis for constructing new complementary strands. With each cycle, the number of copies of the segment of the DNA is doubled. If the system is totally efficient after 20 cycles, more than 1 million copies of the target region will have been made. Often, the PCR reaction mixture is run through 30 cycles, which produces many millions of copies.

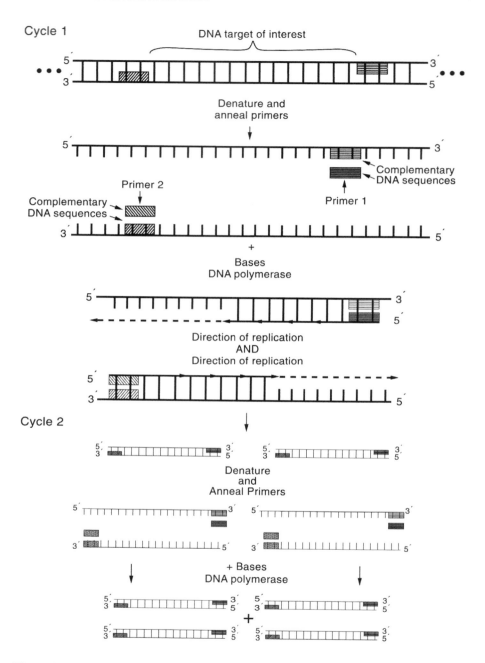

Figure 3.13 The polymerase chain reaction is a technique that allows the production of millions of copies of a specific piece of DNA from as little as one precursor molecule. It is not yet an easy matter to replicate the entire DNA from a cell; it is much easier to do so for defined areas.

This mechanism operates in fruit flies, in which sex determination involves different exon splicing in males and females. Mammals, too, are known to use alternate splice points. In rats, for instance, a DNA sequence is spliced at one junction in the thyroid gland to produce a protein called calcitonin, which regulates calcium levels in the blood. In the rat brain, a different splicing site is used, and the resulting protein acts as a neuropeptide. By making use of these alternate splice points, a single sequence can be used to generate a diversity of products. This "mix and match" approach to protein production could have important consequences for the generation of new proteins with new functions over evolutionary time. As the inheritors of this legacy, modern organisms may have important adaptive abilities that could allow the construction of different products under varying conditions.

To return to the question of defining a gene, the definition at the molecular level is expanded to include not only the DNA that codes for polypeptides or RNA but also introns and sequences like the promoter, which are 5′ to the coding sequence, as well as sequences downstream of the last exon.

Gene Families

Although we talk of genes as being unique sequences, it became apparent early on that some genes are quite similar to others, suggesting both similar functions and a shared evolutionary history.

Gene families were uncovered when researchers went searching for new genes. A coding gene can be located within the mass of a cell's DNA by using a "DNA probe." A probe is a piece of single-stranded DNA with a sequence complementary to part of the gene one is searching for. To search for globin genes, for instance, one might use a probe derived from a β-globin gene exon. If this probe is made radioactive, we have a simple means to trace this molecule. One of the earlier biotechnology techniques used this approach.

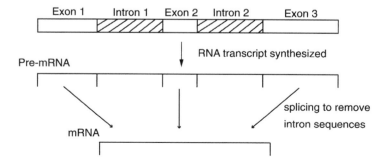

Figure 3.14 Coding sequences in eukaryotic genes are often separated by introns. Introns are transcribed into the pre-mRNA molecule but are removed and the exonic sequences united to form the mRNA.

Let's say that we want to search for DNA sequences similar to that of the β-globin gene. First, human (although it could be from another organism) DNA is cut into many fragments of varying sizes through the use of one or more restriction enzymes. This complex mix is separated by size by placing the mixture in a gel and applying an electric current. The smallest pieces migrate the farthest within the gel; the gel in-

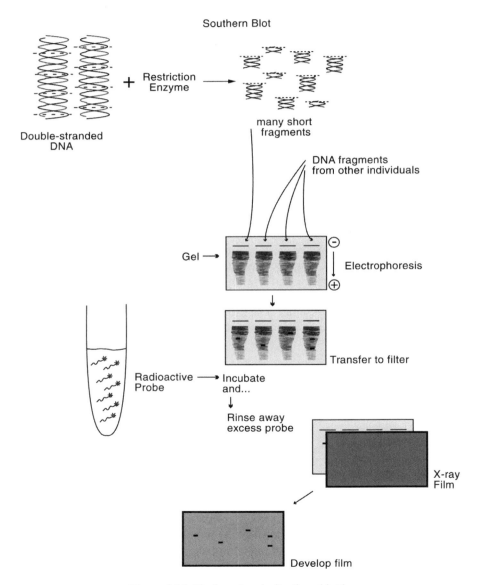

Figure 3.15 The key steps in Southern blotting.

creasingly retards larger and larger pieces. The DNA is transferred onto a filter, fixed in place, and treated so as to make it single-stranded. Then, the probe is constructed.

The probe is made of single-stranded DNA that contains part of the β-globin sequence. One of the bases in the probe sequence is labeled with a radioactive substance, and the radioactive single-stranded probe is incubated with the filter. Where the probe finds its complement, the two DNA strands bind in place. Unbound, excess radioactivity is washed away, and the location of the reaction is detected by layering X-ray film on the filter; where radioactivity is present, the film is darkened, producing bands. If pieces of DNA of known size are run on the same gel, it is possible to estimate the size of each band. This technique is known as Southern blot analysis (Figure 3.15), named after the developer of the technique.

One variable in this process is the stringency or rigor with which one tries to wash off the radioactive probe. If the probe and the gene to which it is bound have a very high level of base matching, then the probe will be tightly bound to the filter, and a very stringent wash would be necessary to wash it away along with the unbound probe. However, if the probe is similar but not identical to the DNA to which it is bound, then it can be washed away easily. The greater the mismatch, the easier it is to dislodge the probe.

Using varying degrees of stringency, accomplished, for instance, by changing the temperature of the wash solution, researchers were able to detect a number of related but not identical human globin genes that are now known to reside on chromosome 11; they are members of the β-family of globin genes. Subsequent determination of the DNA sequences of these genes showed that they have a common evolutionary history. Interestingly, the β-like globin genes are arranged along the chromosome with a polarity; the most 5′ of the genes, called ε, is expressed only during the earliest, embryonic stages of development. The next genes, two γ-genes (in humans, γ^G is the more 5′, λ^A is the more 3′) are expressed later, during fetal development. Further downstream are the globin genes that are expressed after birth, (δ and β). A similar situation was detailed when the α-globin genes were investigated.

Other similarities besides the sequence similarities exist among members of a gene family. They tend to have similar numbers of exons and introns, and the introns often are located in identical places. Such similarities of course are not the result of chance but a common evolutionary history. In addition, it is likely that genes that code for similar proteins will be affected by similar evolutionary constraints.

A process that is responsible for the increase in number of members in a gene family is gene duplication. Segments of the genetic material may be duplicated, producing organisms that have an extra copy of a gene or genes. As long as the organism has one copy of the gene that is performing properly, the "surplus" copy may be relatively free to vary. Mutations and **genetic drift** (see Chapter 4) may ultimately result in an altered DNA sequence that hits on a new, adaptive protein function.

Evolution provides a framework for interpreting the similarities we see among members of a gene family. Just as species have an evolutionary history, and just as one can draw evolutionary trees relating one species to another, one can do the same for gene sequences. After gene duplication, for instance, sequences may begin to diverge. Several factors may affect the rate at which the divergence occurs and the con-

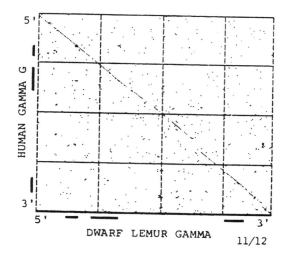

Figure 3.16 Comparison of the human γ^G gene sequence and that of the single γ gene found in another primate, the dwarf lemur.

straints on the divergence. Regardless, it can be a simple matter to demonstrate the similarities between two sequences.

Figure 3.16 demonstrates a marked similarity between γ genes found in humans and in a small nonhuman primate named the dwarf lemur. The long stretches of solid line running along the graph's diagonal indicate areas of the genes that are very similar in sequence (see Box 3.2). Why are they similar? In this case, it is most reasonable to state that it is not due to coincidence. Rather, the similarity reflects a **homology**; the sequence similarity is due to inheritance of the basic structure of the gene from a common ancestral gene.

Are some parts of the modern genes more similar than others, and if so, why? Not surprisingly, one generally finds that exons are more conserved in evolution, whereas intron sequences are freer to accumulate differences. The exons of the α- and β-chains of human hemoglobin, for instance, have a detectable level of sequence homology, demonstrating that these chains arose hundreds of millions of years ago through gene duplication.

As sequences hit upon new adaptive functions, they should experience a rapid rate of evolution. In fact, evidence shows that as the placenta of the higher primates evolved to provide more oxygen during development to a large-brained fetus, the fetal globins underwent rapid evolution.

Other Sorts of DNA: Intergenic DNA, Repetitive DNA, Mini- and Microsatellites

A human body cell contains a tremendous amount of DNA. It is estimated that each nucleus has about 3×10^9 base pairs (bp). Within the genome are sequences of DNA that code for gene products; the genes discussed earlier; and long sequences between the genes, which are referred to as **intergenic DNA**. Making several rough assumptions, some have estimated that a "typical" mammalian gene is ~16,000 bp long, has six or seven exons, and is about five times longer than the mature mRNA that it produces. This DNA is thought to code for the production of about 75,000–100,000 different gene products in humans. Estimating 100,000 genes at 16,000 bp each accounts for about half the genome. The intergenic DNA accounts for the remaining billion or more base pairs. Several assumptions go into making such an estimate, but one thing is clear: The human genome contains a huge amount of coding and noncoding DNA.

It is appealing to assume that the intergenic DNA is debris with no patterning to it—it has sometimes been called "junk DNA"—but evidence indicates that it is not true. Below, I describe some of the components of intergenic DNA.

BOX 3.2   DETERMINING SIMILARITY IN DNA SEQUENCES

Why might one want to determine whether two or more DNA sequences are similar? All of the reasons have to do with some aspect of the sequences' evolution: Are two sequences descended from a common ancestor? Do different parts of a DNA sequence evolve at different rates? Does this tell us something about the way natural selection is affecting the incorporation of mutations into a sequence—that is, are some parts more constrained as to what mutations might be tolerated?

For two very similar sequences, one might be able to just look at the strings of A's, T's, C's and G's and say, "Yes, they are very similar." Other times, the pattern of similarity may be subtle and difficult to detect. In either case, we would like to be able to go beyond saying that a similarity is or is not apparent. We would like to identify parameters such as where the similarities are strongest and the degree to which the two sequences are similar.

Computer algorithms have been developed to help identify sequence similarities. Using one such algorithm, Figure 3.16 illustrates the outcome of a comparison of the human γ^G-gene sequence and that of the single γ-gene found in another primate, the dwarf lemur. First, one specifies exactly how long a segment one wants to compare in the two sequences—this has to do with how similar they have to be to satisfy you that they are indeed so similar that they are likely to be descended from a common ancestor. If we looked at sequences only two bases long, we would find many of short segments that were identical in the two sequences due to chance; these similarities would likely not signify identity due to common ancestry. However, do we want to look at sequences 100 bases long? This sequence is so long that we'd almost never find that two sequences match exactly. Trial and error demonstrates in this case that a comparison of 12 base sequences is reasonable.

Next, do we want to demand that the two sequences be absolutely identical for the 12 bases? Or, are we willing to say that if two 12-base sequences are identical at 11 out of the 12 bases, we will call it a match? Or is a match of 10 out of 12 sufficient? These decisions must be made depending on the particular details of the comparison. For the data shown in Figure 3.16, we chose to require that 11 out of 12 bases in a sequence be identical to score a match.

Now, the sequences being compared are much longer than just 12 bases long. The computer starts with bases 1–12 of the dwarf lemur γ gene, compares that segment with bases 1–12 of the human sequence, and then compares dwarf lemur bases 1–12 with bases 2–13 of the human, with 3–14, and so on to the end of the human sequence. Wherever the comparison matches at least 11 out of 12 bases, the program puts a dot on the graph. Then, the computer starts over, comparing dwarf lemur γ bases 2–13 with bases 1–12, 2–13, 3–14, etc. of the human sequence, again placing dots wherever a match occurs. Against a background of random matches scattered across the graph, a straight line emerges more or less along the diagonal, indicating that the two sequences are very strongly similar.

In the middle of the diagonal, the consistency of matches breaks down. It also breaks down in several other shorter stretches. If you look at the two axes, you see three black lines that indicate the parts of the sequences that correspond to the protein coding areas, the three exons. Separating the three exon sequences are noncoding areas, known as introns. It is not a coincidence that when the sequence of the dwarf lemur first exon is compared with that of the human first exon, resemblance is strong. Likewise, for the other two exon comparisons; the degree of homology is great. When noncoding sequences are compared, the degree of similarity is reduced.

The most likely interpretation of this graph is that natural selection is less tolerant of change in the exon sequences than of change in other areas. If the exon is changed, then the final product of the gene is likely to be altered. This change may be disruptive to the organism and therefore selected against. If an intron is changed, it is not likely to have much, if any, functional significance. Hence, changes in the noncoding areas are more likely to become established in the species' genome. In other words, coding sequences are under stronger selective constraints than noncoding sequences, and noncoding areas can be freer to vary over time.

Pseudogenes

One component of intergenic DNA came to light with the biotechnology revolution, as we obtained the means to rapidly isolate and then determine the DNA sequence of a gene. The β-globin genes were studied early on by using probes designed to isolate the genes, as described earlier. A DNA sequence that bore a striking resemblance to the genes of the family (especially γ^A) but did not correspond to any known protein product was quickly discovered between the human γ^A and the δ genes (Fritsch et al. 1980). Additional investigation (Chang and Slightom 1984) indicated that this sequence was unable to code for protein.

Earlier, I explained that all coding sequences start with the initiator sequence AUG; in this sequence, the initiator is CUA. The sequence also contains a mutation at codon 15 that results in an early termination codon (UGA) and several other de-

STUDYING THE DNA **75**

fects that prevent the sequence from producing a protein. Hence, it was dubbed a **pseudogene** (denoted by the symbol ψ). Additional investigations indicated that this sequence shows significant similarity to a type of globin gene (the η gene) expressed in adult goats; thus, in humans, the gene is variously called the ψβ1 or the ψη gene. Many mammals have pseudogenes in their β-globin gene families.

Globins are by no means the only genes for which pseudogenes are known; in fact most gene families have pseudogenes. Pseudogene members of a gene family may represent the relics of once-functional genes that experienced a mutation that subsequently prevented their expression, although other scenarios are possible.

Repetitive DNA Sequences

Even before the revolution in biotechnology and molecular biology of the 1980s, it was known that the genome contains classes of repetitive DNA sequences. With the development of new techniques, the classes have been further characterized, so many subtypes of repetitive DNA are defined by their size and organization within the genome.

Some of the repetitive DNA is referred to as **satellite DNA**. One type is α-satellite DNA, which in humans consists of a basic 171-bp sequence that is repeated back to back, or in tandem, many times over. The multiple copies of the core sequence repeats are similar but not necessarily identical, and a stretch of tandem repeats can be as much as several million base pairs long. α-Satellite sequences are found localized near the constriction, the centromere, that separates the arms of a chromosome.

Another type of satellite DNA, discovered in the mid-1980s, has come to be called **minisatellite DNA**. The repeated sequence is shorter, ranging from ~15 to 100 bp long, with fewer repeats than α-satellite DNA. A single minisatellite locus may be several thousand base pairs long. Although they are found dispersed throughout the genome, some minisatellites tend to be found near the ends, or **telomeres**, of chromosomes, whereas others are found within introns or just upstream or downstream of coding regions. One well-characterized minisatellite is found just 5′ to the insulin gene, and certain variants are associated with increased risk of **insulin-dependent diabetes**. Because of the structure of the minisatellites, they often are referred to as variable number tandem repeats (VNTRs).

A third class of tandem repeats, **microsatellites**, involve even smaller repeat units. Another name for them is short tandem repeats (**STRs**). Initially, many microsatellites were found to have the dinucleotide CA repeated over and over—as many as several dozen times. Other simple repeating sequences of two, three, or four bases have been detected. Microsatellites appear to be randomly distributed throughout the genome. The CA repeat motif, for instance, appears in many places scattered throughout the genome.

How do scientists identify one set of these alleles as opposed to another? By using PCR. Let's say one finds blood at a crime scene and wants to identify the person who is the source. Short (10–20 bases long) sequences of DNA, primers, have been designed that are complementary to *unique* sequences on either side of the repeat. Thus, the primers flank the section of DNA to be studied. A bit of the crime scene DNA is subjected to repeated rounds of PCR during which only the sequence bounded by the primers is amplified. After 30–40 cycles, there will be millions of copies of this one sequence that can then be analyzed. By using other primer pairs, other sequences can be typed. The alleles identified in the crime scene sample can be compared with those of suspects.

VNTRs and microsatellites have very high mutation rates and yield variation, often at very high levels, in the *number of repeats* at any specific site in the genome. This pattern of variation is very different from the classic thinking about the source and form of genetic variation. For several decades, geneticists thought of variation as arising from point mutations—simple alterations in the sequence of bases. Most simply, a point mutation could take the form of a deletion of a single base, the addition of a base, or the substitution of one base for another. At least in the mind of the geneticist, variation involved a common normal form of a gene and a rare variant form that differed from the normal form in a small but important manner. Sometimes, especially in clinically important cases, a larger alteration might have been caused, for instance, by the deletion of a large section of a gene. Large or small though, variation generated in this fashion was thought to be rare and, in any case, led to a common normal form and a rare variant form of a gene. This model was reasonable for a series of hemoglobin disorders known as **thalassemias** that involved deletions of large segments of the hemoglobin genes.

With the discovery of VNTRs, the thinking about the levels of genome variation changed abruptly. Minisatellites and microsatellites can, by their nature, generate much more variation in a population than point mutations can. Instead of the "normal" sequence and the odd variation caused by, say, a point deletion, the loci of mini- and microsatellites vary in repeat number. A single locus can have a sizable number of equally *un*common alleles, in which case it is called **hypervariable**. This pattern is a far cry from the "usual" pattern of one common and a few rare alleles. For instance, one block of CA repeats, called sAVH-6, is near the end of the long arm of chromosome 7 in humans. Ten different alleles, varying in size from 217 to 235 bp, have been detected in a series of Europeans. The alleles differ in the number of times the sequence CA is repeated. In this group of unrelated Europeans, the alleles varied in frequency from 0.02 to 0.26. Thus, this population is highly heterozygous; very few people inherit two copies of the same-length allele. (Hing et al. 1993).

The VNTR and STR patterns of variation have major implications in population genetics. For these markers, a group of people will have very high levels of heterozygosity. The classic picture developed 25 years ago involved many loci each of which had one, two, or maybe three alleles; the vast majority of people had 2 copies of the common allele. In a diallelic case, even if both alleles are present at equal frequencies, the level of heterozygosity cannot be more than 50%. Now we see that there are genetic markers that might have 10–15 or more alleles, and the most common might have a frequency of only several percent. Any one locus where this pattern holds will have many different genotypes, each present at low frequencies in the population.

The implication of this pattern of variation for the individual is that we are each genetically unique. There are many highly variable VNTRs and STRs. If a person is tested for many of these loci, it is very unlikely that a second person on the face of the planet has exactly the same genotype at all these different loci. This tremendous variability is at the heart of the revolution in forensic (legal) applications of genetics, because it allows for the "individualization" of a DNA sample: the ability to tie a biological sample (blood, sperm, etc.) to one and only one person.

Analysis of these hypervariable loci can be particularly important in understanding genetic similarities between human groups. Genetic similarities can occur for several reasons. They can reflect the operation of identical natural selective forces, migration between the groups, or descent from a recent common ancestor. Mini- and microsatellite loci generally are not thought to be affected by natural selection because they do not appear to have any function. Thus, similarities in frequencies of alleles in different populations can tell us something about the evolutionary history of the groups. If for several of these loci, identical alleles are found in different populations, then it is reasonable to postulate a relationship between the groups. Differences between groups at these loci will largely be the result of genetic drift, not selection.

Although this approach can be very useful, it also presents problems. Assessing the identity of alleles can be difficult; two that appear identical when studied by one technique may turn out to be different when studied more carefully. Nevertheless, analysis of VNTRs can provide for detailed studies of population relationships.

If the single-copy sequences (coding genes and other unique sequences, many of which appear to be functionless) account for about 66–75% of the genome and satellite DNA about 10%, then the remaining 15% or so is composed of two types of repetitive DNA dispersed throughout the genome. These are called long interspersed elements (LINEs) and short interspersed elements (SINEs). Unlike the tandem repeats, in SINEs and LINEs, the repeats are not back to back but scattered throughout the genome. Together, SINEs and LINEs comprise what is called the moderately repetitive class of DNA.

SINEs are sequences about 100–500 bp long that share a common sequence, or variations on that sequence. Members of a SINE family may have less than 85% sequence identity. The best known and most common SINEs are in the Alu family; it alone accounts for about 4–5% of the human genome. The Alu sequence is ~300 bases long, and ~1,000,000 copies are present in a human cell. Examination of Alu SINEs scattered throughout the genome show that they actually occur in subfamilies; members of a subfamily have more similar sequences and presumably a more recent common ancestor. At this time, evidence indicates that SINEs, like the Alu family, may have functional significance.

LINEs may be several thousand base pairs long, up to ~7000 bp. Mammals have a prominent LINE, called L1, that is repeated ~30,000 times. Line sequences vary from repeat to repeat but are fairly homogeneous within a species. LINEs appear to arise from insertion into the genome of a DNA sequence generated via reverse transcription from a mRNA molecule. In reverse transcription, an enzyme (reverse transcriptase) can use a single-stranded mRNA molecule as the basis for synthesizing a DNA molecule—a reversal of the usual direction of synthesis.

Mitochondrial DNA

The above discussion centers on the component of a person's DNA that is found in the nucleus of each cell. However, another form of DNA is present in all cells and follows some of its own rules. This DNA is found in the cell structure called the mitochondrion (plural = mitochondria). Mitochondria produce a chemical called adenosine triphosphate (ATP) that provides the cell with energy, and each cell has several hundred mitochondria. Several characteristics of **mitochondrial DNA** (mtDNA) are different from those of nuclear DNA. For one, it is circular rather than a linear strand. Second, it is rather small. Human mtDNA is ~16,569 bases long (polymorphisms affect the length in any given person), whereas nuclear DNA is billions of bases long. mtDNA bases code for 22 transfer RNAs, 2 other RNAs, and 13 proteins. Unlike nuclear genes, human mitochondrial genes do not have introns.

Three additional features are particularly important for our purposes. First, mtDNA is inherited only from one's mother. Because mitochondria are found in the cytoplasm and because an egg contributes almost all of the cytoplasm to a fertilized egg, mtDNA is transmitted through the matriline. Mother passes it on to her children; her daughters then pass it on as well. Copies passed to a son are reach a dead end, because his mate's mtDNA will be passed to their children. Thus, in evolutionary studies, the mtDNA traces the maternal path of heredity unlike the bilateral path of the nuclear DNA. Second, mutations are incorporated at a much higher rate into mtDNA than into nuclear DNA. Third, mtDNA does not undergo recombination. Combining the second and third features, we find that the differences we observe are only the result of mutations.

Although trying to specify an average mutation rate for the whole nuclear genome is rather simplistic, a broadly correct statement is that mtDNA mutates about 10 times faster than nuclear DNA. This faster rate of mutation is due to the lack of repair mechanisms in mtDNA that are found in nuclear DNA. This feature makes the mtDNA particularly interesting for studying the similarities between groups that share a very recent common ancestor. Over the course of a hundred thousand years, nuclear genes might accumulate very minimal differences—too little to be a good measure of genetic divergence. Over this same time frame, mtDNA will accumulate sufficient alterations to provide assessments of group divergence. Over longer periods of several millions of years and more, however, the mtDNA becomes less useful because in that timeframe it underestimates differences between groups. One site in the mitochondrial genome may have changed multiple times over the generations, but when the DNA sequences from two living organisms are compared, only the current difference can be seen.

One particular region of the mtDNA that has been extensively studied in different human groups is called the **control region**, or D-loop region. This noncoding region evolves quite rapidly and thus is useful for population comparisons.

When investigation of mtDNA divergence of human groups was introduced in the mid-1980s, the data created quite a stir because they indicated a much more recent divergence time than had previously been proposed. Additionally, comparisons of modern mtDNA sequences implied a common ancestor who lived in Africa, a conclusion that was referred to as the "Out of Africa" hypothesis. This conclusion was in sharp contrast to the "regional continuity" view, which postulates major genetic continuity within geographic regions over hundreds of thousands of years of human evolution and little flow of genes between populations who lived in widely separated regions of the world. Even though there were some problems with the early mtDNA data and discussion regarding to the dynamics of human population dispersal and relationships continues, the utility of mtDNA analysis for reconstructing human population history is well-established.

mtDNA can serve another end in studying human evolution: deciphering the interplay between social rules and genetic phenomena. All human groups have rules that regulate one's choice of mates. Often, these rules will result in women or men moving some distance from their place of birth. The mtDNA can trace the movement of genes and, by reflection, the social rules governing mating practices. In situations where women move, their mtDNA will spread to new generations in new places. Male migration will not affect the mtDNA in the new homeland, however. The possibility of differential migration of the sexes must be kept in mind when analyzing the spread of mtDNA types over space.

Knowledge of mating patterns also can be useful when interpreting mtDNA data to trace population history. For instance, a number of mtDNA haplotypes are found widely in Native American populations. Because intertribal female migration is low, the wide spread of mtDNA variants might argue that these variants were present in an original, single group that migrated into the New World from Asia (Forster et al. 1996).

mtDNA is but one component of the human genome. With our increasing ability to manipulate the DNA and with our increasing knowledge of its structure and patterns of variation, the utility of genetic analysis for investigating questions of human biology multiplies many times over.

Y CHROMOSOME: LATEST SEARCH FOR INFORMATION

Returning to the nuclear DNA, 22 pairs of human chromosomes are called **autosomes** to differentiate them from the 23rd pair: the **sex chromosomes**. Although members of autosomal pairs of chromosomes are of equal size and carry alleles for the same traits, the sex chromosomes, called X and Y, are significantly different in size from each other. These chromosomes, as their name indicates, also are involved in determining one's sex. In humans and other mammals, a fertilized egg with two X chromosomes normally develops into a female, and an XY combination develops into a male.

The X chromosome is rather large, but the Y chromosome is very small and contains very few known genes. It is not surprising that many loci found on the X are not present on the Y. Interestingly, areas of similarity and dissimilarity are highly structured. Figure 3.17 depicts a Y chromosome. At the very tips, or telomeres, of its arms are small pseudoautosomal regions, so named because the X chromosome has

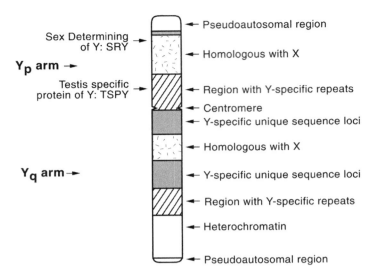

Figure 3.17 A schematic representation of the human Y chromosome. Locations of two genes, *SRY* and *TSPY*, are shown. The short arm, Y_p, is ~15 million base pairs long, and the long arm is ~42 million base pairs long. (Source: After Hammer and Zegura 1996).

homologous or corresponding loci. Genes in these regions of the Y may undergo re-combination. The rest of the Y chromosome is called the Y-specific region; the X has an X-specific region, and there is normally no recombination between these regions of the X and Y chromosomes. Gene mapping has, however, identified regions in the Y-specific region that are homologous to regions found on the human X chromosome. Interestingly, gorilla and chimpanzee Y chromosomes do not contain these sequences, implying that this area has undergone alterations since humans diverged from our closest relatives.

One locus found in the Y-specific region is known as *SRY* (for sex-determining region on the Y). This locus is involved in directing the undifferentiated, embryonic gonad to become testis. In the absence of *SRY* gene product, the gonad starts to develop into ovaries. Between the Y-specific region and the pseudoautosomal region lies the pseudoautosomal boundary. Within the boundary region, the sequence similarity between the X and Y chromosomes abruptly drops from the high levels seen in the pseudoautosomal region.

Females are produced when an X-carrying egg combines with an X-carrying sperm. When a sperm with a Y chromosome combines with an egg, a male is produced. Thus, there is an analogy between the Y chromosome and mtDNA. Whereas the latter traces the matriline—the mother-to-daughter lineage—over time, the Y chromosome is passed from father to son and thus traces the flow of genes through the patriline. This flow can be a very useful aspect for anthropological investigations. Because the Y-specific region does not recombine, the only force that will change it over time is mutation. A mutation that occurs in one male is passed down to all his male lineal relatives. The altered DNA thus can be used as a marker to trace the passage of his Y chromosome through time and space.

Until fairly recently, there were few studies using the Y chromosome for there were very few known Y chromosome variable traits. This has changed recently as researchers have uncovered microsatellites and other highly variable sequences on the Y chromosome. The Y chromosome has been used to investigate the relationships between human populations and the timing of human dispersal. In some cases, the results appear to agree with those gathered through analysis of mtDNA, but it is not always the case.

PATTERNING IN THE GENES: A FUTURE DIRECTION

In this short and simplified introduction to genetics and its significance to biological anthropologists, I have painted a relatively simple picture: DNA directs the production of proteins. My emphasis has been on one aspect of the anthropological application: The genetic variants detected in human populations can be used to trace lines of evolutionary relatedness between populations and other aspects of human population history.

Another line of investigation now promises to revolutionize our understanding of human biology; it involves unraveling the path from the genes to the phenotype. How do the genes, via the proteins, result in our being taller or shorter, well or sick? How do they influence how we deal with the stress of life at high altitude or nutritionally imperfect diets? Why and how do some genes tend to make us to behave in particular ways? We are starting to see how the genes, operating within an environmental circumstance, guide and influence the process by which each of us starts as a single cell and becomes a recognizable organism.

A model for much of this understanding revolves around a series of genes known as homeotic homeobox genes. A homeobox is a highly conserved DNA sequence of ~180 bp that codes a 60-amino acid polypeptide capable of binding to DNA in a specific manner. DNA-binding proteins are capable of activating or repressing target genes and thus are involved in gene regulation.

Homeotic refers to the segmental body plan by which many organisms, from insects to humans, are built. Vertebrates, for instance, have a series of bones (vertebrae) in the back that demonstrate variations on a theme as one traverses from the anterior end of the animal to the posterior. Humans have neck or cervical vertebrae, which are small and mobile; thoracic vertebrae, which are more robust and are connected to ribs; and lumbar and sacral vertebrae, which are heavily built to support weight and not very mobile. Segmentation along this head-to-tail plane is set up early in embryonic life, then refined and detailed as specific genes are turned on or off according to plan. Homeotic genes are involved in segment identity. Very simply put, altering the timing and duration of activation of these genes can cause morphological variation. Humans have four clusters of homeotic homeobox genes, called *Hox* genes, located on four chromosomes, and each cluster contains 9–11 genes.

Surprisingly, it appears that the genes within a Hox cluster are arranged in terms of their spatial expression in the developing organism. The most 3′ genes within a Hox cluster are expressed toward the anterior end of the animal and the more 5′ genes are expressed toward the posterior end. However, the most anterior point where expression begins for one cluster is not the same between clusters. For instance in mice and chickens, the *Hoxb*-3 gene's anterior limit of expression is in the upper neck, whereas the anterior limit of *Hoxc*-6 expression is in the thoracic region. Thus, within each cluster, the Hox genes are expressed in a segmental arrangement. In addition to Hox genes, vertebrates have other genes, unrelated to segment identity, that regulate gene transcription.

Decades ago, King and Wilson (1975) theorized that the shift from an apelike animal to a **hominid** largely involves increasing the size of the brain and decreasing the relative size of the face. They went on to suggest that these alterations would not be very involved. They could be accomplished by altering the timing and activity level of a few genes that affect brain and facial structure: Keep genes affecting growth of certain parts of the brain active a bit longer, and the resulting adult has a larger, more human brain. In this way, they felt that one could resolve the apparent contradiction between genetic data, which showed humans and chimpanzees to be extremely similar, and morphological data, which evidenced some noticeable dissimilarities.

Although the knowledge of regulatory genes was very limited at that time, it is looking more and more like King and Wilson were correct. Cross-species analysis of Hox genes is showing that they can be responsible for major changes in morphology. Although beyond the scope of this chapter, analysis of regulatory genes is clarifying such phenomena as the origin of the foot in evolution. One might imagine that the alterations seen in human evolution, too, would have been contributed to by regulatory genes. The story that is unfolding from analysis of Hox and other regulatory genes is not a simple one at the genetic level. Undoubtedly, many genetic and environmental factors must interact in a highly coordinated fashion to produce a functional individual. Nevertheless, we are starting now to see how these interactions transpire in the developing organism.

CHAPTER SUMMARY

In this chapter, we presented some of the basic genetic processes. We paid particular attention to the molecular mechanisms involved in the fulfillment of DNA's two primary functions: making more DNA and directing the production of proteins.

During the past decade, much information has been uncovered regarding the organization of the DNA. Several classes of DNA and the unique sequences known to code for specific proteins have been elucidated. Some of these classes are hypervariable and, coupled with new techniques such as PCR, allow for the rapid characterization of human samples from very small quantities of material. Such capabilities provide anthropologists with opportunities to both document human variation and investigate its evolution. In the near future, the understanding of regulatory genes should provide insights into modern human adaptations as well as the evolution of modern human morphology.

RECOMMENDED READINGS

Cann RL, Wilson AC, and Stoneking M (1987) Mitochondrial DNA and human evolution. Nature 325: 31–36.

Cavalli-Sforza LL, Menozzi P, and Piazza A (1994) The History and Geography of Human Genes. Princeton, NJ: Princeton University Press

Forster P, Harding R, Torroni A, Bandelt H-J (1996) Origin and evolution of Native American mtDNA variation: a reappraisal. Am. J. Hum. Genet. 59: 935–945

Hammer MF and Zegura S (1996) The role of the Y chromosome in human evolutionary studies. Evol. Anthropol. 5: 116–134.

Hartl DL and Jones EW (1998) Genetics: Principles and Analysis. Sudbury, Mass.: Jones and Bartlett.

Hing AV, Helms C, and Donis-Keller H (1993) VNTR and microsatellite polymorphisms within the subtelomeric region of 7q. Am. J. Hum. Genet. 53: 509–517.

Holland PWH and Garcia Fernandez J (1996) *Hox* genes and chordate evolution. Dev. Biol. 173: 382–395

Johnson RL and Tabin CJ (1997) Molecular models for vertebrate limb development. Cell 90: 979–990.

King M-C and Wilson A (1975) Evolution at two levels in humans and chimpanzees. Science 188: 107–116.

Mourant AE, Kopec AC, Domaniewska-Sobczak K (1976) The Distribution of the Human Blood Groups and Other Polymorphisms, 2nd edition. New York: Oxford Univ. Press.

Schmid CW (1996) Alu: Structure, origin, evolution, significance, and function of one-tenth of human DNA. Prog. Nucleic Acid Res. Mol. Biol. 53: 283–319.

Shubin N, Tabin C, and Carroll S (1997) Fossils, genes and the evolution of animal limbs. Nature (London) 388: 639–648.

Weiss KM (1990) Duplication with variation: Metameric logic in evolution from genes to morphology. Yearbook Phys. Anthropol. 33: 1–23.

Weiss KM (1998) Coming to terms with human variation. Annu. Rev. Anthropol. 27: 273–300.

REFERENCES CITED

Cavalli-Sforza LL, Menozzi P, and Piazza A (1994) The History and Geography of Human Genes. Princeton, NJ: Princeton Univ. Press.

Chang L-YE and Slightom JL (1984) Isolation and nucleotide sequence analysis of the β-type globin pseudogene from human, gorilla, and chimpanzee. J. Mol. Biol. 180: 767–783.

Collins FS, Patrinos A, Jordan E, Chakravarti A, Gesteland R, Walters L, et al. (1998) New goals for the US Human Genome Project, 1998–2003. Science 282: 682–289.

Cranor CF (ed.) (1994) Are Genes Us? New Brunswick, NJ: Rutgers Univ. Press.

Csink AK and Henikoff S (1998) Something from nothing: the evolution and utility of satellite repeats. Trends Genet. 14: 200–204.

Ellsworth DL, Hallman DM, and Boerwinkle E (1997) Impact of the human genome project on epidemiological research. Epidemiol. Rev. 19: 3–13.

Ferraris JD and Palumbi SR (eds.) (1996) Molecular Zoology: Advances, Strategies and Protocols. New York: Wiley-Liss

Flint J, Bond J, Harding RM, and Clegg JB (1996) Minisatellites as tools for population genetic analysis. In AJ Boyce and CGN Mascie-Taylor (eds.): Molecular Biology and Human Diversity. Cambridge: Cambridge Univ. Press.

Forster P, Harding R, Torroni A, and Bandelt H-J (1996) Origin and evolution of Native American mtDNA variation: a reappraisal. Am. J. Hum. Genet. 59: 935–945

Fritsch EF, Lawn RM, and Maniatis T (1980) Molecular cloning and characterization of the human β-like globin gene cluster. Cell 19: 959–972.

Golenberg EM, Bickle A, and Weihs P (1996) Effect of highly fragmented DNA on PCR. Nucleic Acids Res. 24: 5026–5033.

Hammer MF and Zegura S (1996) The role of the Y chromosome in human evolutionary studies. Evol. Anthropol. 5: 116–134.

Hing AV, Helms C, and Donis-Keller H (1993) VNTR and microsatellite polymorphisms within the subtelomeric region of 7q. Am. J. Hum. Genet. 53: 509–517.

Jarne P and Lagoda PJL (1996) Microsatellites, from molecules to populations and back. Trends Ecol. Evol. 11: 424–429.

Jeffreys AJ, Wilson V, and Thein SL (1985) Hypervariable "minisatellite" regions in human DNA. Nature 316: 76–79.

Jorde LB (1995) Population specific genetic markers and disease. In RA Meyers (ed.): Molecular Biology and Biotechnology. New York: VCH Publishers.

Kidd KK, Morar B, Castiglione CM, Zhao HY, Pakstis AJ, Speed WC, Bonne-Tamir B, Lu RB, Goldman D, Lee CY, Nam YS, Grandy DK, Jenkins T, and Kidd JR (1998) A global survey of haplotype frequencies and linkage disequilibrium at the DRD2 locus. Hum. Genet. 103: 211–227.

King MC and Wilson A (1975) Evolution on two levels. Science 188: 107–116.

Krontiris TG (1995) Minisatellites and human disease. Science 269: 1682–1683

Lyons L, Laughlin TF, Copeland NG, Jenkins NA, Womack JE, and O'Brien SJ (1997) Comparative anchor tagged sequences (CATS) for integrative mapping of mammalian genomes. Nature Genet. 15: 47–56.

Mitchell RJ and Hammer MF (1996) Human evolution and the Y chromosome. Curr. Opin. Genet. Dev. 6: 737–742.

Mourant AE, Kopec AC, and Domaniewska-Sobczak K (1976) The Distribution of the Human Blood Groups and Other Polymorphisms, 2nd edition. New York: Oxford Univ. Press.

Poinar HN, Hofreiter M, Spaulding WG, Martin PS, Stankiewicz BA, Bland H, Evershed RP, Possnert G, and Paabo S (1998) Molecular coproscopy: dung and diet of the extinct ground sloth *Nothrotheriops shastensis*. Science 281: 402–406.

Ramkissoon Y and Goodfellow P (1996) Early steps in mammalian sex determination. Curr. Opin. Genet. Dev. 6: 316–321.

Smit AFA 1996. The origin of interspersed repeats in the human genome. Curr. Opin. Genet. Dev. 6: 743–748.

Watson JD and Crick FHC (1953) Molecular structure of nucleic acids. A structure for deoxyribonucleic acid. Nature 171: 964–969

Wilson AF, Elston RC, Siervogel RM, and Tan LD (1988) Linkage of a gene regulating dopamine-β-hydroxylase activity and the ABO blood group locus. Am. J. Hum. Genet. 42: 160–166.

CHAPTER 4

Genetics, Geography, and Human Variation

DENNIS H. O'ROURKE

INTRODUCTION

In Chapter 3, you learned about the fundamentals of heredity via the structure of **DNA**, **protein** synthesis, **genome** organization, and the mechanisms whereby the hereditary material is transmitted to all the cells of the body. The concept of **allele** frequencies was merely introduced. In this chapter, I will examine the mechanisms of evolution that cause changes in allele frequencies and, hence, the genetic structure of populations.

Although in practice, defining a population is rarely a simple or trivial task, a population may be defined as consisting of those individuals of reproductive age who may select each other as mates to produce the next generation. In the case of small, isolated groups of people, the definition of a population is rather straightforward. However, in larger continental areas, where the distribution of individuals is essentially continuous over the landscape, the precise definition of a biological population may be quite difficult. It may encompass cultural and social norms of mate selection, individual movement patterns, and other aspects specific to that population. Nevertheless, once a population is defined, it is a simple matter to identify that population's **gene pool** as all of the **genes** carried by the reproductive members of the population.

Because each individual begins as a hybrid, receiving one genomic complement from each parent at conception, the number of copies of genes (alleles) in a population is twice the size of the population ($2n$, where n is the size of the population). The $2n$ alleles that comprise a gene pool exist in pairs within individuals (**genotypes**) as a result of sexual reproduction. This simple relationship necessitates an initial focus on individuals and genotypes in considering the genetics of populations.

Human Biology: An Evolutionary and Biocultural Perspective, Edited by Sara Stinson, Barry Bogin, Rebecca Huss-Ashmore, and Dennis O'Rourke
ISBN 0-471-13746-4   Copyright © 2000 Wiley-Liss, Inc.

Consider the sample of individuals depicted in Table 4.1. In this hypothetical example, a sample of 400 individuals has been typed for a genetic **marker** with two alternative alleles, A and a, resulting in three possible genotypes: AA, Aa, and aa. Estimating the frequencies of the alleles in this population sample may be done directly by simply counting the relative number of alleles present. For example, each of the 144 AA **homozygotes** carries two copies of the A allele, and each of the 192 **heterozygotes** carries one copy of this allele, for a total of 480 copies of this gene in the sample. Because each individual is **diploid**, the total number of alleles in this example is 800 ($= 2n$) and the relative frequency of the A allele in this sample is $480/800 = 0.6$. Similarly, the number of a alleles in each genotypic class may be counted (e.g., 192 in Aa heterozygotes and 128 in the aa homozygotes) to obtain the frequency of the a allele (0.4). Gene counting is a reliable method to obtain allele frequencies in many instances, but it can be cumbersome and time-consuming. A quicker method of obtaining allele frequencies is also available from the information given in Table 4.1.

Take the frequencies of the three genotypes to be P, H, and Q, respectively, and the allele frequencies may be estimated directly. By convention, allele frequencies are referred to by the letters p and q (and subsequent letters for additional alleles). In Table 4.1, the direct estimate of the A allele frequency is $p = P + H/2$. That is, $A = p = 0.36 + 0.48/2 = 0.6$. This is, of course, the identical allele frequency estimate obtained by gene counting. In similar fashion, the frequency of the a allele may be obtained as $q = Q + H/2$.

An important consequence of the latter method is that it clearly implies a close relationship between the frequencies of genes and genotypes. Indeed, even a casual consideration of the example just described reveals that the frequencies of the homozygote classes are equal to the square of the respective allele frequencies. Thus, the frequency of the A allele is 0.6, and its square ($0.6^2 = 0.36$) is the frequency of its homozygote genotype. The same relationship may be noted for the a allele and its homozygote frequency. If p and q ($= 1 - p$) are taken to represent the two allele frequencies, then a simple relationship is obvious: $p + q = 1$. If this relationship is squared (the binomial expansion), then the resultant terms are $p^2 +$

Table 4.1   Computation of allele frequencies from genotypic data: Gene counting

	Genotype			
	AA	Aa	aa	Total
Number of individuals	144	192	64	400
Frequency	$P = 0.36$	$H = 0.48$	$Q = 0.16$	1
Number of A alleles	288	192	0	480
Number of a alleles	0	192	128	320

Frequency of A allele (p): $480/800 = 0.60$
Frequency of a allele (q): $320/800 = 0.40$

$$p + q = 1.0$$

$2pq + q^2 = 1$. The term $2pq$ in our example is equal to the frequency of the heterozygous genotype (i.e., $2(0.6)(0.4) = 0.48$). This simple relationship, known as **Hardy–Weinberg equilibrium**, is a fundamental principle of population genetics.

Hardy–Weinberg equilibrium represents the relationship of gene and genotype frequencies over time, given specific assumptions. These assumptions include the following (Falconer 1989):

- Parents are a random sample of the population.
- The segregation of alleles to **gametes** is normal.
- All parents are equally fertile.
- All gametes are equally fertile.
- The population is very large.
- Mating between parents is random (panmixia).
- Allele frequencies are equal in males and females.
- All genotypes are of equal reproductive ability.

Stated more simply, these assumptions imply random mating among parents and no forces of evolution operating on the population, where forces of evolution include **(natural) selection**, **mutation**, **gene flow**, and **genetic drift**. I describe the specific actions of these evolutionary forces later. Under such conditions, the stability of the relationship of allele and genotype frequencies permits prediction of both gene and genotype frequencies in subsequent generations by taking account only of gene frequencies rather than genotype frequencies.

Even if the parental generation is not in Hardy–Weinberg equilibrium, one generation of random mating is sufficient to restore the equilibrium between allele and genotype frequencies. This relationship is achieved with no change to the allele frequencies; only the genotype frequencies will have been altered. Hence, in a large **panmictic** population with no forces of evolution operating, the allele frequencies of any particular generation depend on the allele frequencies of the previous generation but not the genotypic frequencies of that generation. Moreover, if all the conditions are met, then the frequency of alleles and genotypes will be maintained over many generations.

The simple but important focus of population genetics is the relationship between genes and genotype frequencies. From this starting point, we can examine deviations from equilibrium. Clearly, nonrandom mating is one mechanism that will result in a deviation from equilibrium of allele and genotype frequencies. Other reasons for the absence of equilibrium are the actions of any of the mechanisms of evolution. These mechanisms not only disrupt the equilibrium but also lead to genetic change in (the evolution of) populations.

In the next section, I will briefly describe population genetic theory to illustrate the effect of these mechanisms on genetic systems.

MECHANISMS OF EVOLUTION

Mutation

A variety of mutational processes and their effects on DNA were described and discussed in Chapter 3. They will not be repeated here. Suffice it to recall that mutations provide the basic material for genetic diversity within populations and species, which is distributed and structured by the mechanisms discussed in the following sections.

Natural Selection

As discussed in Chapter 1, natural selection is the fundamental insight of Charles Darwin. The mechanism by which adaptive change in populations—Darwin's descent with modification—is based on three fundamental premises:

1) All species tend to have more offspring than can survive and reproduce.
2) Organisms vary in their ability to survive and reproduce.
3) Part of the variability in survivorship and reproduction is heritable.

In the modern view, particularly with respect to the genetics of populations, these premises mean that individuals differ with respect to genotypes (genetic variation); genotypes vary with respect to survivorship and reproduction; and we can measure this difference as "**fitness**" of the genotype. More fit genotypes, on average, survive and leave more offspring than less fit genotypes and therefore are considered better adapted to their environment(s) than their less fit counterparts. As a result, alleles associated with the more fit genotype are over-represented in the next generation relative to those associated with less fit genotypes and therefore increase in frequency. From a genetic perspective, natural selection may be defined as the differential reproduction of different genotypes.

The concepts of selection and fitness are complementary. If we define the magnitude of selection against a genotype as "s" (taking values between 0 and 1), then the fitness (f) of this genotype is simply $f = 1 - s$. Although the definition of fitness, and hence selection, appears simple and straightforward in theory, it is more difficult to determine in nature. Because selection may operate at different times throughout the lifespan of an organism, defining the components of fitness for specific genotypes becomes difficult. The traditional method of tackling this problem in nature has been to sample and study age classes within populations.

Selection affects the variation in a population in any of several ways. Figure 4.1 illustrates three modes of selection for a continuously distributed trait. **Directional selection** (Fig. 4.1A) causes the mean of the distribution to be displaced toward higher fitness values. In Figure 4.1B, the individuals near the mean of the distribution have the highest fitnesses. In this case, the mean value of the character does not change over time, but the **variance** of the distribution may decrease. Genetic variation is reduced as a result of selection such that continued selection becomes less and less efficient as the variation on which it operates is diminished. Disruptive

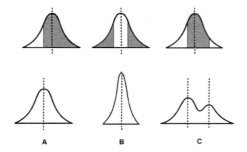

Figure 4.1 Effect of three modes of selection on distribution of continuous trait: directional (A), stabilizing (B), and disruptive (C). Top: Distributions prior to selection. Bottom: Distributions after selection. Dashed lines indicate the mean of each distribution. Shaded areas represent portions of the distribution being selected against. (*Source:* After Price 1996)

selection (Fig. 4.1C) is thought to be a rather rare phenomenon and is not well-known. It implies multiple fitness maxima on the phenotypic scale and may result when a population is subjected to changing or cyclic environments such that different genotypes have a selective advantage at different times. In the extreme, this form of selection may result in speciation.

The simplest and perhaps the most thoroughly examined model is viability selection (i.e., selection resulting from differences among genotypes in survival to reproductive age) on a two-allele **locus**. Consider a single locus with two alleles, A and B. Under Hardy–Weinberg assumptions, the three zygotic genotypes (AA, AB, and BB) will be in the expected frequencies: p^2, $2pq$, q^2. To the degree that these genotypes differentially survive to reproduce, we may describe their fitnesses as f_{AA}, f_{AB}, and f_{BB}, respectively. If we assume an initial population in equilibrium, and with these fitness values, we have

Genotype	AA	AB	BB
Frequency	p^2	$2pq$	q^2
Fitness	f_{AA}	f_{AB}	f_{BB}

These three fitnesses define the ratio of survivorship of the three genotypes to reproduction, so the ratio of surviving, reproducing adults is

$$p^2 f_{AA} : 2pq f_{AB} : q^2 f_{BB} \qquad (4.1)$$

The sum of these three quantities is the mean fitness of the population. The dynamics of these relationships are illustrated with a simple example in Box 4.1 with the well-known example of balancing selection: sickle hemoglobin and malaria.

Whereas the best known example of selection on a relatively simple genetic system in humans is a case of **overdominant selection** (see Box 4.1), most casual presumptions regarding selection presume directionality. That is, selection is presumed

BOX 4.1   BALANCING SELECTION

The elevated frequency of the sickle-cell allele in malarial environments is one of the best known examples of overdominant, or balancing, selection in humans. Assume a population sample of $n = 121$ individuals with the following characteristics:

Genotype	AA	AS	SS
Number observed	100	20	1
Frequency	0.8281	0.1638	0.0081
Fitness	0.90	1.0	0.00

Clearly, the population is in Hardy–Weinberg equilibrium and the allele frequencies in this sample are $A = p = 0.91$ and $S = q = 0.09$. Moreover, from these data, we can demonstrate that the mean fitness (f) is 0.9091.

Because the fitness values are not equal across genotypes, selection is operating on this genetic system. The fitnesses are relative, so the fitness of 1.0 for the heterozygote AS implies that this genotype has the highest reproductive rate, whereas the fitness (i.e., the relative reproductive rate) of the AA genotype is only 90% that of AS. Selection is complete against the SS homozygote. Given the moderate selection against AA, it is of interest to determine the frequency of this allele after a generation of selection. The frequency of an allele after a generation of selection is given by

$$p' = p(pf_{AA} + qf_{AB})/f$$
$$= 0.91[(0.91)(0.9) + (0.09)(1.0)]/0.9091$$
$$= 0.9099$$
$$= 0.91 \tag{4.2}$$

The frequency of the allele is the same after a generation of selection as it was before. Thus, selection in this case has maintained the frequency of both alleles even though both homozygotes have lower fitness than the heterozygote. This form of selection is also known as balancing selection, where the loss of one allele is "balanced" by the loss of the other due to depressed fitnesses for both homozygote classes. It is possible to obtain directly the equilibrium frequency of any allele in such a system, as long as the allele frequencies and fitnesses are known, via the following relationship:

$$p = (f_{AB} - f_{BB})/2(f_{AB} - f_{AA} - f_{BB}) \tag{4.3}$$

The fitness values for each genotype given earlier readily yield the equilibrium frequency of 0.91. This value has already been established, of course, because in this example, the allele frequency did not change from one generation to the next, implying that equilibrium has already been reached. It should be emphasized that this form of selection operates to actually conserve variability, because both alleles are maintained in the population by selection.

This example is a simplified scenario for the distribution of the sickle-cell allele in a population inhabiting a malarial environment. A more extensive discussion of this genetic system and the nature of its interaction with specific environmental and cultural characteristics is presented in Chapter 8.

to change something. In the case of simple single-gene systems, "something" is the frequency of alleles in a genetic system. Unlike the example of sickle-cell anemia and malaria, the nature of allele frequency change under directional selection depends on the presence and strength of **dominance** as well as the selection coefficient. Nevertheless, if the degree of dominance is known, it is possible to compute directly the new expected allele frequency in the next generation, given specific selection regimes. The formulations for such estimates, along with the individual genotype fitnesses and mean fitnesses, are given in Table 4.2.

Genetic Drift

The results of selection presented in Table 4.2 are contingent on several assumptions, including infinite (or at least very large) size and random mating. However, for the vast majority of human history and in many populations today, population size is not only finite but quite small. The realization of the potential effects of small population size led to one of the major debates in the history of modern genetics.

Sewall Wright, one of the three founders of mathematical population genetics, recognized early that small population size meant that the sampling of gametes each generation might result in a departure from the previous generation's frequencies simply by chance. Logically, if population size remained small, the accumulation of these chance departures from the frequencies in each parental generation could, over time, result in substantial departures from the original generation's observed frequencies. Wright considered this random drift of gene frequencies between generations a potentially potent force that could affect the frequencies of genes in populations over time. R.A. Fisher, another of the three founders, fundamentally disagreed. He argued that populations were sufficiently large that such chance phenomena would have little effect, particularly relative to the effects of selection, migration, and mutation. Wright's ideas are so associated with the advocacy of **stochastic** processes in population genetics that random genetic drift is occasionally still referred to as the Sewall Wright effect.

Genetic drift is simply a probability statement regarding the effects of sampling in small, relatively isolated populations. The genetic composition of each generation is a random sample, frequently a small one, of the gametes of the preceding generation. If certain conditions are met (e.g., the assumptions of Hardy–Weinberg equilibrium), then the proportions of alleles from one generation to the next are expected to remain constant. This assumption permits estimation of the proportions of genotypes in one generation from a knowledge of the gene frequencies in the previous one.

In many genetics and statistics texts, this sampling process is illustrated with a large urn filled with marbles of two different colors. Assume that the urn contains equal numbers of black and white marbles. If we repeatedly sample a marble from that urn, record the color, put it back, and then repeat the process, it is obvious that after many such draws, the ratio of black to white marbles in our sample would be 1. That is, we would estimate the frequency of each type as 0.5.

But what if we limited the number of draws to 10? We possibly would draw 5 black marbles and 5 white, replicating the 50/50 ratio in the total group of marbles in the urn. However, we also could have drawn 6 black marbles and 4 white and then estimated the original frequencies in the urn as 0.6 and 0.4 rather than their actual values. The error in estimating the true frequencies of the colored marbles would be due to chance in a small number of samples. As a result of such random sampling, the estimate of the original frequency of black marbles can take any value between 0 and 1 and is clearly related to the number of draws. This sampling scheme is described in probability terms by the rules of binomial sampling, which define the binomial distribution, a well-known entity in statistics. This example is directly relevant to genetics if we simply take the urn of marbles to represent the parental pool of alleles at a biallelic locus and the black and white marbles to represent the proportions of two different alleles at the locus—say, A and a.

To study a concrete and quantitative example, assume the frequency of the A allele in some gene pool is p; therefore, the frequency of the a allele is $q = 1 - p$, such

Table 4.2   Effects of selection on biallelic locus with different conditions of dominance

Mode of inheritance	Genotype frequenices under selection relative to Hardy–Weinberg Equilibrium			Mean fitness after selection	Change in gene frequency
	$AA\ (p^2)$	$Aa\ (2pq)$	$aa\ (q^2)$		
Complete dominance, selection against recessive	1	1	$1-s$	$1-sq^2$	$-sq^2(1-q)/1-sq^2$
Complete dominance, selection against dominant	$1-s$	$1-s$	1	$1-sp^2-2pqs$	$+sq^2(1-q)/1-s(1-q^2)$
No dominance, selection against recessive	1	$1-s$	$1-2s$	$1-spq-sq^2$	$-sq(1-q)/1-2sq$
Heterozygote advantage, selection against homozygotes	$1-s$	1	$1-t$	$1-sp^2-tq^2$	$+pq(ps-qt)/1-p^2s-q^2t$

(*Source:* From Mettler and Gregg 1969; Strickberger 1996.)

that $(p + q) = 1$. The total number of alleles at this locus in the gene pool is n, and i as some arbitrary number of sampled alleles. The probability that the first i of n alleles sampled from this gene pool will be A is $p^i q^{n-i}$. More generally, the probability of sampling i A alleles out of a total of n alleles, irrespective of order, requires multiplying this probability by the number of ways these i A alleles may be taken with the $n - i$ a alleles. This process is equivalent to determining the total number of ways of arranging i things out of n, or

$$\frac{n!}{i!(n-i)!} = \binom{n}{i} \tag{4.4}$$

This is the binomial coefficient, is read n choose i, and gives the probability of i A alleles in n samples as

$$P_i = \binom{n}{i} p^i q^{n-i} \tag{4.5}$$

The mean (**m**) of the binomial distribution is np and the variance (**v**) is npq. It is useful to know that when considering allele frequencies in this context, the n in the above formulation is sometimes given as $2n$, to distinguish the number of individuals in the population (n) from the number of alleles in the gene pool ($2n$).

Because P_i is always >0 for populations in which two alleles coexist (i.e., $0 < p < 1$), the allele frequencies may change from generation to generation without selection simply by chance. Consider a population of five individuals in which the frequencies of two alleles at a locus are each 0.5. What is the probability of obtaining these same frequencies in the next generation? Using the binomial, we can show that this probability is 0.25

$$[\text{i.e., } P_i = \binom{10}{5} (0.5^5)(0.5^5)]$$

Thus, in 75% of cases, the frequencies will change from their initial values.

In the third generation, the probability of having the allele frequencies equal is only ~18%, and after 10 generations, only 5%. Any deviation from equality of frequency, of course, increases the likelihood that one of the alleles will become fixed (at a frequency of 1.0) and the alternate allele lost over time. Thus, the effect of genetic drift is to decrease genetic variation within populations. Because the effect of drift is stochastic, it is independent across populations. Alleles drifting toward **fixation** in one population may well drift to extinction in another. Hence, it also has the effect of increasing genetic variance between populations.

Estimating allele frequencies in some future generation under the action of drift is difficult, because frequencies can drift up or down each generation and because the probability of any particular value must be conditioned on all possible outcomes in every generation. This computational task is exceedingly tedious. Nevertheless, it is possible, via simulation, to graphically illustrate the interactive effects of drift and population size. Figure 4.2 illustrates that when populations of varying sizes are simulated and drift alone is allowed to affect the transmission of allele frequencies to subsequent generations, the effect is most pronounced in smaller populations. Indeed, in small populations, alleles are fixed and lost in relatively few generations; in large populations, the deviation from the "founding" frequencies is minimal.

Gene Flow

Homo sapiens is a highly mobile species, indeed, one of the most mobile of all known vertebrates. The role of migration in the alteration of the genetic structure of populations is particularly well studied. Like most of the genetic effects of the evolutionary mechanisms reviewed so far, the mathematical theory of allele frequency change due to migration, or gene flow, is well developed, and detailed explanations are available (e.g., Crow and Kimura 1970; Cavalli-Sforza and Bodmer 1971; Mielke and Crawford 1980). Sewall Wright noted early the propensity for populations to exchange individuals (and hence genes) with their nearest neighbors (Wright 1943, 1946); that is, populations that live in close proximity exchange members more frequently than populations that are far apart. The genetic effect of this phenomenon is that populations are genetically more similar to nearer groups and genetically more distinct from farther away. Wright referred to this easily demonstrated relationship as **isolation by distance**. Several mathematical treatments of isolation by distance have been developed based on probability models (e.g., Kimura and Weiss 1964; Malécot 1969).

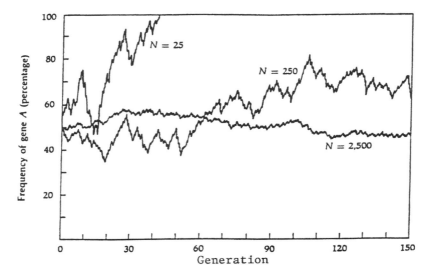

Figure 4.2   Fate of allele frequency in simulated populations of differing sizes, where population size is given by N. All populations were simulated from starting values of 0.5 for the allele frequencies of a biallelic locus. (Source: Bodmer and Cavalli-Sforza 1976)

An obvious question related to migration—in this case, individual migration—is that if one population contributes migrants to another population, how can we estimate the relative contributions of these groups to the contemporary gene pool? The answer to this simple, if fundamental, question was first provided in 1931 (Bernstein 1931). Imagine population *1*, which receives a proportion of its membership (*m*) each generation from population *2*. Then imagine that some genetic marker is at frequency q_1 in population *1* and q_2 in population *2*. Over time (not much time, if *m* is large), the frequency of the marker in population *1* will become more nearly that seen in population *2*. The relationship is given by

$$q'_1 = (1 - m)q_1 + mq_2 \qquad (4.6)$$

where q'_1 denotes the frequency of the marker in population **1** after one generation of such migration. Obviously, this new allele frequency is just the product of the proportion of nonmigrants (original members of population *1*) and their allele frequency plus the product of the proportion of the population that represents migrants and the allele frequency of their source population. If migration remains constant at rate *m* for some time, then the change in allele frequency in population *1* per generation is given by

$$\Delta q = q'_1 - q_1 = -m(q_1 - q_2) \qquad (4.7)$$

This formulation of genetic change as a result of migration presumes unidirectional migration from one population to another. Consequently, the frequency changes in only one population (*1*); the other population (*2*) is assumed to remain static. These simple relationships are most usefully explored where the marker of interest differs in frequency between the two original populations. Only in this way may we see the genetic effects of migration. An example of the computation of migration and admixture between populations is given in Box 4.2.

The relationship between migration and allele frequencies also can be used to assess the contribution of ancestral gene pools to contemporary populations, if some knowledge regarding migration rates is known or can be estimated. An often-cited example is the genetic exchange that has occurred between African Americans and European Americans over the past 400 years. Admixture between these formerly distinct gene pools began with the initial importation of Africans as slaves to the Americas in the 16th century. The genetic effects of this migration, or gene flow, have been more pronounced in the subsequent African American gene pool for at least three reasons.

First, the African American population is substantially smaller than the European American population of North America (~12% of the modern US population), so the relative magnitude of any genetic change due to migration is enhanced and more readily detected. Second, the African American population has remained more isolated from its genetic antecedents in Africa than has the European American population. Finally, traditional social definitions of ethnicity resulted in individuals with any African ancestry, however small, being identified as African American. This, of

BOX 4.2   ADMIXTURE

We may estimate the degree of admixture (**M**) between two gene pools by using the allele frequencies in a contemporary admixed population (q_n), the ancestral population (q_0), and the migrant (or donor) population (Q) as

$$\mathbf{M} = q_n - Q / q_0 - Q \qquad (4.8)$$

Crawford et al. (1976) studied admixture between the indigenous population of the modern state of Tlaxcala in Mexico and the Spanish colonists who arrived in Tlaxcala in the 16th century. Tlaxcalans allied themselves with the Spanish army to help depose their long-time enemies in the western valley, the Aztecs. As a result of this alliance, Tlaxcaltecan history was distinct among native populations of the Americas (Crawford 1976). Considerable admixture between Spanish and Tlaxcaltecans occurred in the administrative center of Tlaxcala City, but little if any occurred in more peripheral Tlaxcaltecan communities.

For the purposes of this example, we will estimate the admixture of Spanish genes into the Tlaxcaltecan gene pool by using a single allele, the Fy^a allele of the **Duffy blood group** system. This allele occurs at a frequency of 0.5261 in contemporary Tlaxcala City and a frequency of 0.7842 in a modern but essentially unadmixed rural community in the same state. The latter value may be taken as an estimate of the ancestral Tlaxcaltecan frequency of the allele. In similar fashion, the ancestral Spanish frequency may be estimated by its current frequency in the Spanish population, 0.3992. The estimate of European admixture into this Amerindian gene pool is simply $(0.5261 - 0.3992)/(0.7842 - 0.3992) = 0.3296$. Thus, approximately one-third of the modern gene pool of Tlaxcala is estimated to be of European origin, and the remaining two-thirds reflect its Amerindian heritage.

Moreover, the complement of the admixture estimate is $(1 - m)^n$, where m is the migration rate estimated as the proportion of alleles introduced by migration (admixture) in each generation, and n is the number of generations such admixture has been occurring. In Tlaxcala, admixture began early in the 16th century. Thus, if we consider that a generation is 25 years, admixture had been occurring for approximately 18 generations by the time the contemporary allele frequency was estimated. Hence,

$$(1 - m)^{18} = 1 - 0.3296 \qquad (4.9)$$

This equation gives an estimate of migration per generation (m) of 2.2%. Although it is based on only a single gene, this estimate is very similar to one based on several genes derived by the original investigators. It presumes that the migration (admixture) has been constant over time. In fact, for Tlaxcala, the majority of direct admixture between native Tlaxcaltecans and Spanish occurred during the first few generations after contact, and the resulting mix of alleles have been segregating in the population since then. Thus, the actual rate of migration of Spanish genes into the Tlaxcalan gene pool was more probably 3–4% per generation, but for a shorter period of time (Crawford et al. 1976).

course, maximizes the estimates of European contribution to the African American gene pool while minimizing the estimates of African contribution to the European American gene pool. Nevertheless, gene flow in both directions may be documented by both historical and molecular genetic research, as in the case of the probable descendants of Thomas Jefferson and Sally Hemings (e.g., Brodie 1974; Gordon-Reed 1997; Foster et al. 1998, 1999; Lander and Ellis 1998; Abbey 1999; Davis 1999).

By using the relationships between migration and gene frequencies defined earlier, we can determine the degree to which genes characterizing the African and European gene pools of several centuries ago contribute to their hybrid descendants today. Because of the asymmetrical effect due to population size and isolation, imposed primarily by the historical conditions of slavery, the question usually posed is what proportion of the contemporary African American gene pool may derive from the European gene pool as a result of gene flow.

In a classic study, Glass and Li (1953) used the frequency of alleles of the **Rh blood group** system to estimate the European American contribution to the contemporary African American population of the United States as ~25%. Such estimates must be taken with considerable caution. First, they typically are estimated from single alleles, a practice notoriously subject to error. Second, this particular genetic system is subject to selection via maternal–fetal incompatibility, so any estimates may be biased due to the action of selection. Third, the ancestors of modern African Americans derive from several source populations and geographic areas of primarily West Africa, and the source populations likely differed genetically. This is undoubtedly true of the European-derived population of the United States as well. Thus, the estimates of the source population frequencies are crude estimates at best. Finally, neither the African American nor the European American population is genetically uniform; both populations demonstrate geographic differences in allele frequencies (Adams and Ward 1973).

Estimation of European admixture in the contemporary African American gene pool depends on estimates of ancestral frequencies and the contemporary population considered. The range of estimates based on single **classical markers** is exceptionally wide (e.g., 0.03–0.41) (Workman 1973). These disparate data demonstrate the lack of precision of such estimates. In a reanalysis, Chakraborty et al. (1992) considered 15 loci simultaneously and estimated the European component of the contemporary African American gene pool to be ~25%. Although the new estimate has the advantage of multiple loci, it, too, depends on the contemporary sample selected (Pittsburgh, Pa.) and the accuracy of the estimates of ancestral frequencies.

Estimates can and have been made for many modern populations worldwide whose recent history is characterized by genetic exchange with previously isolated members of the human community. Even though such estimates may be of heuristic value, or even of some utility (e.g., in gauging disease risk in large segments of national populations), the bulk of genetic variation is found *within* populations rather than *between* them (e.g., Lewontin 1972).

In a reanalysis of this issue, Dean et al. (1994) used 257 **RFLP** (restriction fragment length polymorphism) loci to examine genetic variation in four ethnic groups: African Americans, Native Americans (Cheyenne), Asians (Chinese), and European Americans. All of the loci were **polymorphic** in all four groups, and the analysis indicated that <10% of the variation could be accounted for by differentiation among the four groups. Thus, >90% of the variation observed was the result of within sample variation.

In a similar study, Barbujani et al. (1997) studied the variation of more than 100 molecular markers in 16 populations from around the world. These investigators also found that the vast majority of variation at the molecular level occurred within individual populations, with only one-tenth of the diversity accounted for by differences among continental samples.

GENETIC VARIATION

In much of the foregoing discussion, we assumed that the loci of interest were polymorphic, that is, that there is variation at the locus. Indeed, the study of variation and its patterning is central to evolutionary and anthropological genetics. By convention, a genetic locus is considered polymorphic if at least two alleles at the locus are segregating in the population and both are present at a frequency >1%. Genetic variation at a locus or in a population can be characterized in many ways, but a common method is to define gene diversity (h) as

$$h = 1 - \sum x_i^2 \qquad (4.10)$$

where x_i is the frequency of the ith allele and the summation is over all alleles at the locus. Under the general assumption of random mating within a population, h is also the expected heterozygosity. Thus, the average h over multiple loci is a measure of the heterozygosity of the population and a measure of population variability.

One common assumption of many population genetics models is that each mutation creates a new allele in a gene pool; this assumption is formalized in the mathematical genetic mutation model known as the **infinite allele model**. This model underlies many of the classical models discussed earlier and has guided much work in the study of genetic variation in populations. With the added assumption of selective neutrality, this model predicts the expected heterozygosity of a population at equilibrium as

$$h = \frac{4N_e u}{1 + 4N_e u} \tag{4.11}$$

where N_e is **effective population size** and u is the mutation rate of the gene per generation.

These quantities and models have been used extensively by geneticists to evaluate population variation and to test for neutrality of classical markers. However, the advent of molecular methods raises concerns about their applicability to newer molecular data. For example, h is defined as the expected heterozygosity, which is the probability that two alleles drawn at random from a population are different. For long DNA sequences, where nearly every observed sequence may be different from every other, the value of h will approach 1. Hence, some alternative is required.

For DNA sequence data, an estimate of variation is

$$\Pi = \frac{1}{n(n-1)/2} \sum \Pi_{ij} \tag{4.12}$$

where n is the number of sequences sampled, and therefore $n(n-1)/2$ is the number of possible pairs; Π_{ij} is the number of **nucleotide** differences between the ith and jth sequences; and the summation is over all possible pairs. An alternative formulation useful in some contexts is

$$\Pi = \frac{n}{n-1} \sum x_i x_j \, \Pi_{ij} \tag{4.13}$$

where x_i and x_j are the frequencies of alleles i and j in the sample. This formulation is applicable, for example, to sequence data that may be reduced to allelic states, such as when mitochondrial sequence data are used to define **haplotypes**, whose frequencies may be treated as allelic frequencies.

The DNA sequence analog to the infinite allele model is the infinite sites model, which assumes that for long sequences, each new mutation is novel, occurring at a nucleotide site that has not mutated before. Because degree of polymorphism in DNA sequence data is affected by the length of the sequence, a measure of nucleotide diversity (Nei and Li 1979) that standardizes sequence variation relative to sequence length (L) may be defined as

$$\pi = \Pi/L \tag{4.14}$$

(For additional information about the statistical properties of these quantities and their use in population and evolutionary studies, see Nei 1987 and Li 1997.)

As molecular data accumulate for human populations, such quantities are finding increasing use in human evolutionary studies. For example, Jorde et al. (1995) computed gene diversity for nuclear restriction site polymorphisms (**RSP**s) and short tandem repeats (**STR**s), and Π for **mitochondrial DNA** (mtDNA) hypervariable sequence data in three pooled samples that represented continental populations,

Africans, Asians, and Europeans. Their results indicate that STRs have the highest diversity in worldwide samples, with a mean value of 0.725 ± 0.020; RSPs were characterized by lower diversity (mean $= 0.377 \pm 0.018$); and the mtDNA haplotype data indicate dramatically lower variation, with a mean Π of 0.02 ± 0.001.

These data clearly demonstrate that different types of genetic systems are characterized by different levels of variation, hence different evolutionary content. Certainly, caution must be exercised in making **phylogenetic** inferences from a single system, because it is only one realization of an evolutionary process. Of equal interest in this study was the pattern of variation across the continental samples. The European sample was the most diverse for the RSP markers, and the African sample the least variable. However, for the STR and mtDNA data, African samples were characterized by the greatest degree of variability. For the mtDNA sequence data, Europeans had the least variability. Thus, population history, as well as the nature of the genetic system under study, is of considerable importance in evaluating human genetic variation and evolution.

POPULATION STRUCTURE

Much of the classical theory of allele frequency change presented earlier assumes populations of infinite—or at least very large—size and random mating. Any number of factors operate in human populations to prevent these assumptions from being met. No population is infinite—although some are very large—but most are of small to moderate size, exacerbating the effects of genetic drift. Moreover, for a variety of reasons, most populations are not characterized by random mating. Small size and avoidance of inbreeding ensure nonrandom mating in most human populations. The factors that prevent realization of random mating and other assumptions of equilibrium theory are referred to collectively as population structure. This structure is important in the distribution of allele frequencies in time and space. Thus, mate selection rules, geography, and the interaction of the mechanisms of evolution contribute to the observed distribution of alleles in contemporary populations.

A commonly used statistic in human population genetic studies is **genetic distance**. Genetic distances are quantitative estimates of the degree of similarity or difference between populations based on genetic data. They reflect the degree to which populations share genetic variation and hence are indices of genetic affinity between groups. Several different formulations are used for genetic distances, but space constraints prohibit their review here. The most commonly used genetic distance (D) is that of Nei (1987):

$$D = -\log_e I \qquad (4.15)$$

where I is defined as

$$I = \frac{J_{xy}}{[J_{xx} J_{yy}]^{1/2}} \qquad (4.16)$$

J_{xy} is the probability of randomly selecting two identical alleles from two separate populations (x and y), and J_{xx} and J_{yy} are the probabilities of randomly sampling two identical alleles from the same population. (For discussions of other distance metrics, see Jorde 1985; Weir 1990; Cavalli-Sforza et al. 1994; Goldstein et al. 1995; and Shriver 1995.)

These metrics are important because they not only help describe patterns of genetic variation between populations but also are occasionally used as the basis for additional phylogenetic analyses. In the remainder of this chapter, I will survey some aspects of that patterned variation in several geographic regions and with respect to several problems in human evolution and adaptation.

HUMAN ORIGINS

Although the topic is beyond the scope of this chapter, the role of genetic data and analyses in the study of modern human origins is worthy of a brief note.

As reviewed by Harpending (1994), the current focus of debate regarding the appearance of modern humans is whether the transition occurred early and essentially simultaneously throughout the Old World, such that precursor and modern populations are characterized by regional continuity, or whether modern humans appeared much more recently in a single geographic locale and subsequently migrated from that single point of origin to colonize and replace earlier forms elsewhere. The former model is known as the multiregional, or regional continuity model, and the latter is often referred to as the replacement, or Garden of Eden, model. Each model is associated with specific assumptions and predictions about the nature of modern variation in human populations. Nevertheless, it has proven difficult to unequivocally reject either one.

In a study of 147 mtDNA sequences from around the world, Cann et al. (1987) discovered that the African samples were more variable than the other regional samples. Moreover, construction of a tree from these sequences revealed much deeper branches for the African samples as well as an African root to the tree. The authors concluded that the source population of all modern humans had arisen rather recently (~200,000 years ago) in Africa, lending considerable weight to the replacement model. Since, several questions about the adequacy of analytical methods for mtDNA in phylogenetic analyses have been raised, but new work in both theory and data analyses has continued to appear.

Although classical marker frequencies also have been used to address the questions of human origins, they pose special problems. First, alleles at marker loci are not ordered (Harpending 1994). That is, we don't know whether any particular allele at a locus is more closely or more distantly related to any other allele. Secondly, such markers are coded for in the nuclear genome and are subject to **recombination**. Recombination destroys the genealogical record in genetic data and therefore weakens the phylogenetic inference from such data. Consequently, much of the human origins debate has centered on mtDNA sequence data that is maternally inherited and not subject to recombination. However, mtDNA is a single locus and only one realization of an evolutionary history. Other loci might have a different history. Moreover, with any appreciable number of sequences, the number of trees that can be generated is very large, and determining which tree best reflects the phylogenetic history of the samples is not easily accomplished.

Despite the potential difficulties of these types of genetic data for inferring population histories, numerous studies have followed Cann et al. (1987) in generating trees from mtDNA sequence data from multiple populations. This approach seems reasonable because mtDNA is **haploid** and the diversity seen in contemporary populations is the result of the accumulation of mutations over time. Therefore, the diversity of mtDNA types in a population may be seen as the tips of branches of a tree that ultimately link back to a single mtDNA type. Estimating the time back to that root, the **coalescent** (Kingman 1982a, 1982b; Tavaré et al. 1997), would seem a logical way to discriminate between the two hypotheses of human origins. Unfortunately, estimating the age of the coalescent is dependent on population size (Harpending 1994). Thus, demographic history rather than phylogenetic history is likely to be recovered from human mtDNA sequences.

The theory for this form of inference is based on the distribution of pairwise differences in sequence data (e.g., Rogers and Harpending 1992; Harpending 1994; Rogers 1997). Imagine a population sample of mtDNA sequences. All possible pairwise comparisons can be identified, and for each such comparison, the number of nucleotide differences (e.g., mismatches) can be tabulated. Over all possible pairs, a histogram of differences called the **mismatch distribution** results. The shape of this distribution reveals much about the historical size of the population from which the sequences were drawn. In many cases, the mismatch distribution indicates a relatively small population followed by a population expansion. This finding is consistent over many population samples of sequence data and implies a bottleneck in the relatively recent past for many human groups and is interesting for two reasons. First, like other types of genetic data and analyses, it provides some additional support for the replacement hypothesis, although the data are not entirely inconsistent with the multiregional model. Second, the demonstration of a bottleneck suggests an explanation for the relatively low level of genetic variation that characterizes the human species.

Analyses and inferences of population processes from the mismatch distribution have been criticized for focusing on pairs of sequences when the data are population data, hence ignoring much of the data in hand, and for not taking into account population structure (Marjoram and Donnelly 1994; Harding 1997).

Rogers (1997) evaluated the effect of population structure on the mismatch distribution. He used a geographically structured version of the coalescent algorithm that permitted changes in population size, number of subdivisions over time, and the rate of gene flow between subdivisions. Using simulation, he studied several models of population history. The results indicated that population structure enhances inference from the mismatch distribution and that if mtDNA reflects population growth rather than selection, the multiregional model of modern human origins may be statistically rejected more strongly than before, whereas the replacement model cannot.

These results are unlikely to be the last word on the origin of modern humans. But genetic data and analyses have stimulated considerable interest in reexamining this question from various perspectives. Consequently, researchers have developed and refined concepts in genetic theory—especially with respect to the coalescent and mismatch distributions—and generated much new molecular data that bears on the question of modern human origins. Although the question may not be finally settled, the growing bodies of mtDNA and nuclear molecular markers (Jorde et al. 1998) as well as Y-chromosome data are increasingly more consistent with the recent African origin of modern humans than the model of regional continuity.

GEOGRAPHIC VARIATION AND EVOLUTION IN EUROPE

Europe is undoubtedly the most genetically studied of geographic regions. Geneticists have been documenting patterns of genetic diversity among individual populations of the area for decades, and by the 1970s, it was clear that allele frequencies for many classical markers were not randomly nor uniformly distributed in this geographic space. Rather, allele frequencies often were found to **cline** from areas in the Middle East to Western Europe (Ammerman and Cavalli-Sforza 1984; Sokal et al. 1989). The regular clines of allele frequencies, particularly those of the **HLA system**, across Europe led to the interpretation that the clines were the remnants of an ancient migration.

Archaeologists have hypothesized that the advent of farming technology 10,000 years ago in the Levant was carried to Europe incrementally, not by cultural diffusion but by the migration of farmers from the Middle East. The cline of allele frequencies was taken as confirmation of this historical process, which gave much of the modern structure to the genetics of contemporary Europe (Sokal and Menozzi 1982; Ammerman and Cavalli-Sforza 1984; Sokal et al. 1991).

Cavalli-Sforza and colleagues constructed synthetic gene frequency maps of Europe and the Middle East (e.g., Menozzi et al. 1978) that graphically demonstrated the clinal nature of classical genetic markers in Europe. Although the alleles of the HLA system were the most convincing evidence of such clines, and hence the interpretation of the migration of Neolithic farmers into Europe, they were by no means the only such evidence. Sokal and Menozzi (1982) demonstrated statistically the presence of clines in allele frequency data in this geographic region using **spatial autocorrelation** techniques (see discussion later), providing additional evidence for the interpretation of clines as markers of ancient population movements.

Cavalli-Sforza et al. (1994) summarized the patterns of gene frequency variation in Europe and other regions. They provided the original data as well as synthetic gene frequency maps that graphically illustrate the geographic patterns contained in these classical markers. Synthetic gene frequency maps are typically constructed from **principal component** scores based on the **correlations** between allele frequencies from populations sampled at various geographic locations. Whereas such maps may exhibit spurious geographic patterns if based on databases characterized by much missing data or a regular grid system underlying map surface construction (Sokal et al. 1999a, 1999b; cf. Rendine et al. 1999), maps based on complete data matrices, or single allele frequencies (e.g., Figure 4.3), or that use alternative surface construction techniques also may illustrate informative geographic trends in the original data (O'Rourke and Lichty 1989). However, such maps should always be viewed as only one method of examining patterns in gene frequency data, and geographic trends seen in these maps should be confirmed and analyzed by other statistical methods.

As data have accumulated on newer molecular markers in European samples, the view of allele frequency structure as a reflection of a Neolithic migration has come under renewed scrutiny. Although some of the patterns seen in classical markers are reflected in DNA data (e.g., clinal distribution of the *APOE**4 allele[Lucotte et al. 1997]; Y chromosome markers [Semino et al. 1996]), others are not. However, in a survey of 34 alleles from **micro-** and **minisatellite** markers in more than 300 European populations, Chikhi et al. (1999) found that about one-third of them exhibited clinal distribution across the continent, about the same proportion observed in classical markers by Sokal et al. (1989). Two of the seven loci examined (*APOB* and *F13A*) exhibited variation that appeared to be random, with no geographic structure. Much more molecular data are available on mtDNA variation than on nuclear molecular markers.

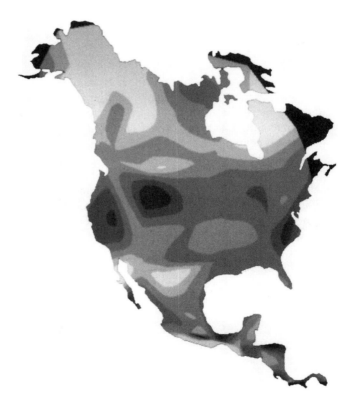

Figure 4.3   Synthetic gene frequency map of the Fyᵃ allele of the Duffy system based on its frequency in 72 native North American populations. Darker shading represents higher frequencies of the allele. See Suarez et al. (1985) and O'Rourke and Lichty (1989) for details of map construction method.

Francalacci et al. (1996) report hypervariable mtDNA sequence data from the population of Tuscany in Italy compared with other populations from Europe (Sardinia, the Basque region, and Britain) and populations from the Middle East. The 49 individuals who made up the Tuscany sample were characterized by 43 distinct haplotypes. Taking the sequence of both **hypervariable regions** together, 1 haplotype was shared by seven individuals, and the remaining 42 were carried by single individuals. Genetic distances between these samples based on the mtDNA **control region** sequences suggest considerable homogeneity among European populations, although the Tuscan sample falls between the other European samples and the Middle Eastern sample used in this study. The mismatch distribution of the Tuscan sequences compared with the other European populations or the Middle Eastern sample also suggests its intermediacy. Despite difficulties in estimation, the authors suggest that the results are consistent with an ancient migration out of the Middle East into Europe, but much earlier than the proposed Neolithic migration, which is postulated to have occurred with the spread of agriculture. Francalacci et al. (1996) claim that the mtDNA sequence data indicate the persistence of the effects of an early population expansion (growth) deriving from outside of Europe, but with an *in situ* increase whose effects remain. These authors conclude that mitochondrial diversity is further enhanced by more recent population expansions.

Similarly, Comas et al. (1996) report on mtDNA sequence data in a sample of 45 unrelated individuals from Turkey. The sequences from the first mtDNA hypervariable region yielded 40 different sequences. By using genetic distances among all sequences to compute a phylogenetic tree, the researchers identified four clusters; one was characterized by an apparent European-specific mutation. Moreover, the estimate of time of divergence of this cluster was 37,000 to more than 100,000 years ago, long before the hypothesized Neolithic expansion but consistent with an earlier hypothesized replacement of Neandertals by modern *Homo sapiens*. Compared with other European populations, the Turkish samples appear to fall between the European and the Middle Eastern samples; a fact these authors also take as evidence of an early expansion of Middle Eastern populations, through Turkey, during the colonization of Europe by modern humans.

In an analysis of a somewhat larger series of European populations, Bertranpetit et al. (1996) also found a low level of genetic diversity, which is consistent with a relatively recent origin of the modern European population. The questions are, how recent, and under what mechanism? These authors reported several hypervariable substitutions that are relatively common in European populations (see Table 4.3). Relative to the Anderson et al. (1981) mtDNA reference sequence, these substitutions occur in roughly 10–20% of Europeans. Some are geographically patterned, whereas others are not. Moreover, the mismatch distributions derived from these data indicate that population expansions in the Middle East preceded those of the European populations and that there is a significant correlation ($r = -0.884$) between the **mean pairwise distance** and distance from the Middle East. The inference is that the movement of modern populations into Europe took several thousands of years. The estimate of the age of these European haplotypes is imprecise because of the ambiguity of the mutation rate (Rogers and Harpending 1992;

Harpending et al. 1993), but estimates between 46,000 and 130,000 years ago are consistent with the observed mismatch distributions.

Sykes et al. (1996) and Richards et al. (1996) also assayed the hypervariable region in various European and Middle Eastern populations and came to similar conclusions. In this study, as in others, sequence diversity is greater in Middle Eastern than in European populations, indicating greater antiquity for the former. Additionally, one of the groups of related haplotypes (a **haplogroup**) defined in this study, characterized by a cytosine-to-thymine (C→T) **transition** at position 16,223 of the Anderson reference sequence (Table 4.3), is estimated to be the oldest of the identifiable European lineage clusters; its estimated divergence time is ~50,000 years. Although it is not clear that all this divergence occurred after European colonization, it is consistent with a lineage identified by Torroni et al. (1994) based on high-density **restriction mapping** that has a minimum age in Europe of 34,000 years. On the basis of these sequence polymorphisms, the more recent lineage divergence times for Europe than the Middle East are given in Table 4.4.

Notably, when the focus of analysis is classical markers rather than molecular ones, certain populations consistently fall outside the general European genetic pattern: Basques, Sardinians, Icelanders, and Lapps (Cavalli-Sforza et al. 1994). Using the molecular data presented by Sykes and colleagues (Richards et al. 1996; Sykes et al. 1996), only the Basques appear similarly divergent based on mean pairwise distances.

In sum, the emerging data derived from mtDNA sequence variation indicate that European lineages derived from Middle Eastern ones several tens of thousands of years ago and that the impact of the Neolithic migration may have been rather less pronounced on the gene pool of Europe than has frequently been assumed. In an expanded analysis of control region sequence data from European and Middle Eastern populations, Sykes (1999) suggests the modern European gene pool derives from three primary colonization events. The first, identified by the early origin haplotype noted above, occurred in the Early Upper Paleolithic Period (~50,000 years before

Table 4.3 Transitions defining European haplogroups and their frequencies

Haplogroup	Definition	Frequency %	Distribution
1	Reference		Clinical, high in west
2	C at 16,311	20.1	No pattern
3	T at 16,294	11.8	No pattern
4	T at 16,270	9.3	Clinical, high in west
5	C at 16,189	20.1	No pattern
6	T at 16,223	17.6	Clinical, high in east
7	C at 16,126	20.8	Clinical, high in east

Haplogroup 1 is defined by the Anderson et al. (1981) sequence, rather than a specific substitution, so no frequency in Europeans is given.
(*Source*: After Bertranpetit et al. 1996)

Table 4.4 Mean pairwise differences and estimated divergence times for European and Middle Eastern mtDNA lineages

Lineage	Europe		Middle East	
	Mean pairwise distance	Divergence time, years BP	Mean pairwise distance	Divergence time, years BP
I	3.05	32,000	4.8	50,000
II	3.59	38,000	5.47	57,500
II A	1.07	11,000	3.85	40,500
II B-W	1.20	12,500		
II B-E	0.57	6,000		
II C	3.46	36,500	4	42,000
III	2.73	29,000	3.03	32,000

Divergence rate used: 1 substitution per 10,500 years.
(*Source*: Sykes et al. 1996)

the present [BP]), and the second, which included the majority of haplotypes identified, took place in the Late Upper Paleolithic Period (11,000–14,000 years BP). There is also evidence from the frequencies and distribution of mtDNA haplotypes of the Neolithic expansion after 10,000 years ago, which may have contributed up to 20% of the current mitochondrial variation observed in modern Europe. Clearly, population movements, incursions, and displacements since the colonization of Europe by modern humans have had an impact on the local genetic structure of populations, but the molecular signal on a continental scale appears to reflect replacement of Neandertals by modern *Homo sapiens*, rather than the Neolithic migration of Middle Eastern farmers (Richards et al. 1996).

Adding credence to this view, Krings et al. (1997) reported that the mtDNA hypervariable sequence from the Neandertal-type specimen differs dramatically from contemporary European sequences. These authors interpret the substantial genetic difference between the Neandertal specimen and modern populations as evidence that that Neandertals were not ancestral to, but replaced by, the modern human gene pool. Molecular variation on the Y chromosome provides additional insights into European population history. Until recently, few genetic markers on the Y chromosome were known. This lack of knowledge was unfortunate because, theoretically, the nonrecombining portion of the Y could serve as the complement to mtDNA markers. Of uniparental inheritance and lacking recombination, such markers would provide a mechanism for tracking variation in paternal lineages, just as mtDNA tracks variation in maternal lineages.

Recently, a wealth of new molecular markers was identified that include a Y-chromosome *Alu* insertion polymorphism (YAP), single nucleotide polymorphisms (SNPs), and microsatellites (Casanova et al. 1985; Hammer 1994; Jobling and Tyler-Smith 1995; Underhill et al. 1996), and reviews of the global variation in these new markers have begun to appear (Hammer et al. 1997; Malaspina et al. 1998).

Malaspina et al. (1998) surveyed the YAP insertion and several Y-chromosome poly-morphisms in populations of Africa, Europe, and western Asia and identified spe-cific regional patterns of variation for these Y-chromosomal markers. For the 23 European populations included in their survey, a pronounced southeast–northwest cline was observed in some but not all markers. Moreover, the estimate of the coa-lescent for these markers, or groups of related markers (networks), resulted in a rel-atively recent origin (Neolithic) for some but an older (Paleolithic) origin for others.

Lucotte and Loirat (1999) recently reported the frequency of Y-chromosome polymorphisms (**DYS**1 haplotypes) from 28 locations in western Europe. The highest frequency of one haplotype of this system (haplotype 15) was found in the Basques. This haplotype exhibits a gradient of decreasing frequencies away from the region of the Pyrenees, in a geographic cline opposite to that typically noted for classical markers across Europe. Lucotte and Loirat (1999) suggested that haplo-type 15 of the DYS1 locus is the ancestral paternal European haplotype and that it represents variation from the pre-Neolithic European gene pool. These recent mol-ecular studies suggest that the origin of the modern European population may not be a simple, single event and that caution should be used in evaluating the age of molecules relative to the age of populations (Hammer et al. 1997; Barbujani et al. 1998; Malaspina et al. 1998; Richards and Sykes 1998). The molecular data avail-able to date for European populations support a Neolithic population migration from the Middle East but also indicate that a substantial portion of the modern Eu-ropean gene pool derives from an earlier, Paleolithic, population. The proportional contribution of each of these prehistoric sources of genetic variation remains sub-ject to debate.

The work of Sokal and colleagues on the relationship between ethnohistorical events and European genetic structure is relevant to this discussion. An ethnohistor-ical database of more than 3,500 records that described the locations and movements in time and space of European populations during the past 4,000 years was compiled to assess the relationship such population movements may have to contemporary gene frequency distributions (reviewed in Sokal et al. 1996, 1997). Although the ge-netic data available for analysis in this case are classical markers rather than molec-ular data, the correlations between geography and history are interesting.

Using a quadrat (grid) system, based on 1° increments in latitude and longitude, these authors computed measures of similarity (distances) among groups of popula-tions in all quadrat pairs based on the ethnohistorical database. These ethnohistorical distances were compared with genetic distances obtained from 26 genetic systems assayed in European populations occupying the same geographic region. With geo-graphic distance held constant, the partial correlation between ethnohistorical and genetic distances in these data is statistically significant. Moreover, the correlation structure is relatively stable. Randomization of the sequence of ethnohistorical events has little impact on the correlation between ethnohistorical and genetic distances, but geography does. Randomizing the region of origin of known population movements reduces the correlation between the distance matrices.

Unfortunately, the ethnohistorical database used does not have the temporal depth to address questions concerning the dispersal of agriculuralists in the Neolithic

Period. Nonetheless, it clearly demonstrates the importance of geography and geographic differentiation between populations in the structuring of genetic variation on a continental scale.

GENE GEOGRAPHY IN THE AMERICAS

Much of the history of human population genetic studies is characterized by investigations of the dynamics of local populations. Nowhere is this characterization more true than in the study of genetic variation of indigenous populations of the Americas.

Beginning with the early studies of Matson (Matson and Schrader 1933; Matson et al. 1936; Matson 1938), who documented the unusually high frequency of the A blood type among Amerind populations of southwest Canada, many investigators surveyed genetic variation in Amerind populations. These populations ranged from the Inuit of the high Arctic to groups at the southern tip of South America. Individual studies typically reported allele frequency data from red cell **antigens** and protein polymorphisms and focused attention on the roles of drift and migration (including admixture from non-Indian gene pools) as determinants of the structure of genetic diversity in small populations. As reports of gene frequency data accumulated, interpretive summaries gave a more continental view of the pattern of genetic variation in the Americas (Spuhler 1979; Salzano and Callegari-Jacques 1988; O'Rourke et al. 1992; Szathmary 1993; Cavalli-Sforza et al. 1994).

North America: Classical Markers

That genetic variation in Amerindians is structured was apparent early, at least with respect to traditional markers. North-to-south clines in allele frequencies have been documented for North American native populations (Suarez et al. 1985), and this geographic structure is conveniently summarized in synthetic gene frequency maps (e.g., Figure 4.3, which reflects the geographic distribution of the Fy^a allele of the Duffy system).

The graphical demonstration of clines in North America can be confirmed by several statistical analytical techniques. One commonly used method is spatial autocorrelation. Autocorrelation simply measures the degree of similarity (correlation) of allele frequencies in a variety of populations residing at various distances from one another (Sokal and Oden 1978a, 1978b). Under the assumptions of isolation by distance, populations that reside in geographic proximity to each other will have significantly positive correlations between allele frequencies, whereas those at increasing distances will show a decrease in the correlation toward zero. In the case of a cline, the decreased correlation between allele frequencies of populations separated by considerable distance will not remain at zero but have a strongly negative correlation.

Figure 4.4 illustrates the generally monotonic decline of the correlation of allele frequencies with distance for the allele frequencies that are mapped in Figure 4.3. The measure of autocorrelation used here is Moran's I (Sokal and Oden 1978a; Cliff and Ord 1981). Such a correlogram is clear evidence for clinal change in allele

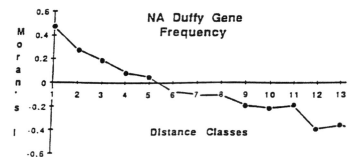

Figure 4.4   Correlogram of spatial autocorrelation coefficients indicating clinal nature of Fy[a] frequencies in native North America (NA).

frequencies with increasing geographic distance. Evidence for admixture with European and African gene pools in Amerind populations of the east coast of North America also are detectable by using these methods, and the amount of genetic admixture is estimable by using methods discussed earlier. Generally, among North American Indian populations, the level of non-Indian admixture averages ~15%, although the range is wide. With respect to the contemporary geographic distribution of allele frequencies, however, inter-Indian gene flow is almost undoubtedly more important than non-Indian admixture (Spuhler 1979).

The origin of the geographical pattern of allele frequency data is the source of continuing investigation. The clinal distributions seen in North America are quite strong for some genetic systems (e.g., ABO, Rh, and P blood groups) (Suarez et al. 1985) but weak or absent for others. These data have been interpreted as possible evidence for natural selection operating via mechanisms correlated with patterns of climatic variation (Ananthakrishnan and Walter 1972; Piazza et al. 1981; O'Rourke et al. 1985) for those systems shown to be correlated with latitude. The interaction of drift and migration may also result in clines for systems that are not associated with corresponding ecological zones.

The clinal structure of red cell antigen frequencies decays in the geographic constraint of Central America, and in this sense, Central America is truly genetically intermediate between North and South America (O'Rourke et al 1992).

South America: Classical Markers

Genetic studies of South America have a long and well known history and three recent summaries of genetic variation in native South America are available (O'Rourke and Suarez 1985; Salzano and Callegari-Jacques 1988; Black 1991). Research by Neel and colleagues (e.g., Neel 1978) in lowland South America revealed a complex pattern of population fission based on kin migration, subsequent population expansion and fusions, and the nature of genetic diversity that resulted from this complex of demographic factors. These studies remain classics of the study of genetic variation in small, relatively isolated populations. Contemporary work continues to characterize individual populations of the region. Belich et al. (1992) report that sequence variation in the HLA-B alleles of the Kaingang and Guarani of southern Brazil are distinct from the alleles found in European and Asian populations. Similarly, Watkins et al. (1992) report new HLA-B alleles in the Waorani of Ecuador deriving from intralocus recombination. The HLA-A and C alleles in these populations, however, are indistinguishable from those found in Europe and Asia. The authors suggest this indicates the ability of rapid evolutionary change in small, isolated populations, and may account for the genetic diversity observed in aboriginal South America. This is similar to the inference of O'Rourke and Suarez' (1985) that the lack of geographic structure among South American populations was consistent with a pattern expected of numerous small populations drifting independently.

In contrast to the case in North America, O'Rourke and Suarez (1985) found no evidence of geographic structure to red cell antigen frequencies in 70 South American populations. Salzano and Callegari-Jaques (1988) did report some clinal distributions for the allele frequencies of some other systems, but generally they too found little geographic structure in the region. Table 4.5 summarizes the relationship between geography and gene frequency for red cell antigen data in native populations of the western hemisphere (O'Rourke et al. 1992). As the entries in Table 4.5 indicate, average genetic distance between populations is greatest in South America and lowest in Central America. F_{st} is a measure of population subdivision, and defines the probability that two alleles sampled at random from a subpopulation will be **identical by descent**. Curiously, although F_{st} values are approximately equivalent in North and South America, the correlation between geographic and genetic distances in the two continents are quite disparate. The correlation between these two distances in North America ($r = 0.337$) is statistically significant and substantially larger than the correspondingly insignificant value for the South American data ($r = 0.128$). In all areas, Central America is intermediate between North and South America.

Table 4.5 Summary statistics for American Indian genetic data

Statistic	North America	South America	Central America
Number of populations	44	70	30
Mean genetic distance	0.055	0.062	0.029
Range	0.003–0.203	0.003–0.275	0.002–0.134
F_{st}	0.902	0.906	0.0517
Correlation: geographic vs. genetic distance	0.337[a]	0.128[b]	0.304[c]

Mantel permutation test significance levels: [a]$p=0.004$; [b]$p=0.020$; [c]$p=0.008$.
(*Source:* After O'Rourke et al. 1992)

Molecular Variation in the Americas

mtDNA Haplogroup Diversity

Despite the increase in importance of molecular markers for population genetic studies, the data on human populations continue to accrue slowly. Classical markers still form the greater source of information regarding genetic variation in most areas. Nevertheless, some inferences regarding molecular variation in Amerind populations are possible.

Initial research on molecular variation in the Americas focused on mtDNA restriction site polymorphisms (RSPs; DNA sequence variations identified by restriction endonucleases, see Chapter 3) and a 9-bp polymorphism. This latter marker occurs between the coding regions for the lysyl transfer RNA gene and the cytochrome oxidase II gene (region V). In most European and African populations, this sequence of **bases** represents a tandem repeat of the 9-bp sequence CCCCCTCTA (where C is cytosine, T is thymine, and A is adenine). However, in many individuals of Asian ancestry, one copy of this repeated sequence has been deleted. Thus, the deleted and repeat forms of the marker are polymorphic in many populations of the Americas and elsewhere.

The 9-bp deletion is polymorphic in most Amerind groups, but its frequency range is 0–100%. Of most interest, however, is its absence in all populations above 55% north latitude, except where its introduction can be attributed to admixture with populations not indigenous to the circumarctic (Shields et al. 1993). The absence of the deletion is not limited to the North American arctic; it also is absent in corresponding regions of Siberia. This peculiar distribution led Cann (1994) to suggest that the presence of the deletion primarily in more southerly and coastal Amerindians may indicate introduction across the Pacific from Polynesia, where the marker is essentially fixed. This suggestion has received little support from additional genetic data because the deletion is not found in the Americas in conjunction with hypervariable sequence variants specific to Polynesia. The position also receives little support from available archaeological evidence in North and South America.

Torroni et al. (1993) performed high-resolution restriction screening on several Amerind populations. In conjunction with the 9-bp deletion, they defined four mtDNA haplogroups that they deemed the original founding haplogroups of all Amerindians. Each haplogroup is defined by the joint occurrence of four markers, but each has a signature marker that is presumptive evidence of its presence in the absence of data for the other three. Haplogroup A is defined by a *Hae*III restriction site at nucleotide position (np) 663 of the Anderson et al. (1981) reference sequence, haplogroup B by the 9-bp deletion in the cytochrome II-tRNA[lys] intergenic region (V), haplogroup C by the absence of a *Hin*cII restriction site at np 13259, and haplogroup D by the absence of an *Alu*I restriction site at np 5176. Recently, a fifth founding haplogroup, X, was confirmed in Native American populations (Smith et al. 1999). Haplogroup X in the Americas is defined by the absence of a *Dde*I restriction site at np 10394, as well as the absence of a *Dde*I site at np 1715 and a C→T transition at np 16278. The relationship of the minimal set of markers and distinct mitochondrial lineages is given in Table 4.6.

Even though the distribution of these lineages in Amerind groups is only beginning to be elucidated, some generalizations are possible. In addition to the absence of haplogroup B from circumarctic populations, this haplogroup is particularly frequent in the southwest of the United States. Haplogroup A is the most common haplogroup among Amerindians—its occurrence is nearly fixed in Athapaskan populations—but its occurrence is low to absent in the southwest United States. It occurs with high frequency in South America. Haplogroups C and D tend to be polymorphic in most but not all Amerind populations, but haplogroup D is particularly frequent on the Columbia plateau in the US northwest, northern California, and the adjacent western Great Basin (Lorenz and Smith 1996). With the exception of haplogroup B, the other three haplogroups common in Amerindians and the Inuit are also found in some indigenous northeast Siberian populations of Chukotka and Kamchatka, although at much reduced frequency (Starikovskaya et al. 1998; Schurr et al. 1999). The documentation of these Amerind hapogroups at low frequency in northeast Siberian populations led Schurr and colleagues (1999) to suggest that they represent evidence of the remnants of the ancient Beringean populations.

mtDNA Sequence Variation

mtDNA hypervariable sequence data are also available for an increasing number of populations, and in general, the sequence data agree with the RSP data. Ward et al. (1991), for example, found four clusters of sequence variants in a single tribal population from the Pacific Northwest (Nuu-Chah-Nulth) that correspond to the four basic haplogroups defined earlier for RSPs and the length polymorphism. Others have also found a similar correspondence, although the sequence support for haplogroup D is not strong. Nevertheless, the extreme sequence diversity of the mtDNA hypervariable region leads to interesting and conflicting estimates for the ages of the lineages and for the timing of origin of native peoples of the Americas.

Table 4.6 Marker Distribution in Amerind mtDNA Lineages

Lineage	Region V 9bp deletion	*Hae* III np 663	*Alu* I np 5176	*Hince* II np 13,259	*Dde* I np 10,394
A	−	+	+	+	−
B	+	−	+	+	−
C	−	−	+	−	+
D	−	−	−	+	+
X	−	−	+	+	−

(*Source*: After Smith et al. 1999)

Y-chromosome Variation

Polymorphisms on the Y chromosome are being identified rapidly (Underhill et al. 1997) and are beginning to shed additional light on patterns of molecular variation in Amerind populations. Several authors have identified a single Y-chromosomal haplotype that is present in a large proportion of American Indian males (Santos et al. 1996; Underhill et al. 1996; Bianchi et al. 1997; Karafet et al. 1997). This haplotype consists of the association of specific markers at three loci. In worldwide population surveys, 23 phenotypic variants associated with molecular variation in the Y-centromeric alphoid block (αh) have been observed. Phenotype II (αh-II) is often associated with allele A of the DYS19 microsatellite locus, and a $C \rightarrow T$ transition in the *DYS199* gene, resulting in the common Amerind haplotype: αh-II/DYS19A/DYS199T (Bianchi et al. 1997; Santos et al. 1996; Karafet et al. 1997).

This haplotype occurs at frequencies >90% in unadmixed Amerind populations of South and Central America and at lower, but appreciable, frequencies in North Amerindians, even with evidence of non-Amerind admixture. It is commonly found in populations of all three language groupings of the Americas advocated by Greenberg (1987). Recently, Bianchi et al. (1998) redefined this haplotype with many more markers and confirmed its prevalence in indigenous populations of the Americas . It is generally absent from Asian populations but has been observed in Siberian Eskimos, Chukchi, and Evens (Karafet et al. 1997), in whom its presence is postulated to be the result of back-migration from northern North America. The preponderance of this haplotype in Native American populations has resulted in it being referred to as the founding Y lineage of the Americas (Pena et al. 1995; Bianchi et al. 1997).

Amerindian Origins

Archaeological evidence for the peopling of the New World has always been controversial. The Clovis culture is widespread in North America by 11,500 years BP, but few sites with earlier dates have been generally accepted. In South America, sites dating to around 11,000–12,000 years BP are numerous, and several have been claimed for considerably greater antiquity. However, only Monte Verde in southern Chile appears to have a convincing case for early occupation (<19,000 years BP).

Linguistic diversity in America has been characterized by Greenberg (1987) as comprising three large groupings: Amerind, which includes all speakers of Indian languages in South America, Central America, and most of North America; the Na-Dene, which includes speakers of Athapaskan languages of the Pacific Northwest; and Eskaleut, which includes the Inupik/Inupiat and Aleut speakers of the circum-arctic. Greenberg et al. (1986) presented data to suggest that this tripartite linguistic classification reflects three waves of migration into the Americas in prehistory and that biological and archaeological data are consistent with this colonization scenario. The proposal has been controversial; Merriwether et al. (1994, 1995) and Merriwether and Ferrell (1996) argued that the ubiquity of multiple founding mtDNA haplogroups in the majority of contemporary Amerind populations suggests a single colonization event.

The nature of mitochondrial inheritance makes possible the estimation of dates of ancestry of related lineages, thus giving rise to the possibility of using a molecular clock to calibrate the colonization of geographic regions for which sufficient molecular data are available. As indicated above, restriction site screening initially yielded only four haplogroups, each of which could be considered a founding lineage. Torroni et al. (1993) argued that this result indicated a reduced number of migrations to the New World and that the founding population(s) must have been characterized by a bottleneck, reducing genetic variability at founding. Subsequent mtDNA sequence analyses (e.g., Ward et al. 1991) indicated that a reduction of genetic variability due to a bottleneck effect at colonization was not necessary to account for contemporary distributions of genetic variation in modern Amerindians.

Most recently, Forster et al. (1996) reanalyzed several hundred Native American and Siberian mtDNA hypervariable sequences to evaluate competing hypotheses regarding the original colonization. They identified at least six mtDNA sequence haplogroups (A1, A2, B, C, D1, and X) for modern Native Americans. The coalescents of these sequences range from 11,300 years ago for the Eskimo, Na-Dene, and Siberian sequences to 21,000–23,000 years ago for the larger aggregate of all North and South American Indians. Thus, the later introduction of Eskimo gene pools and the distinctiveness of the northern North American Na-Dene are supported by these data and analyses. Moreover, the general estimate of first entry into the Americas of 20,000–25,000 years ago in a single migration is consistent with the scenario suggested by Merriwether et al. (1996) and contradicts the long-debated three waves of migration model (see also Szathmary 1993). It is useful to note that the entry date estimated from this reanalysis of mtDNA sequence data agrees with the earliest archaeological evidence from central Alaska.

South American molecular data are equally interesting. South American populations are characterized by a specific D1 subcluster defined by a transition at np 16187. The sequence diversity in this unique South American subgroup, which presumably must have arisen since its entry into South America, indicates a founding event in South America 18,000 years ago. This date, too, is consistent with archaeological dates of early occupation of the South American continent, including the presumed age of Monte Verde (19,000 years BP).

The colonization scenario that results from the analyses of Forster et al. (1996) is that a restricted number of founding lineages entered the Americas between 20,000 and 25,000 years ago, rapidly moved southward, and reached South America by 18,000 years ago. This original colonization was characterized by relatively rapid population expansion of migrating small populations, which countered the effects of genetic drift such that the loss of genetic variation at founding was less than originally hypothesized. Indeed, one strength of this reanalysis is that more realistic parameters for population growth were incorporated into the analyses that led to the estimated coalescent times. The readvance of the last ice age may have reduced population size and, hence, genetic variability in northern North America and Siberia, but not in more southerly latitudes.

After the retreat of the last glacial advance, an expansion of northern, Beringean populations resulted in the distribution of genetic variation seen today. These results confirm many earlier analyses in whole or in part, and are consistent with the conclusions of Szathmary's (1993) excellent summary of the genetics of Native American populations. She argued that migration and colonization should be viewed as a process of **demic** (i.e., population) **expansion** rather than single short-term events; that the original colonists were genetically heterogeneous, at least for mtDNA; and that diversification took place after colonization.

Most recently, Y-chromosome markers have yielded information on Amerind origins. The Native American specificity of the αh-II/DYS19A/DYS199T haplotype and its high frequency in most Amerind populations have led many investigators to see it as the founding paternal lineage in the Americas (Pena et al. 1995; Bianchi et al. 1996, 1997; Santos et al. 1996; Karafet et al. 1997). The predominance of a single haplotype throughout the Americas is inconsistent with multiple migrations of original colonists. Thus, molecular variation on the Y chromosome is generally taken as evidence of a single founding population for the indigenous populations of the Americas, concordant with the single wave of migration scenario advocated by Merriwether et al. (1995, 1996) based on mtDNA data. In a larger analysis of Y-chromosomal variation in 60 populations distributed globally, Karafet et al. (1999) suggested that there are more likely at least two founding haplotypes for Native Americans rather than one (see also Vallinoto et al. 1999).

The question of the geographic location from whence the earliest immigrants to the Americas originated is as contentious as the timing of their entry into the continent. Since the earliest genetic studies of blood group variation, it has been obvious that the indigenous populations of the Americas are of Asian origin. But determining where in Asia has been problematic. Most often, the source populations for early Amerindians have been assumed to be those of extreme northeastern Siberia. This area, after all, is nearest the Beringean area of North America and is the logical place to seek Amerind origins. Moreover, the arctic inhabitants on both sides of the Bering Strait share cultural similarities.

Newer molecular data are beginning to refine our notions of Amerind origins. Kolman et al. (1996) and Merriwether et al. (1996) independently analyzed mtDNA variation in multiple groups of Amerindians and Asians and determined that the greatest similarities between Amerind and Asian populations occurred among groups from Mongolia. The distinctive Amerind mtDNA haplogroups discussed above are curiously absent, or at low frequency, in most Siberian and north Asian populations. However, all four of the primary Amerind haplogroups occur among native populations of Mongolia. Thus, at least for this genetic system, the most likely source for the original entrants to North America is the area in and around Mongolia.

The study of Y-chromosome markers by Karafet et al. (1999) also showed that the paternal lineages most common in the Americas were also found at high frequencies among the Kets and Selkups of southern Siberia, prompting the researchers to suggest a locus of origin of Native American ancestors in the region between the Altai Mountains and Lake Baikal in southcentral Siberia. Santos et al. (1999) also identified the Kets and related populations of the Altai region as possessing the major Y haplotype that defines Native American groups. These authors also argue for a south-central Siberian origin for Native American paternal lineages. Such an interior Siberian source location for early Amerind migrants is consistent with the emerging view of the relatively late colonization of the arctic regions of northeastern Siberia (Goebel 2000) and the possibly recent diversification of Amerind language families (Nettle 1999).

These recent and corroborating results from Y molecular variation also agree with some mtDNA analyses described earlier. The geographic region between the Altai Mountains and Lake Baikal is adjacent to Mongolia, the hypothesized source of Amerind colonists based on analysis of mtDNA variation by Kolman et al. (1995) and Merriwether et al. (1995), although the latter authors advocate a single migration model. Despite these concordances, most of the analyses that can be brought to bear on the question of timing of entry of populations into the Americas are based on genetic variation in mtDNA and Y chromosomal markers.

Although some similarities in distribution and inference are seen, we have reason to expect some differences between mtDNA and Y-chromosomal markers due to different demographic histories for males and females, for example, different migration rates between the sexes, patrilocality, and drift and founder effects. Insufficient data are currently available for other nuclear markers to see whether these inferences will agree with other genetic systems.

ANCIENT DNA ANALYSIS

Although the use of molecular markers has permitted new approaches to broad questions of human evolution on a continental scale, another type of molecular analysis has provided a new perspective on regional population studies: ancient DNA (aDNA) analysis. By the mid-1980s, it was clear that nucleic acids were routinely preserved in prehistoric material, could be accessed and characterized through the development of the **polymerase chain reaction** (PCR), and yield information about the genetic relationships of populations of the past (Pääbo 1985a,b; Pääbo et al. 1989). The field has grown rapidly, and several reviews are already available (Handt et al. 1994a; Herrmann and Hummel 1994; O'Rourke et al. 1996).

Conceptually, aDNA research is straight-forward. Extract remaining **nucleic acids** from prehistoric samples, and using the PCR, amplify fragments bearing molecular markers (e.g., length polymorphisms, RSPs, single-nucleotide polymorphisms, etc.) already demonstrated to be polymorphic and informative in contemporary populations. This conceptual simplicity obscures the practical difficulties of analyzing aDNA.

Unlike molecular screens of contemporary samples, ancient DNA studies are constrained by the quality of the nucleic acids recovered. Virtually all aDNA samples are characterized by degradation in size and nucleotide composition. Recovery of DNA fragments in excess of 500 bp is relatively rare, so studies using aDNA must focus attention on markers that can be scored in PCR products smaller than this size, typically, in fragments <300 bp long. Moreover, the degraded nature of aDNA makes copy number a practical concern. With most ancient samples yielding 1–5% of the DNA expected in modern tissues, mtDNA, with several hundred genomes per cell, has been the genome of choice for aDNA studies. Under optimal conditions, genomic DNA may be accessible and analyzable from ancient samples (e.g., Hummel and Herrmann 1991; Filon et al. 1995; Zierdt et al. 1996), but in most cases, nuclear DNA has proven problematic for reliable amplification from ancient samples. As a result, the majority of aDNA studies conducted to date have focused on mtDNA polymorphisms.

The two primary difficulties of aDNA analyses have to do with contaminants: organic and inorganic contaminants contained in the sample that coextract with the ancient DNA, which act to inhibit the **polymerase**, thus reducing amplification efficiency; and the introduction of modern DNA, which preferentially amplifies in the PCR reaction at the expense of the smaller aDNA fragments. Although contaminants present formidable difficulties for aDNA research, methods have been developed that minimize their effects.

Simply diluting the extracts in sterile distilled and deionized water (ddH$_2$O) or DNA buffer may sufficiently reduce the concentration of coextracted contaminants so that PCR will successfully amplify the target molecules. Increasing the concentration of the polymerase in the PCR reaction, or adding bovine serum albumin (BSA) to bind to possible contaminants, also may increase amplification efficiency and specificity.

In addition to strand breaks that result in the characteristic small size of aDNA, other alterations to the nucleic acids postmortem impede aDNA research. Most ancient nucleic acids also are characterized by damage and alteration to some of the nitrogenous bases or structural sugar molecules. Pääbo (1989) demonstrated that perhaps 10% of the **pyrimidine** residues in aDNA samples may have experienced oxidative damage. Nevertheless, Pääbo (1989), Lindahl (1993), and others (e.g., Tuross 1994) demonstrate that, at least in samples up to a few thousand years in age, nucleic acids may be expected to be preserved in quantities sufficient for analysis. It is also clear that local conditions of interment and preservation determine the quality of aDNA available for analysis in any individual specimen. Typically, cold, dry environments are most favorable for DNA preservation and recovery.

Despite the fact that aDNA studies have been possible for only the past decade, ancient samples already have been studied from Europe, the United Kingdom, Africa, Polynesia, Japan, China, North and South America, and Greenland. Indeed, the literature is growing so rapidly that no comprehensive and detailed survey is possible in a limited space. However, a few representative studies warrant brief discussion.

Handt et al. (1994b) studied the aDNA of "the Tyrolean Ice Man," dated to just over 5000 years ago. Discovered following the retreat of an alpine glacier in the Tyrolean Alps, the preservation conditions seemed favorable for nucleic acid preservation, but the degradation typical of most ancient DNAs characterized this ice-preserved specimen, as well. Handt et al. (1994b) cloned PCR products of the mtDNA hypervariable region and obtained multiple sequences. In addition to documenting exogenous contamination in the sample, the authors nonetheless showed that shorter amplification products circumvented the difficulty of "**jumping PCR**"

(Paabo 1989) and revealed that this individual specimen carried a mtDNA hyper-variable sequence that is most common in contemporary northern European populations. In contrast, Krings et al. (1997) sequenced the mtDNA hypervariable sequence of the Neandertal type specimen from the Neander Valley, and found substantial sequence differences from modern human sequences. The Neandertal sample differed, on average, at 27 nucleotide positions from a collection of nearly 1000 modern human sequences (range = 22–36 substitutions). This is triple the average number of differences observed among the modern samples (mean = 8, range = 1-24), and has been taken as evidence that Neandertals contributed little if any mitochondrial DNA to the modern human gene pool. As noted earlier, this provides additional support to the recent human origins model (Krings et al. 1997; Ward and Stringer 1997).

Hagelberg and colleagues (Hagelberg and Clegg 1993, Hagelberg et al. 1994, 1999) examined ancient mtDNA from a series of Polynesian samples to shed light on the colonization of the Pacific Islands. Archaeological and linguistic data suggest that modern Polynesians derive from an early population of Austronesian speakers of island Southeast Asia that rapidly moved eastward and colonized islands of the Pacific during the archaeologically defined Lapita period (3600–2500 years BP). This "fast-train model" of Polynesian origins is contrasted with other data suggesting that Lapita culture originated in island Melanesia rather than the more westward locale of Southeast Asia. Genetic studies on modern populations have yielded conflicting evidence for these two origin scenarios. Hagelberg and Clegg (1993) and Hagelberg et al. (1994) examined ancient DNA in 21 human skeletal samples from across Oceania to address this question of population movement in prehistory.

Modern Polynesian populations are characterized by a high frequency of the region V deletion (fixed in some populations) and three characteristic substitutions in the hypervariable D-loop region relative to the Anderson et al. (1981) referent: 16,217 (T→C), 16,247 (adenine to guanine [A→G]), and 16261 (C→T). In the ancient Polynesian samples examined by Hagelberg and Clegg (1993), all had the deletion, and of those that could be sequenced for the hypervariable region, all were consistent with two or three of the characteristic Polynesian transitions. Conversely, none of the ancient Melanesian samples possessed the deletion or the characteristic Polynesian transition set. Finally, of six archaeological samples from Fiji and Tonga examined, only one possessed the deletion, and it was by far the youngest in the sample (300 years BP). The deletion was found to be absent in the oldest samples (Melanesian) associated with the Lapita archaeological complex. The inference derived from this study is that the Lapita culture is indeed of Melanesian origin, and has little to do with the Polynesian expansion into the eastern Pacific from a base in the islands of Southeast Asia.

Small populations, and subsequent drift, may be responsible for the fixation of the distinct Polynesian mtDNA lineage observed. As with many studies, some caveats must be offered here. First, relatively few samples were screened relative to the time frame and geographic area of interest, so sampling issues may be quite important to the final inference. Second, although extreme care was taken, undetected modern contaminating DNA cannot be completely excluded as a complicating factor in these analyses. Finally, it is important to remember that mtDNA analyses are based on only a single locus. Inferences of population history based on other genetic systems may well be different than those based on mtDNA. Inclusion of multiple genetic systems from many samples, both ancient and contemporary, rarely leads to simplistic interpretations of prehistory (cf. Jorde et al. 1995; Hagelberg et al. 1999).

Parr et al. (1996) examined mtDNA diversity in the prehistoric Fremont of Utah to assess the biological diversity and possible avenues of population movement in the ancient Great Basin and southwest United States. The Fremont samples studied were excavated from the eastern margin of the Great Salt Lake by investigators who anticipated the future molecular analyses and removed specific skeletal elements from their archaeological contexts on exposure, without cleaning or handling. This technique aided greatly in combating the introduction of modern nucleic acid contaminants. These samples were screened for the four mtDNA markers that define the four primary Amerind lineages (see Table 4.7). Surprisingly, haplogroup A was absent in these samples ($n = 43$), haplogroup B was at high frequency (>60%), and haplogroups C and D were present at low but detectable frequencies. This combination of frequencies was typical of modern groups farther to the south and west of the Fremont area of northern Utah. Indeed, haplogroup A is low to absent in many modern southwestern U.S. populations, just as haplogroup B is found at an exceptionally high frequency.

These data contrast, for example, with the ancient Anasazi who, in an initial examination of relatively few samples ($n = 20$), exhibit a lower deletion frequency and an absence of haplogroup D—despite the facts that the Anasazi and Fremont lived barely 300 miles apart and that the samples overlapped in time (~800–2000 years BP). These contrasts may be extended to ancient Amerind samples in the eastern United States studied by Stone and Stoneking (1993), who found a low frequency of the deletion (10%) and preponderance of haplogroup C; or the results of Kaestle (1997) on ancient samples of the western Great Basin, which are characterized by a low deletion frequency and high frequency of haplogroup D.

These disparate results suggest that prehistoric populations of the Americas were genetically heterogeneous, and like their modern descendants, the variation was geographically structured (O'Rourke et al. 1999, 2000). Continued research on archaeological samples such as those described in this section will undoubtedly continue to shed light on prehistoric population migrations, settlement patterns, and evolutionary dynamics.

Ancient DNA research holds considerable promise in solving a number of longstanding questions in **epidemiology**, as well. Recently, four separate research groups reported evidence of *Mycobacterium tuberculosis* in ancient tissues. The presence of tuberculosis in pre-Columbian populations of the Americas has been controversial, frustrating attempts to discern whether the disease was present prior to European contact or whether it was introduced to Amerind populations via European contact. Unfortunately, the lesion in bone caused by tuberculosis is not specific, and the organism itself does not persist postmortem.

However, Spigelman and Lemma (1993) isolated and identified the DNA of the bacterium from skeletal material in several Old-World samples, and Taylor et al. (1996) and Baron et al. (1996) similarly identified the bacterium in relatively recent bone samples (from tens to a few hundred years old). Salo et al. (1994) extracted, amplified, and sequenced a 123-bp fragment unique to *M. tuberculosis* from naturally mummified tissue obtained from a 1000-year-old Peruvian sample. These data confirm the presence of tuberculosis in the Americas long before European contact and effectively resolves this longstanding debate. It is curious that this study was based on soft tissue samples, because it is the wide experience of many researchers that skeletal material is usually a better source for aDNA that ancient soft tissue.

Finally, Filon et al. (1995) diagnosed β-**thalassemia** from the DNA recovered from the skeleton of an 8-year-old child archaeologically recovered in Israel. Based on archaeological context, the burial was from the Ottoman period, 200–400 years ago. The skeleton evidenced severe porotic hyperostosis, yet the age of the child suggested a mild disease course. Careful extraction, amplification, and sequencing of multiple DNA samples indicated that the individual possessed a frameshift mutation in **codon** 8 (FS8) as a result of a deletion of two thymine bases (noncoding strand), as well as a C → T transition in the second codon of β-globin. The frameshift mutation is polymorphic in thalassemia chromosomes throughout the eastern Mediterranean and reaches a high frequency of 10% in Turkey. According to Filon et al. (1995), this mutation is nearly exclusive to Arabs in modern Israel, and the homozygosity of the marker in the ancient sample is suggestive of **consanguineous mating**. The accompanying transition is polymorphic at a frequency of 13% of modern samples with β-thalessemia.

Without modern medical interventions (transfusions), the frameshift mutation observed in this specimen typically results in early mortality as fetal hemoglobin is replaced by adult hemoglobin. Yet, this individual lived at least eight years. The codon 2 transition present in this sample is part of haplotype IV (Orkin et al. 1982), which in concert with the FS8 homozygosity results in persistence of fetal hemoglobin and a remarkably milder disease course. Thus, the specificity and clinical relevance of aDNA

Table 4.7 Frequency of mtDNA markers in Fremont skeletal samples

	9-bp deletion	*Hae*III np 663	*Alu*I np 5176	*Hinc*II np 13,259
Successful amplifications	41	36	36	31
Number with marker	25	0	34	27
Frequency of marker, %	61	0	94	87

(*Source*: O'Rourke et al. 1999)

research is greater than many imagined until very recently. Other infectious diseases and clinical conditions probably will be studied in antiquity, similarly and profitably.

CHAPTER SUMMARY

This chapter provided a very brief survey of the nature of evolutionary mechanisms and their effects on single genetic markers. It also attempted to place the extensive genetic variation that characterizes contemporary human populations in geographic perspective with a few specific examples:

1. Molecular genetic data have contributed a new perspective on modern human origins. Although the debate is certainly far from over, modern genetic data increasingly strongly suggests that modern humans originated relatively recently and subsequently expanded around the globe.
2. European populations are characterized by relatively low levels of genetic variation, a general feature of our species. The clinal distributions of allele frequencies so clearly demonstrated for many classical markers are not so universal when molecular data are examined. Indeed, the origin of the modern European population may lie well into the Paleolithic Period, with a more modern veneer of clinal variation overlaid as Neolithic farmers moved north and west from the Middle East.
3. Modern populations of the Americas are closely related to Asian populations and may ultimately derive from prehistoric populations inhabiting the region of Mongolia and southern Siberia. North American populations exhibit distinctly different patterns of genetic variability from those on the South American continent when classical markers are examined, but the distinction is not so clear when molecular markers are employed. One marker, the 9-bp deletion, has a peculiar distribution in the Americas, being absent in all populations north of 55 north latitude and essentially polymorphic everywhere else in the Americas. At least in North America, regional patterns of variation in the four or five basic haplogroups are discernible.
4. Ancient DNA is proving useful in directly testing some hypotheses regarding population movements and origins around the world. This approach holds considerable promise for tackling many questions in regional prehistory by examining directly the populations that occupied areas of interest in the not too distant past.

Despite the discussion thus far, much of the interest in human biology and adaptive evolution is not with single neutral markers but with complex quantitative phenotypes. Although the principles of population genetics outlined in this chapter are also the basis for the evolutionary study of **quantitative traits**, a separate body of theory and applications has also been developed for complex traits. They are the subject of Chapter 5.

RECOMMENDED READINGS

Cavalli-Sforza LL, Menozzi P, and Piazza A (1994) The History and Geography of Human Genes. Princeton: Princeton Univ. Press.

Ciba Foundation (1996) Variation in the Human Genome (Symposium 197). New York: John Wiley & Sons.

Graur D and Li W-H (1999) Fundamentals of Molecular Evolution, 2nd edition. Sunderland, Mass.: Sinauer Associates, Inc.

Harding RM (1997) Lines of descent from mitochondrial Eve: an evolutionary look at coalescence. In P Donnelly and S Tavaré (eds.): Progress in Population Genetics and Human Evolution. New York: Springer-Verlag, Inc.

Harpending H (1994) Gene frequencies, DNA sequences, and human origins. Perspect. Biol. Med. 37: 384–394.

Harpending H, Relethford J, and Sherry ST (1996) Methods and models for understanding human diversity. In AJ Boyce and CGN Mascie-Taylor (eds.): Molecular Biology and Human Diversity. Cambridge: Cambridge Univ. Press.

Hartl DL and Clark AG (1997) Principles of Population Genetics, 3rd edition. Sunderland, Mass.: Sinauer Associates, Inc.

Rogers AR (1997) Population structure and modern human origins. In P Donnelly and S Tavaré (eds.): Progress in Population Genetics and Human Evolution. New York: Springer-Verlag, Inc.

Weir BS (1996) Genetic Data Analysis II: Methods for Discrete Population Genetic Data. Sunderland, Mass.: Sinauer Associates, Inc.

Wilson EO and Bossert WH (1971) A Primer of Population Biology. Sunderland, Mass.: Sinauer Associates, Inc.

REFERENCES CITED

Abbey DM (1999) Letter to the Editor: The Thomas Jefferson Paternity Case. Nature 397: 32.

Adams JP and Ward RH (1973) Admixture studies and the detection of selection. Science 180: 1137–1143.

Ammerman AJ and Cavalli-Sforza LL (1984) The Neolithic Transition and the Genetics of Populations in Europe. Princeton: Princeton Univ. Press.

Ananthakrishnan R and Walter H (1972) Some notes on the geographical distribution of the human red cell acid phosphatase phenotypes. Humangenetik 15: 177–181.

Anderson S, Bankier AT, Barrell BG, de Bruijn MHL, Coulson AR, Drouin J, Eperon IC, Nierlich DP, Roe BA, Sanger F, Schreier PH, Smith AJH, Staden R, and Young IG (1981) Sequence and organization of the human mitochondrial genome. Nature 290: 457–465.

Barbujani, G, Magagni A, Minch E, and Cavalli-Sforza LL (1997) An apportionment of human DNA diversity. Proc. Natl. Acad. Sci. U.S.A. 94: 4516–4519.

Barbujani G, Bertorelle G, and Chikhi L (1998) Evidence for Paleolithic and Neolithic gene flow in Europe. Amer. J. Hum. Genet. 62: 488–491.

Baron H, Hummel S, and Herrmann B (1996) *Mycobacterium tuberculosis* complex DNA in ancient human bones. J. Arch. Sci. 23: 667–671.

Belich MP, Madrigal JA, Hildebrand WH, Zemmour J, Williams RC, Luz R, Petzl-Erler ML, and Parham P (1992) Unusual HLA-B alleles in two tribes of Brazilian Indians. Nature 357: 326–329.

Bernstein M (1931) Die geographische Verteilung der Blutgruppen und ihre anthropologische Bedeutung. In: Comitato Italiano per lo Studio dei Problemi della Popülazione. Rome, Italy: Istituto Poligrafico dello Stato.

Bertranpetit J, Calafell F, Comas D, Perez-Lezaun A, and Mateu E (1996) Mitochondrial DNA sequences in Europe: an insight into population history. In: AJ Boyce and CGN Mascie-Taylor (eds.): Molecular Biology and Human Diversity. Cambridge: Cambridge Univ. Press.

Bianchi NO, Bailliet G, Bravi CM, Carnese RF, Rothhammer F, Martinez-Marignac VL, and Pena SDJ (1997) Origin of Amerindian Y-chromosomes as inferred by the analysis of six polymorphic markers. Am. J. Phys. Anthropol. 102: 79–89.

Bianchi NO, Catanesi CI, Bailliet G, Martinez-Marignac VL, Bravi CM, Vidal-Rioja LB, Herrera RJ, and López-Camelo JS (1998) Characterization of ancestral and derived Y-chromosome haplotypes of New World native populations. Am. J. Hum. Genet. 63: 1862–1871.

Black FL (1991) Reasons for failure of genetic classifications of South Amerind populations. Hum. Biol. 63(6): 763–774.

Bodmer W and Cavalli-Sforza LL (1976) Genetics, Evolution, and Man. New York: W. H. Freeman & Co.

Brodie FM (1974) Thomas Jefferson: An Intimate History. New York: Norton.

Cann RL (1994) MtDNA and native Americans: a southern perspective. Am. J. Hum. Genet. 55: 7–11.

Cann RL, Stoneking M, Wilson AC (1987) Mitochondrial DNA and human evolution. Nature 325: 31-36

Casanova M, Leroy P, Boucekkine C, Weissenbach J, Bishop C, et al. (1985) A human Y-linked DNA polymorphism and its potential for estimating genetic and evolutionary distance. Science 230: 1403–1406.

Cavalli-Sforza LL and Bodmer WF (1971) The Genetics of Human Populations. San Francisco: W. H. Freeman & Co.

Cavalli-Sforza LL, Menozzi P, and Piazza A (1994) The History and Geography of Human Genes. Princeton: Princeton Univ. Press.

Chakraborty R, Kamboh MI, Nwankwo M, and Ferrell RE (1992) Caucasian genes in American Blacks: New Data. Am. J. Hum. Genet. 50: 145–155.

Chikhi L, Destro-Bisol G, Pascali V, Baravelli V, Dobosz M, and Barbujani G (1999) Clinal variation in the nuclear DNA of Europeans. Hum. Biol. 70: 643–657.

Cliff AD and Ord JK (1981) Spatial Processes. London: Pion.

Comas D, Calafell F, Mateu E, Perez-Lezaun A, and Bertranpetit J (1996) Geographic variation in human mitochondrial DNA control region sequence: the population history of Turkey and its relationship to the European populations. Mol. Biol. Evol. 13: 1067–1077.

Crawford MH (ed.) (1976) The Tlaxcaltecans: Prehistory, Demography, Morphology and Genetics. Publications in Anthropology, No. 7. Lawrence: Univ. of Kansas Press.

Crawford MH, Workman PL, McClean C, and Lees FC (1976) Admixture estimates and selection in Tlaxcala. In Crawford MH (ed.): The Tlaxcaltecans. Lawrence: Univ. of Kansas Press.

Crow FJ and Kimura M (1970) An Introduction to Population Genetics Theory. New York: Harper & Row.

Davis G (1999) Letter to the Editor: The Thomas Jefferson Paternity Case. Nature 397: 32.

Dean M, Stephens JC, Winkler C, Lomb DA, Ramsburg M, Boaze R, Stewart C, Charbonneau L, Goldman D, Albaugh BJ, Goedert JJ, Beasley RP, Hwang L-Y, Buchbinder S, Weedon M, Johnson PA, Eichelberger M, and O'Brien SJ (1994) Polymorphic admixture typing in human ethnic populations. Am. J. Hum. Genet. 55: 788–808.

Falconer DS (1981) Introduction to Quantitative Genetics, 2nd edition. London: Longman.

Filon D, Faerman M, Smith P, and Oppenheim A (1995) Sequence analysis reveals a β-thalassaemia mutation in the DNA of skeletal remains from the archaeological site of Akhziv, Israel. Nature Genet. 9: 365–368.

Forster P, Harding R, Torroni A, and Bandelt H-J (1996) Origin and evolution of native American mtDNA variation: a reappraisal. Am. J. Hum. Genet. 59: 935–945.

Foster EA, Jobling MA, Taylor PG, Donnelly P, de Knijff P, Mierement R, Zerjal T, and Tyler-Smith C (1998) Jefferson fathered slave's last child. Nature 396: 27–28.

Foster EA, Jobling MA, Taylor PG, Donnelly P, de Knijff P, Mieremet R, Zerjal T, and Tyler-Smith C (1999) Letter to the Editor: Reply. Nature 397: 32.

Francalacci P, Bertranpetiti J, Calafell F, and Underhill PA (1996) Sequence diversity of the control region of mitochondrial DNA in Tuscany and its implications for the peopling of Europe. Am. J. Phys. Anthropol. 100: 443–460.

Glass B and Li CC (1953) The dynamics of racial admixture: an analysis based on the American Negro. Am. J. Hum. Genet. 5: 1–19.

Goebel T (2000) Pleistocene human colonization of Siberia and peopling of the Americas: an ecological approach. Evol. Anthropol. In Press.

Goldstein DB, Linares AR, Cavalli-Sforza LL, and Feldman MW (1995) An evaluation of genetic distances for use with microsatellite loci. Genetics 139: 463–471.

Greenberg JH (1987) Language in the Americas. Palo Alto, Calif.: Stanford Univ. Press.

Greenberg JH, Turner II CG, and Zegura SL (1986) The settlement of the Americas: a comparison of the linguistic, dental, and genetic evidence. Curr. Anthropol. 4: 477–497.

Hagelberg E and Clegg JB (1993) Genetic polymorphisms in prehistoric Pacific islanders determined by analysis of ancient bone DNA. Proc. R. Soc. Lond. B 252: 163–170.

Hagelberg E, Wuevedo S, Turbon D, and Clegg JB (1994) DNA from Ancient Easter Islanders. Nature 369: 25–26.

Hagelberg E, Kayser M, Nagy M, Roewer L, Zimdahl H, Krawczak M, Lio P, and Schiefenhovel W (1999) Molecular genetic evidence for the human settlement of the Pacific: analysis of mitochondrial DNA, Y chromosome, and HLA markers. Phil. Trans. R. Soc. Lond. B 354: 141–152.

Hammer MF (1994) A recent insertion of an Alu element on the y chromosome is a useful marker for human populations studies. Mol. Biol. Evol. 11: 749–761.

Hammer MF, Spurdle AB, Karafet T, Bonner MR, Wood ET, Novelletto A, Malaspina P, Mitchell RJ, Horai S, Jenkins T, and Zegura SL (1997) The geographic distribution of human Y chromosome variation. Genetics 145: 787–805.

Handt O, Höss M, Krings M, and Pääbo S (1994a) Ancient DNA: methodological challenges. Experientia 50: 524–529.

Handt O, Richards M, Trommsdorff M, Kilger C, Simanainen J, Georgiev O, Bauer K, Stone A, Hedges R, Schaffner W, Utermann G, Sykes B, and Pääbo S (1994b) Molecular genetic analyses of the Tyrolean ice man. Science 264: 1175–1778.

Harding RM (1997) Lines of descent from mitochondrial Eve: an evolutionary look at coalescence. In: P Donnelly and S Tavaré (eds.): Progress in Population Genetics and Human Evolution. New York: Springer-Verlag.

Harpending HC (1994) Gene frequencies, DNA sequences, and human origins. Perspect. Biol. Med. 37: 384–394.

Harpending HC, Sherry ST, Rogers AR, and Stoneking M (1993) The genetic structure of ancient human populations. Curr. Anthropol. 34: 483–496.

Herrmann B and Hummel S (eds.) (1994) Ancient DNA. New York: Springer-Verlag.

Hummel S and Herrmann B (1991) Y-chromosome-specific DNA amplified in ancient human bone. Naturwissenschaften 78: 266–267.

Jobling MA and Tyler-Smith C (1995) Fathers and sons: the Y chromosome and human evolution. Trends Genet. 11: 449–456.

Jorde LB (1985) Human genetic distance studies: present status and future prospects. Annu. Rev. Anthropol. 14: 343–373.

Jorde LB, Bamshad MJ, Watkins WS, Zenger R, Fraley AE, Krakowiak PA, Carpenter KD, Soodyall H, Jenkins T, and Rogers AR (1995) Origins and affinities of modern humans: a comparison of mitochondrial and nuclear genetic data. Am. J. Hum. Genet. 57: 523–538.

Jorde LB, Bamshad M, and Rogers AR (1998) Using mitochondrial and nuclear DNA markers to reconstruct human evolution. BioEssays 20: 126–136.

Kaestle FA (1997) Molecular archaeology: an analysis of ancient Native American DNA from western Nevada. Am. J. Phys. Anthropol., Suppl. 24.

Karafet T, Zegura SL, Vuturo-Brady J, Posukh O, Osipova L, Wiebe V, Romero F, Long JC, Harihara S, Jin F, Dashnyam B, Gerelsaikhan T, Omoto K, and Hammer MF (1997) Y Chromosome markers and trans-Bering Strait dispersals. Am. J. Phys. Anthropol. 102: 301–314.

Karafet TM, Zegura SL, Posukh O, Osipova L, Bergen A, Long J, Goldman D, Klitz W, Harihara S, de Knijff P, Wiebe V, Griffiths RC, Templeton AR, and Hammer MF (1999) Ancestral Asian source(s) of New World Y-chromosome founder haplotypes. Am. J. Hum. Genet. 64: 817–831.

Kingman JFC (1982a) On the genealogy of large populations. J. Appl. Probab. 19A: 27–43.

Kingman JFC (1982b) The coalescent. Stochastic Processes Appl. 13: 235–248.

Kimura M and Weiss GH (1964) The stepping stone model of population structure and the decrease of genetic correlation with distance. Genetics 49: 561–576.

Kolman CJ, Sambuughin N, and Bermingham E (1996) Mitochondrial DNA analysis of Mongolian populations and implications for the origin of New World founders. Genetics 142: 1321–1334.

Krings M, Stone A, Schmitz RW, Krainitski H, Stoneking M, and Pääbo S (1997) Neandertal DNA sequences and the origin of modern humans. Cell 90: 19–30.

Lewontin RC (1972) The apportionment of human diversity. Evol. Biol. 6: 381–398.

Li W-H (1997) Molecular Evolution. Sunderland, Mass.: Sinauer Associates, Inc.

Lindahl T (1993) Instability and decay of the primary structure of DNA. Nature 362: 709–715.

Lorenz JG and Smith DG (1996) Distribution of four founding mtDNA haplogroups among Native North Americans. Am. J. Phys. Anthropol. 101: 307–324.

Lucotte G and Loirat F (1999) Y-chromosome DNA haplotype 15 in Europe. Hum. Biol. 71: 431–437.

Lucotte G, Loirat F, and Hazout S (1997) Pattern of gradient of apolipoprotein E allele *4 frequencies in Western Europe. Hum. Biol. 69: 253–262.

Malaspina P, Cruciani F, Ciminelli BM, Terrenato L, Santolamazza P, Alonso A, Banyko J, Brdicka R, García O, Gaudiano C, Guanti G, Kidd KK, Lavinha J, Avila M, Mandich P, Moral P, Qamar R, Mehdi SQ, Ragusa A, Stfanescu G, Caraghin M, Tyler-Smith C, Scozzari R, and Novelletto A (1998) Network analyses of Y-chromosomal types in Europe, Northern Africa, and Western Asia reveal specific patterns of geographic distribution. Amer. J. Hum. Genet. 63: 847–860.

Malécot G (1969) The Mathematics of Heredity. New York: W. H. Freeman & Co.

Marjoram P and Donnelly P (1994) Pairwise comparisons of mitochondrial DNA sequences in subdivided populations and implications for early human evolution. Genetics 136: 673–683.

Matson G and Schrader HF (1933) Blood grouping among the "Blackfeet" and "Blood" tribes of American Indians. J. Immunol. 25: 15–163.

Matson GA (1938) Blood groups and aguesia in Indians of Montana and Alberta. Am. J. Phys. Anthropol. 24: 81–89

Matson GA, Levine P, and Schrader HF (1936) Distribution of the sub-groups of A and the M and N agglutinogens among the Blackfeet Indians. Proc. Soc. Exp. Biol. 35: 46–47.

Menozzi P, Piazza A, and Cavalli-Sforza LL (1978) Synthetic maps of human gene frequencies in Europeans. Science 201: 786–792.

Merriwether DA and Ferrell RE (1996) The four founding lineage hypothesis for the New World: a critical Reevaluation. Mol. Phylogen. Evol. 5: 241–246.

Merriwether DA, Rothhammer F, and Ferrell RE (1994) Genetic variation in the New World: ancient teeth, bone, and tissue as sources of DNA. Experientia 50: 592–601.

Merriwether DA, Rothhammer F, and Ferrell RE (1995) Distribution of the four-founding lineage haplotypes in Native Americans suggests a single wave of migration for the New World. Am. J. Phys. Anthropol. 98: 411–430.

Merriwether DA, Hall W, Vahlne A, and Ferrell RE (1996) mtDNA variation indicates Mongolia may have been the source for the founding population for the New World. Am. J. Hum. Genet. 59: 204–212.

Mettler LE and Gregg TG (1969) Population Genetics and Evolution. Englewood Cliffs, NJ: Prentice-Hall.

Mielke JH and Crawford MH (1980) Current Developments in Anthropological Genetics, Vol. 1: Theory and Methods. New York: Plenum Press.

Neel JV (1978) The population structure of an Amerindian tribe, the Yanomama. Ann. Rev. Genet. 12: 365–413.

Nettle D (1999) Linguistic diversity of the Americas can be reconciled with a recent colonization. Proc. Natl. Acad. Sci. U.S.A. 96: 3325–3329.

Nei M (1987) Molecular Evolutionary Genetics. New York: Columbia Univ. Press.

Nei M and Li W-H (1979) Mathematical model for studying genetic variation in terms of restriction endonucleases. Proc. Natl. Acad. Sci. USA 76: 269-5273.

Orkin SH, Kazazian HH, Antonarkis SE, Goff SC, Boehm CD, Sexton JP, Waber PG, Giardina PJ (1982) Linkage of β-thalassaemia mutations and β-globin gene polymorphisms with DNA polymorphisms in human β-globin gene cluster. Nature 296: 627–631.

O'Rourke DH and Lichty AS (1989) Spatial analysis and gene frequency maps of native North American populations. Coll. Antropol. 13: 73–84

O'Rourke DH and Suarez BK (1985) Patterns and correlates of genetic variation in South Amerindians. Ann. Hum. Biol. 13(1): 13–31.

O'Rourke DH, Suarez BK, and Crouse JD (1985) Genetic variation in North Amerindian populations: covariance with climate. Am. J. Phys. Anthropol. 67: 241–250.

O'Rourke DH, Mobarry A, and Suarez BK (1992) Patterns of genetic variation in Native America. Hum. Biol. 64(3): 417–434.

O'Rourke DH, Carlyle SW, and Parr RL (1996) Ancient DNA: a review of methods, progress, and perspectives. Am. J. Hum. Biol. 8(5): 557–571.

O'Rourke DH, Parr RL, and Carlyle SW (1999) Molecular genetic variation in the prehistoric inhabitants of the eastern Great Basin. In B Hemphill and C Larson (eds.): Understanding Prehistoric Lifeways in the Great Basin Wetlands: Bioarchaeological Reconstruction and Interpretation. Salt Lake City: Univ. of Utah Press.

O'Rourke DH, Hayes MG, and Carlyle SW (2000) Spatial and temporal stability of mtDNA haplogroup frequencies in Native North America. Hum. Biol. 72: 15-34.

Pääbo S (1985a) Molecular cloning of ancient Egyptian mummy DNA. Nature 314: 644-645.

Pääbo S (1985b) Preservation of DNA in ancient Egyptian mummies. J. Arch. Sci. 12: 411-417.

Pääbo S (1986) Molecular genetic investigations of ancient human remains. Proc. Cold Spring Harbor Symp. Quant. Biol. 51: 441–446.

Pääbo S (1989) Ancient DNA: extraction, characterization, molecular cloning, and enzymatic amplification. Proc. Natl. Acad. Sci. U.S.A. 86: 1939–1943.

Pääbo S, Higuchi RG, and Wilson AC (1989) Ancient DNA and the polymerase chain reaction. J. Biol. Chem. 264: 9709–9712.

Parr RL, Carlyle SW, and O'Rourke DH (1996) Ancient DNA analysis of Fremont Amerindians of the Great Salt Lake wetlands. Am. J. Phys. Anthropol. 99(4): 507–519.

Pena SDJ, Santos FR, Bianchi N, Bravi CM, Carnese FR, Roghhammer F, Gerelsaikhan T, et al. (1995) Identification of a major founder Y-chromosome haplotype in Amerindians. Nat. Genet. 11: 15–16.

Piazza A, Menozzi P, and Cavalli-Sforza LL (1981) The making and testing of geographic and gene frequency maps. Biometrics 37: 635–659.

Price PW (1996) Biological Evolution. Fort Worth: Saunders College Publishers.

Rendine S, Piazza A, Menozzi P, and Cavalli-Sforza LL (1999) A problem with synthetic maps (Reply to Sokal et al.). Hum. Biol. 71: 15–25.

Richards M and Sykes B (1998) Letter to the Editor: Reply to Barbujani et al. Amer. J. Hum. Genet. 62: 491-492.

Richards M, Corte-Real H, Forster P, Macaulay V, Wilkinson-Herbots H, Demaine A, Papiha S, Hedges R, Bandelt H-J, and Sykes B (1996) Paleolithic and neolithic lineages in the European mitochondrial gene pool. Am. J. Hum. Genet. 59: 185–203.

Rogan PK and Salvo JJ (1990) Study of nucleic acids isolated from ancient remains. Yearbook Phys. Anthropol. 33: 195–214.

Rogers AR (1997) Population structure and modern human origins. In: P Donnelly and S Tavaré (eds.): Progress in Population Genetics and Human Evolution. New York: Springer-Verlag.

Rogers AR and Harpending H (1992) Population growth makes waves in the distribution of pairwise genetic differences. Mol. Biol. Evol. 9: 552–569.

Saitou N and Nei M (1987) The neighbor-joining method: a new method for reconstructing phylogenetic trees. Mol. Biol. Evol. 4: 406–425.

Salo WL, Aufderheide AC, Buikstra J, and Holcomb, TA (1994) Identification of mycobacterium tuberculosis DNA in a precolumbian Peruvian mummy. Proc. Natl. Acad. Sci. U.S.A. 91(6): 2091–2094.

Salzano F and Callegari-Jacques S (1988) South American Indians: A Case Study in Evolution. Oxford: Claredon Press.

Santos FR, Rodriguez-Delfin L, Pena SDJ, Moore J, and Weiss KM (1996) North and South Amerindians may have the same major founder Y chromosome haplotype. Am. J. Hum. Genet. 58: 1369–1370.

Santos FR, Pandya A, Tyler-Smith C, Pena SDJ, Schanfield M, leonard WR, Osipova L, Crawford MH and Mitchell RJ (1999) The central Siberian origin for Native American Y chromosomes. Am. J. Hum. Genet. 64: 619–628.

Schurr TG, Ballinger SW, Gan Y-Y, Hodge JA, Merriwether DA, Lawrence DN, Knowler WC, Weiss KM, and Wallace DC (1990) Amerindian mitochondrial DNAs have rare Asian mutations at high frequencies, suggesting they derived from four primary maternal lineages. Am. J. Hum. Genet. 46: 613–623.

Schurr TG, Sukernik RI, Starikovskaya YB, and Wallace DC (1999) Mitochondrial DNA variation in Koryaks and Itel'men: population replacement in the Okhotsk Sea–Bering Sea region during the Neolithic. Am. J. Phys. Anthropol. 108: 1–39.

Semino O, Passarino G, Brega A, Fellous M, and Santachiara-Benerecitti AS (1996) A view of the Neolithic demic diffusion in Europe through two Y chromosome-specific markers. Am. J. Hum. Genet. 59: 964–968.

Shields GF, Schmiechen AM, Frazier BL, Redd A, Voevoda MI, Reed JK, and Ward RH (1993) mtDNA sequences suggest a recent evolutionary divergence for Beringian and northern North American populations. Am. J. Hum. Genet. 53: 5449–562.

Shriver MD, Jin L, Boerwinkle E, Deka R, Ferrell RE, and Chakraborty R (1995) A novel measure of genetic distance for highly polymorphic tandem repeat loci. Mol. Biol. Evol. 12: 914–920.

Smith DG, Malhi RS, Eshleman J, Lorenz JG, and Kaestle FA (1999) Distribution of mtDNA haplogroup X among native North Americans. Amer. J. Phys. Anthropol. 110:271–284.

Sokal RR and Menozzi P (1982) Spatial autocorrelations of HLA frequencies in Europe support demic diffusion of early farmers. Am. Nat. 119: 1–17.

Sokal RR and Oden NL (1978a) Spatial autocorrelation in biology 1: methodology. Biol. J. Linn. Soc. 10: 199–228.

Sokal RR and Oden NL (1978b) Spatial autocorrelation in biology 2: some biological implications and four applications of evolutionary and ecological interest. Biol. J. Linn. Soc. 10: 229–249.

Sokal RR, Harding RM, and Oden NL (1989) Spatial patterns of human gene frequencies in Europe. Am. J. Phys. Anthropol. 80: 267–294.

Sokal RR, Oden NL, Wilson C (1991) Genetic evidence for the spread of agriculture in Europe by demic diffusion. Nature 351: 143-145.

Sokal RR, Oden NL, Walker J, DiGiovani D, and Thomson BA (1996) Historical population movements in Europe influence genetic relationships in modern samples. Hum. Biol. 68: 873–898.

Sokal RR, Oden NL, Rosenberg MS, and DiGiovanni D (1997) The patterns of historical population movements in Europe and some of their genetic consequences. Am. J. Hum. Biol. 9: 391–404.

Sokal RR, Oden NL, and Thomson BA (1999a) A problem with synthetic maps. Hum. Biol. 71: 1–13.

Sokal RR, Oden NL, and Thomson BA (1999b) Problems with synthetic maps remain (Reply to Rendine et al.). Hum. Biol. 71: 447–453.

Spigelman M and Lemma E (1993) The use of the polymerase chain reaction to detect Mycobacterium tuberculosis in ancient skeletons. Int. J. Osteoarch. 3: 137–143.

Spuhler JN (1979) Genetic distances, trees, and maps of North American Indians. In WS Laughlin and AB Harper (eds.): The First Americans: Origins, Affinities, and Adaptations. Gustav Fisher: New York, pp. 135–183.

Starikovskaya YB, Sukernik RI, Schurr TG, Kogelnik AM, and Wallace DC (1998) mtDNA diversity in Chukchi and Siberian Eskimos: implications for the genetic history of ancient Beringia and the peopling of the New World. Am. J. Hum. Genet. 63: 1473–1491.

Stone AC and Stoneking M (1993) Ancient DNA from a pre-Columbian Amerindian population. Am. J. Phys. Anthropol. 92(4): 463–471.

Strickberger MW (1996) Evolution, 2nd ed. Sudbury, MA: Jones and Bartlett Publishers.

Suarez BK, Crouse JD, and O'Rourke DH (1985) Genetic variation in North Amerindian populations: the geography of gene frequencies. Am. J. Phys. Anthropol. 67: 217–232.

Sykes B (1999) The molecular genetics of European ancestry. Phil. Trans. R. Soc. Lond. B 354: 131–139.

Sykes B, Corte-Real H, and Richards M (1996) Palaeolithic and Neolithic contributions to the European mitochondrial gene pool. In AJ Boyce and CGN Mascie-Taylor (eds.): Molecular Biology and Human Diversity. Cambridge: Cambridge Univ. Press.

Szathmary EJE (1993) Genetics of aboriginal North Americans. Evol. Anthropol. 1: 202–220.

Tavaré S, Balding DJ, Griffiths RC, and Donnelly P (1997) Inferring coalescence times from DNA sequence data. Genetics 145: 505–518.

Taylor GM, Crossey M, Saldanha J, and Waldron T (1996) DNA from mycobacterium tuberculosis identifies in mediaeval human skeletal remains using polymerase chain reaction. J. Arch. Sci. 23: 789–798.

Torroni A, Schurr TG, Cabell MF, Brown MD, Neel JV, Larsen M, Smith DG, Vullo CM, and Wallace DC (1993) Asian affinities and continental radiation of the four founding native American mtDNAs. Am. J. Hum. Genet. 53: 563–590.

Torroni A, Lott MT, Cabell MF, Chen Y-S, Lavergne L, and Wallace DC (1994) mtDNA and the origin of Caucasians: identification of ancient Caucasian-specific haplogroups, one of which is prone to a recurrent somatic duplication in the D-loop region. Am. J. Hum. Genet. 55: 760–776.

Tuross N (1994) The biochemistry of ancient DNA in bone. Experientia 50: 530–535.

Underhill PA, Jin L, Zemans R, Oefner PJ, and Cavalli-Sforza LL (1996) A pre-Columbian Y chromosome-specific transition and its implications for human evolutionary history. Proc. Natl. Acad. Sci. U.S.A. 93: 196–200.

Underhill PA, Jin L, Lin AA, Qasim Mehdi S, Jenkins T, Vollrath D, Davis RW, et al. (1997) Detection of numerous Y chromosome biallelic polymorphisms by denaturing high-performance liquid chromatography. Genome Res. 7: 996–1005.

Vallinoto ACR, Cayres-Vallinoto IMV, Ribeiro Dos Santos AKC, Zago MA, Santos SEB, and Guerreiro JF (1999) Heterogeneity of Y chromosome markers among Brazilian Amerindians. Am. J. Hum. Biol. 11: 481–487.

Ward RH, Frazier BL, Dew-Jager K, and Pääbo S (1991) Extensive mitochondrial diversity within a single Amerindian tribe. Proc. Natl. Acad. Sci. U.S.A. 88: 8720–8724.

Ward R and Stringer C (1997) A molecular handle on the Neandertals. Nature 388: 225-226.

Watkins DI, McAdam SN, Liu X, Strang CR, Milford EL, Levine CG, Garber TL, Dogon AL, Lord CI, Ghim SH, Troup GM, Hughes AL, and Letvin NL (1992) New recombinant HLA-B alleles in a tribe of South American Amerindians indicate rapid evolution of MHC class I loci. Nature 357: 329–333.

Weir BS (1990) Genetic Data Analysis. Sunderland, Mass.: Sinauer Associates, Inc.

Workman PL (1973) Genetic analyses of hybrid populations. In MH Crawford and PL Crawford (eds.): Method and Theory in Anthropological Genetics. Santa Fe, N.Mex.: New School.

Wright S (1943) Isolation by distance. Genetics 28: 114-138.

Wright S (1946) Isolation by distance under diverse systems of mating. Genetics 31: 39-59.

Zierdt H, Hummel S, and Herrmann B (1996) Amplification of human short tandem repeats from medieval teeth and bone samples. Hum. Biol. 68: 185–199.

CHAPTER 5

Quantitative Variation and Genetics

LYLE W. KONIGSBERG

INTRODUCTION

In Chapters 3 and 4, you learned about the genetic basis for inheritance and the central role of genetic effects in human evolution and history. My task in this chapter is to discuss the genetics of **quantitative trait** variation.

The study of quantitative variation has a very central role in biological anthropology. Almost invariably, discussions surrounding fossil **hominids** have focused on metric aspects of the skeletal remains, and even analyses of **discrete characters** usually must acknowledge that there is an underlying continuum to traits. From a theoretical standpoint, **natural selection** should (and indeed does) operate on many continuous aspects of morphology. In this chapter, I discuss many of the important issues surrounding variation in humans and the quantitative genetic models used to explain this variation. I start with a fairly simple demonstration of how the particulate genetic system first elucidated by Mendel (see Chapter 3) can still be used to explain the genetic control of continuous variation.

CONTINUOUS VARIATION FROM DISCRETE INHERITANCE

Let's start with a simple example of one **locus** with two **alleles** that are completely **codominant,** such that the aa **genotype** adds 0 units to a measurement, the Aa genotype adds 1 unit, and the AA genotype adds 2 units. This system is completely codominant, because the heterozygote is exactly half-way between the two homozygotes. If we assume that the two alleles have equal frequency in a population ($p = q = 0.5$), then the expected proportions of genotypes under **Hardy–Weinberg equilibrium** are 0.25 aa:0.50 Aa:0.25 AA (see Chapter 4). If two loci are independent (i.e., not **linked,** so they obey the principle of **independent assortment**), then the

Human Biology: An Evolutionary and Biocultural Perspective, Edited by Sara Stinson, Barry Bogin, Rebecca Huss-Ashmore, and Dennis O'Rourke
ISBN 0-471-13746-4 Copyright © 2000 Wiley-Liss, Inc.

two locus genotype probabilities are simply the products of the individual locus genotypic probabilities.

If we again assume complete codominance and equal frequencies of the two alleles at the second locus, then we can find the genotypic value and frequency of any particular two-locus genotype. For example, the genotypic value for the *AA,BB* genotype is +4 units, while its expected frequency is 0.25×0.25, or 0.0625. All of the possible genotypes and their frequencies are listed in Table 5.1. Once we sum together genotypes that are indistinguishable (e.g., *AA,bb* has the same genotypic value as *Aa,Bb* [+2]), we find that the ratios of genotypic values are 1:4:6:4:1. For three loci, the comparable ratio is 1:6:15:20:15:6:1; for four loci, it is 1:8:28:56:70:56:28:8:1; and so on. The expected frequencies are symmetrical and can be calculated by using Pascal's triangle or the binomial coefficients described in Chapter 4.

To increase the complexity of this discussion, let's consider a case in which the alleles at each locus are not equally frequent. Figure 5.1 shows a histogram of the expected proportions at each genotypic value in a setting with 20 loci, $p = 0.7$, and $q = 0.3$. For comparison, a **normal distribution** also is shown. Clearly, the distribution of genotypic values approximates the bell-shaped curve of the normal distribution. This will happen even in a case of complete **dominance**, such as when the *AA* and *Aa* genotypes contribute one unit to the measurement, whereas the *aa* genotype contributes nothing. The distribution goes to a normal curve as a direct result of what is known as the central limit theorem, a classic theorem from probability theory and statistics.

From our numerical examples, Figure 5.1, and the central limit theorem, we know that if the number of loci that contribute to a trait is large enough, then the distribution of genetic values will tend toward a normal curve. We might characterize this distribution of genetic values by specifying how dispersed its values are. As a mea-

Table 5.1 Calculation of two-locus genotypic frequencies

Locus 1	Locus 2	Probability	Genotypic Value	Genotypic frequency
aa	*bb*	$0.25 \times 0.25 = 0.0625$	0	0.0625
aa	*Bb*	$0.25 \times 0.50 = 0.1250$	1	$0.1250 + 0.1250 = 0.2500$
Aa	*bb*	$0.50 \times 0.25 = 0.1250$	1	
aa	*BB*	$0.25 \times 0.25 = 0.0625$	2	$0.0625 + 0.2500 + 0.0625 = 0.3750$
Aa	*Bb*	$0.50 \times 0.50 = 0.2500$	2	
AA	*bb*	$0.25 \times 0.25 = 0.0625$	2	
Aa	*BB*	$0.50 \times 0.25 = 0.1250$	3	$0.1250 + 0.1250 = 0.2500$
AA	*Bb*	$0.25 \times 0.50 = 0.1250$	3	
AA	*BB*	$0.25 \times 0.25 = 0.0625$	4	0.0625

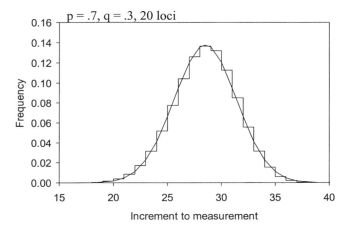

Figure 5.1   Histogram showing the expected frequencies for genotypic values when there are 20 loci, each with an allele frequency of 0.7. The smooth curve is the normal distribution to which the histogram is tending.

sure of "spread," we use the **variance** of the distribution, which is the average of squared values around the average value. In the case of Figure 5.1, the variance of the genetic values could be called an **additive genetic variance**, because the variance is due to the sum of the individual genetic values across all loci. The usual symbol for this variance is V_a. An additional source of variation that we should consider is the environment itself, which may influence the development of quantitative traits (see Chapters 6, 11, and 12). Consequently, even if only a few loci contribute to a quantitative trait, we would expect to see a continuous distribution rather than a discrete histogram of quantitative values. The environment thus has a smoothing effect on the distribution. The usual symbol for variance in a trait due to environmental effects is V_e, for **environmental variance**. If all the variation in a quantitative trait is due to the additive effects of alleles and environmental perturbations, then the **phenotypic** value (p_i) for any individual could be represented as

$$p_i = a_i + e_i \qquad (5.1)$$

where a_i and e_i are the additive genetic and environmental values for the individual.

The additive genetic effects are often referred to as polygenetic effects, because they represent the sum of independent effects across many independent loci. If allelic and environmental effects also are independent, then the phenotypic variance across individuals (i.e., the observed variance of the phenotypic values) is simply the sum of the unobservable additive genetic and environmental variances. This relationship is so central to quantitative genetics that it is summarized in the single term, **heritability**, which is usually represented as h^2. Heritability is defined as the proportion of phenotypic variance (V_p) due to additive genetic effects, or

$$h^2 = \frac{V_a}{V_a + V_e} = \frac{V_a}{V_p} \tag{5.2}$$

When there is dominance at one or more loci, the concept of heritability becomes considerably more complicated. In such a case, an individual's phenotypic value can be due to both additive effects of the two alleles at a locus and nonadditive effects (i.e., dominance). Therefore, we need to rewrite equation 5.1 as

$$p_i = a_i + d_i + e_i \tag{5.3}$$

where d_i represents dominance deviations. Similarly, the heritability could be rewritten from equation 5.2 to include the variances due to both additive genetic (V_a) and dominance (V_d) effects, giving

$$H^2 = \frac{V_a + V_d}{V_a + V_d + V_e} = \frac{V_a + V_d}{V_p} = \frac{V_g}{V_p} \tag{5.4}$$

where V_g is the total genetic variance. This redefinition of the heritability (H^2) is often referred to as "broad sense," whereas the definition from equation 5.2 is referred to as "narrow sense." The broad sense heritability gives an impression of how genetic a trait is but includes the dominance effects that are not transmissible to descendants. Dominance effects are not transmissible because they are properties of the two alleles at each locus. Consequently, dominance effects exist within individuals but cannot be passed to offspring because only one allele per locus is transmitted. The narrow sense definition is generally more useful because it is a measure of the proportion of variance in a trait explained by genetic transmission. I will explain later that narrow sense heritability figures prominently in models of evolution of quantitative traits.

ESTIMATION OF QUANTITATIVE GENETIC PARAMETERS

In the introduction, I suggested that quantitative genetic **parameters** might be important, and in the previous section, I described some relevant parameters. Now, I intend to show how such parameters can be estimated from observed data.

First, consider the notion of **genetic kinship**. Whereas the term kinship can have a rather loose definition in cultural anthropology, in genetics, its definition is quite precise. Genetic kinship is equal to the probability that two individuals share alleles at a locus that are **identical by descent** (or IBD). Identical by descent means that two individuals share copies of the same allele because the copies were passed on to each by a common ancestor. For example, the probability of identity by descent for two full-sibs (i.e., the genetic kinship between two sibs who have the same biological mother and father) is 0.25. Logically, there is half a chance that each parent of the sibs will pass one allele, and half a chance that he or she will pass the other. If the parent passes the same allele to the sibs then this should occur with probability equal to $\frac{1}{2} \times \frac{1}{2} = \frac{1}{4}$, as the probability of these two independent events is just the product of their individual probabilities. It is even possible to calculate an individual's genetic kinship with him- or herself; it is one-half. If we sample an allele twice (with replacement) at a locus, then half of the time we will sample the same (i.e., IBD) allele. In individuals who are inbred (i.e., the offspring of **consanguineous unions**), the probability of IBD rises above one-half because the inbreeding makes it possible for an individual to have two alleles at a locus identical by descent.

The expected additive genetic correlation between pairs of relatives is often referred to as the **coefficient of relationship,** or coefficient of relatedness, and has a central position in much of sociobiological theory (see Hughes 1988, Chapter 2). The coefficient of relationship is the expected proportion of genes that two individuals have in common. For example, the coefficient of relationship for a parent and child is one-half, because the child has received half of his or her genes from that parent.

BOX 5.1 GENETIC CORRELATION AND COVARIANCE BETWEEN RELATIVES

A **correlation** is a standardized measure of the relationship between two variables, such that a perfect negative relationship has a correlation value of -1 and a perfect positive relationship has a value of $+1$. By negative relationship we mean that as the value of one variable increases, the value of the other decreases; in a positive relationship, increases in one variable are associated with increases in the other.

In discussing genetic correlations between relatives, we can ignore negative relationships so that the two extreme cases to consider are correlations of 0 and $+1$. A genetic correlation of zero between two individuals indicates no genetic relationship, and indeed this correlation is expected for unrelated individuals. On a population level, this zero correlation has a simple interpretation. If one of the two individuals has an additive genetic value above the average for a given measurement, we have no particular expectation concerning the second (unrelated) individual. A genetic correlation of 1 between two individuals indicates a perfect genetic relationship and is expected for an individual with him- or herself or between monozygous twins. In twins, the genetic value for one twin is the same as the genetic value for the other twin, so there is perfect agreement. From the example of unrelated individuals and monozygous twins, we can see why expected genetic correlations are never negative. Such a relationship would indicate that if one individual were above the additive genetic mean value, then his or her relative would be below. Mechanically, this relationship would require that genes that code for a larger value in one individual would code for a smaller value in his or her relative.

Although we can use intuition to find the expected additive genetic correlations between unrelated individuals (equal to 0.0) and monozygous twins (equal to 1.0), more formal methods are required to calculate expected correlations between other pairs of relatives. To make these calculations, we return to the notion of genetic kinship. There are several possible ways in which individuals do (or do not) share alleles that are identical by descent. To simplify this discussion, assume that there are no inbreeding loops (i.e., no consanguineous unions). Figure 5.2 shows the possible configurations of alleles for two individuals. The three possible states are that the individuals share both pairs of alleles by descent, that they share a single pair by descent, and that they share no alleles by descent. Unfortunately, the symbols used for expressing the probabilities of these states can be quite confusing. I use Jacquard's notation (Jacquard 1974), where the delta coefficients Δ_9, Δ_8, and Δ_7 correspond to the probabilities of sharing zero pairs, one pair, and both pairs of alleles, respectively. (For an extensive description and discussion of Jacquard's delta notation, see Lynch and Walsh 1998.)

From the delta coefficients, we can determine the probability that an allele selected at random from the same locus in each of two individuals (i and j) will be identical by descent (also known as coefficient of kinship [φ_{ij}]). If the two individuals share both pairs of alleles (which can only occur in bilineal descent), then when we pick one allele from each individual, there is one-half of a chance that the two alleles will be identical by descent. On the other hand, if only one pair of alleles is shared, then there is one-quarter of a chance that when we pick an allele from each of the two individuals they will be identical by descent. Putting this information together, we have

$$\phi_{ij} = \frac{\Delta_{7,ij}}{2} + \frac{\Delta_{8,ij}}{4} \tag{5.5}$$

Genetic kinship values Δ_7, Δ_8, and Δ_9 for several different relationships are listed in Table 5.2 (also see Lynch and Walsh 1998, Table 7.1). These values can be calculated by using the rules of probability applied to genetic transmission through pedigrees.

Although the algebra is too tedious to present here, it is possible to show that the expected additive genetic correlation between two individuals i and j is simply $2\varphi_{ij}$. In the original examples of unrelated individuals and monozygous twins, the genetic kinships within these pairs are 0.0 and 0.5, respectively, leading to correlations of 0.0 and 1.0.

To complete this discussion, consider the expected additive genetic covariance between relatives. A **covariance** is a measure of relationship, like a correlation, but it is not scaled between −1 and +1. It is an average product of deviations around the average value. For example, if the average value for a trait is 10.0, and a mother and her daughter had values of 12.0 and 13.0, then these individuals would contribute 6.0 units (equal to $(12.0 - 10.0) \times (13.0 - 10.0)$) to the covariance between mother–daughter pairs. If we take the expected genetic correlation between individuals as $2\varphi_{ij}$, then the expected genetic covariance between the individuals is $2\varphi_{ij} V_a$. Because additive genetic variance appears in the expected additive genetic covariance between relatives, we can estimate this variance from pedigrees.

Similarly, monozygous twins have a coefficient of relationship of one, because they have all of their genes in common. Dizygous twins (or full sibs) have a coefficient of relationship of one-half. This correlation accounts for only the additive effects of the alleles within loci. If dominance effects also must be accounted for, then the expected genetic correlation between individuals i and j is $2\varphi_{ij} + \Delta_7$, and the expected genetic covariance is $2\varphi_{ij}Va + \Delta_7 V_d$. From Table 5.2, we can see that for, say, monozygous twins, the expected genetic covariance is $V_a + V_d = V_g$.

Estimating Quantitative Genetic Parameters from Fixed Designs

In animal studies, fixed breeding designs are commonly used to estimate quantitative genetic parameters. For example, multiple litters of mice can be reared and measured for a quantitative trait, and then the similarity within sibships (a group of siblings) versus the differences between sibships can be used to estimate quantitative genetic parameters. Such an approach is obviously not feasible with primate species, particularly humans. However, we can use data collected from sibships, or from parents and their offspring (or other relationships), in a post hoc manner. In this section, I present a few of these classical approaches to estimating quantitative genetic parameters.

As noted above, the expected genetic covariance within monozygotic twins is simply equal to the total genetic variance. The increased genetic covariance of monozygotic relative to dizygotic twins can be used to set a rough upper boundary on the broad-sense heritability. Many researchers have published results of twin studies in human genetics. I will not describe methods for estimating heritability from

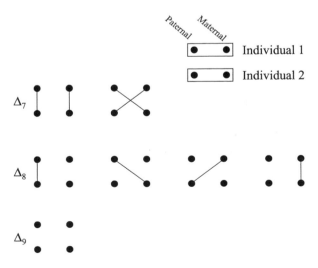

Figure 5.2 Graphical representation of allele sharing between two individuals (after Jacquard 1974, Fig 6.3).

BOX 5.2   ESTIMATION OF HERITABILITY FROM PARENT–OFFSPRING REGRESSION

Regression analysis is often used to estimate narrow sense heritability. Regression refers to the fitting of a straight line that shows the relationship between two variables. For our purposes, the variables will be one measurement on a parent and the same measurement on an offspring. The data will consequently consist of measurements from parent–offspring pairs.

One of the earliest examples of regression analysis is attributed to Sir Francis Galton (1889), who presented a regression analysis of the stature of offspring on the stature of the midparent. Midparent refers simply to the average of the statures of the father and the mother (after increasing the mother's stature to account for sexual dimorphism). Galton presented the midparent's measurements in one-inch increments and then the median value of all adult offspring to midparents in the one-inch interval. His data, listed in Table 5.3, were based on 199 parental pairs and 910 offspring. Figure 5.3 shows the best-fit line for the regression of the offspring medians on the midparents. Regression analysis is now such a common device that it is included in almost all computer spreadsheet packages as well as on many pocket calculators. The best-fit line shown in Figure 5.3 has a slope of 0.685, which is essentially what Galton found (he stated that the slope was 2:3, or 0.667).

How are we to interpret this slope of 0.685? A slope of zero would have a simple interpretation. In this case, the offspring's value bears no relationship to the midparental value, so we know that the heritability for stature is zero. To make any additional interpretation, we must delve into the calculations involved in finding the slope. The slope of the best-fit line for the regression of a variable y on another variable x ($\beta_{y.x}$) is

$$\beta_{y.x} = \frac{\mathrm{cov}\,(x, y)}{V_x} \qquad (5.6)$$

where $\mathrm{cov}(x,y)$ is the covariance between the two variables (see Box 5.1 for a definition of covariance) and V_x is the variance of x. A variance is the average squared distance of each value to the mean; it also can be thought of as the covariance of a variable with itself. Here our x represents midparental values (so I will use the symbol m) and the y are offspring values (for which I will use an o), so we have

$$\beta_{o.m} = \frac{\mathrm{cov}\,(m, o)}{V_m} \qquad (5.7)$$

From Table 5.2, we can see that the expected covariance between one parent and his or her offspring is one-half the additive genetic variance ($V_a/2$); this value is also the expected covariance between the midparent and its offspring [$\mathrm{cov}(m,o)$]. The variance of the midparent values turns out to be one-half the phenotypic variance (V_p; the halving is due to the averaging of the two parental values). Substituting in equation 5.7 gives

$$\beta_{o.m} = \frac{\left(\dfrac{V_a}{2}\right)}{\left(\dfrac{V_p}{2}\right)} = \frac{V_a}{V_p} \tag{5.8}$$

so that the regression slope of offspring on the midparent value is an estimate of the narrow sense heritability (cf. equations 5.8 and 5.2). Consequently, Galton's data suggest a narrow sense heritability for stature of ~0.685.

twin studies here but note that twin studies have a very intuitive feel for estimating heritability. In essence, one expects monozygotic twins to be more similar phenotypically than dizygotic twins, because monozygotic twins are genetic "clones" of one another. To the extent that this expectation is not met, the heritability of the measured trait is lower.

Parent–child pairs can also be used to estimate heritability; however, the estimate is the narrow-sense heritability. This matter is taken up in some detail in the preceding box. Other pedigree units, such as sibships, can be used to make estimates of quantitative genetic parameters. However, I focus on methods that make better use of the many pedigree relationships available when studying humans.

Table 5.2 Probability of identity by descent for two pairs of alleles (Δ_7), one pair (Δ_8), and zero pairs (Δ_9)

Relationship	Δ_7	Δ_8	Δ_9	ϕ
Ego (monozygous twins)	1.0000	0.0000	0.0000	0.5000
Full-sibs	0.2500	0.5000	0.2500	0.2500
Half-sibs	0.0000	0.5000	0.5000	0.1250
Parent–child	0.0000	1.0000	0.0000	0.2500
Grandparent–grandchild	0.0000	0.5000	0.5000	0.1250
Great-grandparent– great-grandchild 0.0000	0.2500	0.7500	0.0625	
First cousins	0.0000	0.2500	0.7500	0.0625
Second cousins	0.0000	0.0625	0.9375	0.0156
Double first cousins 0.0625	0.3750	0.5625	0.1250	

Note: The final column gives the probability of identity by descent when one allele is sampled from each of the two individuals (ϕ).

BOX 5.3   MAXIMUM LIKELIHOOD AND THE ESTIMATION OF QUANTITATIVE GENETIC PARAMETERS

Maximum likelihood is a general method for estimating parameter values within a model. Parametric models specify how variables are distributed and how, potentially, they may relate to other variables. Most college students have dealt with some simple parametric models, such as the bell-shaped curve that is, theoretically, descriptive of their test scores.

The bell-shaped curve (or, more properly, the normal distribution) is characterized by one parameter that measures the central tendency (called the mean or average) and another parameter that measures spread around the mean (called the **standard deviation**). The standard deviation is the square root of the variance. We would estimate these two parameters. It is simplest to calculate the mean and standard deviation directly, but we will consider using maximum likelihood estimation (MLE) here as precursor to estimating quantitative genetic parameters.

The **likelihood** is defined as proportional to the probability of obtaining the observed data conditional on the unobserved parameter values. For example, if we hypothesize that test scores are normally distributed with a mean of 63.7 and a standard deviation of 13.0, then the probability of obtaining a test score of 72.0 is ~0.025. The calculation of this probability is beyond the level of this text, but almost all computer spreadsheet software includes what are known as probability density functions, which give the probability value for a particular observation when one specifies the parameter values.

In this case, we are using a normal distribution with a mean and standard deviation of 63.7 and 13.0, respectively. If we had a collection of test scores from, say, 100 students, the likelihood for the total sample would be simply the product of all the individual likelihoods (probabilities). We can multiply the individual likelihoods because we assume that the student's test scores are independent (i.e., one student scoring above the mean for the test tells us nothing about whether a different student scored above or below the mean).

To formalize this discussion, I should introduce some relatively simple symbols. The total likelihood is expressed by the notation $L(\mu,\sigma|\mathbf{x})$, where L is likelihood; μ and σ are the mean and standard deviation, respectively; and $\mathbf{x}$ represents the collection of all student's scores. The vertical line represents conditioning, with the left side conditional on the (fixed) right side. Consequently, $L(\mu,\sigma|\mathbf{x})$ means that we are finding the total likelihood of the mean and standard deviation conditional on the observed test scores. The total likelihood is

$$L(\mu,\sigma|\mathbf{x}) = \prod_{i=1}^{n} p\ (x_i|\mu,\sigma) \tag{5.9}$$

where p is probability and $\prod$ is the product across all students. This equation also can be written in a logarithmic scale, giving

$$\ln L \ (\mu,\sigma|\mathbf{x}) = \sum_{i=1}^{n} \ln \ (p \ (x_i|\mu,\sigma)) \qquad (5.10)$$

Equation 5.10 gives the total log-likelihood, and if we search across different values for the mean and standard deviation, we can identify the parameter values at which the log-likelihood is maximized. These values would then give us the mean and standard deviation most likely to have produced the observed data under the assumption of a normal distribution. Because these parameter values are the most likely ones to have generated the observed data, they are called maximum likelihood estimates.

The total log-likelihood for pedigree data is more complicated, because we cannot assume independence between individuals. If, for example, we had stature measured on related adults, then the log-likelihood would be

$$\ln L \ (\mu,\mathbf{V}|\mathbf{x}) = -\frac{1}{2}\ln|\mathbf{V}|-\frac{1}{2}(\mathbf{x}-\mu)' \ \mathbf{V}^{-1} \ (\mathbf{x}-\mu) \qquad (5.11)$$

where

$$\mathbf{V} = 2\Phi V_a + \Delta_7 V_d + \mathbf{I} V_e \qquad (5.12)$$

where $\mathbf{V}$ is the phenotypic variance/covariance matrix among the relatives, Φ is a matrix of coefficients of kinship, V_a is additive genetic variance, Δ_7 is a matrix of probabilities that both alleles are identical by descent at any given locus, V_d is dominance variation, $\mathbf{I}$ is an identity matrix, and V_e is environment variance.

Without a background in linear algebra, you may find equations 5.11 and 5.12 quite difficult to fathom. Their gist is that expected relationships between trait values in relatives depend on the pedigree structure (which is observed and provides the coefficients of relationship and dominance) and on unobserved population parameters (V_a, V_d, and V_e). The task here is to scan across possible values for V_a, V_d, and V_e until the maximum log-likelihood is identified. The narrow and broad sense heritabilities can then be found from equations 5.2 and 5.4, respectively.

Estimation of Quantitative Genetic Parameters from Pedigree Data

In general, using only parent–offspring, sibship, or twin data to estimate quantitative genetic parameters in natural populations is quite inefficient. In the lab, one can construct breeding designs to maximize the amount of information, but when working with humans, the data come as observations on pedigrees of varying size and complexity. If, for example, we had data on many unrelated nuclear families, we could use the parent/offspring pairs, but these data would ignore the information within sibships. Furthermore, if we have data from extended families, then we should have information from, for example, grandparent–grandchild pairs. We can incorporate all of these relationships using what is known as the method of maximum likelihood. In Box 5.3, I briefly describe this method within the context of estimating quantitative genetic parameters.

Many quantitative genetic analyses based on human pedigree data have been published. It would be impossible to review all of these analyses here; instead, I offer a few representative examples. Towne et al. (1993) presented an interesting quantitative genetic analysis in that they made use of serial measurements on relatives. They modeled growth in infants within the first two years of life by using a three-parameter model to examine growth in **recumbent length**. They studied 569 infants in 188 families from Ohio and examined the heritability of each of the growth parameters. They estimated the heritabilities for the three parameters at 0.83, 0.67, and 0.78, where the parameters represent the length at birth, a growth rate linear to the logarithm of age, and a rate of change in the growth rate, respectively. It is interesting to compare the heritability for the length at birth (0.83) to the

Table 5.3   Midparent stature and adult offspring stature from Galton's data

Midparent	Adult offspring
64.5	65.8
65.5	66.7
66.5	67.2
67.5	67.6
68.5	68.2
69.5	68.9
70.5	69.5
71.5	69.9
72.5	72.2

(*Source:* Galton 1889)

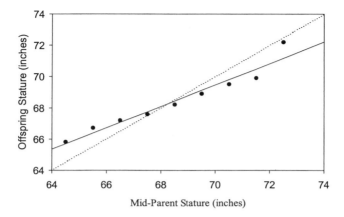

Figure 5.3 Regression line for adult offspring stature on midparental stature (solid line). The dashed line shows a line with a slope of 1.0, and the solid line has a slope of 0.685. (Source: Data from Galton 1889, Table 11)

estimate we obtained for adult stature from Galton's data (0.685). The higher heritability for length at birth is explicable because the disruptive effects of the postuterine environment have yet to occur. This study presents an important point: that heritability levels may vary by life stage.

Weiss (1993) compared narrow sense heritability estimates for **systolic blood pressure** in different populations. Systolic blood pressure is a quantitative trait that has many biomedical implications because of the dire cardiovascular consequences that can develop in individuals who have elevated blood pressure. From the eight populations that Weiss sampled, the average heritability for systolic blood pressure is ~0.30, considerably lower than the heritability we have seen for stature in adults or birth length in newborns. In fact, I will show later that a heritability in the range of 0.30–0.40 is more in line with typical morphological or physiological traits from humans.

In another example, Beall et al. (1997) studied **hypoxic ventilatory response** (HVR; see Chapter 6) among 320 Tibetans and 542 Bolivian Aymara and used the pedigree relationships to estimate narrow sense heritabilities of 0.345 and 0.220, respectively. They also found that the resting ventilation heritability was slightly lower than the HVR heritability in the Tibetans (equal to 0.317) and was considerably lower in the Aymara (equal to 0.055). Their analysis also accounted for household effects, that is, the shared environmental factors due to individuals residing in the same house.

MULTIVARIATE QUANTITATIVE GENETIC ANALYSIS

In the previous section, I briefly described how a single trait could be analyzed within a quantitative genetic framework. In this section, I take up the issue of multivariate analysis, in which we consider two or more traits. Equation 5.12 can be generalized so that we consider not only covariances among relatives, but also co-variances among traits (within and between relatives). The importance of this type of analysis is that it allows us to look at the inheritance of, say, not only body size (as with stature) but also body shape. Thus, the **allometric** analyses discussed in Chapter 9 also can be assessed from a quantitative genetic standpoint.

Multivariate quantitative genetic analyses can be used to access **pleiotropy,** which is the effect of genes on different phenotypes. As an example of such pleiotropic effects, Beall et al. (1997) found an additive genetic correlation of 0.67 between resting ventilation and HVR among the Tibetans. These data suggest that there are genes that have effects on both resting ventilation and ventilation under hypoxic conditions. The additive genetic correlation between a trait measured under two different environmental conditions can be used to examine environment-by-genotype (or, more properly, environment-by-polygenotype) interaction. If the additive genetic correlation between two traits is 1.0 (the maximum possible value), then there is no interaction between the environment and additive genetic effects, because individuals are simply shifted a certain amount in their phenotypic values.

QUANTITATIVE GENETIC MODELS FOR THRESHOLD TRAITS

It may at first appear contradictory to include discrete traits in a chapter on quantitative variation, but many traits that we might think of as discrete are more or less continuous. For example, Lake et al. (1997) present a genetic analysis of recurrent apthous stomatitis (RAS; more commonly known as mouth ulcers). Whereas they had respondents classify the frequency of their RAS and that of their relatives as "frequent," "sometimes," "rarely," and "never," this categorization is arbitrary and fuzzy.

Because one component to RAS frequency appears to be psychological stress (Pedersen 1989), respondents may have difficulty classifying the frequency of RAS events. Bouts of RAS can increase and decrease in frequency throughout an individual's lifetime, possibly in tandem with the amounts of stress they are experiencing. Even traits that we consider as truly discontinuous and stable throughout life can reflect an arbitrary dichotomy on a continuous scale. Such is the case for handedness. Even though we think of ourselves and others as left- or right-handed, this dichotomy is based largely on the hand with which we write. This classification does not admit that in some individuals, the asymmetry is less absolute; it uses a single behavior to classify all hand preferences. Finally, many conditions of extreme medical importance, such as alcoholism, diabetes, and most psychological disorders, are often spoken of in dichotomous terms when we know in fact that they have an underlying continuity.

Some traits that are truly discrete in humans have relatively simple monogenic bases, but many are better accommodated within what is known as the **threshold model**. In the threshold model, it is assumed that some continuous underlying distribution, often referred to as the trait **liability**, is dichotomized by a threshold. For example, if the trait of interest is **cardiac ischemia** (a decidedly discrete event), then we might assume a continuously distributed risk factor. Once one crosses over a boundary (i.e., threshold), then the trait is expressed. We generally can measure quantitative traits that are highly correlated with the unobservable total risk factor, but the total risk itself is unobservable. All that we bear witness to are the actual endpoints, which in the case of cardiac ischemia would be coronary events. Individual risk factors that we might be able to measure include the low-density lipoprotein cholesterol (LDL-C) serum levels, the high-density lipoprotein cholesterol (HDL-C) to LDL-C ratio ("good" to "bad" cholesterol), or blood pressure. However, ultimately, we (and our physicians) are interested in the total risk to heart disease.

To simplify the threshold model, we usually assume that the liability has a standard normal distribution. Robertson and colleagues (Robertson and Lerner 1949; Dempster and Lerner 1950) showed that it is possible to score threshold traits as either a "0" (absent) or a "1" (present), calculate the narrow sense heritability on this observed scale by using any of the methods discussed earlier, and then calculate the heritability on the underlying scale by using a simple approximation. Falconer (1965) gave a method for calculating the heritability of discrete traits that uses the increased incidence of a trait in the offspring of parents who also have the trait. In an important series of papers, Cheverud and Buikstra (1981a, 1981b, 1982) applied this method to cranial discrete traits from pedigreed rhesus macaques. They found that the discrete traits generally had higher narrow sense heritabilities than did **craniometric** variables from the same crania. Sjøvold (1984) applied the threshold model in a study of cranial discrete traits by using pedigreed human crania from a 19th-century charnel house in Hallstatt, Austria. He used father–son and mother–daughter comparisons to estimate the narrow sense heritability for 20 discrete traits. The heritabilities for these traits were not very high. The average heritability was only ~0.28, and the median was even lower, at ~0.22.

The cranial discrete trait studies are of anthropological interest because we often use these types of discrete traits in making taxonomic assignments to fossils and in assessing biological distances between prehistoric human groups. Threshold models also have been applied extensively to more biomedically relevant discrete traits. I will discuss such applications later as related to locating major genes that have effects on discrete medical conditions.

COMPLEX SEGREGATION ANALYSIS

I already have discussed traits that appear discontinuous yet are continuous on an underlying scale. In this section, I consider traits that are continuous but have an underlying discrete causality. Such would be the case in a classic Mendelian model, in which a single locus determines the trait value and then environmental effects "spread-out" or "smear" individual trait values. If the distance between the mean trait values for the genotypes is large and the effects from the environment are small, then it may be possible to observe data from many individuals as a mixture of distributions. This type of analysis is often referred to as a commingling analysis, because we observe a commingled series of distributions. Commingling analyses are a very indirect way to suggest the presence of simple Mendelian inheritance, because the mixture of distributions could be caused by environmental effects. A more powerful method for examining the Mendelian hypothesis is **complex segregation analysis** (CSA). Jarvik (1998, 942) introduced a summary and description of CSA with the welcome message, "I will describe the CSA methods with record brevity, facilitated by the absence of any equations." The reader should consult Jarvik as well as references therein for further information, because our discussion here must be relatively brief and nontechnical.

Segregation analyses are typically done by using the method of maximum likelihood. The parameter values to estimate are average trait values for the genotypes, allele frequency, environmental variance, and three transmission probabilities. The transmission probabilities represent the probabilities that AA, Aa, and aa parents would transmit an A allele. Under the Mendelian hypothesis, these transmission probabilities are 1.0, 0.5, and 0.0. Likelihood methods can be used to compare the estimated transmission probabilities with their expected values; if the values are close to their expected values, evidence for simple Mendelian inheritance is good.

An important point from CSA is that we cannot directly observe the genotypes (*AA, Aa,* or *aa*) in our analysis; we can only make probabilistic statements about which genotypes individuals may carry. For example, if our analysis suggests that the *aa* genotype gives the highest trait value and that it is completely recessive, then a mother, father, and their offspring all with relatively high values would give a high probability that both parents have the recessive genotype. This probability calculation also has to include the allele frequency estimate, because a high frequency for the recessive allele would increase the frequency of recessive genotypes.

The CSA can be extended to include polygenetic effects. In this case, there is a major Mendelian **locus**, as in the CSA described earlier, but polygenetic heritability also must be estimated. This heritability represents the proportion of phenotypic variation explained by additive effects at loci after accounting for the major locus. As a consequence of allowing for major locus transmission, this residual polygenetic heritability must be lower than the heritability one would calculate in a traditional quantitative genetic analysis that ignored the possibility of major locus transmission. Beall et al. (1994), who worked on high-altitude physiology, used CSA to demonstrate major gene and polygenic transmission for saturation of arterial hemoglobin (SaO_2) among Tibetans. They studied 188 individuals in 11 pedigrees and 9 singletons and measured saturation of arterial hemoglobin by using a noninvasive technique. After allowing for the effect of age, they found evidence for a major locus with dominance and a polygenetic heritability of 0.253. This heritability is much lower than the value of 0.409 that they obtained by using a traditional quantitative genetic analysis that ignored the possibility of major locus transmission.

In a recent study of bone mineral density (BMD) among Chuvasha and Turkmenian populations, Livshits et al. (1999) conducted a bivariate (two-trait) CSA of compact and spongy bone. Measurement of BMD has important biomedical health implications because it is a proximate variable to **osteoporosis**. Livshits and coworkers measured BMD for compact bone and spongy bone read from hand X-rays.

They found evidence for a single major gene with pleiotropic effects on both compact and spongy BMD, and in most instances, they were able to reject the hypothesis that a polygenic background accompanied the major gene effects.

LINKAGE AND QUANTITATIVE TRAIT LOCI

The above section on complex segregation analysis and major genes leads to the problem of how to locate genes in the **genome** once they have been identified in pedigree analyses. It is beyond the scope of this chapter to detail the methods used to search for major genes that affect quantitative traits in humans. Elston (Elston 1980, 1990, 1998) has reviewed many of the relevant methods, and Vieland (1998) has reviewed alternative procedures. Rogers et al. (1999) have specifically addressed the role of quantitative trait mapping in both humans and non-human primates within an anthropological context. I briefly discuss **linkage** analysis because it is a method that is under considerable use today and because its results are frequently reported in the popular press. This method nicely combines molecular genetic assays (such as those described in Chapter 4) with the quantitative genetic model; consequently, at this point we have some of the requisite methodological and analytical tools to discuss linkage analysis for **quantitative trait loci** (QTLs).

Almasy and Blangero (1998) showed that equation 5.12 can be generalized to include information from **marker** loci (i.e., loci that have been assayed). If relatives share marker **haplotypes** that are tightly linked to a QTL, then the probability that they will have identical by descent alleles at the QTL is increased. By using this relationship, it is possible to screen a QTL (i.e., a statistically defined but unlocated locus) against several marker loci spread throughout the genome. The probabilities that relatives who share marker loci will also share QTL alleles is a function of the **recombination** rate and therefore reflect the physical distance of a QTL to a known genetic location. The strength of a putative linkage is usually assessed by using a measure known as a **LOD score**. LOD stands for log-odds and is the base-ten logarithm comparing the likelihood for the estimated recombination rate against a recombination rate of 0.5 (no linkage). For example, a LOD score of 2 indicates that the observed data are 100 times more likely to have arisen under the estimated recombination rate (i.e., if the marker and QTL are linked) than under no linkage (recombination rate of 0.5). A LOD score of 3 indicates that the data are 1000 times more likely to have arisen if the QTL is linked to the marker. LOD scores of 2 are generally taken as suggestive of linkage, whereas scores of 3 or above are taken as strong evidence for linkage.

To date, QTL linkage analysis has been successful in several applications, notably, a study that pinpointed a QTL affecting serum leptin and fat mass to **chromosome** number 2 (Commuzzie et al. 1997) and a study that showed linkage of a **non-insulin-dependent diabetes mellitus** QTL to the short arm of the 10th chromosome (Duggirala et al. 1999). Both of these studies were undertaken among Mexican Americans, a group for whom obesity and diabetes pose considerable health risks. Abel and Dessein (1998) reviewed CSA and linkage studies for **infectious diseases** and noted the evidence for linkage of leprosy, malaria, and mycobacterium susceptibility to various marker loci. Their work serves to contrast the **candidate gene method** with the genomewide searches that are now possible thanks to the large panel of **polymorphic** sites known for the human genome. The candidate gene method relies on selecting markers that are at or near loci whose biochemical effects might be expected to affect the quantitative phenotype. Although the candidate gene method provides a powerful tool for screening possible linkages, it also runs the risk of missing linkages. Comuzzie and Allison (1998) discuss both the candidate gene approach and the genome scanning approach in "the search for human obesity genes." They note the successful identification of QTL linkages from both methods of analysis.

ARE THE ASSUMPTIONS OF THE QUANTITATIVE GENETIC MODEL CORRECT?

In the introduction to this chapter, I noted that if many loci had equal effects on a quantitative trait, then the trait would show a normal distribution. This model posits that many genes contribute to the formation of a trait, an assumption that may be difficult to justify in light of our discussion of major genes in the previous sections. In this section, we examine whether the assumptions of the polygenic model are correct. We can approach the problem in two ways. First, we can address the question of the number of loci that affect quantitative traits. Under the polygenic model, the number of loci should be quite large. Second, we can examine whether loci have equal contributions to traits, or whether a few major loci dominate the genetic landscape. If the latter is the case, then one of the major assumptions of the polygenic model has been violated.

The estimates of the number of loci across many different taxa are conflicting. Because the number of QTLs is usually assessed by crossing inbred lines, far fewer estimates exist from humans and other primates than from other species. Lande (1981) suggested that skin reflectance in humans is only affected by a small number of QTLs. Roff (1997, 23) carefully considered many previous studies of the number of QTLs, and concluded the following:

> Whether the number of loci determining quantitative traits is in the tens or hundreds (or even thousands) remains to be shown. Nevertheless, we can conclude that the data support the hypothesis of numerous loci determining quantitative traits and, hence, to suppose that the most fundamental assumption of quantitative genetic theory is correct.

As shown in the previous sections, whether or not major loci affect quantitative traits in humans has been extensively examined. The existence of such major loci has very obvious biomedical relevance if the trait of interest is, say, serum cholesterol or sodium level, adiposity, or blood pressure. Unfortunately, traits that are of more direct anthropological interest have received less attention. Given the time and cost involved in the search for major genes that affect biomedically relevant phenotypes, it can only be hoped that some form of **anthropometric** survey can be included with such studies. The cost of collecting anthropometric data is minuscule compared with the monetary and time costs involved in collecting many other quantitative pheno-types. In any event, with the general paucity of complex segregation and linkage analyses for many morphological (anthropometric) traits, it is currently impossible to guess how well the traditional quantitative genetic model fits the extensive anthro-pometric databases in existence. In the coming years, it will be necessary to examine whether the traditional size and shape variables from human anthropometry (and craniometry) are influenced by QTLs with substantial effects. If such loci exist, then the use of (polygenic) quantitative genetic models that exclude major gene effects will be seriously called into question.

IS THERE AN "AVERAGE HERITABILITY?"

Now that we have the tools for estimating heritability, we need to consider how (and why) heritabilities might differ between traits and across populations. Cheverud (1988) suggested after examining phenotypic and genetic variances, co-variances, and correlations among many mammalian taxa that average heritability was ~0.35. It is important to note, though, that he was referring specifically to morphological traits in mammals and not behavioral or **life history** traits. As I ex-plain later, these latter traits should, on average, have lower heritabilities than do morphological traits.

Konigsberg and Ousley (1995) specifically addressed the question of variation between trait heritabilities in humans. For their analysis, they used anthropometric data collected in pedigrees from five native North American tribes. This data had originally been collected under Franz Boas' direction for the 1893 World Columbian Exposition (Jantz et al. 1992). Konigsberg and Ousley found in their analysis of 12 anthropometric traits that the average heritability was ~0.40, near Cheverud's 0.35. Other studies in humans have suggested higher heritabilities, but extenuating circumstances usually explain the higher heritabilities. For example, Towne et al. (1993) showed that the heritability of infant stature is greater than that of adult stature, because body length at birth is less affected by environmental per-turbations. (This argument does not apply to body weight at birth, which is under substantial effects from the maternal environment.) Another trait in humans with a high heritability, the total finger ridge count (i.e., the sum of the ridge counts from fingerprints), can be explained by the fact that summing across fingers reduces the special environmental variation unique to the individual fingers (see Falconer and Mackay 1996, 173). The heritability for individual finger ridge counts is much lower than for the total, though the heritabilities by finger are still substantially higher than 0.35, presumably because ridge counts are unaffected by the postnatal environment.

In addition to looking at heritabilities for different traits within the same popula-tion, we can also look at the heritabilities for the same quantitative trait across dif-ferent populations. After drawing a comparison of systolic blood pressure heritabilities across populations, Weiss (1993, 105) noted that "for many traits, such as blood pressure, the heritability is similar across populations that live in presumably different environmental circumstances ….” He surmises that this similarity may be due to “a balance among evolutionary forces,” a point that he returns to in a later chapter of his book.

Although the question of whether there is an average heritability for morphometric traits remains unanswered, we can make some generalities about heritability. First, narrow sense heritability is simply the proportion of additive genetic variance out of the total phenotypic variance. As such, there is nothing particularly mechanistic about heritability, and it can change in a population simply as a result of changing levels of environmental variation. For example, if the environmental variation for a trait decreases, then the heritability will increase. Another important point, which I will discuss in a later section, is that evolutionary forces can affect heritability. For example, we know that **genetic drift** leads to decreased genetic variation within populations (see Chapter 4) and increased genetic variation between populations. Similarly, genetic drift leads to decreased additive genetic variation within populations and increased additive genetic variation between populations. Consequently, the heritability of a trait will decrease within a population subject to extreme genetic drift.

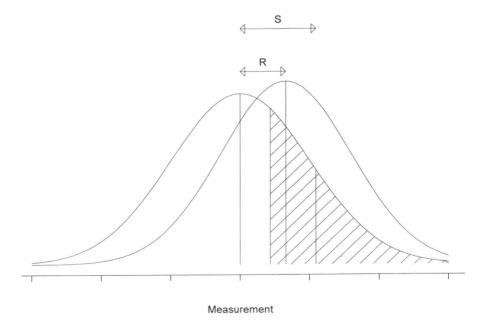

Measurement

Figure 5.4   Graphical representation of the response to selection. The shaded region represents the proportion of individuals in the parental generation that reproduce, whereas the curve shifted to the right represents the offspring generation. Vertical lines give (from left to right) the preselection mean in the parental generation, the mean of the offspring, and the mean of the selected (breeding) parents.

Another consideration in assessing narrow sense heritabilities is that the heritability cannot exceed the repeatability of a trait in question. Repeatability of a trait refers to the extent to which repeat measurements of a trait will reproduce previous measurements. The repeatability of morphometric traits is generally quite high, so heritability may also be high. In contrast, the repeatability of behavioral traits tends to be lower, and as a result, the upper limit for heritability of behavioral traits is lowered. Some behavioral traits, although they are highly repeatable, may still be relatively ill-defined. For example, I.Q. and GRE scores may be highly repeatable (and hence maybe even heritable, though this is doubtful). However, whether I.Q. measures intelligence and whether the GRE predicts future performance in graduate school remains an open question. When an observer measures traits such as aggression or altruism, measurement error may be considerable, so a priori, the heritability is unlikely to be terribly high.

EVOLUTION AND QUANTITATIVE TRAITS

It will not be possible to fully explore the role of evolutionary forces in shaping quantitative trait variation within the scope of this chapter. I only hope to sketch some of the relevant information for selection, drift, and migration. (For a full account of the role of evolution in quantitative trait variation, see Roff 1997).

Selection on Quantitative Traits

The relationship between **directional selection** (often referred to as truncation selection in the animal breeding literature) and evolution of a quantitative trait is quite straightforward. We start with the assumption that a normally distributed trait is selected, such that only some proportion of the population survives to reproduce. From the amount of selection and the heritability of the trait, it is possible to find the mean of the measurement in the next generation. In its simplest form, the change in the mean measurement, also known as the response to selection, is

$$R = h^2 S \qquad (5.13)$$

where S is the difference between the mean of the selected population and the total (preselection) population, R is the response (i.e., the difference between the offspring mean and the parental [preselection] mean), and h^2 is the narrow sense heritability (see Figure 5.4).

It is immediately apparent from equation 5.13 that if a trait is not heritable, it will show no response to selection, no matter how intense the level of selection. Selection will also lead to a reduction in the additive genetic variance, and as a consequence the phenotypic variance will also be decreased. Figure 5.4 shows an example of one generation of directional selection, where only about one-third of the parental population reproduces, and where the heritability is 0.6. The figure shows both the increase in the mean for the offspring generation (note that the response to selection is

60% of the selection) and the reduction in overall variance. Because the additive genetic variance has decreased while the environmental variance has remained the same, the narrow sense heritability must also decrease in the next generation.

The relationship between the response to selection and the narrow sense heritability is so critical that it has been encapsulated in a single theorem known as Fisher's Fundamental Theorem (after Sir R. A. Fisher). In essence, Fisher's theorem states that the rate of evolution by natural selection is proportional to the additive genetic variance. Because we know that selection will decrease the additive genetic variance, we can infer that the rate of evolution by selection will diminish over time. This effect has long been known to animal breeders, who can make substantial gains or improvements to their stock by selective breeding in early generations. With continued selection in later generations, they find a diminishing return, because genetic variation has been removed through selection. Fisher's theorem can explain the general observation that measures of **fertility** generally have very low heritabilities and that most life history traits also display low heritabilities. The reason for this observation is that natural selection has all but eliminated the genetic variation in fertility, which, as one of the main components of fitness, is under intense selection.

Multivariate Selection

One of the more fascinating areas of research in quantitative genetics involves selection on two or more traits that are correlated. Arnold (1994) has given a very readable account, while the classic reference is Lande and Arnold (1983). Lande and Arnold show that the response to selection on two or more traits can be written as

$$\Delta \bar{z} = \mathbf{G} \mathbf{P}^{-1} \mathbf{s} \tag{5.14}$$

where $\Delta \bar{z}$ is the change in the vector of mean phenotypes after one generation of truncation selection, $\mathbf{s}$ is the vector of direct selection coefficients on each trait, and $\mathbf{G}$ and $\mathbf{P}$ are the t by t additive genetic and phenotypic variance/covariance matrices among the traits, where t is the number of traits. Equation 5.14 is quite similar to equation 5.13. The matrix $\mathbf{G}\mathbf{P}^{-1}$ is the multivariate generalization for narrow sense heritability, $\Delta \bar{z}$ is analogous to R, and $\mathbf{s}$ is analogous to S.

An interesting implication from equation 5.14 is that selection for, say, an increase in size of one part of the body, can lead to a decrease in the size of another part. It can occur if additive genetic covariance is negative, because the correlated response to selection for the larger size for one part will lead to a decrease in the other part. Unfortunately, whereas we know a considerable amount about the $\mathbf{P}$ matrix (indeed, studies of stature estimation from long bone lengths are but one example of estimating phenotypic covariances among measurements), we know relatively little about the $\mathbf{G}$ matrix. What we do know from empirical studies, as well as from theoretical arguments (Atchley 1984; Cheverud 1984; Houle 1991; Konigsberg and Ousley 1995) is that most additive genetic covariances are positive. This consequence follows logically from how we assume evolution by natural selection operates. For negative genetic covariances to form, there would had to have been selection to increase the size of one part at the same time that there was selection to decrease the size of another part. Such massive selection on shape, rather than size, is difficult to conceive within an adaptive framework. It is unfortunate that we know so little about the structure of additive genetic covariation. Without knowledge of these covariances, we could entirely miss the selective patterns implied by the fossil hominid record.

Drift, Migration, and Population Structure Analysis from Quantitative Traits

Drift and migration (**gene flow**) have the same effects on quantitative traits as they do on single locus genes. Drift leads to increased additive genetic variance between demes and decreased additive genetic variance within demes. As a result, the within-deme narrow sense heritability will decrease over time as additive genetic variance is eroded. Similarly, migration has a homogenizing effect on between-group differences in quantitative traits but increases variances within demes. Because of the relationships of drift and migration to the structure of additive genetic variance, it logically follows that quantitative traits can be used to assess population structure. However, considerable debate over the utility of quantitative traits in population structure analyses has ensued (Felsenstein 1986; Lewontin 1986; Rogers 1986). The resolution of this debate largely lies in theoretical work by Rogers and Harpending (1983), in which they show that a single completely heritable quantitative trait contains the same amount of information as a single biallelic marker locus (see also Williams-Blangero and Blangero 1992). However, a central problem in using any trait for population structure analysis is that the trait should be selectively neutral (i.e., not be under direct selection or genetically correlated with a trait that is under selection). It is generally more difficult to argue that quantitative traits are selectively neutral, as much of the study of past human evolution has focused on morphological evolution by natural selection.

Lande (1976) presented a method for contrasting selection and drift when examining temporal change within or between lineages. He showed that if the value for the narrow sense heritability of a trait can be assumed and the amount of evolutionary time (in generations) is known, then equation 5.14 can be solved to find the proportion of individuals who would have to be selected against each generation to effect the observed phenotypic change. This model assumes no drift. In an alternative model, he assumed that there was no selection, and then found the **effective population size** at which there is only a 5% chance that drift could have given rise to the observed phenotypic change. Sciulli and Mahaney (1991) and Christensen (1998) applied these types of analyses to examine human dental size reduction. They suggest that the level of reduction in human tooth size is indicative of selection, because the effective population sizes necessary for drift to explain the reduction are far too small. Lofsvold (1988) described a method for extending Lande's work to cover multivariate evolution. Konigsberg and Blangero (1993) used his method to examine the level of craniometric difference between the Moriori and the Tolai. They found that the level of phenotypic difference between the two extant groups was so great that independent drift (within populations) could not explain the differentiation. Furthermore, they suggested that there would have to have been extreme selection on the founding populations to give rise to the level of observed phenotypic differentiation.

Wright's Shifting Balance Theory

Our discussion of evolution by natural selection and of the effects of drift and migration on quantitative traits has greatly simplified the field by separately considering the evolutionary forces. A more satisfying approach to the evolution of morphology can be found in Sewall Wright's theory, referred to as the Shifting Balance Theory, in which he considers all evolutionary forces together. Wright (1977) very succinctly described his shifting balance process as consisting of three phases. An example of the first two phases of Wright's theory is illustrated in Figure 5.5. In the first phase, random genetic drift can lead to deviations in allele frequencies (and, consequently, to the means of quantitative traits). In the second phase, selection can increase the average fitness of a population; and in the third and final phase, colonization (i.e., migration) can spread the new adaptation. Obviously, there must initially be **mutation** for variant alleles to exist. Wright's model therefore incorporates mutation, migration, selection, and drift.

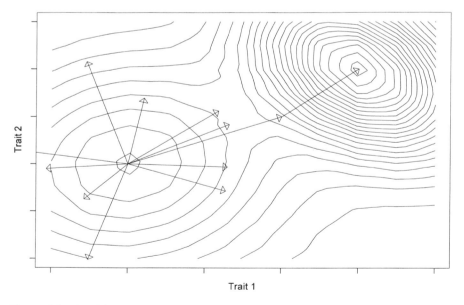

Figure 5.5   Graphical representation of the first two phases in Wright's shifting balance theory. The fitness peak to the lower left represents the initial location of a population, whereas the radiating arrows represent genetic drift away from this peak. Selection will return the population to the initial peak, except in the one case where a fitness valley has been crossed. Here, selection takes the population to the second, and higher, peak.

One point I have not adequately covered is that natural selection can only maximize **fitness**, and it only acts locally. Wright used the analogy of topography in discussing the path of a population across a fitness surface. Because evolution by natural selection can only increase fitness, if a population is perched on a fitness peak, natural selection cannot operate to bring the population to some higher fitness peak, because the population cannot go "downhill" by natural selection. However, as drift leads to random changes in allele frequencies (and hence in quantitative trait means), drift can take populations across fitness "valleys." Once a population has crossed a fitness valley, natural selection will take the population "uphill" to the new local maximum fitness. Thus, drift can be thought of as a random search that sometimes takes a population across a low fitness valley. Consequently, if the fitness surface itself does not change and there is no genetic drift, the population cannot evolve further after natural selection has acted to bring it to a local maximum fitness.

We might ask how important a force genetic drift was in hominid evolution. Thompson (1986, 102) noted that "the history of bottlenecks in population size, and the resulting few contributors to a current small population, is probably typical of the history of the small isolated groups in which much of human evolution has occurred." If this is the case, then drift was likely a potent force in our evolutionary history.

CHAPTER SUMMARY

In this chapter, we have covered a wide range of topics, including traditional quantitative genetic, multivariate quantitative genetic, complex segregation, and quantitative trait locus linkage analyses. In addition, we have merely scratched the surface of evolutionary models applicable to quantitative traits. These analyses and models are difficult to comprehend, and you may wish to consult some of the references listed at the end of this chapter.

In truth, human genomes and their functioning are incredibly complex in determining quantitative trait variation, and the addition of environmental variation only complicates the picture. The complexity of the analyses and models briefly presented here pales in comparison to what must be the true complexity of the systems that underlie quantitative trait variation in humans.

RECOMMENDED READINGS

Abel L and Dessein AJ (1998) Genetic epidemiology of infectious diseases in humans: design of population-based studies. Emerg. Infect. Dis. 4: 593–603.

Arnold SJ (1994) Multivariate inheritance and evolution: a review of concepts. In CRB Boake (ed.): Quantitative Genetic Studies of Behavioral Evolution. Chicago: Univ. of Chicago, pp. 17–48.

Comuzzie AB and Allison DB (1998) The search for human obesity genes. Science 280: 1374–1377.

Elston RC (1998) Methods of linkage analysis—and the assumptions underlying them. Am. J. Hum. Genet. 63: 931–934.

Falconer DS and Mackay TFC (1996) Introduction to Quantitative Genetics, 4th edition. Essex, U.K.: Longman.

Lynch M and Walsh B (1998) Genetics and Analysis of Quantitative Traits. Sunderland, Mass.: Sinauer Associates.

Roff DA (1997) Evolutionary Quantitative Genetics. New York: Chapman and Hall.

Rogers J, Mahaney MC, Almasy L, Comuzzie AG, and Blangero J (1999) Quantitative trait linkage mapping in anthropology. Yrbk. Phys. Anthropol. 42:127–151.

Thompson EA (1986) Pedigree Analysis in Human Genetics. Baltimore: Johns Hopkins Univ. Press.

Weiss KM (1993) Genetic Variation and Human Disease. New York: Cambridge Univ. Press.

Wright S (1977) Evolution and the Genetics of Populations: Experimental Results and Evolutionary Deductions. Chicago: Univ. of Chicago.

REFERENCES CITED

Abel L and Dessein AJ (1998) Genetic epidemiology of infectious diseases in humans: design of population-based studies. Emerg. Infect. Dis. 4: 593–603.

Almasy L and Blangero J (1998) Multipoint quantitative-trait linkage analysis in general pedigrees. Am. J. Hum. Genet. 62: 1198–1211.

Arnold SJ (1994) Multivariate inheritance and evolution: a review of concepts. In CRB Boake (ed.): Quantitative Genetic Studies of Behavioral Evolution. Chicago: Univ. of Chicago, 17–48.

Atchley WR (1984) Ontogeny, timing of development, and genetic variance-covariance structure. Am. Natur. 123: 519–540.

Beall CM, Blangero J, Williams-Blangero S, and Goldstein MC (1994) Major gene for percent oxygen saturation of arterial hemoglobin in Tibetan highlanders. Am. J. Phys. Anthropol. 95: 271–276.

Beall CM, Strohl KP, Blangero J, Williams-Blangero S, Almasy LA, Decker MJ, Worthman CM, Goldstein MC, Vargas E, Villena M, Soria R, Alarcon AM, and Gonzales C (1997) Ventilation and hypoxic ventilatory response of Tibetan and Aymara high altitude natives. Am. J. Phys. Anthropol. 104: 427–447.

Cheverud JM (1984) Quantitative genetics and developmental constraints on evolution by selection. J. Theor. Biol. 110: 155–171.

Cheverud JM (1988) A comparison of genetic and phenotypic correlations. Evol 42: 958–968.

Cheverud JM, and Buikstra JE (1981a) Quantitative genetics of skeletal nonmetric traits in the rhesus macaques on Cayo Santiago. I. Single trait heritabilities. Am. J. Phys. Anthropol. 54: 43–49.

Cheverud JM and Buikstra JE (1981b) Quantitative genetics of skeletal nonmetric traits in the rhesus macaques on Cayo Santiago. II. Phenotypic, genetic, and environmental correlations between traits. Am. J. Phys. Anthropol. 54: 51–58.

Cheverud JM, and Buikstra JE (1982) Quantitative genetics of skeletal nonmetric traits in the rhesus macaques of Cayo Santiago. III. Relative heritability of skeletal nonmetric and metric traits. Am. J. Phys. Anthropol. 59: 151–155.

Christensen AF (1998) Odontometric microevolution in the Valley of Oaxaca, Mexico. J. Hum. Evol. 34: 333–360.

Comuzzie AB and Allison DB (1998) The search for human obesity genes. Science 280: 1374–1377.

Commuzzie AG, Hixson JE, Almasy L, Mitchell BD, Mahaney MC, Dyer TD, Stern MP, et al. (1997) A major quantitative trait locus determining serum leptin levels and fat mass is located on human chromosome 2. Nature Genetics 15: 273–275.

Dempster ER and Lerner IM (1950) Heritability of threshold characters. Genetics 35: 212–236.

Duggirala R, Blangero J, Almasy L, Dyer TD, Williams KL, Leach RJ, O'Connell P, and Stern MP (1999) Linkage of type 2 diabetes mellitus and age of onset to a genetic location on chromosome 10q in Mexican Americans. Am. J. Hum. Genet. 64: 1127–1140.

Elston RC (1980) Segregation analysis. In JH Mielke and MH Crawford (eds.): Current Developments in Anthropological Genetics. New York: Plenum, pp. 327–354.

Elston RC (1990) Models for discrimination between alternative modes of inheritance. In D Gianola and K Hammond (eds.): Advances in Statistical Methods for Genetic Improvement of Livestock. New York: Springer-Verlag, pp. 41–55.

Elston RC (1998) Methods of linkage analysis—and the assumptions underlying them. Am. J. Hum. Genet. 63: 931–934.

Falconer D (1965) The inheritance of liability to certain diseases, estimated from incidence among relatives. Ann. Hum. Genet. 29: 51–76.

Falconer DS and Mackay TFC (1996) Introduction to Quantitative Genetics, 4th edition. Essex, U.K.: Longman.

Felsenstein J (1986) Population differences in quantitative characters and gene frequencies: a comment on papers by Lewontin and Rogers. Am. Nat. 127: 731–732.

Galton F (1889) Natural Inheritance. London: Macmillan and Co.

Houle D (1991) Genetic covariance of fitness correlates: what genetic correlations are made of and why it matters. Evol. 45: 630–648.

Hughes AL (1988) Evolution and Human Kinship. New York: Oxford Univ. Press.

Jacquard A (1974) The Genetic Structure of Populations. New York: Springer-Verlag.

Jantz RL, Hunt DR, Falsetti AB, and Key PJ (1992) Variation among North Amerindians: Analysis of Boas's anthropometric data. Hum. Biol. 64:435–461.

Jarvik GP (1998) Complex segregation analysis: uses and limitations. Am. J. Hum. Genet. 63: 942–946.

Konigsberg LW and Blangero J (1993) Multivariate quantitative genetic simulations in anthropology with an example from the South Pacific. Hum. Biol. 65: 897–915.

Konigsberg LW and Ousley SD (1995) Multivariate quantitative genetics of anthropometric traits from the Boas data. Hum. Biol. 67: 481–498.

Lake RIE, Thomas SJ, and Martin NG (1997) Genetic factors in the aetiology of mouth ulcers. Genet. Epidemiol. 14: 17–33.

Lande R (1976) Natural selection and random genetic drift in phenotypic evolution. Evol. 30: 314–334.

Lande R (1981) The minimum number of genes contributing to quantitative variation between and within populations. Genetics 99: 541–553.

Lande R and Arnold SJ (1983) The measurement of selection on correlated characters. Evol. 37: 1210–1226.

Lewontin RC (1986) A comment on the comments of Rogers and Felsenstein. Am. Nat. 127: 733–734.

Livshits G, Karasik D, Pavlovsky O, and Kobylianski E (1999) Segregation analysis reveals a major gene effect in compact and cancellous bone mineral density in 2 populations. Hum. Biol. 71: 155–172.

Lofsvold D (1988) Quantitative genetics of morphological differentiation in Peromyscus. Evol. 42: 54–67.

Lynch M and Walsh B (1998) Genetics and Analysis of Quantitative Traits. Sunderland, Mass.: Sinauer Associates.

Pedersen A (1989) Psychological stress and recurrent aphthous ulceration. J. Oral Pathol. Med. 18: 119–122.

Robertson A and Lerner IM (1949) The heritability of all-or-none traits: viability in poultry. Genetics 34: 395–411.

Roff DA (1997) Evolutionary Quantitative Genetics. New York: Chapman and Hall.

Rogers AR (1986) Population differences in quantitative characters as opposed to gene frequencies. Am. Nat. 127: 729–730.

Rogers AR and Harpending HC (1983) Population structure and quantitative characters. Genetics 105: 985–1002.

Rogers J, Mahaney MC, Almasy L, Comuzzie AG, and Blangero J (1999) Quantitative trait linkage mapping in anthropology. Yrbk. Phys. Anthropol. 42:127–151.

Sciulli PW and Mahaney MC (1991) Phenotypic evolution in prehistoric Ohio Valley Amerindians: natural selection versus random genetic drift in tooth size reduction. Hum. Biol. 63: 499–511.

Sjøvold T (1984) A report on the heritability of some cranial measurements and non-metric traits. In GN van Vark and WW Howells (eds.): Multivariate Statistical Methods in Physical Anthropology. Boston: Reidl, pp. 223–246.

Thompson EA (1986) Pedigree Analysis in Human Genetics. Baltimore: Johns Hopkins Univ. Press.

Towne B, Guo S, Roche AF, and Siervogel RM (1993) Genetic analysis of patterns of growth in infant recumbent length. Hum. Biol. 65: 977–989.

Vieland VJ (1998) Bayesian linkage analysis, or: how I learned to stop worrying and love the posterior probability of linkage. Am. J. Hum. Genet. 63: 947–954.

Weiss KM (1993) Genetic Variation and Human Disease. New York: Cambridge Univ. Press.

Williams-Blangero S and Blangero J (1992) Quantitative genetic analysis of skin reflectance: a multivariate approach. Hum. Biol. 64: 35–49.

Wright S (1977) Evolution and the Genetics of Populations: Experimental Results and Evolutionary Deductions. Chicago: Univ. of Chicago.

CHAPTER 6

Human Adaptation to Climate: Temperature, Ultraviolet Radiation, and Altitude

CYNTHIA M. BEALL AND A. THEODORE STEEGMANN, JR.

INTRODUCTION

Evolutionary theory reasons that successful populations have adapted to or are adapting to their environments. Climate is a pervasive and consistent aspect of the physical environment. Changing slowly on a geological time scale, it acts as an agent of **natural selection** and results in well-adapted populations. Climate varies widely across geographic space, and thus, evolutionary theory predicts that populations will differ biologically in ways that reflect their history of adapting to local climatic stress. The human body is sensitive to many aspects of climate. In this chapter, we describe adaptations to three widely studied, biologically stressful, and highly variable climatic features: temperature, ultraviolet solar radiation, and altitude above sea level.

COLD AND HEAT

Cold Adaptation

Functioning in a cold climate requires maintaining internal (**core**) temperatures high enough to maintain mental and physiological processes as well as muscle and joint mobility to complete tasks. Feet and especially hands are vulnerable: If they get too cold, then they will stop functioning altogether (because movement and dexterity are lost) and may even freeze. Human populations adapt biologically with an intricate set of thermoregulatory responses to maintain internal body temperature. Indigenous populations exposed to extreme cold, such as arctic winters, may have evolved some

Human Biology: An Evolutionary and Biocultural Perspective, Edited by Sara Stinson, Barry Bogin, Rebecca Huss-Ashmore, and Dennis O'Rourke
ISBN 0-471-13746-4 Copyright © 2000 Wiley-Liss, Inc.

unique biological protections. Survival under cold stress also requires behavioral adaptations, and indigenous populations have traditional practices to protect themselves from cold.

Cold Injury and Death

Cold adaptation may reveal itself most clearly when it fails, for example, in death by **hypothermia** (low body temperature) or in tissue destruction by **frostbite** (freezing). These events prove that cold is dangerous to people. Responses by the body in an effort to protect itself, such as changes in **hormonal** or blood flow levels, also mark the presence of cold stress.

Our well-being is most threatened by hypothermia (due to heat loss outrunning heat production or gain) because it is frequently lethal. Internal (core) temperature is maintained in all humans at ~37.0–37.6 °C (98.6–99.6 °F), as measured orally. If core temperatures drops 1 °C (1.6 °F), then the thermoregulatory system compensates by increasing heat production (metabolism). When the deep body temperature drops to 35 °C (95.0 °F), the system responds by **shivering** to produce more heat. At 34 °C (93.2 °F), judgment begins to fail (depriving us of one of our main self-defenses), and loss of physical coordination soon follows. At this point, the cold-stressed victim is in desperate trouble. At 31–32 °C (87.8–89.8 °F), shivering fails, stupor sets in, and only outside intervention will help. The degree of loss that turns a normally functioning person into one falling toward death is only 6 °C (11 °F)— a very slim margin indeed.

Accidental hypothermia in unimpaired adults is fairly uncommon, suggesting that humans are in fact well adapted to the cold. However, occasional incidents of hypothermia occur among skiers, hikers, hunters, sailors, climbers, swimmers, indigents, and the elderly (Danzl et al. 1989, 35). Because hypothermia cases are not centrally registered, we cannot offer exact figures. However, Danzl and Pozos (1987) summarized information from 12 North American areas (with Alaska at one extreme and Florida at the other). They reviewed 428 hypothermia cases collected over a 2-year period, apparently in the mid-1980s. Cases were classified as severe if core temperature was <32.2 °C (90.2 °F) and mild if above that temperature. Of the 428 patients, 73 died (17.1%), mostly those with the lowest body temperatures.

Sixty-nine of the 428 cases were from Florida. Even in warm climates, water temperatures are often <25 °C (77 °F) and present hypothermia risk. As water temperatures fall even lower, immersion becomes extremely dangerous, as illustrated by the 1912 *Titanic* disaster. The steamship sank in –2 °C (28 °F) water. Hundreds of people who escaped into the sea with flotation vests had died when help arrived 2 hours later. In 1963, the cruise ship *Lakonia* caught fire and sank in 17.8 °C (64 °F) water. Of the 200 who escaped with life preservers, 120 died before rescue; hypothermia apparently was a major cause of death (Steinman and Hayward 1989). Doyle (1998) estimates that 600–700 hypothermia deaths in the United States annually occur among alcoholics, the poor, the homeless, and others who are without the usual defenses. Cold season deaths (all causes) are most common in the very young and in elderly adults (Kilbourne 1997).

The primary defense against hypothermia is behavioral, that is, avoiding situations in which body heat is lost faster than it can be produced. Every culture teaches a set of practices about how to do that, including use of light, layered clothing; shelter; food; heat sources; avoidance of alcohol; and caution in dangerous situations. Biological defenses such as body fat and muscle, possibly body build, capacity to shiver, and **acclimatization** all conserve heat.

Frostbite is fatal less often than hypothermia, but it can cripple and may cause increasingly severe symptoms as the victim ages. Frostbite serious enough to cause functional tissue loss is seen in civilian populations, but not frequently. For example, in the northern United States, Cook County Hospital (Chicago) recorded only 843 cases of frostbite over a 10-year period, and the rate for all of Finland was only moderately higher (Smith et al. 1989, 102). Even the Mt. Blanc (Switzerland) winter recreation area averages only 66 cases per year with the feet involved in 55% of the cases, the hands in 46%, and the face in 17% (Foray 1992).

The best information about frostbite comes from soldiers in cold weather combat, a situation in which it often is impossible to exercise self-protective behavior. Even under those conditions, not everyone is frostbitten. Frostbite risk to U.S. soldiers increases with age, smoking, previous injury, "race" (especially tropical ancestry), malnutrition, fatigue, and enlisted status (Hamlet 1988, 434). Frostbite risk decreases with training, experience, and good equipment. The classic illustration is Schuman's (1953) analysis of 716 frostbite injuries to American soldiers during the Korean War (winter 1950–51). Detailed examinations of smoking, age, rank, and several other factors showed that African American heritage (Negro was the term used in the original report) was the single greatest risk factor. Consider the following findings:

Subgroup comparison	Strength of the finding, (χ^2)
"Southern Negro" (raised in warm climate) more risk than "Southern White" (raised in warm climate)	112.5
"Northern Negro" (raised in cold climate) more risk than "Northern White" (raised in cold climate)	40.1
"Northern Negro" (raised in cold climate) more risk than "Southern White" (raised in warm climate)	22.7

(*Source:* Schuman 1953, 396)

Whether the subgroups were both raised in warm or cold areas of the United States, African American soldiers still suffered significantly more frostbite than their European American counterparts. African Americans from colder areas also suffered more cold injury than European Americans from warmer regions, suggesting that adjustment to cold stress during growth (**developmental acclimatization**) may not compensate completely for lack of genetic defenses.

One interpretation is that natural selection for frostbite resistance occurred in the European ancestral population through generations of cold exposure, but not in

Africans. Regardless of their tropical evolutionary history, African Americans may show stronger **sympathetic nervous system** response to stress than people of European ancestry (Calhoun and Oparil 1995). Because the sympathetic nervous system is one of the channels through which surface blood vessels are restricted, any stress (including cold) that triggers a sympathetic response will shut down those vessels, reduce the flow of warming blood to the surface, and increase the risk of cold injury. A better known expression of this sympathetic vasoconstriction is elevated blood pressures in African Americans (Reed 1993). The importance of this point is that people of African ancestry may show heightened sensitivity to cold injury not because of their tropical origins but as a result of some unrelated event in their evolutionary history that affected sympathetic response. The thermal evolutionary biology of this issue is far from clear, but some evidence from Asia indicates that people native to northern areas may have some special cold protection (So 1975, and later in this chapter [Figure 6.6]).

Some of the fundamental physiological responses to cold by humans are less obvious than frostbite injury. Cold causes the blood vessels of the extremities to constrict (most severely in those least well adapted to cold); increases **hematocrit** levels (i.e., the volume of red blood cells in the blood, expressed as a percentage; a higher hematocrit results in "thicker" blood); and raises blood pressure and metabolic rate (Burton and Edholm 1955; Vogelaere et al. 1992). These responses create more work for the heart and may worsen coronary artery disease. In northern industrial countries, acute coronary problems rise as daily temperature falls, especially during snow-shoveling season (Houdas et al. 1992). The lungs may be affected as well. The nose and mouth warm and humidify inhaled air so that when the air reaches the lungs, it is 37 °C (98.6 °F) and fully saturated with water.

But sometimes, the upper respiratory system fails to keep pace, and cold, dry air reaches the lungs, which may exacerbate chronic bronchitis (persistent inflammation of the larger passages conveying air to and within the lungs), constrict pulmonary vessels, and even lead to pulmonary edema (accumulation of fluid in lung tissues and air spaces). Regnard (1992), reviewing these problems, noted that with chronic cold stress, bronchial walls thicken and decline in function (he called this syndrome "Eskimo lung"), and Shephard and Rode (1996) found this problem in arctic populations.

Respiratory problems often worsen during the cold season. Momiyama (1975; Figure 6.1) showed that in Japan, the number of deaths from pneumonia and bronchitis has been as much as six times higher in winter than in summer. The seasonal contrast remained as the overall death rate declined in postwar Japan. Figure 6.1 raises several important adaptive issues. Seasonal cold is accompanied by change in diet, sunlight, humidity, and the amount of time indoors and closely exposed to others. Consequently, the seasonal variation in **mortality** shown in Figure 6.1 may reflect factors other than cold, even though many are cold-related. The decline over time in bronchitis and pneumonia mortality may reflect improvements in nutrition, building heating, and health care in Japan, but the Japanese author attributed the decline to home heating improvements alone.

Certainly, relatively few directly cold-induced injuries and deaths occur each year compared with those due to **parasitic, infectious,** or nutritional diseases. Why? Because biological and behavioral adaptations suffice to avoid cold injury in most circumstances. Cold is an ancient stress, and adaptive responses are well-honed, whereas diseases and malnutrition that depend on population density are relatively new stresses in the long span of human existence.

Thermoregulation (Cold and General)

Think of the human body as a machine that breaks down if it is too hot or too cold. It gains heat (or increases **heat load**) in several ways. The more oxygen and energy our cells burn (metabolism), the hotter they become; cells burn "hottest" during exercise, shivering, and release of hormones such as epinephrine and in response to infection. Heat load also rises if we eat hot food, contact hot surfaces (**conduction**), are exposed to hot air (**convection**), or receive **infrared** waves (**radiation**). We lose heat in the absence of these conditions by the same channels whenever the environmental (**ambient**) temperature is below body temperature. We also rid ourselves of heat by special mechanisms such as shunting blood to the surface and sweating. Deep body temperature has little opportunity for heat compensation, because cells function only within narrow temperature ranges.

The body keeps itself at functional temperatures by **thermoregulation**. This process rests on an ancient, intricate physical and physiological control system that

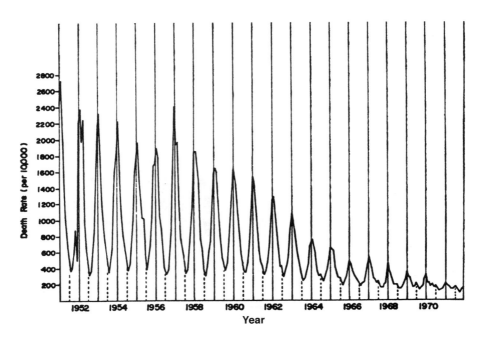

Figure 6.1   Seasonal data from Japan showing pneumonia and bronchitis deaths, 1952–1970. Notice the differences between winter and summer rates and the steady overall decline in rates, which was attributed to improvements in household heating. (*Source:* Momiyama 1975, 146)

we share with most other mammals. Werner (1987) identified four subsystems that work together to ensure thermal equilibrium:

1. *Passive traits*, such as **subcutaneous fat** (just below the skin) and muscle tissue, provide insulation without changing biologically on exposure to temperature change. Body proportions belong in this category, too, as do varying numbers of **capillaries** in the skin itself. These anatomies do not have to "do" anything to be effective.
2. *Receptors* are scattered widely in tissues. They are input **neurons** in the skin and at deeper levels that are triggered when tissue temperatures get too low or too high.
3. *Information processors* or central neurons are specialized cells in the brain's **hypothalamus** and cerebral cortex. They receive incoming warning signals from receptors and also sense brain temperature directly. Below a **set point** (a temperature threshold at which control cells are stimulated to act), the brain initiates physiological and behavioral actions that produce and conserve heat.
4. *Effectors* carry neural and hormonal messages that cause blood vessels to constrict (conserve heat), muscles to start shivering, respiration and metabolism to rise (produce heat), make sweating begin (lose heat if the body is hot), and start self-defensive behavioral responses such as clothing adjustment.

Thermoregulation is a classic example of a self-regulating system managed by complex feedback loops. The feedback system can adapt. The set points at which the

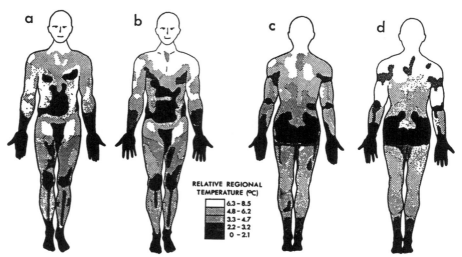

Figure 6.2   Regional differences in body surface temperatures of people immersed in 45.6 °F (7.5 °C) water for 15 minutes. a and c, nonexercise patterns; b and d, after swimming. (Source: Hayward et al. 1973, 710)

body initiates response are adjustable. Cold acclimatization lowers the core temperature at which shivering starts; heat acclimatization lowers the core temperature at which sweating begins (Brück 1980). Paradoxically, cold acclimatization set point adjustment allows the deep body to undergo more cooling before shivering is started. However, this process may save metabolic energy that would be used by shivering without endangering the body.

Adjustment of set points by heat acclimatization is more complex. The thermal system is kept in "emergency mode" and starts its cooling system (sweating) at the earliest hint of heat stress. Intermittent heat exposure actually lowers the deep body temperature at which sweating begins. However, the results of studies of continuous heat exposure disagree, and some researchers have found that the set point increases, so the body is hotter before protective action kicks in. This finding is analogous to that of increased toleration to cold after acclimatization (Werner 1987). The general conclusion is that heat acclimatization keeps the body at work somewhat cooler; however, at best, set points can be adjusted up and down only ~1 °C (1.8 °F) from unacclimatized levels. These adjustments are limited because cells will function properly only within narrow temperature ranges. In mammals, the amount of core temperature variation on which natural selection has to operate is very small.

Active physiological responses to cooling are as follows. The first is **vasoconstriction**. Neurons and cold sensors within surface blood vessels cause the vessels to narrow and reduce blood flow, allowing the superficial body "shell" to cool off while keeping much of the heated blood deep in the body. Figure 6.2 is a striking illustration that shows how hands, feet, and some areas over fat deposits may become quite chilled. This response may persist, even to the point of local cold injury, to protect vital organs. Second, a small drop in deep body temperature brings on shivering (i.e., involuntary muscle contraction for the purpose of producing heat rather than work). Shivering can raise the metabolic rate above the resting level by a factor of 4. Heat is also produced through a third channel, nonshivering thermogenesis, which is involuntary heat production that results from hormonal action on adipose tissue. Fourth, exercise is a behavioral means of heat production. Toner et al. (1985) reported that subjects doing leg exercise in cool water for 1 hour raised their deep body temperature an average of 0.3 °C (0.5 °F) (compared with a loss of 0.5 °C (0.9 °F) when they did not exercise); even in cold 18 °C (64 °F) water, exercise reduced the rate of body temperature loss. However, exercise heating is limited by individual endurance and food energy resources, and some types of whole body exercise such as swimming may actually increase heat loss.

Subcutaneous fat, which provides major passive protection from cold, insulates deeper tissues. Whales and seals survive in freezing water wrapped in a blanket of blubber. Fat protects people, too. Figure 6.3 shows that people with thicker subcutaneous fat (measured as **skinfolds**, i.e., thickness of double folds or "pinches" of skin and subcutaneous adipose tissue) can tolerate colder temperatures before increasing heat production. Critical temperature as used in Figure 6.3 means the specific air or water temperature at which basal (resting) body heat production is no longer sufficient to keep one's core temperature steady, and heat production is increased through shivering, exercise, etc. "Skin thickness" indicates the thickness range of just the two

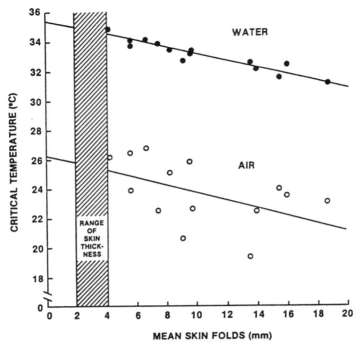

Figure 6.3   The environmental temperature level at which the body begins to produce compensatory heat (or the "lower critical temperature") decreases as subcutaneous fat thickness increases. People with more body fat respond more slowly to cold before producing compensatory heat. (Source: Smith and Hanna, 1975, 97)

skin layers of a skinfold. Values above the skin thickness range reflect amounts of subcutaneous adipose tissue. Experiments in which men and women were immersed into 0 °C (32 °F) water for 30 minutes—an extremely severe test—supported these results. Hayward and Eckerson (1984) found that people with thicker skinfolds lost less core temperature than the others.

Veicsteinas' 1982 experiments demonstrated that muscle is even more insulative than adipose tissue, although the benefit is seen mostly at rest or low levels of exertion. These tests measured tissue temperatures below the skin surface in water-cooled humans. Muscle was analyzed as a metabolically inactive, passive insulative layer without regard to its additional function as a heat producer, and offered good insulation simply because it was often thicker than the overlying fat layer. This insight is significant for understanding variation among human populations. Contemporary cold climate populations tend to have relatively large muscle masses, and our Pleistocene ancestors and relatives were heavily muscled (see discussions by Ruff 1994 and Holliday 1997). We may speculate that our ancestors had substantial muscle mass not only to do their strenuous tasks but also because it provided thermal protection. They lived a life of constant exposure, lacking the elegant cultural protections of the modern world, and many tropical areas can impose a good bit of seasonal cold exposure. For instance, Scholander et al. (1958) found that Australian Aborigines traditionally slept nude during winter desert nights at environmental temperatures of 0 °C (32 °F).

The head and neck present humans with special heat conservation difficulties in the cold, partly because they are exposed but also because they do not show severe vasoconstriction. Steegmann (1979) showed that the faces of Native North Americans working at –20 to –35 °C (–4 to –31 °F) remained relatively warm. Although this response may provide protection from frostbite, it also leads to hypothermia due to heat loss. In water immersion experiments, Wade (1979) demonstrated that ~25% of total body heat loss was from the head and neck. Cold climate populations typically compensate for this using a broad range of head and neck insulators.

Finally, people with greater physical fitness produce more heat when exposed to cold (Bittel 1992). Although they may also lose more heat because they tend to carry less insulative body fat, this ability to generate more heat and better endure work could be the difference between life and death in a cold emergency.

Cold Acclimatization and Tolerance
Acclimatization responses begin as the newborn first experiences cold and can lead to substantial changes during the growth period. In addition to developmental acclimatization, acclimatization occurs throughout the life cycle, particularly as the seasons change.

LeBlanc (1975) examined how this mechanism works in adults, over a period of a few months. He tested a group of Canadian soldiers who were exposed to severe winter cold between November and April. By midwinter, their response to a standardized cold test showed less shivering, a lower metabolic rate, and reduced deep body temperature (although in other studies, deep body temperature stayed level). LeBlanc's results confirmed those of Davies and Johnston (1961), who also discovered that heat production fell less rapidly than shivering.

These results imply that a strong (if uncomfortable) emergency boost in metabolism from shivering is replaced in part by a less active type of thermogenesis. That is, shivering decreases, but hormonal control of metabolism increases. Clearly, this takeover is based on greater oxidation of carbohydrates and lipids in the tissues. The mechanisms by which it takes place are not clear, but the hormone **norepinephrine** may play a role (Vallerand and Jacobs 1992; Budd et al. 1993). By whatever mechanisms, the acclimatized person tolerates low internal temperatures better, burns less metabolic fuel, and feels more comfortable in the cold. As cold exposure and acclimatization continue, a better tolerance for cold-induced pain develops, partly because the emergency vasoconstriction lessens. Higher hand temperature not only relieves cold pain but also increases dexterity—another critical life-and-death issue (Ellis et al. 1985).

Variation in Adaptive Responses Within Populations
In any group, some people resist cold stress effectively and some do not; the range of physiological responses typically shows the classic bell-shaped curve. If those with the

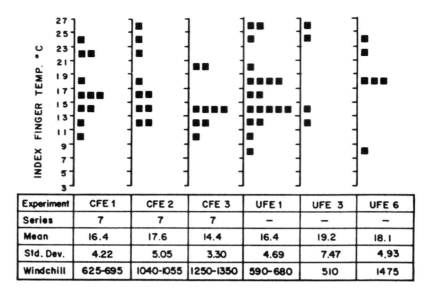

Experiment	CFE 1	CFE 2	CFE 3	UFE 1	UFE 3	UFE 6
Series	7	7	7	−	−	−
Mean	16.4	17.6	14.4	16.4	19.2	18.1
Std. Dev.	4.22	5.05	3.30	4.69	7.47	4.93
Windchill	625-695	1040-1055	1250-1350	590-680	510	1475

Figure 6.4 Variations in individual Ojibwa (native North American) finger temperatures are shown for outdoor controlled field experiments (CFEs) and normal outdoor work in uncontrolled field experiments (UFEs). The severity of natural cold conditions is described in the bottom row, where 510 is fairly mild and 1250 is severe cold. (*Source:* Steegmann 1977, 358)

poorest responses die or become disabled in a crisis and if those failures have some genetic basis (**heritability**), then there is potential for natural selection by cold and the maintenance or improvement of biological adaptation to cold over generations.

As an example of variation in cold resistance, consider how well indigenous people working in the cold maintain their finger temperatures (see Box 6.1). The Native American Ojibwa of Northern Ontario, Canada, grow up and work in one of the world's coldest inhabited areas. The temperature of winter work settings is commonly –20 to –35 °C (–4 to –31 °F). Field experiments measured hand temperatures in six different work places during or immediately after work. These data are realistic because they come from active workplaces rather than laboratories. Hand temperatures are influenced by exertion level, insulation use, and by other factors such as food intake, which the men were free to control. Hand temperatures are functionally meaningful measures of adaptation because manual dexterity fails between 16 and 12 °C (61 and 54 °F) and tactile sensation usually disappears at ~12 °C (54 °F) (Lockhart et al. 1975). Figure 6.4 (Steegmann 1977) shows two sets of finger temperatures. Controlled field experiments were measurements taken after men had been snowshoeing at their own pace for ~45 minutes and then standing without exercising for another 25 minutes, and uncontrolled field experiments were measurements taken as men did their normal winter work, such as chopping wood or cutting ice blocks. Windchill is an environmental stress index that combines temperature and wind speed; values of 1200–1500 indicate bitterly cold conditions. Despite differing exposure conditions, mean finger temperatures did not vary greatly, and individual temperatures were normally distributed. However, even among these men with genetic and acclimatizational adaptations, some showed finger temperatures in the range of 12–16 °C (54–61 °F), where dexterity and sensitivity are lost. But none of the men were close to frostbite, and in fact, frostbite was uncommon in this group.

Variation in cold resistance is also found among individuals who are not acclimatized. This variation is illustrated by experimental hand cooling, which causes surface

---

BOX 6.1 FINGER TEMPERATURE MEASUREMENT

We assume that the ability to keep our hands warm in cold settings is adaptive. Higher tissue temperatures not only defend against cold injury but also maintain manual dexterity. Hand cooling tests are used to measure variation in cold response among individuals and for comparing different populations.

The best thermal measuring instrument for field experiments is the electric resistance thermometer (telethermometer). It operates on the principle that temperature change increases or decreases electrical resistance in metals in a linear fashion. The thermistor (sensing element, applied to the skin) is in circuit, and a change in resistance to current maintained through the circuit is compared with an internal reference resistor. The ratio between the two resistances is continuously converted to temperature measures and is shown on the instrument face. Older instruments had analog dials (the gray box in the Figure 6.5), and present models are digital display (the black box) (both instruments made by Yellow Springs Instrument Company).

Test Procedure

1. The laboratory ambient (surrounding) air temperature should be warm enough that the subject is comfortable and not experiencing hand cooling due to body chill or cold air before the test; 25 °C (75 °F) is about right. The subject should be in a quiet, resting state for at least 30 minutes before testing. Accuracy of the telethermometer is checked against a laboratory-grade mercury thermometer accurate to 0.1 °C (0.18 °F), an example of a general procedure called calibration.

2. Using water and crushed ice, a hand bath is prepared and kept at 5 °C (41 °F). This temperature is used in our laboratory because it is sufficient to test cold response but not so cold as to cause cold injury, extreme pain, or fainting. This experience is, nevertheless, challenging for most subjects; they should be in good health and should not have high blood pressure or other conditions that would subject them to risk from the experiment.

3. With hypoallergenic surgical tape, small surface thermistors are applied to the middle finger of each hand, 5 mm (0.2 in.) above the nail base. Care must

be taken to apply the tape so that it does not constrict blood flow (because warm blood from the body core is the primary means of finger warming).

4. The subject is seated, and finger temperatures are recorded for 5 minutes, or until the temperature stabilizes. The stable temperatures (usually taken at 30-second or 1-minute intervals) are averaged to get preimmersion temperature.
5. The "cold hand" is then immersed in the chilled water bath up to the lower end of the radius. Water should be stirred regularly to prevent thermal layering and the temperature maintained at 5 °C (41 °F) by adding ice. A thermistor may be added to the stirring rod as a monitor. Temperatures of both the chilled and unchilled (contralateral) hand are recorded at least every minute. An alternative procedure favored by Japanese investigators cools only one finger, rather than the entire hand, and is somewhat less stressful.
6. At 30 minutes, the cooled hand is withdrawn from the water and patted dry with a towel. Temperatures of both fingers continue to be recorded during a 10-minute recovery period.

The mean temperature of the finger during 30 minutes of cooling is an index of cold resistance, as is percentage of recovery to preimmersion temperature at 5 and 10 minutes after cooling. Other analyses, such as finger temperature increase during cold-induced vasodilation, also may be used. The contralateral (uncooled) finger temperature indicates the amount of central nervous system response (degree to which both hands vasoconstrict when one is cooled) or whether there is more of a local tissue response.

blood vessels to constrict and finger temperatures to plummet (because the fingers are deprived of warm blood). However, many people override vasoconstriction. Their skin blood vessels undergo cycles of **cold-induced vasodilation** (CIVD; increased diameter of the blood vessels) that rewarm fingers and maintain function. Acclimatization resulting from repeated experience with cold enhances CIVD, but the response has genetic origins as well (reviewed in Steegmann 1975).

Genetic causation is demonstrated by experiments done on 21 young men of European ancestry who grew up in Hawaii without significant exposure to cold. Because Europeans have a long history of residence in glacial and, more recently, in seasonally cold temperate climates, it is assumed that they underwent natural selection by cold in their homelands and carry "cold adaptation" **genes**. However, European Americans who are born and raised in tropical Hawaii are not strongly acclimatized to cold. If they show specific adaptations to cold, we assume that these responses are genetically based. In the Hawaii experiments, subjects' hands were cooled for 70 minutes in 0 °C (32 °F) air moving at 4 feet per second. The sample was then divided into two groups: according to whether CIVD waves allowed the individuals to regain 0–7 °C (32–44.6 °F) in finger temperature over the 70 minutes (low CIVD responders) or more than 7 °C (high CIVD responders). The results (Figure 6.6) (as well as statistical comparison) showed that high responders finished the test with warmer fingers. Figure 6.6 illustrates the variation in final finger temperatures (2.8–12.4 °C or 35–54 °F) and in CIVD, a mechanism that protects humans from cold injury. This result is particularly interesting because it suggests that Europeans do have some genes for cold resistance that are expressed even in groups not acclimatized to cold and that there is genetic variation in the population. Furthermore, it implies that genetically based geographic differences in cold resistance may exist in different human populations, but these experiments did not compare any individuals of tropical ancestry.

Variation in survival times of unprotected humans in freezing (0 °C) water is also likely. Obviously, obtaining this kind of data is difficult, but Hayward and Eckerson (1984) estimated survival times based on rates of deep body temperature loss by experimental subjects in cold water. The investigators extrapolated from individual core

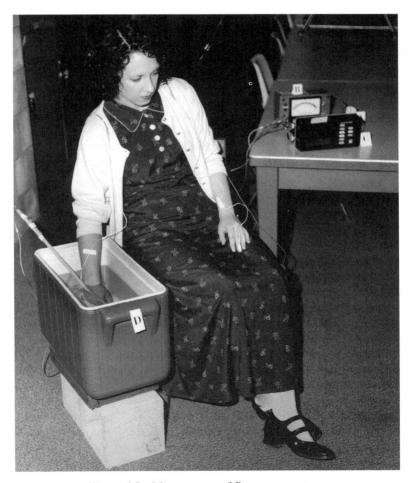

Figure 6.5 Measurement of finger temperature.

cooling curves to the time required to reach 30 °C (86 °F), at which temperature death usually occurs. Most would survive 60 minutes, but at 75 minutes, 25% would be dead; at 90 minutes, 50%; at 120 minutes, 75%; and all would perish within 150 minutes. This extrapolation corresponds well to survival rates in the *Titanic* and other disasters. The distribution curve was fairly normal, even though there were more late survivors (after 2 hours) than early fatalities (less than 1 hour). Factors such as body fat, muscle mass, and physical fitness probably cause the individual differences.

Adaptive Responses Between Populations: Biogeography

Biogeographic "rules" are statements about how biological traits in widely distributed warm-blooded species vary in relation to climatic differences. Because humans span the globe, human biologists hypothesize that adaptations may be expressed as **clines** or as **correlations** between climatic factors such as temperature and biological traits, reflecting adaptations to temperature.

Finger cold response among East Asians ranging from subarctic Northern China to the steaming jungles of Indonesia is one example. Yoshimura and Iida (1952) developed a protection index that combines average finger temperatures and CIVD responses during 30-minute finger immersion in ice water. A score of 3 is weak, and 9 is strong. Figure 6.7 shows a cline of mean protection index values that probably reflects differences in both genetics and acclimatization. If data from other groups were added to the figure, it might become less clear, but the north-to-south change in index values is quite evident.

The Asian cold response series may reflect a general human arctic-to-tropic pattern of clinal variation. Meehan's classic experiment (Meehan 1955) compared

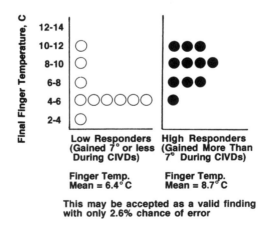

Figure 6.6   Final finger temperatures of men of European ancestry born and raised in Hawaii, exposed to moving 0 °C air for 70 minutes. Low responders showed either no or few CIVD waves and finger temperature gains of 7 °C (12.6 °F) or less. High responders showed more capacity to dilate their skin blood vessels, consequently gained more temperature, and completed experiment with significantly higher temperatures. (*Source:* Steegmann, unpublished data)

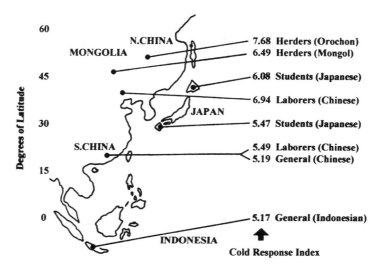

Figure 6.7   Values are indicators of resistance to water finger-cooling experiments: 9 indicates strong resistance to cooling, and 3 indicates poor resistance. Subarctic natives working outdoors (Orochons) are best protected, and tropical Indonesians were less able to keep their fingers warm. (*Source:* Yoshimura and Iida 1952; Kondo 1969; Toda 1975)

Athabascan Indians and Inuit (Eskimos) from the Arctic with European Americans and African Americans from the lower 48 United States. These groups represent populations that are presumably highly acclimatized and genetically well-adapted (Native Americans); moderately acclimatized and moderately genetically adapted (European Americans); and moderately acclimatized, probably not genetically adapted (African Americans). After 30 minutes of hand immersion in 0 °C (32 °F) water, only 5% of the Arctic natives showed 0 °C (32 °F) finger temperatures compared with 20% of the European Americans and 60% of the African Americans— results confirmed in other studies. Meehan assumed that a higher proportion with fingers above freezing indicated a better adapted population. Natural selection for cold resistance has doubtless been operating since humans first braved lethal cold. It explains why long-term Arctic residents do much better than those of tropical origins and the Europeans, moderately cold adapted, fall in between. However, we have done much more testing for cold resistance among cold climate peoples than with tropical natives. We need to conduct major cold resistance investigations in Australia, South and Southeast Asia, and Africa before we can draw firm, species-wide conclusions about clines that indicate physiological adaptation to cold.

Physical characteristics also vary with mean annual temperature. Ruff (1994) showed that cold climate populations have wider bodies than those from the tropics (Figure 6.8). This observation is an expression of **Bergmann's Rule**, that in a widespread, warm-blooded species, populations in colder climates are generally larger than those in temperate and warm climates. This distribution results in more body

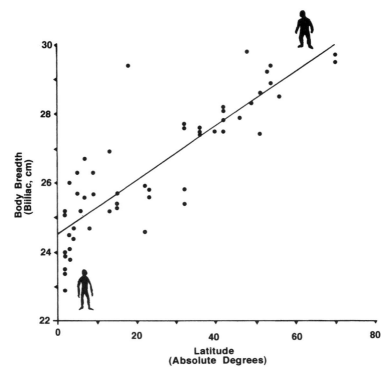

Figure 6.8 The increase in human body breadth as latitude increases and climatic surroundings get colder. Body breadth is estimated using the biiliac diameter (maximum breadth across the pelvic bones). (*Source:* Redrawn from Ruff 1994)

weight (i.e., metabolic heat-producing potential) per unit of body **surface area** (i.e., heat loss potential) in the cold. A perfectly spherical human would, according to thermal physics, conserve heat best. Ruff's data show a very strong positive correlation of latitude to body width ($r = 0.886$). They may represent the result of natural selection acting over many generations, if those with slender builds (small body breadth) were more likely to die from hypothermia or be crippled from frostbite, leaving survivors to pass on a higher proportion of relatively stocky body genes.

However, physiological test results do not paint a clear picture of the assumed advantages of a larger build. For instance, Toner et al. (1986) and Glickman-Weiss et al. (1993) compared subjects with similar subcutaneous body fat but different muscle mass and relative body surface area. The expectation was that the individuals who had relatively low body surface area for their weight (that is, were relatively more round in shape) would better conserve deep body heat when immersed in cool or cold water. In neither set of experiments was this response strongly or clearly observed. In contrast, Sloan and Keatinge (1973) studied a group of children swimming in a cold pool, and discovered that subcutaneous fat was their primary physical defense against cooling. In this situation, low relative surface area played a secondary but measurable role in cold resistance. These experiments may be stressful for subjects and are problematic because of small sample sizes, only moderate differences between individuals with relative low and high surface area, and a very complex cold response physiology. Clearly, this response is not a matter of simple physics. Maybe the ability to reduce heat loss through vasoconstriction, which decreases circulation to the skin and outer shell of tissue, reduces the influence of relative surface area in heat loss, or maybe the two sets of traits interact. This problem is clearly also of anthropological importance, awaiting experimental verification or refutation.

A second human biogeographic pattern is **Allen's Rule**: Cold climate populations tend to have short arms and legs relative to their height. Such body proportions may reduce heat-radiating surface area, functionally paralleling Bergmann's Rule. The correlation between mean annual temperature and relative leg length is –0.62, meaning that the colder the climate becomes, the shorter the legs relative to body length. Climatically induced differences in body build may have been more extreme in the past. Holliday (1997) and Trinkaus (1981) have demonstrated that Neandertals of glacial Europe had "hyperpolar" body proportions, even more extreme than those of modern Inuit.

Relationships between human body shape and climate are not simple and straightforward. Rather, they are apparently products of multiple forces acting simultaneously. Certainly, heat and cold are expected to play a role. However, these actions may go beyond genetic patterning that is produced by natural selection. In experimental animals, relative leg length may be shortened during growth in cold, probably by direct reduction of extremity tissue metabolism rates. This mechanism may also play a role in reducing human leg length.

Other studies support the importance of nongenetic factors in producing Bergmann's Rule. Newman and Munro (1955) discovered that European Americans from the colder parts of the country had lower surface area for their weight than southerners. Because U.S. whites are originally from Western Europe, we must assume that these New World differences in body form are produced during growth by nutritional or thermal factors. They cannot be genetic adaptations to the new environment in so short a period of time. Furthermore, clines in recent populations show weaker climate–body size relationships than in the past (Katzmarzyk and Leonard 1998). One possible explanation is that nutrition is improving worldwide, and especially among tropical peoples. Warm climate populations are consequently catching up in height and weight to those from colder regions. It will be interesting to see whether larger tropical people are also getting relatively wider. If not, Ruff's findings on body width will be sustained, as will the inference of potential thermal advantages.

A final cline, known as Thomson's Rule, involves the human nose, which becomes relatively more narrow as the climate becomes cooler and dryer. Using a world sample, Weiner (1954) reported a –0.82 correlation between the nasal index (i.e., a narrowness indicator: nose width divided by nose height) and vapor pressure (the actual amount of water suspended in air). Because this correlation is higher than others (e.g., between nose shape and simple latitude), it suggests that low humidity is the driving force. Carey and Steegmann (1981) also discovered that the human nose protrudes from the face more where the climate is dry and cold ($r = -0.72$). We infer that natural selection by cold, dry air has narrowed the nose and increased its protrusion. Cold air will hold little water vapor, and the respiratory system must remain warm and moist to function and to resist tissue damage. The logic of the relationship between climate and nose shape is that a relatively narrow, beak-like nose creates more turbulence of incoming air and more contact between inhaled air and the hot, moist nasal passage walls. This shape then increases the capacity of the nose to heat and humidify air (Proctor and Andersen 1982). Although we know that inspired air reaches nearly deep body temperature and 100% humidity by the time it gets to the lungs, no experimental data show that a narrow nose is better at achieving this state. (For a well-balanced, critical review of this issue, see Franciscus and Long 1991.)

Behavioral Cold Adaptations

By far, the greatest of specialized human cold adaptations are cultural and behavioral. Body chilling is uncomfortable, and the most immediate response is usually to do something to reduce the distress. Fire, tailored clothing, tents, sleds, and good foraging skills are essential for surviving winters in cold areas of the planet. Cultures vary widely in the specific responses, and the following tale illustrates a few methods used by the northern Algonkians of Canada. Here, a newcomer's education to subarctic survival begins.

> Two men leave the cabin, a fog of warm air with them instantly banished by the stunning cold. Both wear windproof cloth parkas and soft, well insulated moosehide moccasins and mittens. One kneels to bind on his snowshoes while the other slips into his bindings with casual skill.
>
> Seeing the stranger having trouble with the unaccustomed task, Joseph crouches beside me and fastens the ties in an Algonkian hitch. Moving with precision and speed, his bared hands are exposed but a moment. Without a word we move down a trail into the bush where his trapline lies waiting. I am surprised by the ease of snowshoeing—only later to recognize that our trail has been cleared through the dense forest and is kept packed down by occasional travelers. Joseph sets an easy pace, glancing back now and then to be sure I am keeping up. I know he can easily travel 35 miles a day this way so an even pace is important to him as well. Since we are not camping this trip, he carries only his rifle and an old canvas bag slung over one shoulder. The miles go by in silence. Sheltering trees protect us from the wind and we cool down only when he stops to check his traps. I have dressed too warmly and I can feel the sweat on my back. In the Ojibwa way, my training proceeds in silence.

Coming down through a stand of mature spruce trees, Joseph motions for a halt and with a light hatchet from his bag, cuts a number of spruce boughs. They are layered into an oblong island in the snow. Joseph kicks out of his snowshoes and along with the rifle, sticks them butt down in the snow. I take mine off as well, floundering for a moment as I sink to my waist in the powder. The spruce bough mat however bears up our weight as Joseph builds a fire on top of the boughs. He props a small pan in the flames to thaw some ice he collected as we passed a frozen stream. As we wait for the water to heat, our clothes, moccasins and mitts throw off wisps of steam, drying gently; we hunker down on the mat, eased by the rest. I soon realize how thirsty I am. While we carried no water, which would freeze in minutes, I see now that the boiling water is for tea. I wait, anxious to observe what exotic trail food my mentor has in his sack. The bag of Planters peanuts and the Snickers candy bar he offers never tasted so good, and soon we are off again. Rested and refreshed, with dry moccasins, it seems only a middling passage before the cabin reappears, peaceful by the lake's edge.

(Source: Adapted from experiences by A. T. Steegmann, Jr.,
Weagamow, Ontario, January 1974)

From observations like these and many others, anthropologists have extracted advice on cold climate survival, some of which follows.

1. If it is too cold, just stay inside. Otherwise, use natural shelter for wind protection or make a windbreak, particularly when not exercising. Fire, especially with a heat reflector, compensates for lower heat production at rest. Stop periodically to make shelter and heat rather than getting exhausted trying to make it home.

2. Clothing should be lightweight, layered, windproof, and not restrictive of movement. Vulnerable hands and feet need special attention, even if no more than a simple fabric or leather shell stuffed with wild grass. Also cover your head and neck, and keep everything dry.

3. Move at a rate fast enough to produce body heat but not so fast as to dampen clothes with sweat. Do not get dehydrated, and do replace burned energy. Shephard and Rode (1996) point out that traditionally, Inuit (Eskimo) ate a lot of fish and marine mammals, high in polyunsaturated fats. These fats gave them great amounts of energy and were not of the type that produce cardiovascular diseases.

4. Emergencies, particularly falling into freezing water, are the ultimate threat to life. Have a plan. Travel with others. In a crisis with cold water, the person who is soaked should exchange part of their clothes with another. This may slow body cooling enough to allow you to build a fire or reach shelter.

Nelson's (1969, 1973) observations on life in cold regions are particularly rich sources on this topic.

General Heat Adaptation

It seems appropriate to begin this section with another narrative—this one on how easy it is to discover your limits.

> Arching the valley wall stands a hot, still Philippine sky. Below, a lone jeepney jolts its way over the "first class bad road" accompanied by occasional complaints from pigs carried in back. Flanking the road, rice paddies steam in the mid-day quiet, farmers having already retreated to their homes until cooler evening. An anthropologist, the sole outsider here, watches the landscape pass. Along the top of the valley there rests a bed formed by an ancient lava flow. Caves have made a space below the hard crust and it occurs to me that people may have sought shelter there from time out of mind. Kiko, my Filipino colleague and co-investigator, also looks up and we exchange one of those "another frontier" comments.
>
> Although we are here to do a pilot growth study, time is put aside for a trip to the caves. Setting out in the cool morning, we hike back along the road a few kilometers to reach a little settlement. Along the way, a rice farmer harrows his paddy fields behind the straining bulk of his carabao and we stop to watch. It is already hot and I wonder how the man and his draft animal tolerate such heavy work. Sun burns into my shirt making the sweat flow, sticking cloth to skin, and blocking evaporation. There is no shade on the road. It was still a cold, late spring in Buffalo when we left and neither of us is ready for such a sudden shift to the wet, lowland heat.
>
> Kiko hires some local guys to guide us through the forest up to the caves. They treat it as a lark, carrying only their light, loose clothing. And, each straps on a talibang, the long iron knife used for all kinds of tasks. As we start climbing, our guides clear a trail—traveled only occasionally and now overgrown. I watch the economy of motion they have learned already as young men, and see that they are not dripping sweat despite the work.
>
> The first little sign of trouble is steam on the inside of my glasses. I stop to wipe my face with somewhat shaky hands, but I feel no sense of distress from the exercise or the heat. Finally I put the glasses in a snap pocket. Through a little clearing and up perhaps another 200 meters, we see the smaller cave. Climbing the slope I feel my heart pounding way faster than the grade would merit and my skin feels like boiled meat. I have to stop, not from exhaustion

but rather, distress. The rest does little to take the edge off of my heart rate. "I'm not sure I can make to the big cave, Kiko." It seems an understatement. We are nearly at the small cave entrance and we climb into the welcome cool. It is too low to stand but you can scramble in bent over. That seems hard with muscles getting weak and uncooperative, so I just drop on a low mound of earth. The guides are chatting with Kiko, discussing a bat they have caught. I can hardly attend to what is going on. After a while, I feel better, but as we move out to make the next climb, I realize my heart is already beating wildly. Kiko tells the guides we have to start down. He is feeling the heat too, but not so much.

The descent is easier. Near the trail end, a stand of coconut trees comes along. Our guide knocks some nuts down and takes the ends off with his knife. How could we have not brought a canteen? The delicious, sharp coconut milk helps restore order and by the time we are back down to the road, is seems an almost normal hike.

(*Source:* Adapted from experiences by A. T. Steegmann, Jr., Sorsogon, the Philippines, May 1988)

Normal human core temperature is very hot—about 37–37.6 °C (98.6–99.7 °F)—and close to the highest tolerable temperatures of 40–42 °C (104–107.6 °F), above which organs begin to fail, tissues hemorrhage, and the irreversible slide toward death begins (Sminia et al. 1994). Heat stress is produced by not only external environments but also metabolism. Humans are heavy-duty metabolic furnaces that produce great quantities of heat during physical work and even while seated or sleeping. Cellular metabolism is boosted 13% for each 1 °C rise in body temperature above 37 °C (Hubbard and Armstrong 1988). The challenge with heat stress is to keep deep body temperature low enough that the brain, heart, and liver continue to operate. Heat is probably one of the most dangerous stresses induced by physical climate. Work helps the body keep warm at low temperature but adds to the stress in a hot environment. It is true that our species is of tropical origin and that we function well in the heat. However, that function depends on some remarkable adaptations, such as sweating and learning to recognize the limits past which we stray into danger. We survive in heat only if we keep our bodies below the ever-present lethal ceiling.

The press reported more than 600 deaths in Chicago during a severe heat spell in 1995. A Beijing heat wave in 1743 AD reportedly killed 11,000. Mortality rates escalate beyond a certain threshold. For example, as the Shanghai daily high temperature rose from 34 to 38 °C (93 to 100 °F), the mortality rate rose from 100 to 300 people a day. But even in more normal conditions, summer heat takes its toll. Although high temperatures threaten the elderly in particular, some other groups, such as athletes, industrial workers, and military personnel are also at special risk of heat death (Yarbrough and Hubbard 1989).

Thermoregulation (Heat)

For humans, of all the processes for removing excess body heat and keeping deep tissues acceptably cool, sweating is the most important. In fact, humans are among the champion sweaters of the animal kingdom (Newman 1970). By conduction and **vasodilation**, heat moves from the body core to the skin. There, evaporation of 1 liter (about 1 quart) of sweat takes away 580 kcal of heat, and humans can sweat up to 4 liters an hour at the extreme. Sweat is produced by glands distributed densely over the body surface. However, for sweating to work well, deep body heat must surface at a high rate. This process is aided by the second cooling process: blood vessels that bring up the blood and rush it through the skin. An average person resting in the heat can push 8 liters of blood through the skin a minute and the dilation (widening) of surface blood vessels begins at the same time sweating begins (Yarbrough and Hubbard 1989). The hot skin also will radiate infrared energy away from the body.

Serious threats to survival in the heat are heat exhaustion and heat stroke. Heat exhaustion is marked by weakness, fatigue, frontal headache, impaired judgment, dizziness, faintness, thirst, nausea, vomiting, and muscle cramps. It sometimes strikes when the victim gets up from a seated or prone position, and fainting (syncope) may occur. With heat exhaustion, core temperatures are seldom >40 °C (104 °F) and the central nervous system is not seriously impaired. Usually, heat exhaustion results from dehydration and salt loss from sweating. It happens because sweating rapidly drains body water and because humans almost never voluntarily drink enough fluids to keep up with sweat loss. Dehydration reduces sweat rate and blood pressure, raises core temperature and heart rate, and inhibits normal cerebral function. Dehydration may lead to heat stroke, but recovery usually is spontaneous with rest and replacement of salt and fluids (Yarbrough and Hubbard 1989). Heat exhaustion and heat stroke, although separated in clinical diagnosis, are considered to be on a physiological continuum.

Heat stroke is a serious illness that will cause death fairly rapidly. Heat stroke victims typically survive the immediate crisis, but 70% die within 24 hours due to hemorrhage and disruption of cellular metabolic machinery. Sometimes, there are no warning symptoms; the thermoregulation system fails suddenly, causing collapse. Classic signs are body temperatures >41 °C (105.8 °F), dry skin, and disruption of the central nervous system as indicated by delirium or coma. Most heat stroke fatalities are among infants, the aged, and the poor during severely hot and humid weather; however, exertional heat stroke may be brought on during exercise in industrial workers and athletes. Heat stroke results from falling central venous pressure and depleted **plasma volume** (volume of the fluid portion of the blood) due to heavy sweating and the movement of fluids into muscles as they fight to retain electrolytes. This process reduces blood flow to the skin as muscle circulation and thermoregulation compete and core temperature rises. Finally, blood coagulates and the brain is damaged, leading to death (Werner 1993). Given the seriousness of this life threat, it is odd that we have not yet been able to detect genetically based differences in heat stroke resistance, comparing peoples from hot and cold regions. Either heat stroke resistance is not genetically based (removing it from the action of natural selection), or it is genetically based and found in all populations.

Heat Acclimatization

Although our knowledge of genetic factors in **heat resistance** is rudimentary, capacity for heat acclimatization is better explored. Like a high state of physical fitness, acclimatization also allows more physical work output while keeping a lower body temperature. One of the great advantages of full heat acclimatization is that one gets better at estimating the time to stop exercising at the beginning of thermal overload.

Working in heat for 10 days to 2 weeks produces better heat resistance due to a set of physiological acclimatizations. These acclimatizations result in a lower core temperature, less heat storage, a lower threshold at which sweating and surface vasodilation begin, and reduced heart rate and metabolic rate. Sweat becomes better distributed over the body (especially in arms and legs) and carries less sodium. Although the heart beats more slowly, it pumps more with each stroke, and blood plasma volume increases as well. Increased plasma volume is the key adaptation because avoiding dehydration is essential. There is no evidence that lifelong acclimatization to heat produces better response than short-term adjustment (Werner 1993). The adrenal hormone aldosterone plays some role, especially by increasing the reabsorption of sodium as sweat moves up the tubules that connect sweat glands to the body surface.

Heat acclimatization certainly makes people feel better subjectively and increases the capacity to maintain cooler body temperatures while working in heat. However, as Henane (1981) noted, acclimatization brings down body heat content only 1.4–3.0 °C (2.5–5.4 °F), depending on workload. As a result, most of our self-protection is left to our biological defenses, such as a high body surface area-to-weight ratio and long legs to keep our bodies away from the hot surface. Technical and behavioral safeguards are also essential (see later).

Variation in Heat Adaptation Within Populations

A huge amount of information and everyday experience shows that some people in any group tolerate heat poorly. Infants and the elderly, those with excess body fat, females, and people in poor states of physical fitness show less **heat tolerance** than others (Hanna and Brown 1979). Performance of subjects in a stressful 4-hour heat exercise test points up "normal" heat tolerance differences with some drama:

> When the 20 unacclimatized Caucasians in South Africa were exposed to the test, 10 failed to finish. Of those, 5 were at the point of collapse with rectal temperatures below 40 °C (104 °F). The Frenchmen in the Sahara Desert were, in general, better. Only two failed to finish [out of 15] …. Four [out of 15] Arabs failed to complete the tasks. [One was in distress; a second showed poor motivation.] Two Arabs withdrew at the end of the third hour. Both were near the collapse point; one vomited and the other had a rectal temperature of 39.7 °C (103 °F)—the highest recorded in the Arab group.
>
> (*Source:* Wyndham et al. 1964, 1052)

Variation in Heat Adaptation Among Populations

Because sweating is a primary human heat adaptation, one might hypothesize that the people best adapted should have more sweat glands (or at least have them better distributed). However, despite years of anatomical and physiological investigation, it seems that no human population has a clear genetic advantage in sweat production or control. All humans are born with about the same numbers of sweat glands, and "bringing them into production" is largely a matter of exposure to heat (Hanna and Brown 1979). The presence of large numbers of sweat glands and our ability to activate them seems to be a hallmark of our tropical origins.

The variation in human body shape—Bergmann's and Allen's Rules—shows that the hot tropics are inhabited by relatively thin people, that is, with narrow bodies and more surface area for a given amount of body weight. Good experimental evidence shows that this body shape reduces heat stress (Epstein et al. 1983; Ruff 1994) because it provides a large platform for heat radiation and sweat evaporation. Shvartz et al. (1973) demonstrated that after completing identical exercise in humid heat, men with relatively high surface areas for their weight stored less body heat than those with lower surface areas. Whether this response is a genetic adaptation to heat stress or a fortuitous byproduct of the generally poorer nutritional status of tropical populations is not clear. Moderate undernutrition produces smaller, thinner people, which could be an advantage in heat without being caused by heat selection. A classic study by Roberts (1953) found a fairly strong correlation of –0.60 between body weight and mean annual temperature. A recent replication used only data published since 1970 and found a correlation of only –0.26. The smaller correlation was due to a larger gain in body size in the tropical populations in recent decades relative to the temperate and cold populations (Katzmaryk and Leonard 1998).

Heat resistance is a physiological term usually defined as the ability to keep deep body temperature below dangerous levels when exposed to heat. Heat tolerance, more subtle in usage, implies greater capacity to function normally even when body core temperatures approach the upper limits. Tolerance is typically indicated by statements from subjects about how they feel and on the observer's sense of whether the subject is struggling to continue (even if the deep body temperature is the same as others who are not distressed).

Heat resistance comparisons among ethnic groups have produced complex results. Wyndham's experiments subjected people to extreme heat stress: 4 hours of moderate exercise at 34 °C (93 °F) and 90% relative humidity. Subjects were all acclimatized to heat (although there may have been differences in physical fitness).

Group	Ending Rectal Temperature	
	°C	°F
African Bantus	38.0	100.66
South African Whites	38.3	101.00
South African Bushmen	38.9	102.24
French Soldiers	39.0	102.36
Australian Aborigines	39.1	102.45
North African Arabs	39.2	102.54

(*Source:* Strydom and Wyndham 1963)

The North African Arabs and South African Bushmen, who had the thinnest body builds, did not keep the lowest core temperatures. However, the Australian Aborigines and the South African Bushmen, who were used to rigorous outdoor life in hot climates, tolerated the tests well, suggesting that core temperature was not the sole factor determining function in heat.

In an experimental test of Americans of African and of European ancestry, Riggs and Sargent (1964) matched the groups for body weight. The European Americans were taller, giving them a little more relative body surface area. The two groups came from areas of the United States with equivalent summer temperatures (a control for developmental and seasonal acclimatization). Core temperatures were nearly identical at test completion (African American = 38.0 °C [100.4 °F]; European American = 38.1 °C [100.6 °F]). However, signs of poor tolerance (irregular pulses, failure to finish due to distress, and clinical dehydration) were limited to those of European ancestry. The European Americans may or may not have been less acclimatized. However, the African Americans tolerated the heat better subjectively, and their sweat rate was also lower, an important finding because these groups were better matched in most respects than the Strydom and Wyndham series. Overall, it seems that tropical origin may not be a striking advantage in the heat when core temperature is the sole criterion of heat adaptation. However, for core temperatures <39 °C (at which people in real situations tend to operate), individuals with tropical ancestry may tolerate work with less heat distress.

Duncan and Horvath (1988) tested ethnic Chinese, East Indians, and Malays, all of whom had been born and raised in hot, humid Malaysia. The subjects exercised at ~50% of their maximum **aerobic** capacity for 2 hours in 34.9 °C (94.8 °F) heat with a relative humidity of 87%—a very stressful challenge. The following physiological patterns were recorded at the end of 2 hours:

Ethnic group	Heart rate, beats/minute	Core temperature °C	°F	Sweat loss, ml/hour/m^2 of body surface area
Malay	137.8	38.4	101.4	322
East Indian	150.3	38.4	101.4	335
Chinese	155.0	38.7	101.8	380

(*Source:* Duncan and Horvath 1988)

All three groups were acclimatized to heat and performed well (moderate core temperatures and completion of the 2-hour test). However, the Malays who worked at a little higher percentage of their maximum work capacity (i.e., they were closer to their peak capacity and produced more body heat than the other ethnic samples) maintained a significantly lower heart rate. This is a good marker of less heat strain.

Interestingly, the Malays have the longest evolutionary history in wet heat and the Chinese the shortest. So, the Malays have a hint of advantage in tropical origins among equally well-acclimatized groups, but the results of all studies considered are far from clear. This issue is still a major research area in anthropology, especially in terms of interactions among ethnicity, body build, physical fitness, nutrition, behavioral patterns of pacing and work scheduling, and evident tolerance for work in heat. Industrial ergonomists, who study factors relating to human work efficiency, have made substantial progress on this topic in first world workers. An entire volume of the journal *Ergonomics* is devoted to this problem (Volume 38, pages 1 to 192, 1995).

Cultural Adaptation to Heat

Heat stress usually elicits behavioral responses, just as cold stress does. The general rules for reducing cold stress apply to heat, such as avoidance behavior, special shelter, and clothing. Roosters awaken tropical villagers early, and it is good that they do. Strenuous work should be done before the heat of the day, in the evening, or during cooler periods. Scheduling activities to avoid heat is a major defense against it, and periods of midday rest (siestas) reduce metabolic heat buildup. Staying out of the sun and avoiding overheated places also help, so trees, umbrellas, and open but roofed areas are common in hot zones. Houses and clothing (especially in sunny areas) are insulated to resist heating, but clothing is often scanty in shady places such as forests or is designed to evaporate sweat in climates with a lot of sun exposure.

Moving air—ventilation—is probably one of our most important cultural heat adaptations; it enhances sweating as a primary cooling mechanism. However, early human evolution occurred in the tropics under heat stress conditions; therefore, heat adaptation is ancient (Newman 1970). Even now there may not be much directional natural selection by heat stress (Hanna and Brown 1979) because mortality, at least, strikes many people beyond reproductive age or results from behavior that is not genetically driven. The relation of individual heat resistance to fertility is an extraordinarily interesting research frontier, particularly in view of the heat sensitivity of human sperm.

SOLAR RADIATION

The effect of solar radiation on human biology depends on the radiation wavelength and the chemical composition of the cells of directly exposed organs: the eyes and the skin. The wavelength spectrum of solar radiation that reaches the earth includes invisible infrared (>750 nanometers), **visible** (400–750 nanometers), and invisible **ultraviolet** (UV) (<400 nanometers) wavelengths. A biological effect occurs when a **photochemical reaction** (one initiated by electromagnetic light waves or particles) is initiated in a cell that contains chemical compounds called **chromophores**, which absorb light at the incident wavelength. If a cell does not contain chromophores, then the radiation is transmitted. Considerable scientific attention has focused on UV radiation and chromophores in human skin.

UV Solar Radiation

UV radiation accounts for ~9% of the total solar energy output; its intensity at the earth's surface varies daily, seasonally, and with latitude and altitude. It is highest at noon, in the summer, at the equator, and at higher altitudes. In addition, UV radiation intensity is decreased irregularly by factors such as cloud cover and air pollution, and increased irregularly by reflective surfaces such as water and snow. UV radiation is subdivided into **ultraviolet A** (UV-A) in the 315- to 400-nanometer range and **ultra-violet B** (UV-B) in the 280- to 315-nanometer range.

Figure 6.9 illustrates seasonal and latitudinal variation in solar radiation intensity at the lower and upper wavelengths of the UV-B range. The difference between the y-axes of the two panels illustrates that the intensity at the lower wavelength is several degrees

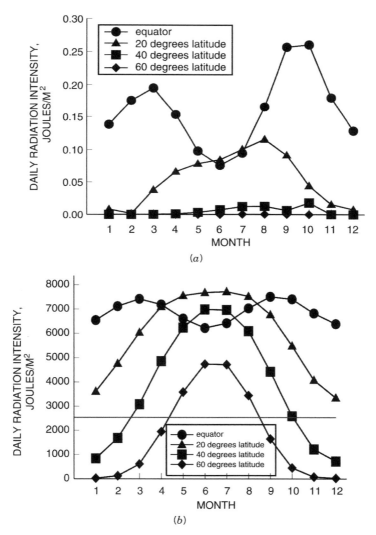

Figure 6.9   Seasonal and latitudinal variation in solar radiation intensity at the lower (a; 290 nanometers) and upper (b; 315 nanometers) ends of the UV-B range. (*Source:* Data from Johnson, Mo and Green 1974, 180, 183)

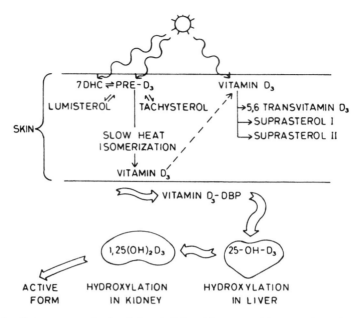

Figure 6.10   Cutaneous synthesis of vitamin D3 and its removal from, or degradation in, the skin. (*Source:* Webb 1993, 188; reprinted with permission from the author and Plenum Publishing Corporation)

of magnitude lower than at the higher wavelength. For example, during September and October, when the UV-B radiation intensity at the lower wavelength (Fig. 6.9a) is ~0.25 joules/m^2/day, it is ~7500 joules/m^2/day at the higher wavelength (Fig. 6.9b).

UV-B radiation is most abundant during the summer months, and the seasonal contrast is more marked at the higher wavelength. For example, the highest monthly intensity during June and July in Cleveland, Ohio, at 41° north latitude, is about 10 times that during December and January at the 315-nanometer wavelength. Daily UV-B radiation intensity is higher at lower latitudes. For example, the highest monthly intensity in Cleveland, Ohio, is about the same as the lowest monthly intensity in Quito, Ecuador. These comparisons illustrate that UV-B radiation reaching the earth's surface varies widely. The high extreme of UV-B radiation occurs at the lower latitudes during the summer months and the low extreme of UV-B radiation occurs at the higher latitudes during the winter months.

This climatic variation is relevant to human biology because UV-B initiates a vital biochemical process in the skin: vitamin D synthesis. Vitamin D is essential for maintenance of the skeleton because it promotes intestinal absorption of dietary calcium and maintenance of serum calcium levels (Holick 1994a, 1994b). Vitamin D synthesis is initiated when the chromophore 7-dehydrocholesterol (7DHC) in the cells of the **epidermis**, the outermost layer of the skin, absorbs **photons** (particles of radiant energy) in the UV-B range of 290–315 nanometers and photoconverts to

previtamin D. Previtamin D is thermally converted into vitamin D3 or photocon-
verted back to 7DHC or to inactive lumisterol and tachysterol (Figure 6.10). Vitamin
D3 is bound to vitamin D-binding protein for removal from the skin into the bloodstream for subsequent transport to the liver and kidneys, where additional biochemical reactions produce active vitamin D.

UV-B–initiated synthesis has been the main source of vitamin D throughout human evolution and remains the sole source for most of humanity today, because dietary sources of vitamin D such as oily fish and fish liver oil are uncommon.
High-intensity UV-B radiation does not lead to vitamin D overproduction because
7DHC photoconversion to previtamin D3 reaches an equilibrium at about 10–15%
conversion. As a result, previtamin D and vitamin D3 levels do not increase indefinitely with higher UV-B intensity. In contrast, low-intensity or no UV-B radiation
results in little or no production of previtamin D3, possible vitamin D deficiency,
and adverse consequences for skeletal health.

The horizontal line in Figure 6.9b approximates the minimum radiation intensity
necessary to initiate photoconversion of 7DHC to previtamin D3 at the most abundant wavelength in the effective range. Daily radiation intensity is below this threshold for months at a time at some latitudes. This **"vitamin D winter"** may be as short
as 1 month at 25° north or south latitude and as long as 8 months at 60° north or south
latitude. All of Europe and Canada and substantial parts of Asia have at least 4
months of vitamin D winter each year when adequate vitamin D must be obtained
from body fat stores or diet to prevent a vitamin D deficiency.

Evidence shows that vitamin D levels of healthy people of all ages covary with
UV-B radiation intensity, that is, tend to be lower at higher latitudes than at lower lat-

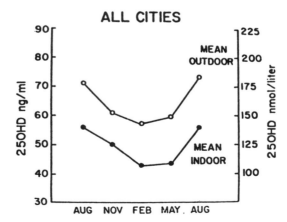

Figure 6.11   Average blood levels of 25-OH vitamin D (the principal circulating form of vitamin D and used as an indicator of vitamin D status) at four seasons of the year in healthy adult
middle-aged males working in Boston, Detroit, Seattle, Des Moines, and Palm Beach. (*Source:*
Neer 1985, 17; reprinted with permission from the author and the Annals of the New York
Academy of Sciences)

itudes and lower during the winter than during the summer. For example, the mean umbilical cord blood vitamin D levels of Chinese babies born at 40 and 47° north latitude were only about one-third the level of their counterparts born at 22 and 30° north latitude. Similarly, the mean blood vitamin D levels of men in Boston and Seattle (42 and 47° north latitude, respectively) were one-half those of men in Palm Beach (25° north latitude) (Neer 1985). Vitamin D levels vary seasonally at these latitudes, too. Chinese babies born in the spring, after maternal vitamin D winter, had lower vitamin D levels in umbilical cord blood than those born in the fall, after maternal summertime UV-B exposure. Winter blood levels of vitamin D among men working indoors and outdoors in five U.S. cities were ~20% lower than during the summer (Figure 6.11). A winter/summer contrast of similar magnitude was observed among elderly women in Maine (46° north latitude) (Rosen et al. 1994).

Latitudinal and seasonal variation in indicators of calcium **homeostasis** reveal functional consequences of variation in UV-B and vitamin D levels. One measure is the number of **centers of ossification** (points at which bone formation occurs) in the wrist. These centers indicate calcium deposition in the growing bones of the wrist. Wrist ossification centers were more frequent in southern Chinese newborns than in northern newborns and were most frequent among babies born in the fall. Calcium retention was better in middle-aged Norwegian men during the summer than in the winter because of better vitamin D–mediated calcium absorption (Neer 1985). The elderly Maine women lost bone mineral during the winter, not the summer (Rosen et al. 1994). The incidence of hip fractures among US adults over 65 increases significantly during winter at all latitudes (Holick 1995).

These examples illustrate that seasonal variation in UV-B is associated with—probably causally—seasonal variation in mean blood levels of vitamin D in males

---

BOX 6.2   SKIN COLOR MEASUREMENT

Our skin has major adaptive functions, such as protection and heat regulation. One protective function is blocking or transmitting ultraviolet radiation. Transmission is regulated partly by skin pigments.

Skin pigmentation, which produces what we perceive as skin color, can be quantified by use of a reflected light meter, or photoreflectometer. As the amount of skin pigment increases, the amount of light that the skin reflects decreases. The photoreflectometer is used with color filters to estimate different pigments and different amounts of pigments. For example, filters in the red part of the spectrum produce a red light beam from the instrument. These waves are absorbed by the red pigments in the skin (especially hemoglobin in blood vessels) so that only nonred pigments such as melanin determine the amount of reflectance. On the other hand, with a green filter, the skin will "look" darker, and larger amounts of both hemoglobin and melanin will decrease reflectance. Use of both filters in sequence allows us to estimate overall skin reflectance and to separate out the effects of hemoglobin and melanin.

Measurement

In Figure 6.12, an anthropologist (A. T. Steegmann, Jr.) holds a search unit (which produces filtered light and senses its reflectance) against the volunteer's forehead. He is careful to hold the unit flush to the skin without pressure and keeps back stray strands of hair (which reduce reflection). The forehead, a site exposed to sun, is usually compared with the unexposed inside of the upper arm to estimate the amount of tanning. Percent reflectance is read from the dial on top of the instrument.

Photoreflectometer (Photovolt Corp., Model 610)

The instrument works as follows. In the tubular search unit body, a bulb produces light. Two condenser lenses direct that light through a glass filter. Part of the outgoing parallel beam strikes a compensating photocell, and the rest hits the skin to be diffusely reflected back to a second (measuring) photocell. Because the two photocells are calibrated to give a 100% reflectance reading on a completely white reflection surface, the difference between the current from the two cells when measuring any other surface is read by galvanometer. The voltage is expressed on the instrument dial as percent reflectance. The reflectometer works on standard alternating current but is probably more accurate and easier to use on battery power.

With each change of filter, the machine is recalibrated, usually using an enamel plaque of known reflectance. Each type of reflectometer has its own set of filters to transmit at various wavelengths. The filters supplied with the machine shown in Figure 6.12 have peak light transmission values of 430 nanometers (blue), 535 nanometers (green), and 650 nanometers (red). The red value is close to that of hemoglobin.

Figure 6.13 summarizes results obtained with another widely used machine, the EEL (Evans Electroselenium, Ltd.). It uses a set of nine filters, including one called the 609 filter, which has a peak light transmission value of 685 nanometers. This wavelength is in the range associated with the largest differences in percent reflectance depicted in Figure 6.13 and has been especially meaningful for quantifying variation in skin color and skin chromophores.

and females from birth to old age at latitudes as low as 22° north and that functional skeletal consequences result. Evolutionary logic suggests that individuals at the low end of the normal range of variation of vitamin D synthesis or blood vitamin D levels would be at higher risk of these skeletal consequences and their associated higher risk of **morbidity** and mortality. Traits that enhance vitamin D status would be selected for and selection would be strongest in environments where vitamin D status would be most stressed by low UV-B radiation.

Melanin, Vitamin D Synthesis, and UV-B Radiation

The major trait influencing vitamin D synthesis is the amount of epidermal **melanin**. Melanin is a chromophore in the epidermis that absorbs solar radiation across a wide range of wavelengths. Solar radiation that enters the skin is either absorbed by the chromophores or transmitted. Absorption in the visible range accounts for skin color. Melanin competes with 7DHC for absorption in the UV range and is the major factor influencing the transmission of UV radiation in the epidermis and **dermis** (i.e., the outermost vascular layer of the skin, just below the epidermis). About 29% of UV-B solar radiation penetrates to the basal layer of the epidermis and the dermal vasculature of "Caucasians" compared to only 7% among "Blacks" (Kaidbey et al. 1979). (Quotation marks here and elsewhere indicate terms used by the original authors to denote the populations from which the study samples were drawn).

A portion of transmitted radiation is also reflected back through the skin. The percentage of incident radiation reflected back through the skin is a measure of the amount of melanin in the epidermis: the more melanin, the more absorption and the less reflectance of solar radiation. This measurement, called the **percent skin**

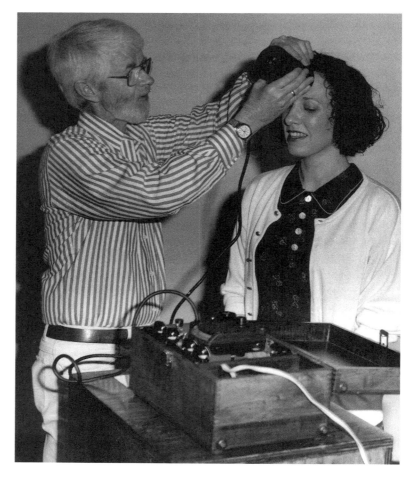

Figure 6.12 Measurement of skin reflectance.

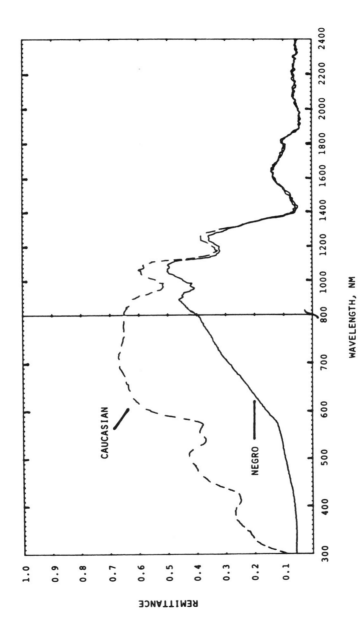

Figure 6.13   Skin reflectance of "dark Negroid" and "fair Caucasian" skin. (*Source:* Anderson and Parrish 1982, 166; reprinted with permission from the authors and Plenum Publishing Corporation)

reflectance, is used to quantify the amount of epidermal melanin (see Figure 6.12). Figure 6.13 illustrates the lower percent skin reflectance of "dark Negroid" skin in the visible (400–750 nanometers) and the UV radiation range, including UV-B (<315 nanometers), compared with "fair Caucasian skin."

A cline among human populations from heavily to lightly melanized skin corresponds to increasing latitude. The cline is exhibited by indigenous populations of three major geographic areas: Africa and Europe, Asia and the Pacific, and the New World. Figure 6.14 presents the percent skin reflectance of indigenous populations throughout the world. All the populations with percent skin reflectances <35% are found within 20° of the equator; populations with intermediate percent skin reflectances (40–60%) are found in the middle latitudes roughly 20–40° from the equator; and populations with the highest percent skin reflectances (>60%) are found above 40°north or south latitude. The populations with the lowest UV-B radiation intensity and longest vitamin D winters have the least melanin.

Melanin is synthesized in specialized cells called **melanocytes** that are interspersed with basal cells in the basal layer of the epidermis, at the junction of the epidermis and dermis (Figure 6.15). Figure 6.16 illustrates that lightly melanized populations have fewer melanocytes. The figure represents a trend rather than a precise relationship because it is a compilation of many studies, most of which did not report skin reflectance or latitude of the study sample. Populations 1–6 are Europeans in Europe and their descendants in the United States and Australia, all indigenous to high latitudes, where populations have relatively high skin reflectances (as indicated in Figure 6.14). Although there is variation among these six samples, as a group, they have fewer melanocytes than samples from populations native to latitudes within 15–30° of the equator (populations 7–10), where populations have low skin reflectances. Samples 11

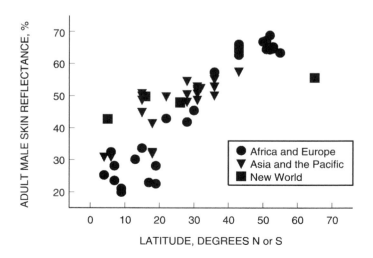

Figure 6.14   Skin reflectance increases with increasing latitude. (*Source:* Data for the 609 filter [dominant wavelength 685, thought to best measure melanin content in skin] from Robins 1991, Table 7-2)

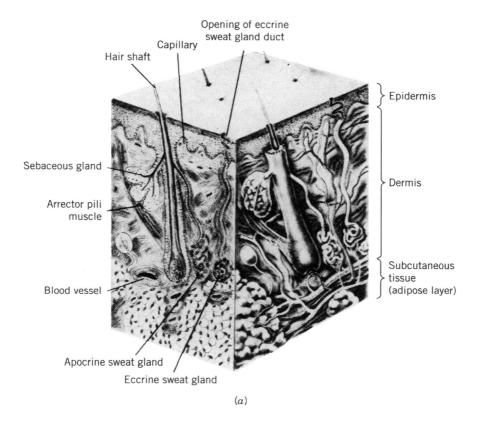

Hair shaft

Capillary

Opening of eccrine
sweat gland duct

Sebaceous gland

Arrector pili
muscle

Blood vessel

Apocrine sweat gland

Eccrine sweat gland

Epidermis

Dermis

Subcutaneous
tissue
(adipose layer)

(a)

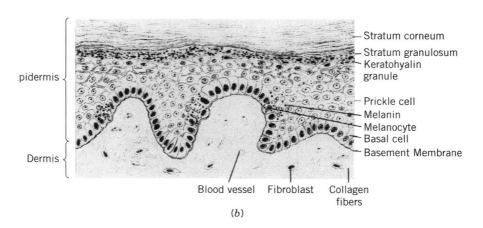

Epidermis

Dermis

Stratum corneum

Stratum granulosum

Keratohyalin
granule

Prickle cell

Melanin

Melanocyte

Basal cell

Basement Membrane

Blood vessel   Fibroblast   Collagen
fibers

(b)

Figure 6.15 Diagrammatic cross section of normal skin (a) and epidermis (b). (*Source:* Parrish 1982, 5, 6; reprinted with permission from the author and Plenum Publishing Corporation)

and 12, with the most melanocytes, are Asian and Native American samples of unspecified geographic origin. This figure suggests that the common textbook assertion that the number of melanocytes is not different between populations is inaccurate, at least for the forearm, where these measurements were made.

Each melanocyte synthesizes melanin into small organelles called **melanosomes,** which are secreted into some three dozen neighboring skin cells called **keratinocytes.** Complexes of small melanosomes aggregate in organelles called lysosomes in the keratinocytes of lightly melanized "white" skin. Larger, single melanosomes are distributed throughout the keratinocytes of heavily melanized "black" skin. Keratinocytes are continuously formed and lost. As keratinocytes move from the basal layer to the surface of the skin, melanosomes and other organelles degrade, and the keratinocytes of the outermost layer are shed.

Melanin competes with 7DHC for UV-B photons: the more melanin, the less 7DHC photoconversion to previtamin D3 from a given solar radiation dose. Laboratory studies provide evidence that individuals with little melanin are likely to maintain better vitamin D, calcium status, and skeletal health in environments with low UV-B radiation intensity because more photons penetrate to the lower layers of the epidermis where 7DHC photoconversion to previtamin D3 occurs. The result is that previtamin D3 and vitamin D can be synthesized at lower UV-B intensity and after shorter exposure. For example, a single brief exposure caused a nearly 10-fold increase in serum vitamin D in two "lightly pigmented Caucasians" but no change in three "heav-

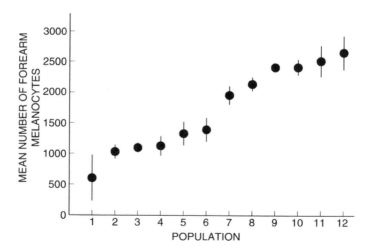

Figure 6.16   Number of forearm melanocytes in populations indigenous to high and low latitudes. Populations 1 and 5 are Australian, 2 and 3 are British (from the United Kingdom), 4 is Swedish, 6 is American "White," 7 is African American, 8 is Australian Aborigine, 9 and 10 are Solomon Islanders, 11 is Native American, and 12 is Asian. (*Source:* Compiled from Szabo 1954; Staricco and Pinkus 1957; Mitchell 1963; Szabo 1967; Mitchell 1968; Szabo 1969; Garcia 1977; Rosdahl and Rorsman 1983; Scheibner et al. 1988)

ily pigmented Negro volunteers" (Clemens et al. 1982). When "Caucasian" and "Black" skin was exposed to the same UV light source, Caucasian skin began synthesizing previtamin D3 much sooner—in 2 minutes, compared with 16 minutes in one study. In another study, the time to maximum previtamin D3 formation was 30–60 minutes, compared with 180–210 minutes (Holick et al. 1981).

Exposing "White" and "Black" volunteers to the same experimental UV light source over 6 weeks caused both to increase blood levels of vitamin D. However, the "White" volunteers consistently synthesized more: Their blood levels were 30–130% higher than those in the "Black" volunteers (Brazerol 1988). This pattern of faster and greater vitamin D synthesis would be advantageous at high latitudes because lightly melanized individuals could continue to synthesize vitamin D in the months of low-intensity UV-B radiation, before and after vitamin D winter. Furthermore, actual UV radiation intensity is reduced by cloud cover, indoor and shaded activity, and clothing. Individuals with the ability to respond quickly to brief UV-B stimulus— lightly melanized individuals—would have more favorable blood levels of vitamin D and better skeletal health.

In contrast, heavily melanized individuals would synthesize less vitamin D in such environments and would be at risk of short- or long-term vitamin D deficiency and its sequelae. For example, heavily melanized populations indigenous to low latitudes who have migrated during the past few centuries to high latitudes with low UV radiation are at increased risk of poor vitamin D status. The blood levels of vitamin D in African American children and children from the Indian subcontinent living in the United Kingdom are lower than those of indigenous children, despite equal exposure to UV radiation. The former children are at the low end of a range of variation in blood levels of vitamin D, centered around lower means than those of children indigenous to the high latitudes.

Vitamin D is essential for calcium homeostasis, which in turn is essential for growth of the skeleton in length, volume, and weight as well as for **skeletal maturation**, modeling into its definitive form, and continuous modeling and repair throughout the life cycle. The pathological consequences of inadequate production of vitamin D are **rickets** in children and **osteomalacia** in adults. These deficiency diseases represent the "tip of the iceberg" of more frequent subclinical deficiency that would probably be of equal evolutionary significance under a regime of **directional selection**.

During growth and development, vitamin D deficiency is called rickets. Low blood levels of vitamin D indicate serum rickets and can lead to inadequate calcification of bone matrix and soft bones. Normal bone growth and remodeling are compromised, resulting in large, knobby joints and stunted linear growth. In severe cases, the body weight of the standing child can cause deformities in the softened legs, pelvis, and spine. Muscle pain and weakness make it difficult for children to support their own body weight and move normally. Children with rickets have a higher mortality rate than other sick children. Although many children survive rickets, they do so with greater risk of skeletal deformity that can limit their ability to perform physical work. Women with rachitic pelvic deformities have a higher frequency of difficult delivery as well as maternal and **neonatal** death. These severe outcomes are clearly maladaptive. Moderate disease would also be disadvantageous because muscle weakness and skeletal deformity could limit everyday function and, among women, successful childbirth.

Adult serum rickets may lead to osteomalacia, a decalcification and softening of existing bone that is a risk factor for bone loss, bone fracture, and higher mortality rates. Bone pain, tenderness, muscle weakness, **anorexia**, and weight loss accompany osteomalacia. Women with poor vitamin D status give birth to vitamin D–poor newborns and produce milk that contains low vitamin D levels. Thus, poor vitamin D status can lead to increased risk of morbidity and mortality and lower completed fertility. All other things being equal, populations with high melanin concentrations who live under low UV radiation would be at higher risk of having poor vitamin D status than those with low melanin concentrations.

This evidence is consistent with the hypothesis that natural selection acted to produce and maintain the cline in melanin concentration (Figure 6.14) by directional selection for low melanin production in higher latitudes. The fossil record indicates that early **hominids** lived within 20° latitude of the equator under intense, year-round UV-B radiation for millions of years before moving into the middle and high latitudes. The early hominids probably were heavily melanized. Modern great apes (orangs, gorillas, and chimpanzees) have heavily melanized epidermis, which suggests that the early pongids and hominids inherited this trait from their common ancestor. Subsequent northward migration of *Homo erectus* and *Homo sapiens* exposed hominids to less intense UV-B radiation, vitamin D winter, and natural selection for traits that lead to a better vitamin D status.

Directional selection would have favored **genotypes** with less melanin that could respond quickly and vigorously to low and intermittent doses of UV-B radiation and thereby maintain vitamin D and calcium levels and skeletal integrity. Rickets and osteomalacia (from mild to severe) would have been selective factors against high melanin concentrations in high latitudes. Individuals in the lower range of vitamin D synthesis would have been at high risk of poor skeletal health and function and have higher mortality and lower fertility rates.

This explanation is contrary to the conclusion of a review of the literature, in which Robins (1991) argued against the importance of UV radiation and vitamin D synthesis as factors selecting for light melanization and skin color at high latitudes. Instead, Robins argued for the importance of cold temperatures and cold injury as selective factors: "The least likely backdrop from which rickets would emerge was the open-air, hunter-gatherer lifestyle of naked or semi-naked *Homo* on the Eurasian and North American landscapes during the Pleistocene" (Robins 1991, 207).

Robins favored the frostbite hypothesis for the evolution of lightly melanized skin. His arguments are not convincing. First, he treated rickets as a categorical variable (i.e., individuals are either rachitic or healthy). Instead, rickets is a biomedically defined condition based on a cutoff value in a continuous distribution of blood levels of vitamin D and on skeletal manifestations.

There is no reason to assume that natural selection respected the boundaries defined by modern biomedicine and that it acted on one side of the boundary and not the other. In a population with a mean and standard deviation of blood levels of vitamin D or bone health and other things being equal, individuals with darker pigmentation are going to be in the lower ranges of that distribution and at higher risk of morbidity and mortality and lowered fertility. Natural selection for better vitamin D status would have favored less melanin production directionally proportional to the level of vitamin D production rather than in the all-or-nothing fashion assumed implicitly by Robins. Second, his assumption that early hominids at high latitudes were unclothed and thus continuously exposed to UV-B radiation is unwarranted. The highest mean monthly temperatures at the latitude of London (~52° north) are 26–27 °C (79–81 °F) during July and August, and the next highest mean monthly temperature is 17 °C (62 °F). Thus, the mean annual temperature in northern Europe was not compatible with nakedness or near-nakedness, even during the summer, when UV radiation is most intense. Third, Robins argued inconsistently that the early hominids were naked and producing plenty of vitamin D because of the mild climate and were simultaneously subject to selection by cold stress. Finally, Robins cited two lines of evidence to support the frostbite hypothesis for the evolution of light melanization at high latitudes (Robins 1991, 208–209): increased susceptibility to cold injury among "Negroids" compared with "Caucasoids" and of pigmented compared with nonpigmented skin of piebald guinea pigs.

With respect to the first line of evidence, scientists demonstrated in the 1950s that the increased susceptibility to cold injury among "Negroids" was due to different vascular responses to cold (e.g., Meehan 1955); that is, there was no causal relationship with melanin. With respect to the second line of evidence, the cited experiments have no apparent relevance to humans. Guinea pig skin was frozen with Freon or liquid nitrogen and then thawed, and the damage was assessed (Post et al. 1975). The temperature of Freon is –30 °C (–22 °F), and the temperature of liquid nitrogen is –210 °C (–346 °F). The temperature for Freon was rarely encountered and that for liquid nitrogen was never encountered in the natural hominid environment. Human skin freezes at much higher temperatures than those experimental ones, and the experimental temperature extremes are not relevant to human evolution. Thus, the frostbite hypothesis for the evolution of light melanization at high altitudes was not supported by the cited evidence. The UV-B radiation and vitamin D hypothesis is much stronger because the biological links among UV-B, vitamin D, skeletal health, fertility, and mortality are both plausible and supported by empirical evidence.

It is useful to consider the factors favoring maintenance of heavy melanization in the lower latitudes. Early hominids at low latitudes with high UV-B stress would have benefited from the high melanin concentration; therefore, stabilizing selection would have maintained this ancestral characteristic. One cost of this adaptation was greater absorption of solar radiation and a greater heat load under direct solar radiation; however, this cost was outweighed by the benefits.

Several lines of evidence indicate why high-melanin-producing genotypes would have been favored under high UV-B stress. Lightly melanized populations indigenous to high altitudes who have migrated to low latitudes during the past few centuries offer a natural experimental test of the hypothesis that they have lower fitness in an environment with high UV-B intensity. The population of Australia, at 15–30° south latitude, is descended mainly from people who came during the past few centuries from the British Isles, at 50–60° north latitude. Australia has the highest incidence of skin cancer in the world. Figure 6.17 illustrates that the 1970 US population, largely composed of descendants of Europeans from 50° north latitude or higher, exhibited a latitudinal gradient of skin cancer incidence and mortality that paralleled the monthly dose of UV-B. On the whole, skin cancer mortality rates are not very high in the United States today; however, the current fatality rate of more than 50% for advanced skin cancer would have prevailed prior to developing the sophisticated medical treatments available now in the developed countries. Thus, selection could have eliminated genes for skin cancer susceptibility.

More heavily melanized populations have less skin cancer at the same latitudes. For example, the incidence of nonmelanoma skin cancer (i.e., cancer of the basal and squamous cells of the epidermis [Fig. 6.15]) in Texas was 5, 22, and 113 cases per

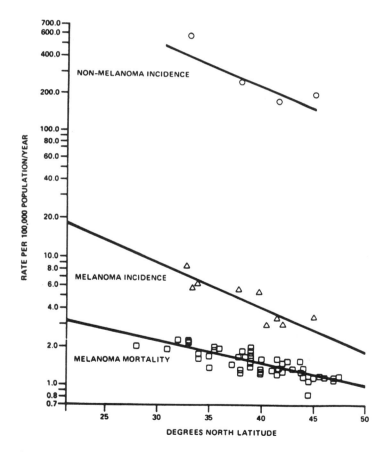

Figure 6.17 Latitudinal gradient of skin cancer incidence and mortality in the US population in 1970. (*Source:* Fears et al. 1976, 462, Copyright © 1976 American Public Health Association; reprinted with permission from the American Public Health Association)

10,000 population per year among Nonwhite, Hispanic, and Anglo residents, respectively (Weinstock 1993). The importance of melanin in decreasing the risk of skin cancer is emphasized further by the example of albinos, who do not synthesize melanin, in the tropics. These individuals inevitably develop fatal skin cancer by their 20s or 30s. This evidence indicates that a higher risk of skin cancer would have acted to keep the frequency of genes for light melanization low. Exposure to UV-B would have resulted in higher mortality rates throughout adulthood and selection against mutations for less melanin.

The link between skin cancer and UV-B radiation is **DNA** damage. The sequence of events is well understood for nonmelanoma skin cancer. UV radiation mutates the p53 tumor suppressor gene (inhibits cancerous cell growth and division) and then promotes unrestrained growth of the damaged cells (Leffell and Brash 1996). Furthermore, UV-B can suppress the immune system by depleting the number of specialized immune system cells called Langerhans cells in the epidermis. This suppression can result in inadequate response to malignant cell growth and to other stresses on the immune system (Vermeer and Hurks 1994).

In addition to causing skin cancer and depleting the immune system, the photodegradation (i.e., photochemical conversion to less complex compounds) of nutrients such as vitamin D and folates in the skin and dermal circulation has potentially pathological consequences. UV can photodegrade vitamin D3 into other compounds before its removal from the skin. Hypothetically, this process could cause vitamin D deficiency even if synthesis were adequate (Webb et al. 1988). Folates are also degraded by UV-A radiation in the 360-nanometer range. Folate deficiency is associated with elevated intrauterine mortality. Heavy melanization hypothetically decreases the risk of nutrient photodegradation by high UV radiation intensity, but this interesting possibility has not been studied. Thus, several factors explain why genes for low melanin synthesis would have had lower fitness at low latitudes.

High epidermal melanin content is protective against photodegradation of nutrients, immune system depletion, and DNA damage in environments with intense UV-B radiation. Low epidermal melanin content protects against vitamin D deficiency and its sequelae in environments with low UV-B radiation.

Vitamin D Winter, Lactase Persistence, and Calcium Metabolism

Another trait becomes important in populations exposed to prolonged vitamin D winter when UV-B radiation is absent and melanin is irrelevant. The genetic **polymorphism** for **lactase persistence** appears to be another means of maintaining or enhancing calcium homeostasis in populations under severe low UV-B stress.

The intestinal enzyme **lactase** hydrolyzes **lactose**, the **disaccharide** in milk, into **monosaccharides** that can be absorbed into the bloodstream, where their calories are available for metabolism. From birth to around 5 years of age, almost all individuals in all human populations produce lactase and can absorb the sugars from lactose. In most populations, the production of lactase decreases by ~90% around 5 years old, and thereafter, individuals do not hydrolyze lactose or absorb the sugars. The exceptions occur in a few populations, mainly of northern European residence or origin, in which most individuals maintain high levels of lactase throughout the life cycle. This trait is inherited as an **autosomal dominant** trait at a locus on **chromosome** 2, and the **phenotype** is called lactase persistence.

The recessive phenotype is called **late-onset lactase deficiency**. Drinking milk may cause the person with this trait to suffer a variety of symptoms, such as abdominal bloating and pain, flatulence, and diarrhea because the undigested lactose is metabolized by bacteria in the large intestine that produce hydrogen. This condition is termed **lactose intolerance**. Not all late-onset lactase deficiency phenotypes suffer symptoms of lactose intolerance after drinking milk, partly because chronic exposure (e.g., 10 days) to lactose results in natural selection and adaptation of the bacteria in the large intestine. Regardless of symptoms of lactose intolerance, individuals who do not produce lactase do not absorb the sugars (see also Chapter 9).

The geographic coincidence of populations with very high prevalence of lactase persistence and those with low UV-B exposure and long vitamin D winters above 40° north or south latitude is illustrated in Figure 6.18. Two lines of evidence, neither completely conclusive, indicate that individuals with lactase-persistent phenotypes would be favored in such environments because they would have better calcium status. One line suggests that lactase-persistent phenotypes absorb calcium better in the presence of lactose than in the presence of other sugars. Thus, calcium status would be enhanced by drinking lactose-rich milk rather than consuming cheese or yogurt, in which the lactose has been broken down into simple sugars by bacteria. Another line of evidence suggests that lactase-persistent phenotypes generally absorb calcium better than late-onset lactase deficiency phenotypes. Natural selection would favor the lactase persistence gene under low UV-B conditions because it confers an advantage when calcium homeostasis is precarious.

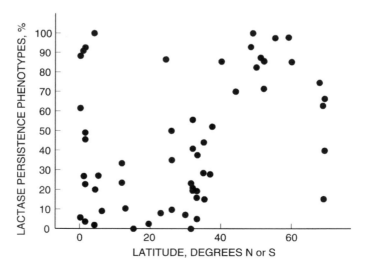

Figure 6.18   Percent lactase persistence and latitude. (*Source:* Compiled from data in Durham 1991, 264).

Figure 6.18 illustrates that the populations above 50° latitude have 75% or more lactase persistence phenotypes, whereas those within 30° latitude have <25% lactase persistence phenotypes. There are two exceptions to this generalization. First, a few equatorial populations have high prevalence of lactase-persistence phenotypes. They are African pastoralists who rely on milk for a substantial proportion of their calories at some times of the year. Under those conditions, a caloric benefit would accrue to lactase-persistence phenotypes who can metabolize the lactase sugar. Second, a few very high latitude populations have a low prevalence of lactase-persistence phenotypes. They include Eskimos and Lapps. The Eskimo diet contains dietary vitamin D sources in fish; therefore, natural selection would not operate so strongly on biological traits that enhance calcium status. The reason for the Lapps' low prevalence of lactase persistence is not known. However, calcium intake is low and calcium homeostasis is poor in both Eskimos and Lapps (Shephard and Rode 1996).

Prolonged vitamin D winters appear to have selected for a high frequency of the allele for lactase persistence in indigenous populations where it enhances calcium status.

Acclimatization to Seasonal Variation in UV-B Radiation Intensity

Populations at medium and high latitudes are seasonally exposed to high-intensity UV-B and have adaptive responses. Sunburn, until recently thought to be a solely pathological response to a high dose of UV radiation, is now viewed as a result of the skin's adaptive response. Under certain conditions, skin cells with UV-damaged DNA initiate cell death or **apoptosis**. The dead cells are responsible for the skin peeling that follows sunburn (Kamb 1994). Peeling and potential compromise of the protective barrier function of the skin is a cost of this adaptive response. The greatest sunburn efficiency is at the 290- to 320-nanometer wavelength, which is very similar to that for previtamin D3 photoconversion. The smallest dose of UV radiation necessary to produce sunburn (the **minimal erythemal dose** [MED]) varies widely. The more melanin, the less chance of sunburn from a given UV dose and the higher UV dose necessary to elicit sunburn. One study reported that the MED was 13 times higher in "Black" than "Caucasian" study participants (Kaidbey et al. 1979). The implication is that individuals with more melanin sustain less DNA damage and have fewer cells undergoing apoptosis.

Another adaptive response to UV-caused DNA damage is tanning—an increase in melanin content on exposure to UV radiation. UV can disrupt normal bonds between DNA **nucleotide bases**, which then form new bonds between adjacent **pyrimidine** bases called pyrimidine dimers that can result in mutations. The dimers may be eliminated by a process called nucleotide excision repair, which replaces the disrupted bases with intact bases and generates a fragment of DNA that contains the damaged nucleotides. These fragments are a signal to increase melanogenesis (Eller et al. 1994). The increase in melanin increases the absorption of UV radiation and limits further DNA damage. Thus, suntanning is simultaneously a sign that DNA damage has been incurred and repaired and an adaptive response against further damage.

These observations suggest why poor suntanning ability is a risk factor for skin cancer: It is an indicator of failure to repair UV-caused DNA damage.

Tanning is initiated in the course of repair of UV-damaged DNA. However, the increase in melanin itself has a protective effect because it increases the absorption of damaging UV-B. After repeated sun exposure, the number of melanocytes, their branches, and melanin synthesis all increase. Skin thickness also increases. These responses all decrease the transmission of UV-B to the basal epidermis and dermis. However, tanning does not achieve the same melanin level as inheritance of genes for high melanin production (tanning only increases the MED two to three times) and thus provides less protection against UV-B than genes for high production.

Behavioral Adaptations to UV Radiation

The major behavioral adaptation to low UV-B radiation intensity is a diet containing vitamin D. It would be predicted to occur at the highest latitudes with the least UV radiation. For example, in very high latitudes, inclusion of fatty fish in the diet allows residence despite a very long vitamin D winter. Nowadays, some industrial countries fortify foods with vitamin D. Historically and prehistorically, however, human behavior was more likely to exacerbate low UV-B levels. The actual UV radiation exposure may be well below the potential exposure indicated in Figure 6.9 because cultural practices such as wearing clothing, spending time indoors, and burning biomass fuels that produce particulate pollution can decrease skin exposure.

Both high- and low-intensity UV-B radiation may be stressful. Geographic, **epidemiological**, and clinical data indicate that heavy melanin is better adapted to high-intensity UV at low latitudes, whereas little melanin is better adapted to low-intensity UV at high latitudes. Data also indicate that the lactase-persistence phenotype is probably better adapted than the late-onset lactase-deficiency phenotype under conditions of prolonged absence of UV radiation.

HIGH-ALTITUDE HYPOXIA

Plentiful at sea level, oxygen becomes increasingly scarce at higher altitudes. Although all populations are exposed to solar radiation and temperature stress, only those living at great elevations above sea level—usually taken to be 2500 meters (8250 feet) or higher—are exposed to hypobaric **hypoxia**. The term hypoxia refers both to reduced oxygen in ambient air and reduced physiologically available oxygen compared with sea level. High-altitude hypoxia results from the shorter column of atmosphere at higher altitudes, which compresses the air less and results in lower barometric pressure—hence the term hypobaric—and fewer molecules in a volume of air. Barometric pressure decreases exponentially with altitude. At 1700 meters (~5600 feet) in Denver, barometric pressure is 83% of the sea level value; at 4000 meters (13,200 feet) on the Andean or Tibetan plateaus, it is 60% of the sea level value; and at 5000 meters (16,500 feet) in some of the highest permanent camps, it is only 55%

of sea level barometric pressure. Oxygen makes up ~21% of the air at all altitudes; therefore, lower barometric pressure results in fewer oxygen molecules in each breath. Hypoxia is a severe physiological stress because cellular metabolic processes continuously use oxygen, which is not synthesized or stored.

Failure to adapt to hypoxia may have serious consequences. Some acutely exposed visitors do not adapt and become ill with acute mountain sickness (AMS), a temporary, adverse reaction to hypoxia that usually occurs within a day of arrival. The symptoms of headache, nausea, anorexia, vomiting, fatigue, and breathlessness usually subside after a few days. However, normal physiological values and physical performance are not restored until descent to sea level. The prevalence of AMS increases with altitude and degree of hypoxic stress: 9, 13, 34, and 52% of mountaineers studied in the Swiss Alps at 2850, 3050, 3650, and 4550 meters (~9400, ~11,000, ~12,000, and 15,000 feet) had three or more symptoms of AMS (Maggiorrini et al. 1990). About 1–2% of those exposed to altitudes of 3600 meters (~12,000 feet) or more develop severe intolerance known as high-altitude pulmonary edema (HAPE; accumulation of fluid in the lungs) and/or high-altitude cerebral edema (HACE; accumulation of fluid in the brain). In addition, some well-adapted high-altitude residents lose their adaptation and become ill with chronic mountain sickness (CMS). The symptoms include headache, breathlessness, bone pain, fatigue, insomnia, and confusion. AMS and CMS can be fatal, and both can be alleviated by descent to sea level.

A physiological measure of hypoxic stress is the percentage of arterial hemoglobin that is saturated with (carrying) oxygen (SaO_2). Figure 6.20 illustrates the gradual decrease in SaO_2 from nearly 100% at sea level to as low as 70% at 5500 meters (18,200 feet). The carotid body, a tiny organ just above the bifurcation of the common carotid artery in the neck, senses changes in SaO_2 (Ward et al. 1989) and initiates physiological homeostatic processes to enhance oxygen transport.

Oxygen transport includes the processes of acquiring, carrying, delivering, and using oxygen. The respiratory system acquires oxygen by moving air into the **alveoli** of the lungs, where oxygen diffuses into the blood. Oxygen combines with the hemoglobin in the red blood cells and is carried in the blood flow, circulated by the heart and blood vessels. Oxygen diffuses from red blood cells in the capillaries into the surrounding tissues, where cells metabolize the oxygen for energy. Many homeostatic mechanisms are engaged to maintain adequate oxygen transport under hypoxic stress. The specific mechanisms and their relative strength and efficacy vary depending on length and age of exposure and population of origin.

Oxygen Transport and Acclimatization by Sea Level Natives

Sea level natives acutely exposed to high-altitude hypoxia during sojourns of a few days to a few weeks have been widely studied. Initially, SaO_2 falls abruptly and then increases over 2–3 weeks (although not to sea level values). For example, a group of young men had SaO_2 >95% at sea level that fell to an average of 81% during their first day at 4300 meters (14,190 feet) and rose to an average of 88% by the 19th day of exposure (Reeves et al. 1993). Such acclimatizations are achieved

---

BOX 6.3   PERCENT OXYGEN SATURATION OF ARTERIAL HEMOGLOBIN MEASUREMENT

The percentage of arterial hemoglobin that is saturated with (carrying) oxygen (SaO_2) is a measure of hypoxic stress. This chapter describes altitude, intrapopulation, and interpopulation variation in SaO_2. SaO_2 can be measured with a finger pulse oximeter such as the one pictured in Figure 6.19, which was used in a Tibetan nomad camp at 4850 meters (~16,000 feet). The measurement is noninvasive and easy to obtain from people of all ages. Earlier techniques for measuring SaO_2 required arterial blood samples and generally were applied only to small samples of young adult males studied in laboratory settings.

Pulse oximeters such as this Criticare Model 501+ (Criticare Systems, Inc., Waukeshau, WI) work by measuring the transmission of two wavelengths of light, red and infrared, through the finger (or earlobe or, in the case of infants, the big toe or foot). The amount of red light transmitted through the finger measures SaO_2. More-oxygenated blood is brighter red and transmits more red light. Less-oxygenated blood is darker, absorbs more and transmits less red light. Transmission of infrared light through the finger is unaffected by SaO_2 and is used as a reference for the variable transmission of red light. The sensor on this Tibetan nomad boy's finger emits light on one side of the finger and measures the amount transmitted to the other side. The anthropologist (C. M. Beall) is holding the oximeter unit, which houses electronics to run a computer program to convert the transmittance into SaO_2, which is displayed on the face of the oximeter. The SaO_2 reading is an average over five to six heart beats that is updated as SaO_2 fluctuates.

SaO_2 is measured under standardized conditions such as sleeping, exercising, or quietly sitting. For example, one study protocol averaged measurements taken at 1-minute intervals for 10 minutes in each of four activities (awake, feeding, quiet sleep, and active sleep) recorded from infants ranging from 6 hours to 4 months of age. Another study protocol averaged SaO_2 taken at 10-second intervals for 1 minute in quietly seated, awake individuals 5–79 years of age.

The noninvasive, portable oximeter has substantially expanded our understanding of adaptation to high altitudes by enabling rapid and inexpensive measurement of many people of all ages. As technological developments yield similar equipment to measure additional traits, our knowledge of the physiology of populations living in different climates will improve tremendously.

---

by homeostatic changes in oxygen transport processes. Oxygen acquisition is enhanced by immediate increase in the depth and rate of breathing: the **hypoxic ventilatory response**. Oxygen carrying is enhanced by higher heart rate and larger volume of oxygenated blood leaving the heart per contraction. This response diminishes after a week or so at altitude. Oxygen carrying is also enhanced by increasing the concentration of hemoglobin, the oxygen-carrying molecule. This increase is accomplished at first by a decrease in plasma volume and later by increased production of hemoglobin-containing red blood cells. Oxygen diffusion is enhanced by opening inactive capillaries and increasing tissue perfusion with blood. Oxygen metabolism is affected differently at low and high extremes. **Basal metabolic rate** (BMR; the minimum energy expenditure for maintenance of respiration, circulation, body temperature, and other vegetative functions) is elevated partly due to the increase in breathing (Huang et al. 1984). For example, men exposed to 4300 meters (14,190 feet) on Pikes Peak had a 17–27% elevation in BMR during a 2-week exposure (Butterfield et al. 1992).

At the other end of the energy demand spectrum, metabolic rate measured during maximum physical work is markedly decreased because hypoxia limits the capacity to meet the higher metabolic requirements, despite the various adaptations. The actual amount of reduction varies. For example, physically fit individuals experience a greater reduction than sedentary individuals. One study reported a decrease of 27% among runners acutely exposed to 4,000 meters compared with a decrease of 14% among sedentary males (Buskirk 1976). Visitors to high altitudes cannot achieve their previous maximum physical work capacities at sea level, even after weeks or months of high-altitude exposure.

Oxygen Transport by High-Altitude Natives

Indigenous high-altitude native populations have been studied intensively in the Andean and Tibetan Plateaus and their surroundings. Quechua and Aymara Indians have lived at altitudes above 2500 meters (8250 feet) on the Andean Plateau for ~11,000 years (Aldenderfer 1999). Tibetans and the related populations of Sherpas and Bods

Figure 6.19 Measurement of SaO$_2$.

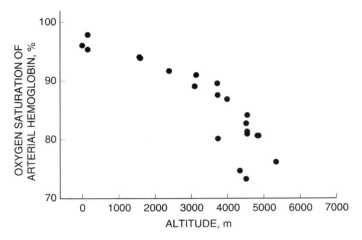

Figure 6.20   The decrease in SaO2 with altitude. (Source: Compiled from data in Altman and Dittmer 1972, 1893–1987)

have lived above 2500 meters (8250 feet) on the Tibetan Plateau for more than 4000–5500 years, although the exact length of habitation is not known (Ehrich 1992).

Recent migrant populations to high altitudes include Europeans and Han Chinese. Europeans have been migrating to the Andean Plateau during the 450 years since the Spanish Conquest. Individuals of European descent have lived in the Colorado Rockies for the past 100 years or so at altitudes >2500 meters. Han Chinese began migrating to the Tibetan plateau during the past few decades. From an evolutionary perspective, the indigenous Andean and Tibetan highlanders are populations whose **gene pools** have had time to be modified by natural selection to increase the frequency of traits enhancing oxygen transport, whereas European and Han high-altitude resident populations have not. Similarities among indigenous and migrant highland populations presumably reflect the general human capacity to acclimatize to lifelong high-altitude hypoxia, whereas differences may reflect population-specific adaptations or traits unrelated to high-altitude adaptation.

Andean and Tibetan highlanders partially offset hypoxic stress with several enhancements of oxygen acquisition, which include a higher rate and depth of breathing relative to sea level residents. For example, resting ventilation of Andean highlanders at 4540 meters (~15,000 feet) is more than one-third higher than that of their sea level counterparts at sea level (Hurtado 1964). However, more detailed studies of ventilation in which a trait called the hypoxic ventilatory response (HVR) is measured have produced unexpected results. The general term HVR refers to the increase in breathing on exposure to hypoxia. Formal measurement of HVR entails inducing controlled hypoxia, usually by rebreathing exhaled air, and measuring the increase in ventilation in terms of liters of air per minute relative to the increase in hypoxic stress over a range of hypoxic stress. Surprisingly, high-altitude Andean natives exhibit very little increase in ventilation on exposure to such additional experimentally-induced hypoxia. This "blunted" HVR also develops in Colorado high-altitude natives and residents of European descent after 20–25 years of exposure. However, Tibetan highlanders have HVRs similar to those of sea level residents. The similarity between indigenous Andean and migrant Colorado high-altitude natives and long-term residents suggests that this is the usual human response to chronic hypoxic stress. The contrast with the Tibetans suggests that the latter have unique capacity to maintain HVR under chronic hypoxic stress.

High-altitude natives also enhance oxygen acquisition with larger lung volumes than their sea level counterparts. Total lung capacity is the sum of **vital capacity** (VC) and **residual volume** (RV). Large VCs could have functional advantages, including more oxygen molecules per breath and a greater surface area for their diffusion into the bloodstream. As Figure 6.21 indicates, the increase is ~25% relative to sea level in several samples of indigenous populations. VCs of Andean highlanders are consistently been found to be substantially larger than those of sea level natives. VCs of Tibetan samples sometimes have been found to be larger than those of sea level natives, although some studies find no difference between high-altitude native Tibetans and sea level Europeans (e.g., Hackett et al. 1980). Larger VCs may be advantageous for maintaining SaO$_2$ during heavy exercise. For example, Tibetans have larger VCs than Han Chinese migrants to Tibet and tend to maintain SaO$_2$ during exercise, whereas the Han tend to lose SaO$_2$ (Sun et al. 1990). If this correlation represents a causal relationship,

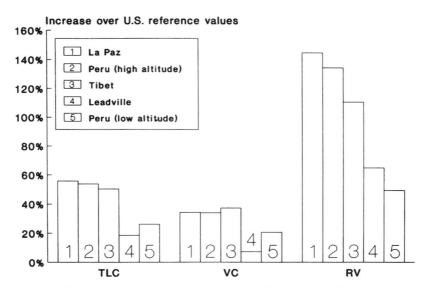

Figure 6.21   Comparison of percentage increases in TLC (Total Lung Capacity), vital capacity (VC), and residual volume (RV) of high altitude natives and sea level natives above U.S. low-altitude reference values. (*Source:* Greksa 1994, 495, Copyright © 1994 John Wiley and Sons; reprinted with permission from Wiley-Liss, Inc., a subsidiary of John Wiley and Sons, Inc)

then it would enable sustained high levels of physical work in daily life without incurring additional transient hypoxic stress.

RV is the volume of air that remains in the lungs after a complete expiration. Figure 6.21 indicates that Andean and Tibetan highlanders have more than double the RV predicted by sea level reference values and that the percent increase over sea level in RVs is much greater than for VCs. Figure 6.21 also demonstrates that Andean lowlanders have larger lung volumes than European sea level natives, suggesting that this population has large lung volumes irrespective of altitude. Large RVs represent larger oxygen stores, which can prevent or lessen additional decreases in SaO_2 during daily activities such as sleep, when ventilation is normally lower and more irregular, or during tasks such as lifting, which often are accompanied by pauses in breathing such as grunts or breath holds (Strohl et al. 1986). If SaO_2 were to fall frequently during recurring activities, hypoxic stress would increase. Thus, the larger RV may function in a similar manner to the larger VC to enable heavy work in daily life while maintaining SaO_2.

Oxygen carrying in the blood is a function of hemoglobin concentration and SaO_2. Oxygen carrying is often enhanced at high altitudes by an increase in hemoglobin concentration over sea level values. The amount of the increase varies. Andean highlanders have ~1.5 grams/deciliter more hemoglobin concentration than Tibetan highlanders at the same altitudes. Higher hemoglobin concentration is beneficial only to the point at which blood viscosity becomes so high that circulation is impeded, SaO_2 falls, and hypoxic stress increases (at ~18 grams/deciliter (Winslow and Monge 1987). Andean male samples average around 18 grams/deciliter at ~3700 meters, whereas Tibetan male samples do not reach this value until nearly 5000 meters. In fact, some Tibetan samples at 3300- to 4100-meter altitudes have mean hemoglobin concentrations no different from sea level reference values.

Elevated hemoglobin concentration is not essential for successful life at high altitude, and some researchers question whether this response has any physiological value (Winslow and Monge 1987, 203). Oxygen carrying is also enhanced by higher SaO_2 within the limits imposed by high-altitude hypoxia. SaO_2 varies considerably about the mean at any one altitude. For example, at 4000 meters in Tibet, the SaO_2 is 89 ±4% (mean and standard deviation); and at 5000 meters, it is 84 ±3%. Evidence indicated a genetic contribution to this variation among Tibetans (see later).

Considering oxygen use, some evidence of adaptation exists at the level of cellular metabolism. For example, heart and brain metabolism may preferentially use **glucose** over fats as a fuel source. Glucose is more oxygen-efficient than fat (Hochachka et al. 1996). Oxygen use measured by energy metabolism is normal among Andean and Tibetan high-altitude natives in the sense that BMR and maximal physical work capacity do not differ from normal sea level values. In this sense, oxygen transport adaptation is complete.

However, oxygen demands are increased under regularly occurring conditions such as growth and pregnancy. Growth is slow at high altitude. Indigenous high-altitude populations are generally much shorter and lighter than their sea level counterparts throughout the life cycle. Interpreting these facts as a response to hypoxia is difficult, because the same pattern is characteristics of undernutrition and there is evidence of undernutrition in some indigenous samples. One way to separate the effects of under-nutrition and altitude has been to study upper socioeconomic status European children who have migrated to high-altitude in the Andes. Such children are unlikely to be malnourished. They are slightly shorter and lighter than children at sea level, suggesting that high-altitude hypoxia does slow growth somewhat (Stinson 1982; Greksa et al. 1984). Similar results have been found among preschoolers in the US plains and mountain states (Yip et al. 1988). In contrast to overall body size measured by height and weight, growth in chest size is accelerated among Andean but not Tibetan highlanders (Beall 1981). Because larger size is not attributable to undernutrition and because there is a correlation between chest dimensions and lung volumes among Andean highlanders, this morphological characteristic may reflect morphological adaptations in the lung. However, no direct evidence supports this theory.

High-altitude pregnancy and fetal growth increase maternal oxygen demands and elicit adaptations to enhance oxygen transport to the fetus. Success in meeting these demands can be measured by birth weight, which is an important determinant of infant survival. Mean birth weight decreases with increasing altitude (Figure 6.22). Because relatively low birth weights are associated with higher infant mortality rate, heavier babies are more likely to survive than lighter babies. Andean and Tibetan women bear heavier babies than European and Han women at the same altitude, which suggests better oxygen transport to the fetus. One way this is achieved during high-altitude pregnancy is by a rise in ventilation from its already higher baseline and by an increase in HVR. The effect is to increase SaO_2 a few percent and to decrease maternal, and presumably fetal, hypoxic stress (Moore 1990). Andean and Colorado women with larger HVR increases during pregnancy give birth to heavier neonates. This trait has not been measured in pregnant Tibetan women, among whom a different trait has been investigated. Tibetan women redistributed blood flow during pregnancy to favor uterine circulation. This redistribution also would enhance oxygen delivery to the fetus.

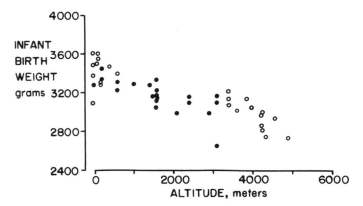

Figure 6.22   Birth weight decreases with increasing altitude. (*Source:* Moore and Regensteiner 1983, 292, © 1983 Annual Review of Anthropology; reprinted with permission from the authors and from Annual Reviews, Inc.)

High-altitude Andean and Tibetan natives are distinct from sea level natives at sea level in several aspects of oxygen transport, including higher average ventilation, larger average lung volume, and smaller body size. They are similar to sea level natives in average BMR and average maximal physical work capacity. However, the two high-altitude populations are different in some ways. For example, the Andean HVR is blunted (whereas the Tibetan HVR is normal) and the Andean hemoglobin concentration is higher than the Tibetan. Many approaches have been used to determine whether these quantitative differences in traits for oxygen transport are mainly influenced genetically, by environmental exposure during development, or by the immediate environment.

Genetic and Developmental Aspects of Oxygen Transport

The quantitative genetic approach uses data collected from biological relatives to ascertain whether there is familial patterning of these traits and to calculate heritability (h^2) to evaluate the relative strength of genetic and random environmental influences (see Chapter 5). A significant genetic basis for the traits must be established to evaluate the potential for genetic adaptation. The h^2 of numerous oxygen transport traits is moderate to high at high altitudes. For example, the h^2 of hemoglobin concentration was 0.86 in one Tibetan and 0.87 in one Aymara sample at ~4000 meters, whereas the h^2 of HVR was 0.35 in the Tibetan and 0.22 in the Aymara sample. Thus, there is potential for natural selection on both traits in both populations. Whether natural selection has acted to produce the existing genetic variation and the observed mean values cannot be determined from these studies. Knowledge of the quantitative effects of the alleles and the allele frequency of the traits in high- and low-altitude populations would be necessary.

Allele frequency is known for one trait for which Tibetan populations have a major gene: SaO_2 (Beall et al. 1994). The average SaO_2 of Tibetan nomads at 4850–5450 m (~16,000–18,000 feet) with the autosomal dominant allele is 84% compared with an average of 78% for those without. The allele frequency of the allele for higher SaO_2 is 0.56. The average SaO_2 of Tibetan villagers at 3900–4065m (~12,900–13,400 feet) with the dominant allele is 88% compared with an average of 83% for those without. The higher mean genotypic values of the villagers reflects the lower altitude, whereas the similar-sized difference between genotypes in the two sites reflects the influence of the dominant allele in both. The functional consequence of the higher mean is equivalent to living at ~1000 meters (3300 feet) lower altitude. As additional data become available from other Tibetan populations, it will be possible to determine whether an altitudinal cline exists for this gene. Andean high-altitude populations do not exhibit genetic variation in this trait, thus revealing another difference between the two populations.

Another approach to understanding the genetic basis of high-altitude adaptive traits is to compare indigenous highlanders who might have genetic adaptations with European migrants who presumably have no genetic adaptation to high-altitude hypoxia. This method has been used in the Andes to test the developmental adaptation hypothesis (DAH), which reasons that highland populations are distinctive due to hypoxic exposure during development that causes structural and functional modifications of those normal processes. The DAH proposed that lifelong high-altitude residents adapt during growth and development by preferential delivery of scarce oxygen to oxygen transport systems rather than to other systems, such as the musculoskeletal. This hypothesis specifically argues that natural selection has not acted on the gene pool of high-altitude natives to enhance their ability to survive and reproduce under hypoxic stress. It reasons that lifelong exposure is sufficient to elicit the responses characteristic of highlanders: Natives have complete developmental adaptation, migrants vary according to their age at arrival, and sojourners have no adaptation.

The DAH has been tested most thoroughly with a comparison of European children 9–20 years of age living at high and low altitude in Bolivia (3600 and 4000 meters [~11,900 and 13,200 feet]). These healthy, well-nourished children were from a population without ancestral exposure to high-altitude, and thus their responses to high altitude represent general human biological adaptive capacities. This study assessed oxygen acquisition by VC, energy metabolism by maximal physical work capacity, and musculoskeletal development by body size and maturation (Greksa 1990). It found that VC was slightly greater in the highland samples than the lowland. However, the enhancement was much less than in Figure 6.21, and oxygen metabolism measured by maximum physical work capacity was lower at high altitudes prior to adolescence and equal to sea level values after adolescence. These data suggest that developmental events during the adolescent period are essential to achieving complete adaptation (in the sense of equivalence of high- and low-altitude values). Growth and maturation were slightly delayed prior to adolescence. This study provides some support for the DAH in that it found changes in the direction of similarity to Andean highlanders and offered evidence that adolescence is a critical period for attaining full adaptation.

However, the magnitude of the effects on oxygen acquisition and musculoskeletal development was much smaller than reported for high-altitude native populations. This could result if natural selection has acted to increase the magnitude of the general adaptive capacity exhibited by the European children. Despite these morphological contrasts, oxygen delivery as measured by maximal physical work capacity did not differ after adolescence.

Evidence indicates three modes of adaptation to high-altitude hypoxia: (1) acclimatization among visitors, (2) genetic adaptation in SaO_2 and HVR among Tibetans, and (3) developmental adaptation in VC and maximal work capacity among Andean and European populations.

FUTURE DIRECTIONS

This description of adaptation to climate is simplified, particularly with respect to age and sex differences. Relatively few studies included women, children, and the elderly. It was often assumed implicitly that young adult males adequately represented the adaptive responses of the entire population. However, age and sex differences in body size and composition, neuroendocrine function, and women's simultaneous adaptation to climatic stress as well as the recurrent stresses of pregnancy and lactation indicate the need for specific investigation of various subgroups. For example, high-altitude pregnancy has been studied, with interesting results that have improved general understanding of adaptation to hypoxia. The metabolic and vascular responses involved in thermoregulation and gestation could hypothetically compete or complement one another and warrant study as well. Age differences throughout the life cycle modulate the effectiveness of some adaptive responses. For example, the epidermal concentration of 7DHC declines steadily from childhood onward, and the number of melanocytes declines throughout adulthood. Therefore, the effectiveness of adaptation to UV radiation probably varies with age. Future research that includes broader age–sex samples would result in more realistic understanding of the complexity of adaptation.

Climatic change is also an issue for human biology. Global warming as a consequence of industrial emission of carbon dioxide and other greenhouse gases has the potential to increase the degree and geographic extent of heat stress (as well as infectious disease stress). At the same time, aerosol cooling as a consequence of particulate pollution, reflecting solar radiation back into space, is occurring in some areas such as eastern Asia, eastern North America, and Europe. The net effect of global warming and aerosol cooling will vary geographically and depend on pollution patterns. Ozone depletion due to photochemical reactions with pollutants in the atmosphere has the potential to increase UV-B radiation intensity reaching the earth's surface. It could decrease stress from low UV-B radiation at high latitudes and increase stress from high UV-B radiation intensity at low latitudes.

Other areas for future research include investigation into

1. the neuroendocrine integration of adaptation (e.g., the hormonal control of metabolism or of melanin synthesis),
2. the quantitative and biochemical genetics of biological variation within and between populations (e.g., melanin concentration or SaO_2), and
3. the consequences of biological variation for other adaptive domains, such as differential mortality and fertility rates.

CHAPTER SUMMARY

In this chapter, we have reviewed how humans adapt to the wide range of climatic conditions in which they live. Several acclimatizations occur in individuals exposed to extremes of heat, cold, ultraviolet radiation, and high altitude. Populations with a long history of exposure to extreme cold may have evolved some unique biological protections, but this evolution is less clear for populations exposed to extreme heat. Variation in body size and shape may also be related to temperature. The evidence is consistent with the hypothesis that natural selection acted to produce and maintain the cline in skin color. Adaptations to high altitude include acclimatizations, genetic adaptations, and developmental adaptations.

Several themes emerge from this discussion of adaptation to climate. One is the importance of behavior in modulating exposure (and thus selective pressure) to heat, cold, and UV radiation. Another is the contrast in adaptive responses between newly exposed populations and those with long histories of exposure to cold, UV radiation extremes, and altitude. At the same time, little is known for certain about the mechanisms underlying the various forms of adaptation to climate in human populations. There is enormous opportunity for additional important fieldwork investigating the human population biology of residents of climatic extremes.

RECOMMENDED READINGS

Beall CM, Blangero J, Williams-Blangero S, and Goldstein MC (1994) A major gene for percent of oxygen saturation of arterial hemoglobin in Tibetan highlanders. Am. J. Phys. Anthropol. 95: 271–276.

Beall CM, Brittenham GM, Macuaga F, and Barragan M (1990) Variation in hemoglobin concentration among samples of high-altitude natives in the Andes and the Himalayas. Am. J. Hum. Biol. 2(6): 639–651.

Frisancho AR (1993) Human Adaptation and Accommodation. Ann Arbor: Univ. of Michigan Press.

Greksa LP (1990) Developmental responses to high-altitude hypoxia in Bolivian children of European ancestry: a test of the developmental adaptation hypothesis. Am. J. Hum. Biol. 2(6): 603–612.

Hochachka PW, Clark CM, Holden JE, Stanley C, Ugurbil K, and Menon RS (1996) ^{31}P magnetic resonance spectroscopy of the Sherpa heart: a phosphocreatine/adenosine triphosphate signature of metabolic defense against hypobaric hypoxia. Proc. Natl. Acad. Sci. 93: 1215–1220.

Houston C (1992) Mountain sickness. Sci. Am., October: 58–66.

Jung EG and Holick MF (eds.) (1993) Biological Effects of Light 1993. Berlin, N.Y.: Walter de Gruyter and Co.

Leffell D and Brash D (1996) Sunlight and Skin Cancer. Sci. Am., July: 52–59.

Marriott BM and Carlson SJ (eds.) (1996) Nutritional Needs in Cold and High-Altitude Environments. Washington, D.C.: National Academy Press.

Nelson RK (1973) Hunters of the Northern Forest. Chicago: Univ. of Chicago Press.

Niermeyer S, Yang P, Shanmina, Drolkar, Zhuang J, and Moore LG (1995) Arterial oxygen saturation in Tibetan and Han Infants born in Lhasa, Tibet. N. Engl. J. Med. 333(19): 1248–1252.

Pandolf KB, Sawka MN, and Gonzalez RR (eds.) (1988) Human Performance Physiology and Environmental Medicine at Terrestrial Extremes. Indianapolis: Benchmark Press.

Ruff CB (1994) Morphological adaptation to climate in modern and fossil hominids. Yearbook Phys. Anthropol. 37: 65–107.

Rutstrum C (1968) Paradise Below Zero. New York: Macmillan Company.

Ward MP, Milledge JS, and West JB (1989). High Altitude Medicine and Physiology. Philadelphia: Univ. of Pennsylvania Press.

Webb AR, Kline L, and Holick MG (1988) Influence of season and latitude on the cutaneous synthesis of vitamin D: exposure to winter sunlight in Bolton and Edmonton will not promote vitamin D synthesis in human skin. J. Clin. Endocrin. Metab. 67(2): 373–378.

Young AR (ed.) (1993) Environmental UV Photobiology. New York: Plenum Press.

REFERENCES CITED

Aldenderfer M (1999) The Pleistocene/Holocene transition in Peru and its effects upon human use of the landscape. Quarternary International 54: 11–19.

Altman PL and Dittmer DS (1972) Biology Data Book, 2nd edition. Bethesda, Md., Federation of American Societies for Experimental Biology.

Anderson RR and Parrish JA (1982) Optical properties of human skin. In JD Regan and JA Parrish (eds.): The Science of Photomedicine. New York: Plenum Press, pp. 147–193.

Beall CM (1981) Optimal birthweights in Peruvian populations at high and low altitudes. Am. J. Phys. Anthropol. 56: 209–216.

Beall CM, Blangero J, Williams-Blangero S, and Goldstein MC (1994) A major gene for percent of oxygen saturation of arterial hemoglobin in Tibetan highlanders. Am. J. Phys. Anthropol. 95: 271–276.

Bittel J (1992) The different types of cold adaptation in man. Int. J. Sports Med. 13 (Suppl 1): S172-S176.

Brazerol WF, McPhee JA, Mimouni F, Specker BL, and Tsang RC . (1988) Serial ultraviolet B exposure and serum 25 hydroxyvitamin D response in young adult American Blacks and Whites: no racial differences. J. Am. Coll. Nutr. 7(2): 111–118.

Brück K (1980) Basic mechanisms in longtime thermal adaptation. In Z Szelenyi and M Szekely (eds.): Contributions to Thermal Physiology. Budapest: Akademiai Kiado, pp. 263–273.

Budd GM, Brotherhood JR, Beasley FA, Hendrie AL, Jeffery SE, Lincoln GJ, and Solaga AT (1993) Effects of acclimatization to cold baths on men's responses to whole-body cooling in air. Eur. J. Appl. Physiol. 67: 438–449.

Burton AC and Edholm OG (1955) Man in a Cold Environment. London: Edward Arnold.

Buskirk ER (1976) Work performance of newcomers to Peruvian highlands. In PT Baker and MA Little (eds.): Man in the Andes: A Multidisciplinary Study of High Altitude Quechua. Stroudsburg, Pa.: Dowden, Hutchinson, Ross, Inc., pp. 283–299.

Butterfield GE, Gates J, Fleming S, Brooks GA, Sutton JR, and Reeves JT (1992) Increased energy intake minimizes weight loss in men at high altitude. J. Appl. Physiol. 72(5): 1741-8.

Calhoun DA and Oparil S (1995) Racial differences in pathogenesis of hypertension. Am J. Med. Sci. 310 (Suppl. 1): S86–S90.

Carey JW and Steegmann AT Jr (1981) Human nasal protrusion, latitude, and climate. Am. J. Phys. Anthropol. 56: 313–319.

Clemens TL, Henderson SL, Adams JS, and Holick MF (1982) Increased skin pigment reduces the capacity of skin to synthesise vitamin D. The Lancet, January 9: 74–76.

Danzl DF and Pozos RS (1987) Multicenter hypothermia survey. Ann. Emerg. Med. 16: 1042–1055.

Danzl DF, Pozos RS, and Hamlet MD (1989) Accidental hypothermia. In PS Auerbach and EC Geehr (eds.): Management of Wilderness and Environmental Emergencies, 2nd edition. St. Louis: Mosby, pp. 55–76.

Davies TRA and Johnston DR (1961) Seasonal acclimatization to cold in man. J. Appl. Physiol. 16: 231–234.

Doyle R (1998) Deaths from excessive cold and excessive heat. Sci. Am. 278: 26.

Duncan MT and Horvath SM (1988) Physiological adaptations to thermal stress in tropical Asians. Eur. J. Appl. Physiol. 57: 440–444.

Durham WH (1991) Coevolution: Genes, Culture, and Human Diversity. Stanford, Calif.: Stanford Univ. Press.

Ehrich RW (ed.) (1992) Chronologies in Old World Archaeology. Chicago: Univ. of Chicago Press.

Eller M, Yar M, and Gilchrest B (1994) DNA damage and melanogenesis. Nature 372: 413–414.

Ellis HD, Wilcock SE, and Zaman SA (1985) Cold and performance: the effects of information load, analgesics and rate of cooling. Aviat. Space Environ. Med. 56: 233–237.

Epstein Y, Shapiro Y, and Brill S (1983) The role of surface area-to-mass ratio and work efficiency in heat intolerance. J. Appl. Physiol. 54: 831–836.

Ergonomics 38: 1–192, 1995.

Fears TR, Scotto J, and Scheidman MA (1976) Skin cancer, melanoma and sunlight. Am. J. Pub. Health 66: 461–464.

Foray J (1992) Mountain frostbite: current trends in prognosis and treatment, from results concerning 1261 cases. Int. J. Sports 13(Suppl 1) S193–S196.

Franciscus RG and Long JC (1991) Variation in human nasal height and breadth. Am. J. Phys. Anthropol. 85: 419–427.

Garcia RI, Mitchell RE, Bloom J, and Szabo G (1977) Number of epidermal melanocytes, hair follicles, and sweat ducts in skin of Solomon Islanders. Am. J. Phys. Anthropol. 47: 427–434.

Glickman-Weiss EL, Nelson AG, Hearon CM, Goss FL, Robertson RJ, and Cassinelli DA (1993) Effects of body morphology and mass on thermal responses to cold water: revisited. Eur. J. Appl. Physiol. 66: 299–303.

Greksa LP (1990) Developmental responses to high-altitude hypoxia in Bolivian children of European ancestry: a test of the developmental adaptation hypothesis. Am. J. Hum. Biol. 2: 603–612.

Greksa LP (1994) Total lung capacity in Andean highlanders. Am. J. Hum. Biol. 5: 491–498.

Greksa, LP, Spielvogel H, Paredes-Fernandez L, Paz-Zamora M, and Caceres E (1984) The physical growth of urban children at high altitude. Am. J. Phys. Anthropol. 65: 315–322.

Hackett PH, Reeves J, Reeves CD, Grover RF, and Rennie D (1980) Control of breathing in Sherpas at low and high altitude. J. Appl. Physiol. 49(3): 374–379.

Hamlet MP (1988) Human cold injuries. In KB Pandolf, MN Sawka, and RB Gonzalez (eds.): Human Performance Physiology and Environmental Medicine at Terrestrial Extremes. Indianapolis: Benchmark, pp. 435–466.

Hanna JM and Brown DE (1979) Human heat tolerance: biological and cultural adaptations. Yearbook Phys. Anthropol. 22: 163–186.

Hayward JS, Collis M, and Eckerson JD (1973) Thermographic evaluation of relative heat loss of man during cold water immersion. Aerospace Med. 44: 708–711.

Hayward JS and Eckerson JD (1984) Physiological responses and survival time prediction for humans in ice water. Aviat. Space Environ. Med. 55: 206–212.

Henane R (1981) Acclimatization to heat in man: giant or windmill, a critical appraisal. In Z Szelenyi and M Szekely (eds.): Contributions to Thermal Physiology. Budapest: Akademiai Kiado, pp. 275–284.

Hochachka PW, Clark CM, Holden JE, Stanley C, Ugurbil K, and Menon RS (1996). ^{31}P magnetic resonance spectroscopy of the Sherpa heart: a phosphocreatine/adenosine triphosphate signature of metabolic defense against hypobaric hypoxia. Proc. Natl. Acad. Sci. U.S.A. 93: 1215–1220.

Holick MF (1995) Environmental factors that influence the cutaneous production of vitamin D. Am. J. Clin. Nutr. 61(Suppl 1): 638S–645S.

Holick MF (1994a) Vitamin D: new horizons for the 21st century (McCollum Award Lecture, 1994). Am. J. Clin. Nutr. 60: 619–630.

Holick MF (1994b). Sunlight, vitamin D, and human health. In EG Jung and MF Holick (eds.): Biologic Effects of Light 1993. Berlin: Walter de Gruyter and Co., pp. 3–15.

Holick MF, MacLaughlin JA, and Doppelt SH (1981) Regulation of cutaneous previtamin D photosynthesis in man: skin pigment is not an essential regulator. Science 211: 590–593.

Holliday T (1997) Postcranial evidence of cold adaptation in European Neandertals. Am. J. Phys. Anthropol. 104:245–258.

Houdas Y Deklunder G, and Lecroart J-L (1992) Cold exposure and ischemic heart disease. Int. J. Sports Med. 13(Suppl l.): S179–S181.

Huang SY, Alexander JK, Grover RF, Maher JT, McCullough RE, McCullough RG, Moore LG, Weil JV, Sampson JB, and Reeves JT (1984) Increased metabolism contributes to increased resting ventilation at high altitude. Respir. Phys. 57: 377–385.

Hubbard RW and Armstrong LE (1988) The heat illnesses: biochemical, ultrastructural, and fluid electrolyte considerations. In KB Pandolf, MN Sawka, and RB Gonzalez (eds.): Human Performance Physiology and Environmental Medicine at Terrestrial Extremes. Indianapolis: Benchmark pp. 305–359.

Hurtado A (1964) Some Physiologic and Clinical Aspects of Life at High Altitudes. In L Lander and JH Moyer (eds.): Aging of the Lung. Perspectives. New York: Grune and Stratton, pp. 257–278.

Johnson FS, Mo T, and Green AES (1976) Average latitudinal variation in ultravaiolet radiation at the earth's surface. Photochem. Photobiol 23: 179-188.

Kaidbey KH, Agin PP Sayre RM, and Kligman AM (1979) Photoprotection by melanin: a comparison of black and Caucasian skin. J. Am. Acad. Dermatol. 1: 249–260.

Kamb A (1994) Sun protection factor p53. Nature 372: 730–731.

Katzmarzyk PT and Leonard WR (1998) Climatic influences on human body size and proportions: Ecological adaptations and secular trends. Am J Phys Anthropol. 106:483-503.

Kilbourne EM (1997) Cold environments. In EJ Noji (ed.): The Public Health Consequences of Disasters. New York: Oxford, pp. 270-286.

Kondo S (1969) A study on acclimatization of the Ainu and the Japanese with reference to hunting temperature reaction. J. Fac. Sci. (U. Tokyo) III 4: 254–265.

LeBlanc J (1975) Man in the Cold. Springfield, Ill.: Thomas.

Leffell D and Brash D (1996) Sunlight and skin cancer. Sci. Am., July: 52–59.

Lockhart JM, Kiess HO, and Clegg TJ (1975) Effect of rate and level of lowered finger temperature on manual performance. J. Appl. Physiol. 60: 106–113.

Maggiorini MBB, Walter M, and Oelz O (1990) Prevalence of acute mountain sickness in the Swiss Alps. BMJ 301: 853–855.

Meehan JP (1955) Individual and racial variations in vascular response to cold stimulus. Milit. Med. 116: 330–334.

Mitchell RE (1963) The effect of prolonged solar radiation on melanocytes of the human epidermis. J. Invest. Dermatol. 41: 199–212.

Mitchell RE (1968) The skin of the Australian aborigine; a light and electronmicroscopical study. Aust. J. Dermatol. 9: 314–328.

Momiyama M (1975) Seasonal variation in mortality with special reference to thermal living conditions. In H Yoshimura and S Kobayashi (eds.): Physiological Adaptability and Nutritional Status of the Japanese. Tokyo: Univ. of Tokyo Press, JIBP Synthesis, vol. 3, pp. 136–147.

Moore LG and Regensteiner JG (1983) Adaptation to high altitude. Annu. Rev. Anthropol. 12: 290–296.

Moore LG (1990) Maternal O_2 Transport and fetal growth in Colorado, Peru, and Tibet high-altitude residents. Am. J. Hum. Biol. 2(6): 627–637.

Neer RM (1985) Environmental light: effects on vitamin D synthesis and calcium metabolism in humans. Ann. N.Y. Acad. Sci. 453: 14–20.

Nelson RK (1969) Hunters of the Northern Ice. Chicago: Univ. of Chicago Press.

Nelson RK (1973) Hunters of the Northern Forest. Chicago: Univ. of Chicago Press.

Newman R (1970) Why man is such a sweaty, thirsty, naked animal: A speculative review. Hum. Biol. 42: 12–27.

Newman RW and Munro EH (1955) The relation of climate and body size in U.S. males. Am. J. Phys. Anthropol. 13: 1–17.

Parrish JA (1982) The scope of photomedicine. In JD Regan and JA Parrish (eds.): The Science of Photomedicine. New York: Plenum Press, pp. 3–17.

Post P, Daniels F, and Binford R (1975) Cold injury and the evolution of "white" skin. Hum. Biol. 47: 65–80.

Proctor DF and Andersen I (1982) The Nose: Upper Respiratory Physiology and the Atmospheric Environment. Amsterdam: Elsevier Medical.

Reed WL (1993) Health and Medical Care of African-Americans. Westport, Conn.: Auburn House.

Reeves JT, McCullough RE, Moore LG, Cymerman A, and Weil JV (1993) Sea-level PCO_2 relates to ventilatory acclimatization at 4,300 m. J. Appl. Physiol 75: 1117–1122.

Regnard J (1992) Cold and the airways. Int. J. Sports Med. 13(Suppl.): S182–S184.

Riggs SK and Sargent F (1964) Physiological regulation in moist heat by young American negro and white males. Hum. Biol. 36: 339–353.

Roberts DF (1953) Body weight, race, and climate. Am. J. Phys. Anthropol. 11: 533–558.

Robins A (1991) Biological Perspectives on Human Pigmentation. Cambridge: Cambridge Univ. Press.

Rosdahl I and Rorsman H (1983) An estimate of the melanocyte mass in humans. J. Invest. Dermatol. 81(3): 278–281.

Rosen CM, Morrison T, Zhou H, Storm D, Hunter SJ, Musgrave K, Chen T, Liu W, and Holick MF (1994) Seasonal effects of sunlight on bone mass in elderly women. In EG Jung and MF Holick (eds.): Biologic Effects of Light 1993. Berlin: Walter de Gruyter and Co., pp. 16–27.

Ruff CR (1994) Morphological adaptations to climate in modern and fossil hominids. Yearbook Phys. Anthropol. 37: 65–107.

Scholander PF, Hammel HT, Hart JS, LeMessurier DH and Steen J (1958) Cold adaptation in Australian Aborigines. J. Appl. Physiol. 13: 211-218.

Scheibner A, McCarthy WH, and Nordlund JJ (1988) Age and seasonal variation in melanocyte distribution in normal human epidermis. In AM Kligman and Y Takase (eds.): Cutaneous Aging. Tokyo: Univ. of Tokyo Press, pp. 201–211.

Schuman LM (1953) Epidemiology of frostbite: Korea, 1951–52. In Cold Injury—Korea 1951–1952. Ft. Knox, Army Medical Research Laboratory Report 113, pp. 205–568.

Shephard RJ and Rode A (1996) The Health Consequences of Modernization: Evidence from Circumpolar Peoples. Cambridge: Cambridge Univ. Press.

Shvartz E, Saar E, and Benor D (1973) Physique and heat tolerance in hot-dry and hot-humid environments. J. Appl. Physiol. 34: 799–803.

Sloan REG and Keatinge RW (1973) Cooling rates of young people swimming in cold water. J. Appl. Physiol. 35:371-375.

Sminia P, van der Zee J, Wondergem J, and Haveman J (1994) The effect of hyperthermia on the central nervous system. Int. J. Hyperthermia 10: 1–30.

Smith DJ Jr, Robson MC, and Heggers JP (1989) Frostbite and other cold-induced injuries. In PS Auerbach and EC Geehr (eds.): Management of Wilderness and Environmental Emergencies, 2nd edition. St. Louis: Mosby, pp. 101–118.

Smith RM and Hanna JM (1975) Skinfolds and resting heat loss in cold air and water: temperature equivalence. J. Appl. Physiol. 39: 93–102.

So JK (1975) Genetic, acclimatizational, and anthropometric factors in hand cooling among North and South Chinese. Am. J. Phys. Anthropol. 43: 31–38.

Staricco RJ and Pinkus H (1957) Quantitative and qualitative data on the pigment cells of adult human epidermis. J. Invest. Dermatol. 29: 33–43.

Steegmann AT Jr (1975) Human adaptation to cold. In A Damon (ed.): Physiological Anthropology. New York: Oxford, pp. 130–166.

Steegmann AT Jr (1977) Finger temperature during work in natural cold. Hum. Biol. 49: 349–362.

Steegmann AT Jr (1979) Human facial temperatures in natural and laboratory cold. Aviat. Space Environ. Med. 50: 227–232.

Steinman AM and Hayward JS (1989) Cold water immersion. In PS Auerbach and EC Geehr (eds.): Management of Wilderness and Environmental Emergencies, 2nd edition. St. Louis: Mosby, pp.77–188.

Stinson S (1982) The effect of high altitude on the growth of children of high socioeconomic status in Bolivia. Am. J. Phys. Anthropol. 59: 61–73.

Strohl KP, Cherniack NS, et al. (1986) Physiological bases of therapy for sleep. Am. Rev. Respir. Dis. 34: 791–802

Strydom NB and Wyndham CH (1963) Natural state of heat acclimatization of different ethnic groups. Fed. Proc. 22(3), part 1: 801–808.

Sun SF, Droma TS, Zhang JG, Tao JX, Huang SY, McCullough RG, McCullough RE, Reeves CS, Reeves JR, and Moore LG (1990) Greater maximal O_2 uptakes and vital capacities in Tibetan than Han residents of Lhasa. Respir. Phys. 79: 151–162.

Szabo G (1954) The number of melanocytes in human epidermis. Br. Med. J., May 1: 1016–1017.

Szabo G (1967) The regional anatomy of the human integument with special reference to the distribution of hair follicles, sweat glands, and melanocytes. Phil. Trans. R. Soc. London 252: 447–485.

Szabo G (1969) The biology of the pigment cell. In EE Bittar (ed.): The Biological Basis of Medicine. London: Academic Press, vol. 16, pp. 59–91.

Toda Y (1975) Thermal adaptability of Indonesians. In H Yoshimura and S Kobayashi (eds.): Physiological Adaptability and Nutritional Status of the Japanese A: Thermal Adaptability of the Japanese and Ainu. Tokyo: Univ. of Tokyo Press.

Toner MM, Sawka M, Holden WL, and Pandolf KB (1985) Comparison of thermal responses between rest and leg exercise in water. J. Appl. Physiol. 59: 248–253.

Toner MM, Sawka MN, Foley ME, and Pandolf KB (1986) Effects of body mass and morphology on thermal responses in water. J. Appl. Physiol. 60: 521–525.

Trinkaus E (1981) Neandertal limb proportions and cold adaptation. In CB Stringer (ed.) Aspects of Human Evolution. London: Taylor and Francis, pp. 187–224.

Vallerand AL and Jacobs I (1992) Energy metabolism during cold exposure. Int. J. Sports Med. 13(Suppl 1): S191–S193.

Veicsteinas A, Ferretti G, and Rennie DW (1982) Superficial shell insulation in resting and exercising men in cold water. J. Appl. Physiol. 52: 1557–1564.

Vogelaere P, Savourney G, Deklunder G, Lecroart J, Brasseur M, Bakaert S and Bittel J (1992) Reversal of cold induced hemoconcentration. Eur. J. Physiol. Occup. Med. 64: 244–249.

Vermeer BJ and Hurks M (1994) The clinical relevance of immunosuppression by UV irradiation. J. Photochem. Photobiol. B Biol. 24: 149–154.

Wade CE, Dancany S, and Smith RM (1979) Regional heat loss in resting man during immersion in 25.2 °C water. Aviat. Space Environ. Med. 50: 590–593.

Ward MP, Milledge JS, and West JB (1989) High altitude medicine and physiology. Philadelphia: Univ. of Pennsylvania Press.

Webb AR (1993) Vitamin D synthesis under changing UV spectra. In AR Young . (ed.): Environmental UV Photobiology. New York: Plenum Press, pp. 185–202.

Webb AR, Kline L, and Holick MF (1988) Influence of season and latitude on the cutaneous synthesis of vitamin D: Exposure to winter sunlight in Bolton and Edmonton will not promote vitamin D synthesis in human skin. J. Clin. Endocrinol. Metab. 67(2): 373–378.

Weiner JS (1954) Nose shape and climate. Am. J. Phys. Anthropol. 12: 1–4.

Weinstock MA (1993) Ultraviolet radiation and skin cancer: epidemiological data from the United States and Canada. In AR Young (ed.): Environmental UV Photobiology. New York, Plenum Press, pp. 295–343

Werner J (1987) Thermoregulation mechanisms: adaptations to thermal loads. In S Samueloff and MK Yousef (eds.): Adaptative Physiology to Stressful Environments. Boca Raton, Fla.: CRC Press, pp. 17–25.

Werner J (1993) Temperature regulation during exercise: an overview. In CV Gisolfi, DR Lamb, and ER Nadel (eds.): Exercise, Heat, and Thermoregulation. Perspectives in Exercise Science and Sports Medicine. Dubuque, Iowa: Brown and Benchmark, , vol. 6, pp. 49–79.

Winslow RM and Monge C (1987) Hypoxia, Polycythemia, and Chronic Mountain Sickness. Baltimore, Md.: Johns Hopkins Univ. Press.

Wyndham CH, Metz B, and Munro A (1964) Reactions to heat of Arabs and Caucasians. J. Appl. Physiol. 19: 1051–1054.

Yarbrough BE and Hubbard RW (1989) Heat-related illness. In PS Auerbach and EC Gheer (eds.): Management of Wilderness and Environmental Emergencies, 2nd edition. St. Louis: Mosby, pp. 119–143.

Yip R, Binkin NJ, and Trowbridge FL (1988) Altitude and childhood growth. J. Pediatr. 113: 486–489.

Yoshimura H and Iida T (1952) Studies on the reactivity of vessels to extreme cold. Part II: Factors governing the individual difference of the reactivity, or the resistance to frostbite. Jpn. J. Physiol. 2: 177–185.

CHAPTER 7

The Epidemiology of Human Disease

LISA SATTENSPIEL

INTRODUCTION

Epidemic. Disease. For most people, these words conjure up feelings of discomfort and fear. But for a few, the words generate so much interest that they are driven to investigate how, when, where, and why diseases occur. These individuals are called **epidemiologists**, and their discipline is called **epidemiology**.

The field of epidemiology arose from attempts to understand and control the great epidemics that were rampant in Western Europe until the 20th century. The discipline now includes the investigation of not only infectious and epidemic diseases but all kinds of human diseases. This broadened scope has resulted in various approaches to studying and understanding human disease.

MAJOR CONCERNS AND APPROACHES IN EPIDEMIOLOGY

One major focus of epidemiologists is the determination of the causes of a particular malady. In modern Western cultures, where epidemiologists take a biomedical approach to disease, this process consists of answering several primarily biological questions about the disease, such as,

"Is it an **infectious disease** or a **noninfectious disease**?"

"What are the genetic determinants of this disease, if any?"

"Is the disease caused by extrinsic (environmental) factors, or is an internal infectious organism (**pathogen**) responsible?"

Non-Western cultures, however, sometimes have different views of what causes disease and even what constitutes disease itself. For example, the Ogori, an African people studied by Gillies (1976), believe that childhood illnesses are due to difficulties in

Human Biology: An Evolutionary and Biocultural Perspective, Edited by Sara Stinson, Barry Bogin, Rebecca Huss-Ashmore, and Dennis O'Rourke
ISBN 0-471-13746-4 Copyright © 2000 Wiley-Liss, Inc.

the emergence and identification of the shadow soul, which is normally transmitted through reincarnation of a recently deceased person. Insanity and tuberculosis are thought to be caused by magical medicine. Many other diseases are attributed to causes rooted in a person's personal social, moral, or ritual situation.

Epidemiology is a population-centered discipline and focuses especially on interactions among the **host**, the **agent**, and the **environment**. In this endeavor, epidemiologists study the biology and the behavior of both humans (the host) and any pathogens (agents) that cause the disease, as well as the characteristics of nonliving agents and environmental factors that may influence the development and manifestations of the disease.

Some epidemiologists center their research on specific characteristics of causal factors that have been identified, especially if the causal factors are biological organisms such as **bacteria, viruses**, or worms. For example, if a disease is caused by a **parasitic** worm, epidemiologists might be interested in understanding the life cycle of the worm and how it is passed from one person to another or whether the worm can survive in other species. They also might be interested in how human behaviors influence the life cycle of the worm or how characteristics of the environment, such as temperature or amount of rain, affect the survival and reproduction of the parasite or the ease with which it spreads to humans.

Another interest of epidemiologists is the identification and study of attributes of the human host that affect the introduction and persistence of a disease in an individual. These attributes include demographic characteristics such as age, sex, ethnicity, and economic status; biological characteristics such as degree of **immunity**, **susceptibility**, and nutritional status; and behavioral characteristics such as the nature of risk behaviors, cultural traditions that influence disease transmission and persistence, and psychological factors that influence the manifestation of disease.

Epidemiologists also are interested in how the environment influences the nature and persistence of a disease. They might study the relationship between electromagnetic fields near high-intensity power lines and the development of cancer, the effects of environmental degradation on the evolution of new viruses, or how the widespread use of antibiotics has led to the development of antibiotic-resistant strains of bacteria.

Many epidemiologists collect the information they need through observational studies within a population. They identify members of a population to study and then observe whether those individuals experience particular risk factors (e.g., smoking, living near a toxic waste site, etc.) and whether those individuals have or develop the disease they are studying. Other epidemiologists use experimental approaches to conduct their research. However, because disease (defined as the absence of health, or the opposite of ease) is not beneficial to the human host, most people won't allow researchers to experiment on them. Notable scientific advances sometimes are attained when people take part (usually willingly, but sometimes by force) in experimental studies. For example, in 1900–1, the Walter Reed Commission dramatically illustrated that yellow fever is a mosquito-borne disease. The disease developed in volunteers bitten by mosquitoes that had recently fed on people in the early stages of the disease but did not develop in volunteers bitten by mosquitoes that had fed on uninfected people. More commonly, though, experimental epidemiology involves clinical trials, in which medical procedures, control measures, and other treatments are tested for effectiveness.

Epidemiologists are often interested in describing and understanding observed patterns of disease occurrence. This work overlaps with and uses some of the same methods as that of medical geographers. For example, epidemiologists and medical geographers may superimpose a map of infectious disease distribution on a climate map of the same area to see whether there is a link between the disease and climate. They may plot the cases of leukemia within a region to determine whether there is any clustering of the disease in space and time, or they may study how an infectious disease spreads throughout a human population distributed in space.

The concerns of epidemiologists overlap with those of anthropologists when they study how other cultures determine disease causation and attempt to control disease spread. Their concerns also overlap when they study the effects of cultural factors on the development and manifestations of different diseases. For example, an epidemiologist may observe that a certain kind of **food-borne illness** occurs less commonly in one subpopulation than in another. Anthropological study of the cultures of the two subpopulations may reveal that this difference is due to dietary practices, such as not eating raw meat or preparing the food in a way that kills an organism implicated in the outbreak.

Only a relatively small proportion of epidemiologists combine an anthropological focus with their public health orientation; epidemiologists generally have stronger training in human biology and public health. Although public health-oriented epidemiologists and anthropologically oriented epidemiologists use many of the same methods, they may have significantly different approaches to epidemiological questions. These approaches are clearly illustrated in many descriptive epidemiological studies, in which a researcher studies the amount and distribution of disease according to who has it, where it is found, and when it occurs. For example, when a public health-oriented epidemiologist considers attributes of a person, he or she usually looks at broad categories such as age, sex, ethnic group, social class, occupation, or marital status—all of which are defined prior to the study. These kinds of data are routinely collected in censuses, at the outset of medical studies, and in many other situations, so it is reasonable to set up a study along such lines. Anthropologically oriented epidemiologists, on the other hand, also may use these categories, but more often, they will study the culture first to determine attributes that the natives consider essential elements of their culture. Anthropologically oriented epidemiologists assume that natives are active participants and may consider variables such as lifestyle or food preparation techniques, which may be ignored by public health-oriented epidemiologists.

The two kinds of epidemiologists also differ in their approach to studies of environmental aspects of a disease. Public health-oriented epidemiologists tend to focus on the physical features of the environment and group-related behaviors, such as natural boundaries, political subdivisions, urban–rural differences, and international comparisons.

Anthropologically oriented epidemiologists tend to be more interested in how a culture's response to disease changes when the physical or social environment changes. In addition, because of their focus on both individual behaviors and the culture's view of its structure and the diseases that affect it, anthropologically oriented epidemiologists are often more aware of views about medicine and medical care that do not reflect the Western biomedical paradigm (which tries to find a discrete cause for every malady). This awareness may help to explain how the members of an unfamiliar culture have tried to deal with a disease—knowledge which may, in turn, help the epidemiologists determine effective control measures that will be accepted by the population.

In practice, the majority of anthropologically oriented epidemiologists do not differ much in orientation, methods, and approach from public health–oriented epidemiologists. However, with the increasing contact among people from many different cultural backgrounds, there is a great need for more anthropologically oriented epidemiologists with strengths in anthropological approaches as well as a thorough knowledge of public health approaches.

BASIC EPIDEMIOLOGICAL CONCEPTS

Terminology

As is true of most scientific disciplines, epidemiologists use several specialized terms, the most important of which are infectious disease and noninfectious disease. Infectious or **communicable diseases** are caused by specific infectious agents or their toxic products; these agents or toxins are transmitted from one person to another, directly or indirectly. Noninfectious diseases include most kinds of cancer (as far as we know), genetic diseases, nutritional diseases, allergies, and diseases caused by other environmental factors. Although the procedures overlap, people studying infectious diseases use different methods and ask different questions from people who study noninfectious diseases. One person rarely focuses on both infectious and noninfectious diseases (except, perhaps, if studying interactions between nutritional status and infectious diseases).

Regardless of whether a disease is infectious or noninfectious, when it is present in a population at a relatively constant (usually low) level at all times, it is referred to as **endemic**; when the number of cases in a fairly localized area suddenly increases above the expected or normal level for a short time, the disease is said to be **epidemic**; and when the number of cases occurring worldwide suddenly increases, the disease becomes pandemic. The total number of cases of a disease in a given population during a particular time interval is referred to as the **prevalence**, whereas the number of *new* cases during a particular time interval is referred to as the **incidence**. An epidemic or pandemic occurs when the incidence of a disease suddenly increases, whereas an endemic disease shows no change in incidence over time. High prevalence rates can occur in all three situations, as can low overall prevalence rates. (Thus, it is possible to talk about an epidemic occurring even though there may only be a few cases, if the underlying average prevalence rates are near zero.)

Several special terms are associated with the progression of an infectious disease within an individual from the time a **susceptible** person (i.e., one at risk for infection) becomes infected until that person either recovers or dies. Figure 7.1 illustrates this process, which can be divided into several stages of varying length depending on the particular disease. At the onset of infection, a person enters both the incubation period (i.e., time from infection to the development of symptoms) and the **latent period** (i.e., length of time between infection and the ability to infect someone else); the incubation period is usually somewhat longer than the latent period, which usually ends before symptoms occur. It is difficult to control the spread of many infectious diseases because transmission of the infectious organism can and often does occur before anyone is aware of illness.

The end of the latent period signals the beginning of the infectious period, and the end of the incubation period signals the beginning of the period of symptomatic illness. The infectious period lasts until the person can no longer transmit the disease, which may be before symptoms have subsided. For many but not all infectious diseases, a person may enter a state of temporary or permanent immunity at the end of the infectious period.

Death may occur at any time during this process; the chance of death depends on the disease, the underlying health of the infected person, and contributing environmental factors (e.g., quality of health care, sanitation).

The Nature and Classification of Disease Agents

At the most basic level, human diseases can be divided into two categories: those caused by infectious organisms or particles and those with noninfectious causes. Table 7.1 shows the major kinds of diseases within these two categories and examples of each.

Classes of matter that are capable of causing infectious diseases in humans include bacteria, rickettsiae, viruses, prions, protozoa, fungi, and metazoa. In some of

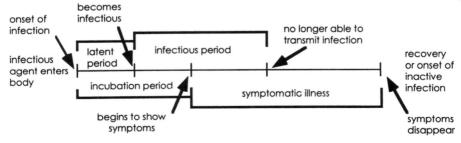

Figure 7.1   The progression of an infection in an individual. This scenario assumes that the person survives the stages of active infection. Death from an infectious disease can occur at any stage of infection and depends on the infectious organism, the underlying health of the infected person, and environmental conditions.

these classes, the majority of species are parasitic; that is, they depend on the host for their own nourishment and survival, to the detriment of the host. In other classes, only a few species are parasitic.

Bacteria are microorganisms that lack a true nucleus and usually have their **DNA** in a single molecule. Rickettsiae are sometimes classified as a kind of bacteria but are more frequently considered to be a separate kind of microorganism. They have a bacteria-like structure but are obligate intracellular parasites of mammals, which means that during at least some stages of the life cycle all kinds of rickettsiae *must* live within the cell of a mammal and draw on that animal's resources. Viruses also are obligate intracellular parasites, but they have a much simpler structure than rickettsiae. It consists of a **protein** coat surrounding a molecule of either DNA or **RNA**, but not both, with little other internal structure. Viral DNA or RNA can be either single or double stranded. In addition, viruses are not capable of independent metabolic activities; rather, they take over the energy production in a host cell to facilitate their replication. For this reason, biologists usually consider viruses to be nonliving.

Prions are also nonliving particles capable of causing infectious diseases. They are pieces of protein that lack any genetic material of their own but can be encoded in the host's genetic material and can be transmitted from person to person. **Protozoa** are single-celled organisms that lack a true cell wall (therefore, they are more like animals than like plants). Fungi, including yeasts, molds, and mushrooms, are multicellular decomposers and are distinguished from other organisms by a characteristic growth pattern, spore formation, and lack of chlorophyll. Metazoa are multicellular

Table 7.1   Major causes of human disease and examples of each type

Cause	Examples
Infectious disease	
Bacteria	Strep throat, tuberculosis, syphilis, ulcers?
Rickettsiae	Typhus, Rocky Mountain spotted fever
Viruses	Measles, herpes, warts, some cancers
Prions	Kuru, Creutzfeld–Jakob disease
Protozoa	Giardia, trypanosomiasis, malaria
Fungi	Histoplasmosis, athlete's foot, ringworm, yeast
Metazoa	Schistosomiasis, hookworm, tapeworm
Noninfectious disease	
Nutrition	Cirrhosis of the liver, rickets, scurvy, obesity
Poisons	Carbon monoxide, ergotism, arsenic, lead
Allergens	Hay fever, sinusitis, poison ivy
Metabolic disorders	Neonatal jaundice, Parkinson's disease
Hormonal disorders	Hypothyroidism, diabetes, giantism
Genetic diseases	Tay–Sachs disease, sickle-cell anemia
Psychological factors	Posttraumatic stress syndrome, anorexia nervosa
Physical factors	Skin cancer, altitude sickness, bends

animals that possess tissue differentiation. Metazoa that cause disease in humans include most of the common worms that are ubiquitous in tropical regions and occasionally are found in more temperate regions as well. Because of their large size and relatively complicated structures and complex life cycles, metazoa are commonly referred to as **macroparasites**. The other six classes of infectious organisms are referred to as **microparasites**.

Noninfectious diseases also have many causes, the most common of which are given in Table 7.1, with examples of related diseases. In many cases, a particular noninfectious disease is due to interactions between two or more of these causes. It is also often the case that the causes of a particular noninfectious disease have not been or cannot be determined using modern scientific techniques. Nevertheless, many diseases have been related to environmental factors.

Nutritional diseases occur as a result of either too much or too little of a particular nutrient or nutrients. Scurvy results from a deficiency of Vitamin C; **rickets** results from a deficiency of Vitamin D. Cirrhosis of the liver can result from an excess of alcohol (or from other non-alcohol-related causes), whereas obesity is sometimes considered to be a disease related to an overabundance of nutrients in general. Diseases that result from poisons include botulism, which occurs when foods become contaminated with a toxin produced by the bacterium *Clostridium botulinum,* and ergot poisoning ("St. Anthony's Fire"), which results from eating grain contaminated with ergot fungus. Allergens are substances that stimulate an immune response in the body and lead to local or generalized (systemic) body reactions. Sometimes, byproducts of metabolism or inappropriate amounts of **hormones** can lead to disease in humans. For example, hypothyroidism, which results from an inadequate amount of thyroid hormone, is one of the most commonly treated medical conditions in the United States today. Diseases such as Tay–Sachs disease and sickle-cell anemia are due to simple genetic traits. Many diseases are influenced by psychological factors, and for some, such as posttraumatic stress syndrome, psychological factors are the primary cause. Physical factors, such as sudden changes in atmospheric pressure or ionizing radiation, are implicated in some kinds of human disease.

The existence and severity of both infectious and noninfectious disease is influenced by many risk factors. Several of these factors and examples of what they influence are given in Table 7.2.

Modes of Transmission of Infectious Diseases

The major problem facing all living organisms is how to guarantee that at least some of the genetic material from one generation is passed to the next generation. If the organism happens to be a human parasite, then this transmission of genetic material can occur only if the parasite finds a human host in which to spend all or part of its life cycle. However, all living organisms have a fixed life span, so a given parasite cannot spend an infinite amount of time in a single host; it must develop ways to spread from one host to another. Parasitic organisms use many different strategies, called modes of transmission, to accomplish this feat.

Table 7.2 Factors that influence exposure, susceptibility, or response to disease agents

Factors	Examples
Host factors	
Genes	Predisposition to diabetes, heart disease, or breast cancer
Age	Development of heart disease or Alzheimer's disease
Sex	Development of premenstrual syndrome or breast cancer
Ethnicity	Parasites from raw fish, hemolytic disease of the newborn
Economic status	Malnutrition, diseases related to inadequate sanitation
Cultural factors	Malnutrition resulting from food taboos, limitations on use and availability of health resources
Physiological state	Pregnancy, fatigue, stress, nutritional state
Prior immunologic experience	Hypersensitivity, vaccinations, maternal antibodies, natural immunization or resistance
Underlying level of health	AIDS patients' susceptibility to tuberculosis and other infections
Environmental factors	
Physical	Climate, soil type, water quality
Biological	Nature of vegetation, types of food available, prevalence of alternate hosts for infectious diseases, density of infectious disease vectors
Social	Human population density, structure of human communities, level of sanitation

Because of the nature of modes of transmission, it is useful to divide them into two basic types: direct modes, whereby transmission occurs directly from one primary host to another, and indirect modes, whereby an intermediate host or agent is needed to effect transmission from one primary host to another primary host. Table 7.3 shows the major kinds of direct and indirect transmission and gives examples of each.

The most familiar mode of transmission of human diseases is respiratory transmission. Infectious organisms are spread through the air when a person coughs, sneezes, or even breathes. Because the infectious organisms are usually carried in droplets of moisture released from the body, this mode of transmission is often referred to as **droplet transmission**. Respiratory transmission can occur only if a susceptible person is in close enough contact with an infectious person to breath in the infectious droplets released by the infectious person. This happens much more easily when large numbers of humans live in close association with one another on a regular basis; hence, droplet transmission increased significantly in importance after humans settled down and began living in larger, more sedentary communities. Many of the familiar childhood diseases, such as measles, mumps, and influenza, are spread by this means.

Several diseases of public health importance, especially gastrointestinal illnesses and some macroparasites, are spread by **fecal–oral transmission**. In these diseases, the infectious stage of a parasite is excreted in the feces of humans or other animals. During the course of daily activities, susceptible humans come into contact with these feces and (usually) inadvertently introduce the infectious organisms into their mouths, by which means they enter the body and cause illness.

Many of the most feared human diseases are spread through sexual contact, including HIV, gonorrhea, syphilis, and genital herpes. Sexually transmitted disease organisms usually require a warm, moist environment for survival and do not live long outside the human body.

Some diseases can be transmitted from a mother to her unborn offspring, either during pregnancy, when the pathogen crosses the placental barrier, or at the time of delivery. This type of transmission is called vertical (or congenital) transmission because the disease spreads from one generation to the next. Often, these pathogens occur commonly in older children and/or adults, where they may range from mild to relatively severe in their effects. However, when they attack a fetus or newborn infant, who does not have a fully developed immune system, they may cause much more serious illness or even death of the infant. For example, if a woman contracts rubella (German measles) during the first trimester of pregnancy, her infant may be born with Multiple Congenital Rubella Syndrome, a disease characterized by blindness, deafness, heart problems, and mental retardation.

Table 7.3   Modes of transmission of human infectious diseases

Mode	Example
Direct transmission	
Respiratory	Influenza, chicken pox, measles
Fecal–oral	Giardia, hepatitis A, rotavirus, pinworm
Sexual	Syphilis, gonorrhea, HIV, genital herpes
Vertical (Congenital)	Rubella, syphilis, HIV, toxoplasmosis
Direct physical contact	Yaws, diphtheria, chicken pox, herpes simplex
Indirect transmission	
Vehicle-borne	
Water-borne	Cholera, hepatitis, typhoid fever
Food-borne	Botulism, salmonellosis, tapeworm
Soil-borne	Hookworm, tetanus, histoplasmosis
Needle sharing[a]	Hepatitis B, HIV, HTLV-I(Human T-lymphotropic virus, Type I)
Vector-borne	
Mosquitoes	Malaria, yellow fever, dengue fever
Ticks	Rocky Mountain spotted fever, Lyme disease
Fleas	Bubonic plague, murine typhus fever
Lice	Louse-borne typhus fever, trench fever
Flies	Trypanosomiasis, leishmaniasis, onchocerciasis
Other	Chagas disease
Complex cycles	Schistosomiasis, Guinea worm, hydatid disease

HIV, human immunodeficiency virus, HTLV-I, human T-lymphotropic virus, Type I.
[a]Includes both vaccinations and intravenous drug use.

Direct physical contact with infectious lesions can spread some diseases from one person to another. Yaws, pinta, and endemic syphilis are diseases caused by bacteria closely related to (and in the case of endemic syphilis, indistinguishable from) the species that causes venereal syphilis. They are very common worldwide and are transmitted principally through direct skin contact with infectious lesions.

Indirect modes of transmission include mechanisms that require an intermediate agent to facilitate transmission between human hosts. The intermediate agents can be either inanimate vehicles or living **vectors**. Inanimate vehicles include objects and materials such as needles (through vaccination or intravenous drug use); water, which can carry an infectious organism from one place to another; soil, in which infectious organisms can live while waiting for an unsuspecting host to come along; or food, which is frequently shared among many humans.

Figure 7.2 illustrates the transmission cycle of schistosomiasis, a water-borne disease with an interesting and complex life cycle that involves an intermediate

Figure 7.2 Life cycle of schistosomiasis. (a) Eggs produced by the sexual stage leave humans via urine or feces. (b) Eggs that reach the water shed their shells and hatch into free-swimming miracidium. (c) Miracidium must locate snails of the appropriate species and genotype within 1 day. (d) Miracidia multiply asexually through two stages into thousands of cercariae. (e) Free-swimming cercariae live for up to 2 days. (f) Cercariae swim until human skin of suitable warmth and smell is encountered. (g) Cercariae penetrates skin and becomes wormlike schistosomule. (h) Schistosomule migrates to lung, sometimes producing cough, then appears in portal system of liver, where it reaches sexual maturity and mates. (i)Worm pairs migrate to blood vessels lining lower small intestine and large intestine (or bladder) and lay eggs. Some eggs work their way into the interior of the intestine and are excreted into the water with feces or urine (Benenson 1995) .

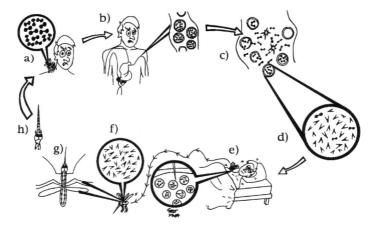

Figure 7.3   Life cycle of malaria. (a) Susceptible human is bitten by mosquito carrying malaria zygotes (known as sporozoites). (b) Trophozoites invade human liver and multiply asexually. (c) Schizonts leave liver and enter red blood cells, where they duplicate asexually to form merozoites. (d) Some of the red blood cells rupture and release the merozoites and their toxic byproducts into the bloodstream. (e) The clinical symptoms of malaria appear. Some of the merozoites develop into sexual gametocytes, which may mature but which cannot undergo sexual reproduction within the human host. Mosquitoes bite human who carries mature gametocytes in blood. (f) Malaria zygote develops within mosquito. (g) Zygote moves from intestinal tract of mosquito into bloodstream. (h) Zygotes, now sporozoites, settle in salivary glands of mosquito. (*Source:* Adapted from Wood 1979)

snail host. Water is the vehicle responsible for carrying the schistosomiasis pathogen from one host to another. Living vectors usually are arthropods, which include, for example, mosquitoes, ticks, fleas, and lice. Figure 7.3 illustrates the transmission cycle of malaria, a vector-borne disease transmitted from one human to another by a mosquito. The pathogens that cause malaria are not carried mechanically from one human to another by the mosquito; rather, like many other vector-borne disease agents, they complete some of their life stages within the vector itself.

HISTORICAL ATTEMPTS TO UNDERSTAND HUMAN DISEASE

Because disease is a part of all cultures, humans have probably always pondered its nature and causes. However, without written records, it is impossible to really know what people thought about disease causation and treatment in prehistoric times. Modern epidemiology is primarily a product of intellectual thought dating from the classical civilizations of Greece and Rome, even though Greek and Roman medicine almost certainly reflected knowledge passed down from many previous generations and populations.

The most important concept derived from classical Greek thought can be traced to the writings of Hippocrates, who believed that all matter was made up of four elements: earth, air, fire, and water, which possessed the corresponding qualities of cold, dry, hot, and moist. These elements were carried in the body by four humors: phlegm, yellow bile, blood, and black bile (Fox et al. 1970). Hippocrates believed that disease resulted from a disturbance in constitution, which included not only bodily humors but also the climate and weather. The goal of Hippocratic physicians was to restore health by restoring the proper balance of the humors through methods designed to aid the natural healing capacities of the body.

Although Hippocrates' ideas about bodily humors may not seem such an obvious part of modern medicine, they form the base from which modern medicine is derived. In fact, they were the predominant Western view of disease causation for many centuries. For example, Cipolla (1992) told the story of Giovan Battista Cartegni da Bagnone, a 17th-century Italian doctor. Dr. Cartegni was asked by the Florentine Magistracy to investigate abnormal levels of morbidity in the area of Marciana, Castello di Cascina, Pontedera, and Ponte di Sacco during a three-day period in March 1610. Cartegni observed that "the winter had been 'cold beyond measure and snowy.' During the cold weather the heads [of the peasants] had become filled with humours. These humours should have fallen 'from the head to the throat' but this did not happen during the winter months" (Cipolla 1992, 36). Dr. Cartagni believed that because of the intense cold "the phlegmatic humours were 'frozen and unable to move' and for that reason had remained blocked in people's heads." When the warmth of the spring arrived, the humours began moving again and dropped down to the throat and chest, where they caused the inflammation of the lungs that he observed.

Hippocrates and his followers also contributed other ideas to the development of modern Western medicine. They stressed the importance of the environment and climate as determinants of health, were concerned about the interplay between diet and disease, recognized different susceptibility to disease at different ages, and attempted to minimize explanations for disease based on magic or other supernatural forces (Moore et al. 1980).

Besides Hippocratic ideas, modern epidemiology is built around the concept of **contagion**, or the transmission of infection by direct contact between individuals, droplet spread, or contact with inanimate objects. Although ideas about this concept had been around for a long time, the first person to formalize its definition was Girolamo Fracastoro. Fracastoro is best known for his poem about syphilis, *Syphilidis, sive morbi Gallici, libri tres* (*On Syphilis, or the French Disease, in Three Books*), but he also developed the idea of contagion in his 1546 book *De contagione et contagiosis morbis et eorum curatione* (*On Contagion and Contagious Diseases and their Cure*). Fracastoro believed that diseases were transferred from person to person via invisible particles present in the air. In fact, the Latin word *contagio* means "contact" or "touching."

The concept of contagion was also influenced by European (especially Italian) attempts to understand and control bubonic plague. When discussing plague, "people spoke of an ill-defined but universally recognised 'corruption and infection of the air,' which degenerated into highly poisonous 'sticky' miasmas that killed the person they infected, either by inhalation or by contact. … the 'corruption and infection of the air' could be caused by an inauspicious conjunction of the stars, vapours rising from marshy water, the eruption of volcanoes, foul and filthy conditions or exhalations arising from 'rebus et corporibus putridis et corruptis' [decomposed and corrupted matter and bodies]" (Cipolla 1992, 4). Miasma was a commonly used but poorly defined term that referred to particles present in low-quality air that were thought to be passed from infected persons to susceptible persons to cause contagious illnesses.

Probably the most important building block for modern epidemiology is the **germ theory of disease**. This theory was developed and tested by Pasteur in his bacteriological research of the 1860s. Pasteur showed that disease could be produced after introduction of a **virulent** microorganism (or germ) into the body. (His work would have been impossible without microscopes, which had come into common use in the late 1600s.)

Throughout the latter half of the 19th century, scientists attempted to understand the transmission of infectious diseases and control epidemics. John Snow, one of the most influential early epidemiologists, conducted a landmark field study that clearly demonstrated that contaminated water (i.e., some kind of particle carried in water) was responsible for the spread of cholera. During his study, Snow plotted all deaths from cholera on a map of a large area of London, along with the primary water sources for the region. Two different companies supplied water, with distributions that overlapped in an area that included the Broad Street pump. Snow was able to conclusively trace the cholera to contaminated water coming from one company and, in particular, to the Broad Street pump itself. He then convinced the authorities to remove the handle of the pump, which stopped the epidemic (Barua 1992, Longmate 1966, Snow 1965).

Although Snow was able to prove that something in the water was causing the cholera outbreak, he was not able to isolate the real cause. This determination had to wait until Robert Koch, a German physician, isolated the microorganism *Vibrio cholerae* in 1883 and proved its link to the disease. (The bacillus was first observed by Filippo Pacini in Italy in 1854, but he did not clearly demonstrate that it caused cholera [Barua 1992]). Koch's work solidified the role of Pasteur's germ theory in modern Western epidemiology and medicine. He derived a series of postulates that are still used to verify that a particular microorganism is the cause of an infectious disease. A similar series of postulates was proposed by a German anatomist, Jakob Henle, in the 1840s (Risse 1993); however, because this fact is not well-known, the postulates continue to be called Koch's postulates. The postulates are as follows:

1. The disease microorganism must be present in every case of the disease.
2. The microorganism must be grown in pure culture derived from the diseased host.
3. The same disease must result when a sample of this culture is injected into a new host.
4. The microorganism must be recovered from the body of the new host.

These postulates were developed to handle bacteria, which are relatively easy to isolate and culture, and the postulates are an effective way to verify a bacterial cause for disease. However, viruses are difficult to isolate and grow in culture, so it has been problematic to use Koch's postulates to verify that an illness is caused by a virus. This situation has become less of a problem as laboratory techniques for isolating viruses have become more sophisticated.

METHODS USED BY EPIDEMIOLOGISTS TO STUDY DISEASE IN HUMAN POPULATIONS

Collection of Epidemiological Data

To understand the causes of disease and its impact in human populations, one must have descriptive information about what kinds of diseases are present, how frequently they occur, how often people die from a particular disease, which people are at highest risk, etc. These kinds of data derive from many sources.

At the core of epidemiology is the analysis of **vital statistics** on **morbidity** (illness) and **mortality** (death). Modern epidemiology was not possible until people began regularly collecting such data. This practice began in Western Europe in the 1500s and only recently has reached many regions of the world. In Europe, registration of births and deaths was originally associated with church membership, and many of the best historical demographic and epidemiological records are found in parish records. (This is still true in many parts of the world.) One of the earliest government requirements for the collection of vital statistics was a 1532 English ordinance that required parish priests to compile weekly "Bills of Mortality," which gave the numbers and causes of deaths. This ordinance arose out of attempts to control the major epidemics of bubonic plague that were ravaging Europe at the time.

In 1662, an Englishman named John Graunt began to analyze data recorded in the London Bills of Mortality to help identify reasons for both the overall numbers of deaths and patterns of death by different causes. Yet, despite his and other scholars' work (most notably Edmund Halley, the discoverer of Halley's Comet), registration of births and deaths was not compulsory in England until 1836. Fortunately, largely as a consequence of British colonialism and the establishment of the World Health Organization (WHO) and the United Nations, statistics on births and deaths are regularly collected in most parts of the world today. In addition, information is often recorded regarding causes of death and the number of cases of serious infectious and noninfectious illnesses.

Other major sources of data similar to vital statistics are large-scale population **surveys** and **surveillance** by organizations such as the US Centers for Disease Control (CDC) and the WHO. These organizations keep detailed weekly records of cases of numerous infectious and noninfectious diseases. The CDC focuses primarily (but not exclusively) on diseases within the United States, whereas the WHO keeps track of reportable diseases throughout the world. Standardized forms are used in most parts of the world, so an epidemiologist doing research on diseases on a small Pacific island will find data on cases of the same diseases that are reported by authorities in the United States, central Africa, and many other places.

Vital statistics and surveillance data are usually collected by government authorities, religious authorities, or nurses or doctors at local medical clinics who have treated patients with the particular illnesses. In general, these data are not collected by epidemiologists, who may also have little or no control over the types and quality of data collected. The nurses, doctors, religious authorities, and government workers are responsible for recording the data on the proper forms and then submitting them to the proper higher authorities. Depending on the importance a particular region attaches to this information relative to other issues facing them, the data may or may not be recorded accurately. Consequently, the quality of these sources of epidemiological and demographic data varies substantially. In addition, health care is grossly inadequate in many regions of the world, and there may not be anyone available to record **demographic** and epidemiological events. Diagnoses of causes of illness and death may be inaccurate, and the numbers of births and deaths may be underreported. Because of these and other concerns, at the outset of any epidemiological study that uses such data, the quality of the data must be carefully assessed.

Epidemiologically oriented clinical trials provide data that have been generated for a specific epidemiological purpose. An epidemiologist usually has been involved in such a project from the beginning and has had input into which data are collected and how the study is designed. Because of the controlled nature of the studies, these data may be of more uniform quality than census and health data collected by a government organization. However, if the study design focused on a subpopulation rather than the entire group, the sample population may not be representative of the entire population.

Clinical trials are usually more expensive and time-consuming than large population surveys for the collection of demographic and health data. Furthermore, although data collected in clinical trials might be useful for the study for which they were intended, they may not be generalizable to other studies unless similar questions are being addressed.

Descriptive Methods for the Analysis of Epidemiological Data

The first step in most epidemiological studies is to use descriptive statistical methods to analyze the data collected in the study. This process usually involves the calculation of simple **ratios** and other statistics that allow the researcher to describe patterns of illness and mortality within the study population. Inferences from these analyses then can be used to help the researcher determine future, more sophisticated directions to take in the study.

Several descriptive statistical measures are used routinely in epidemiological studies. They can be divided into two categories: those that describe the levels and risk of morbidity in a population, and those that describe the levels and risk of mortality.

Measures of Morbidity

The levels of morbidity in a population are usually determined by measuring the incidence and prevalence of a disease. Recall that the incidence of a disease is the number of new cases of a disease occurring during a particular time interval, whereas the prevalence is the total number of existing cases of a disease in a population during a particular time interval. Note that both prevalence and incidence refer to a particular population and a particular time interval.

Incidence is the number of transitions from a healthy state to a diseased state among the individuals in a population. A given person can go through this transition only once for any particular episode of disease. Hence, incidence measures the population's level of risk for the disease. Prevalence, however, does not take into account when a person developed a disease. It represents a snapshot of a population at a particular time and measures how many people have the disease at that time, regardless of how long they have had it. Because the duration of disease usually varies, prevalence is not a measure of risk.

Two measures of prevalence are commonly used in the medical and epidemiological literature. Point prevalence measures the prevalence at a particular time and is the measure described earlier. Period prevalence measures how many people have had a disease at any time during a defined time interval (e.g., years or days).

Prevalence and incidence rates are calculated by dividing the numbers of cases by the population at risk (see Box 7.1). Prevalence rate is defined to be the total number of cases of a disease in a given population during a particular time interval divided by the number of people at risk for the disease at the midpoint of the time interval. Incidence rate is defined to be the number of new cases of a disease in a particular population during a specified time interval divided by the number of persons exposed to the risk of developing the disease. The denominator is usually assumed to be the population at the midpoint of the time interval considered.

Measures of Mortality

One of the reasons disease is feared is that the consequences of some disease will be the ultimate cause of death for almost everyone. Humans therefore have a natural interest in the frequency or rate of deaths due to disease. Epidemiologists measure these mortality rates in many different ways. The simplest mortality rate, the **crude mortality rate**, is based on only the total number of deaths within a population during a specified time period. As with prevalence rates, the denominator for the crude mortality rate is usually assumed to be the population at the midpoint of the time interval considered (see Box 7.1).

Often it is desirable to divide the population into categories and look at the mortality rates within a category. Commonly used categories include sex, age, occupation (especially for deaths due to occupational risks), and kind of disease. These category-specific mortality rates are calculated similarly to the general mortality rate: The numerator is the number of deaths within a category, and the denominator is the number of people within that category at risk of dying. Because these rates focus on subgroups within a larger population, the denominator does not include the entire population; rather, it includes only the subgroup of interest. (See Box 7.1 for an example of how to calculate **age-specific mortality rates**. Analogous calculations are used for other category-specific rates.)

For chronic but often fatal diseases, such as most forms of cancer, epidemiologists may be interested in determining the probability of death within a certain time after diagnosis of the disease. Such a rate is called a case fatality rate and is used both to measure the severity of a disease and to measure the benefits of new therapies for the disease. Because case fatality measures the probability of death, more severe diseases have a higher case fatality rate, and beneficial therapies lower the case fatality rate. The population of interest (numerator) for the case fatality rate is the number of people who die from a particular disease within a specified time period. The population at risk is the total number of people who have had that disease for the time period involved. The formula used to calculate the case fatality rate is given in Box 7.1.

The proportion of total deaths that are due to a specific cause is called the **proportionate mortality ratio** (PMR). Diseases that generate a higher PMR cause death more commonly than diseases with a lower PMR (see Box 7.1). Because the PMR is a proportion, it can be used to compare the impact of a particular disease in different populations. For example, if the PMR for heart disease in population A is 50% (i.e., half of all deaths are due to heart disease) but only 40% in population B, then the impact of the disease is higher in population A. The total number of cases in population B may exceed the total number of cases in population A if population B is larger than population A, but the impact remains higher in population A.

Age Standardization of Epidemiological Rates

Different populations differ in **age distribution** (i.e., the proportion of people at each age within a population). This difference makes it difficult to compare populations, because differences in epidemiological rates, such as morbidity or mortality rates, might be due to differences in the numbers of individuals of a particular age rather than real differences in the impact of death or a disease. **Age standardization** allows epidemiologists to directly compare health statistics in populations that differ in their age distributions.

Consider population 1, which has many more infants and children than population 2. Population 1 is likely to have many more cases of childhood illnesses like measles than population 2, not because population 1 is less healthy, but because population 1 has more people at risk. To compare the impact of measles in the two populations, one must first remove the effects of differences in the age distributions of the two populations. Epidemiologists use two methods of age standardization for epidemiological rates: an indirect method and a direct method (see Box 7.1).

The Design of Epidemiological Studies

Describing the occurrence and distribution of disease and mortality in human populations is not the sole interest of epidemiologists. After the distribution of a disease has been described, many epidemiologists shift to a search for causal associations between that disease and environmental risk factors. Epidemiologists usually use a two-stage process to isolate the causes of disease. First, they attempt to determine

BOX 7.1   COMMON DESCRIPTIVE STATISTICS USED IN EPIDEMIOLOGICAL ANALYSES

Epidemiologists use various statistical measures and techniques in their research, some of which are general measures and techniques used in many disciplines and some of which are specific to epidemiology. Most of the measures described in this box are ratios, proportions, probabilities, or **rates**. These four measures are very closely related and are calculated similarly.

A ratio is simply one number divided by another. Ratios are used to express the frequency of some characteristic relative to some other characteristic.

A proportion measures how frequently a particular characteristic occurs in the entire population relative to all other characteristics in the population. It is a special kind of ratio in which the numerator is included in the denominator and is calculated by using the formula

$$\text{proportion} = \frac{\text{No. of times a characteristic appears in a population}}{\text{No. of times it could have appeared}} \tag{7.1}$$

The denominator of this fraction is the population at risk for having the characteristic and is usually the entire population being studied.

A probability is the limit (ultimate value) of a ratio (or proportion) as the sample size gets infinitely large:

$$\text{proportion} = \frac{\text{No. of times a characteristic appears in a population}}{\text{Size of the population being sampled}} \tag{7.2}$$

From this formulation, it should be clear that a probability is the proportion of times a characteristic appears in a population when the population is infinitely large. Because of this, probabilities are often estimated by the proportion within the population of the characteristic being studied.

A percentage reflects the number of times a given characteristic would appear if the sample population were 100. Percentages are calculated by multiplying probabilities or proportions by 100.

Rate refers to the proportion of events that occur during a specified time interval. Rates are distinguished from proportions and probabilities by the time element, a practice that is not followed in epidemiological usage as rigorously as it could be. The denominator must consider the number of people at risk during a particular time period; this number is related to the length of the time interval considered. Thus, the time interval must be considered explicitly in the calculation of a true rate.

A common unit used in the calculation of a rate is a person-year; 1 person-year is equal to 1 full year in the life of 1 person. A true rate would consider, for example, the number of cases of disease per person-year of observation; a denominator consisting of a simple count of the number of people exposed to a disease at some time during a year would specify a proportion rather than a rate, because those people may not have been exposed the entire year. Unfortunately, several measures are commonly referred to as rates when in fact they are ratios or proportions.

The general formula to use in calculating rates is

$$\text{rate} = \frac{\begin{array}{c}\text{No. of times an event happens}\\ \text{(or characteristic appears) in a population during a particular time}\end{array}}{\text{No. of persons at risk for that event throughout the time period}} \quad (7.3)$$

A common number that appears in the denominator of epidemiological rates is midyear population, which is used as an estimate of the number of person-years lived by members of a population during a particular 1-year period. This number is the average number of persons living throughout the year (or the average number of persons at risk throughout the year), assuming that deaths are spread out evenly throughout the year. To see this, let the midyear population size be N and the total number of deaths during the year be $2x$, with x deaths occurring in each half of the year. Then, the population at the beginning of the year is $N + x$ and the population at the end of the year is $N - x$. Because deaths are distributed evenly throughout the year, the average population is the average of the beginning and end of the year populations, or

$$\frac{1}{2}[(N + x) + (N - x)] = N \quad (7.4)$$

Calculation of the common rates, ratios, proportions, and probabilities used in epidemiology primarily involves determining both the number of people who express the desired characteristic and the nature and number of the population at risk (the hard part). The most important of these measures in epidemiology are incidence and prevalence rates, mortality rates, and the proportionate mortality ratio.

Incidence and Prevalence Rates

The incidence rate measures the number of new cases of a particular disease in a particular time period, relative to the number of people at risk for infection. The point prevalence rate (technically, a ratio) measures the total number of cases of a disease relative to the total population, not only the new cases. Period prevalence measures the total number of cases of a disease during a specified time period relative to the number of persons at risk for the disease during that time period.

$$\text{incidence rate} = \frac{\begin{array}{c}\text{No. of new cases of a disease}\\ \text{occuring during a specific time period}\end{array}}{\begin{array}{c}\text{No. of persons at risk for developing}\\ \text{the disease during that time period}\end{array}} \quad (7.5)$$

$$\text{point prevalance rate} = \frac{\begin{array}{c}\text{No of cases of a disease present}\\ \text{in a population during a specific time period}\end{array}}{\begin{array}{c}\text{No. of persons who might potentially}\\ \text{have the disease at that time}\end{array}} \quad (7.6)$$

$$\text{period prevalance rate} = \frac{\begin{array}{c}\text{No. of cases of a disease present}\\ \text{in a population during a specific time period}\end{array}}{\begin{array}{c}\text{No. of persons at risk for having the}\\ \text{disease during that time period}\end{array}} \quad (7.7)$$

Rates of Mortality

The crude mortality rate relates the number of actual deaths in a population during a specific time period to the total number of people who could have died during that time period. Because most estimates of such rates are small decimals, epidemiologists usually multiply this rate by 1000 and express the mortality rate as "rate of death per 1000 population." (This also may be done for prevalence and incidence measures.) In mathematical terms, this definition is expressed as

$$c.m.r. = \frac{\begin{array}{c}\text{No. of deaths during}\\ \text{a particular time period}\end{array}}{\begin{array}{c}\text{Total population size at the}\\ \text{middle of the time period}\end{array}} \times 1000 \quad (7.8)$$

The crude mortality rate does not take into account any heterogeneity in the population caused by factors such as age or sex. However, it is possible, and often desirable, to express mortality as a function of age, sex, marital status, ethnic group, or another group characteristic of interest. To do so, both the numerator and the denominator must be adjusted to reflect the more narrowly defined number of deaths and the population at risk for those deaths. For example, the age-specific mortality rate would be calculated as follows:

$$a.s.m..r. = \frac{\text{No. of deaths of people aged } x \text{ during a particular time period}}{\text{Total population aged } x \text{ at the middle of the time period}} \times 1000 \qquad (7.9)$$

The case fatality rate is defined as the number of people who die of a disease divided by the number of people who have the disease:

$$c.f.r. = \frac{\text{No of deaths due to a particular disease in a given time interval}}{\text{No. of cases of the disease}} \qquad (7.10)$$

Most often, case fatality is used to estimate the severity of a disease within a time interval beginning at the time of diagnosis. In this common situation, the denominator would be the number of cases of the disease diagnosed one time interval ago. For example, a 1-year case fatality rate would determine the death rate among diseased individuals within the first year of diagnosis.

Proportionate Mortality Ratio

The proportionate mortality ratio measures the proportion of total deaths that are due to a specific cause. Calculation of this ratio results in a decimal number between 0 and 1, so it is usually multiplied by 100 and expressed as a percentage:

$$PMR = \frac{\text{No. of deaths due to a particular cause during a given time period}}{\text{Total number of deaths during that time period}} \times 100 \qquad (7.11)$$

Age Standardization of Epidemiological Rates

Different populations usually have different age distributions. Because the chances of death and illness are strongly influenced by age, two populations may differ in their rates of morbidity and mortality simply because they differ in the numbers of people of different ages. For example, a population with large proportion of infants and young children will have more cases of childhood diseases than a population with a larger proportion of adults. To compare the overall impact of childhood diseases on the two populations, one must standardize the age distribution so that it is similar in the two populations. This adjustment will guarantee that any observed differences are due to the disease rather than differences in the numbers of people at risk for the disease.

Two methods of age standardization are in general use. The direct method involves taking the age-specific death rates from the two populations in question and applying them to a third "standard" population, such as the population of the United States in 1996. The age-specific death rates for *each* study population are multiplied by the numbers of people in respective age categories in the standard population to provide estimates of the number of deaths that would occur in the standard population if people died at the rates observed in the study populations. These estimates are then divided by the total size of the standard population to produce age-adjusted mortality rates that would represent the study populations if they had the same age structure as the standard population (see Table 7.4).

Calculations for the indirect method of age standardization are performed in the opposite direction. The age-specific death rates in the standard population are multiplied by the number of people of a particular age in each of the study populations to give the number of deaths expected in each age group. For each study population, the expected numbers of deaths in all age categories are added to give the total number of deaths expected in that study population. The total number of deaths *observed* in that study population is then divided by the number of deaths *expected* to produce the standardized mortality ratio (SMR). If the SMR is below 1, then fewer deaths were observed than expected; if the SMR is above 1, then more deaths were observed than expected (see Table 7.4). The SMR can be multiplied by the standard population's crude death rate to give an indirect standardized mortality rate.

whether there is a statistical association between a particular characteristic and a disease. Next, they analyze the observed pattern of statistical associations to try to determine the role of the characteristic in causing the disease (Lilienfeld and Stolley 1994). The methods used to determine the statistical associations fall into two broad categories: those based on group characteristics and those based on individual characteristics.

Studies that focus on group characteristics often compare mortality and/or morbidity experience from a given disease in different population groups. Observed patterns of disease are related to differences and similarities in local environment, personal living habits, or genetic composition. Although groups are recognized as being formed of individuals, variability among individuals in these factors is not stressed.

Studies that focus on individual characteristics try to answer two questions:

Do persons with the disease have the characteristic more frequently than those without the disease?

Do persons with the characteristic develop the disease more frequently than those who do not have the characteristic?

The combined results from these two kinds of studies often lead to greater understanding of the causes of the disease being studied.

To determine whether there is a statistical association between particular characteristics and the development of a disease, epidemiologists must carefully design the structure of their studies. Four basic study designs are used: randomized trials, **cohort studies**, **case–control studies**, and **cross-sectional** studies.

Randomized Trials

In randomized trials, unlike in other epidemiological studies, individuals are allocated to groups randomly rather than with set criteria in mind. It is assumed that because the individuals were allocated randomly, the different groups do not vary in their underlying characteristics. The groups are then treated differently, and reactions to the treatment are observed. Because of the underlying assumption, significant observed differences in the outcomes of the treatments are attributed to the treatments themselves (Figure 7.4).

Sophisticated statistical methods have been developed to guarantee that the individuals within the study population are truly randomized by treatment group and that the sample sizes of the treatment groups are large enough to provide confidence in the results. Discussion of these topics is beyond the scope of this chapter; however, interested readers are encouraged to consult an epidemiological textbook such as Gordis (1996) or Lilienfeld and Stolley (1994).

Because they minimize the degree of bias in allocating a population to different treatment groups, randomized clinical trials provide the strongest statistical basis for associations between risk factors or treatments and disease. However, because many epidemiological and medical studies involve health risks, it may be impossible for ethical or practical reasons to randomize the study population into treatment or exposure groups. Consequently, epidemiologists frequently depend on well-designed cohort or case studies to provide evidence for causal associations.

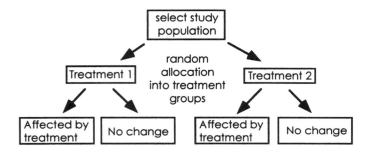

Figure 7.4   Basic study design for randomized trials.

Table 7.4 Age standardization of epidemiological and demographic rates

A. Data from two hypothetical study groups and a hypothetical standard population[a]

	(1)	(2)	(3)	(4)	(5)	(6)
Age (years)	No. in group 1	Group 1 age-specific mortality rates per 1000 people	No. in group 2	Group 2 age-specific mortality rates per 1000 people	No. in standard population	Standard population age-specific mortality rates per 1000 people
0–20	200,000	30	500,000	40	400,000	50
21–40	300,000	10	200,000	10	250,000	20
41–60	300,000	10	200,000	20	250,000	10
>60	200,000	50	100,000	30	100,000	20

[a] Total population in each group and the standard is 1,000,000 people, and each population experiences a total mortality of 100 persons per 1000 population. The crude mortality rate is found by adding elements in the age-specific mortality columns for each group.

B. Direct method of age standardization

Age (years)	Standard pop. (from Column 5 above)	Mortality rates of group 1 per 1000 people (from column 2 above)	Expected no. of deaths in group 1 (standard population times group 1 mortality divided by 1000)	Mortality rates of group 2 per 1000 people (from column 4 above)	Expected no. of deaths in group 2 (Standard population times group 2 mortality divided by 1000)
0–20	400,000	30	12,000	40	16,000
21–40	250,000	10	2500	10	2500
41–60	250,000	10	2500	20	5000
>60	100,000	50	5000	30	3000

Age-adjusted mortality rate (total expected mortality in a group divided by total standard population times 1000): Group 1 = [(12,000 + 2,500 + 5,000)/1,000,000] × 1000 = 22.0 per 1000 population; Group 2 = [(16,000 + 2,500 + 5,000 + 3,000)/1,000,000] × 1000 = 26.5 per 1000 population.

C. *Indirect method of age standardization*

Age (years)	Mortality rates of standard population (from column 6 above)	Observed no. of deaths in group 1 (group 1 mortality rates times group 1 population/1000)	Expected no. of deaths in group 1 (standard mortality rates times group 1 population/1000)	Observed no. of deaths in group 2 (group 2 mortality rates times group 2 population/1000)	Expected no. of deaths in group 2 (Standard mortality rates times group 2 population/1000)
0–20	50	6000	10,000	20,000	25,000
21–40	20	3000	6000	2000	4000
41–60	10	3000	3000	4000	2000
>60	20	10,000	4000	3000	2000

SMR (standardized mortality ratio) = (Total observed no. of deaths in a group)/(Total expected no. of deaths in a group) Group 1 = (6000 + 3000 + 3000 + 10,000)/(10,000 + 6000 + 3000 + 4000) = 0.957; Group 2 = (20,000 + 2000 + 4000 + 3000)/(25,000 + 4000 + 2000 + 2000) = 0.879.

Indirect standardized mortality rate (SMR × crude death rate of standard population) Group 1 = 0.957 × 100 = 95.7; Group 2 = 0.879 × 100 = 87.9

Cohort Studies

Cohort refers to any designated group of people (or other organisms) who share characteristics of interest to a researcher. For example, all the people born in a certain year represent an age cohort. In a cohort study design, the epidemiologist divides the study population into two or more cohorts.

The most common kind of cohort study consists of two groups; one consists of individuals exposed to a factor thought to be related to the disease being studied, and the other consists of nonexposed individuals. The epidemiologist then follows these two groups over time and notes how many people develop the disease in each group (Figure 7.5). If the factor that forms the basis of the exposed group is related to the disease in question, then more people within that group will develop the disease than in the unexposed group. Various statistical tests are used to determine whether the incidence in the exposed group differs significantly from the incidence in the unexposed group. The most common of these tests are discussed later and in Box 7.2.

A major problem with cohort studies is that the study population must be followed over a long time to see whether the disease will develop or not. Because of the time periods involved, there are many kinds of cohort studies, all of which observe exposed and unexposed populations at different times.

One kind, a concurrent cohort study (also concurrent prospective study or **longitudinal** study), identifies a group of people who have not been exposed to the risk factor and who do not show symptoms of the disease. The epidemiologist follows this population for a period during which some will become exposed to the risk factor; then, the initial population is divided into exposed and unexposed groups. At this time, all of the people remain healthy. The epidemiologist continues to follow them for a specified period to see who develops the disease.

A second kind of cohort study, a retrospective cohort study (also historical cohort study or nonconcurrent prospective study), bypasses the first (and sometimes the second) step and shortens the time needed for the study by determining exposure to a risk factor (and possibly disease development) from interviews or medical records at the time the study begins.

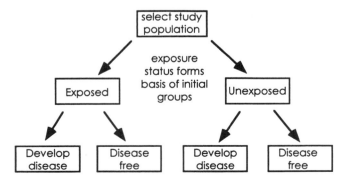

Figure 7.5   Basic study design for cohort studies.

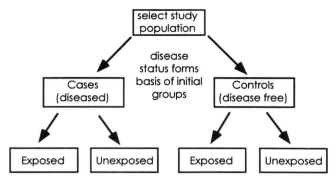

Figure 7.6   Basic study design for case–control studies.

One of the best known cohort studies is the Framingham Heart Study, which identified a cohort of more than 5000 residents of the town of Framingham, Massachusetts, who were between 32 and 60 years of age and were free of heart disease when the study began in 1948. These people were followed over the next 20 years to see whether they developed heart problems and which of several risk factors they possessed (e.g., high blood pressure, smoking, obesity, high cholesterol, etc.). This study continues to have a major impact on how heart disease is studied and on what is known about the risk factors for the development of heart disease and the chances of actually developing the disease when a person does or does not possess particular risk factors.

Case–Control Studies

In case–control studies, the epidemiologist identifies ahead of time a group of people who have the disease of interest (the **cases**) and a group of people who do not have the disease (the **controls**). Prior exposure to suspected or known risk factors is ascertained at the beginning of the study for both cases and controls (Figure 7.6). This setup is opposite that of cohort studies, in which exposure forms the basis of group formation and the presence of disease is determined in the exposed and nonexposed groups separately. In a case–control study, if a particular risk factor is related to the development of the disease, then a history of exposure to the risk factor should occur significantly more often in cases than in controls.

The epidemiologist in charge of a case–control study is responsible for deciding how many cases and controls to include in the study. This decision is based on many different factors, which include cost, difficulty of identifying exposure to a risk or presence of the disease, and whether the disease is common or rare. The decision is often not based on the relative frequency of cases and controls in the general population, and because of the selection process, the proportion of cases in the study population may be much higher than the proportion of cases in the general population. Consequently, case–control studies generally provide no information about the prevalence of a disease in the larger population. However, they do provide some of the most valuable information on whether particular risk factors are linked to the disease.

BOX 7.2   MEASURES OF RISK USED IN EPIDEMIOLOGICAL STUDIES

Absolute Risk

The **absolute risk** to individuals of developing a particular disease is measured by the incidence rate for that disease. (See Box 7.1 for details about how to calculate this risk.)

Relative Risk

When epidemiologists conduct an observational study, they are often interested in determining the strength of the association between the disease and a particular risk factor. This association is measured as the **relative risk**.

$$RR = \frac{\text{incidence rate of the disease in the group exposed to a particular risk}}{\text{incidence rate of the disease in an unexposed group}} \qquad (7.12)$$

Relative risk can be calculated directly for cohort studies, which have the following structure:

	Disease develops	No disease	Total
Exposed	a	b	$a + b$
Not exposed	c	d	$c + d$

Of all the people who have been exposed to the risk factor, some will develop the disease (a) and some will not (b); of all the people who have not been exposed to the risk factor, some will develop the disease anyway (c) and some will not (d). The incidence rate of disease in the exposed group is $a/(a + b)$, whereas that in the unexposed group is $c/(c + d)$. Thus,

$$\text{Relative risk} = \frac{\dfrac{a}{a+b}}{\dfrac{c}{c+d}} \qquad (7.13)$$

A relative risk <1 occurs when the rate of a particular disease is lower in people exposed to the risk than in unexposed people. This kind of result suggests that the risk factor may confer some protection from the disease. A relative risk >1 implies a positive association between the risk factor and development of the disease, with the strength of the association increasing as the relative risk increases. It is important to remember, though, that association does not necessarily mean there

is a causal link between the particular risk factor and the disease. More sophisticated statistical methods are needed to substantiate a causal link.

Odds Ratio

The **odds ratio** is a measure of relative risk that is used in case–control studies, in
which comparisons are made between a group of people who have the disease in
question (cases) and a disease-free group (controls).

A case–control study considers four subgroups: cases who experience the risk
factor, controls who experience the risk factor, cases who do not experience the
risk factor, and controls who do not experience the risk factor. This structure can
be represented as

	Cases	Controls
Exposed	a	b
Not exposed	c	d
Total	$a + c$	$b + d$

The odds of an event can be defined as the ratio of the number of ways an event
can occur to the number of ways it does not occur. When considering a group of
people with a disease, the number of ways disease resulted from exposure to a risk
factor is equal to the number of people who have the disease and were exposed to
the risk factor (a). The number of ways disease did not result from exposure to the
risk factor is equal to the number of people who were not exposed but got the disease anyway (c). Thus, the odds that a case was exposed to the risk factor is a/c.
Similarly, the odds that a control was exposed to the risk factor is b/d. The odds ratio is defined as the ratio of the odds (or probability) that the cases were exposed
to the odds that the controls were exposed:

$$OR = \frac{\frac{a}{c}}{\frac{b}{d}} = \frac{ad}{bc} \tag{7.14}$$

An odds ratio of <1 occurs when the rate of disease is lower in cases than in
controls. This kind of result suggests that the risk factor may confer some protection from the disease. An odds ratio >1 implies a positive association between the
risk factor and development of the disease, with the strength of the association increasing as the odds ratio increases. It is important to remember, though, that association does not necessarily mean there is a causal link between the particular
risk factor and the disease. More sophisticated statistical methods are needed to
substantiate a causal link.

Attributable Risk

Attributable risk measures how much of the incidence of a disease can be attributed to a particular risk factor or cause compared with all other causes. It can be calculated for the group exposed to a particular risk factor or for the entire population. To determine the attributable risk, measurements are needed of both the **background risk** (i.e., the incidence of the disease in people who are not exposed to the risk factor) and the overall risk to the group in question (exposed group or entire population).

The background risk is the same for both kinds of attributable risk. It is subtracted from the risk to the group in question; this difference measures the amount of the overall risk that derives specifically from the risk factor under consideration and not from other causes. The proportion of the total risk that this difference contributes is the attributable risk for the group in question.

$$
\begin{array}{c} \text{Attributable} \\ \text{risk for the} \\ \text{exposed group} \end{array} = \frac{\text{(incidence in exposed group)} - \text{(background risk)}}{\text{incidence in exposed group}} \tag{7.15}
$$

$$
\begin{array}{c} \text{Attributable} \\ \text{risk for the} \\ \text{total} \\ \text{population} \end{array} = \frac{\text{(incidence in total population)} - \text{(background risk)}}{\text{incidence in total population}} \tag{7.16}
$$

Epidemiologists spend considerable effort in choosing cases and controls for their studies. Most often, controls are matched with cases for several characteristics, so that the two groups are as similar as possible with regard to everything except the exposure factors under study (and, of course, presence of disease). Matching of cases and controls helps to guarantee that any significant associations between the exposure factors and the presence of disease in the cases are not a byproduct of other differences between the cases and the controls.

Cross-Sectional Studies

A fourth kind of study commonly used by epidemiologists is a cross-sectional study. This type of study is like taking a snapshot of a population at a particular time. A sample population is selected that reflects the composition of a larger group, and then risk factors and disease status are simultaneously assessed among all members of the sample group (Figure 7.7).

A cross-sectional study provides a measure of the prevalence of a disease within the sample population but does not provide any information on the duration of the disease. It is used to help identify possible risk factors for a disease. However, because there is no way to identify people who possessed the risk factor but may have died before the time of the study, cross-sectional studies generally do not help in determining whether an observed association between the risk factor and disease was, in fact, a causal association. Epidemiologists usually rely on cohort or case–control studies to determine the presence of causal associations.

Methods Used to Help Estimate an Individual's Risk for a Disease

One of the major goals of epidemiological studies is to determine whether an excess (or possibly reduced) risk for disease is associated with a particular exposure factor or treatment. Epidemiologists have devised several measures to assess the risk for disease.

Absolute Risk

One simple measure of risk is the incidence of a disease within a population, also known as the absolute risk. Recall that incidence is the proportion of people in a population who have newly developed a disease (see Box 7.1). Although this measure indicates how likely a person at random is to develop a disease in a given time period, it does not take into account any risk factors that may be associated with the disease in the general population. In addition, if the absolute risk is determined only among individuals known to be exposed to the risk factor, then there is no information about the risk of developing the disease without exposure to the risk factor.

Consider, for example, a population of 200 people, of whom 50 have heart disease, 80 are smokers, and 40 of the heart disease cases are smokers. The absolute risk for heart disease in the population, then, is 50/200, or 25%. However, the risk for heart disease among smokers is 40/80, or 50%, and the risk among nonsmokers is (50–40)/(200–80), or 8%. In this example, it appears that smoking increases the risk for heart disease, but we have no way to estimate the size of that effect and whether it is real or a statistical artifact, especially because the sizes of the smoking and non-smoking groups are different.

Epidemiologists have devised numerous measures to compare the absolute risk of different groups, such as cases and controls. These measures take two basic forms: those based on the ratio of two risks and those based on the difference between two measures. Relative risk and the odds ratio are based on the ratio of two measured risks; attributable risk is based on the difference between two measures.

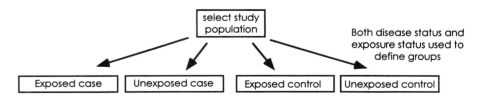

Figure 7.7   Basic study design for cross-sectional studies.

Relative Risk

Used to assess the strength of an association between a particular risk factor and the development of a disease, relative risk is the ratio of the calculated risk in the exposed group to the calculated risk in the unexposed group (see Box 7.2). If these risks are equal, then the relative risk will be 1. If the risk in the exposed group is greater than the risk in the unexposed group the relative risk will be >1, and if the risk in the unexposed group is greater than the risk in the exposed group, the relative risk will be <1. Relative risks can be calculated directly for cohort studies because the study population is divided at the outset into exposed and unexposed groups.

Odds Ratio (Relative Odds)

Relative risk cannot be calculated directly in a case–control study because exposure to the risk factor does not provide the basis for group definition. Epidemiologists use a different measure of risk, the odds ratio, to compare estimates of risk in a case–control study.

"Odds" are another way to talk about probability; the odds that something will happen is equivalent to the probability that something will happen. Odds are determined by dividing the number of times the event did happen by the number of times it could have happened. (Suggestion: Look at Box 7.1 and confirm that the incidence rate is a measure of the odds of developing a particular disease.)

In a case–control study, the odds ratio compares the odds that a case is exposed to the risk factor to the odds that a control is exposed to the risk factor. (See Box 7.2 for an explanation of how to calculate this ratio.) It is interpreted similarly to the relative risk. If exposure to the risk factor is not associated with the disease, the odds ratio will be 1. If exposure to the risk factor increases the rate of disease, the odds ratio will be >1. If exposure to the risk factor decreases the rate of disease, the odds ratio will be <1.

The odds ratio always can be used as a measure of association in case–control studies. In some cases, it also can be used to estimate the relative risk. Three conditions must be met in such cases (Gordis 1996):

1. The study cases are a representative sample of all people in the population with the disease, with respect to exposure to the risk factor.
2. The study controls are a representative sample of all people in the population without the disease, with respect to exposure to the risk factor.
3. The disease being studied is relatively rare.

Attributable Risk

In addition to determining whether there is an association between a risk factor and a disease and assessing the strength of that association, epidemiologists are interested in how much of the incidence of a disease can be attributed to a particular risk factor or cause. This quantity, called the attributable risk, can be calculated for the exposed population or for the total population.

If a risk factor is a causal factor in the development of a disease, then the frequency of the disease in an unexposed population will be lower than the frequency of the disease in the exposed population. However, for most complex diseases, usually a few cases arise even in the absence of exposure to the risk factor. Thus, some risk for development of the disease is independent of exposure to the risk factor. This risk, called the background risk, affects both exposed and unexposed persons. Epidemiologists calculate the background risk from data on disease in unexposed persons and the total risk from data on disease in exposed persons. The total risk in exposed persons is the background risk plus any additional risk due to exposure to the risk factor. Thus, the difference between the total risk in exposed persons and the background risk can be used to calculate the amount of risk that can be attributed to exposure to the risk factor itself. The attributable risk for the exposed group is the proportion of the total risk that this difference contributes (see Box 7.2). It is the percentage that the risk for exposed persons could be reduced if the risk factor were eliminated as a cause of disease.

The attributable risk for a population is used to estimate how much the total risk in a population could be reduced if a risk factor were eliminated and answers questions about the total impact of an intervention on a community. This quantity is found by subtracting the background risk from the incidence in the total population and expressing it as a proportion of the risk to (or incidence in) the total population (see Box 7.2).

After epidemiologists have identified potential risk factors, identified whether there is an association between the risk factors and development of the disease, and determined the strength of that association and its impact on cases of the disease, they try to prove that the observed association is a causal association. That is, the risk factor is not only associated with the disease; it is partially or fully responsible for the development of the disease. A discussion of methods used to verify a causal association is beyond the scope of this chapter, but interested readers are encouraged to consult epidemiology textbooks such as Gordis (1996) or Lilienfeld and Stolley (1994).

CASE STUDY: THE EPIDEMIOLOGY OF DIABETES IN NATIVE AMERICAN AND CANADIAN POPULATIONS

Diabetes is a serious metabolic disorder that is widely distributed throughout human populations. The most common form of the disease is **non-insulin-dependent diabetes mellitus** (NIDDM), which is responsible for 85–90% of all cases of the disease (Bennett and Knowler 1984).

Although the disease is present in only ~2% of the Caucasian population, prevalence rates are much higher in many Native North American groups, ranging from <3% in Athapaskan-speaking tribes of Alaska to >50% in older adult Pima Indians in Arizona (Mouratoff et al. 1969; Statistics Canada 1981; Butler et al. 1982; Knowler et al. 1983; Tan et al. 1983; Klein et al. 1984; Young et al. 1990). These prevalence rates vary considerably in different geographic regions. In addition, the frequency of the disease in some, but not all, Native North American populations has increased markedly (see Szathmáry 1994 for a review of these changes), possibly because of the effects of modernization and the adoption of a Western lifestyle. In this section, I describe the research of Young et al. (1990), who tried to determine the factors (both genetic and environmental) that led to the present geographic pattern of diabetes prevalence and the reasons for recent changes in the prevalence rates in some regions. (See also Chapters 9 and 13 for a more detailed discussion of NIDDM.)

The work of Young et al. was stimulated by the work of James Neel, who tried to explain the high prevalence of NIDDM (Neel 1962). In his original hypothesis, known as the **thrifty genotype hypothesis**, Neel proposed that diabetics possessed genes that allowed for a rapid **insulin** response in the presence of sugars in the blood (usually derived from foods). This insulin response would allow excess blood sugar to be converted to fat and stored by the body for later use if food were scarce. Neel argued that a rapid insulin response would be a selective advantage in environments with a "feast or famine" resource base, that is, where food alternated between being superabundant and scarce. The insulin response would allow people to make maximum use of the nutrients available during the feast phase, then have those nutrients available in the form of body fat that could be converted to energy during famine.

Neel postulated that the same genes that were advantageous under feast or famine conditions would be disadvantageous when food was readily available year-round. A continuous food supply would lead to obesity and excess insulin levels, eventually resulting in the development of diabetes. Neel later adapted his hypothesis to reflect changing ideas about the pathophysiology of diabetes, but it remains one of the leading hypotheses used to explain high rates of diabetes in Native North Americans. These adaptations and other hypotheses are described more fully in Szathmáry (1994).

One problem with the thrifty genotype hypothesis is that it seemed to apply well to the environment of the American Southwest and the high-carbohydrate, agriculture-based diet there but less well to other environments. This discrepancy prompted Young and colleagues to study a different group of populations: the aboriginal peoples of Canada.

Young et al. (1990) included in their study all communities that submitted official reports to local Indian health service units in Canada. These communities represented 76% of the total population of Inuit and on-reserve registered Indians in Canada. The actual study populations consisted of all reported cases of diabetes in each of the communities that submitted reports. These communities were grouped in several different ways and then analyzed for differences in diabetes prevalence among the larger groups.

One of the unusual aspects of this study is that cultural factors (e.g., **tribe, culture area**, and **language family**) were explicitly considered when determining how to combine different communities for statistical analysis. Language is a reasonable predictor of genetic relationships among groups (Spuhler 1972). Populations within the same culture area are similar in several cultural characteristics, such as subsistence system, sociopolitical organization, and religious systems; such similarities may or may not be due to genetic relationships. However, they may indicate common environmental features that could influence susceptibility to and prevalence of diabetes. Tribal designations represent modern sociopolitical ties and are correlated with genetic and cultural similarities among populations. Both the degree of geographic isolation (e.g., remote, rural, or urban) and the geographic location (latitude and longitude) of a community were considered during analysis.

Young et al. (1990) used standard epidemiological techniques to analyze the data. To compare prevalence rates of communities with different numbers of people in each age cohort, the researchers first used the direct method to calculate age-standardized prevalence rates (see Box 7.1) for the administrative regions, which usually corresponded to the provincial or territorial boundaries in Canada. Crude prevalence rates were then used to compare linguistic families, culture areas, and geographic locations. In addition, for some analyses, a summary measure for an entire community was said to constitute a "case," and statistical analyses were performed on a data set that consisted of the crude diabetes prevalence rate associated with each "case."

The variables (e.g., culture area and language) that were used to group communities together for statistical analysis were not independent. Thus, Young et al. (1990) explored several approaches to representing the cultural, biological, and environmental regions occupied by native Canadian populations. (Interested readers are encouraged to consult the original article for details.) Results of the study show that prevalence rates of diabetes in Canada are influenced by language family, culture area, latitude, longitude, and degree of geographical isolation. The crude prevalence rates vary from 0.3% in the Arctic region (3 of every 1000 people have diabetes) to >4% in the Northeast culture area. Prevalence rates are higher in southern latitudes; this difference is also reflected in the rates within the Athapaskan language group, among whom the prevalence was 3.4% in the plains region but only 0.6% in the subarctic. However, this north–south gradient was not observed in the Algonkian language family. Instead, the populations who live in the plains region had a lower prevalence (1.9%) than subarctic (2.6%) and northeast (4.3%) communities within the language group. Results from this language group illustrate an observed west–east gradient for diabetes prevalence, with increasing frequencies in the eastern regions, although this gradient is not as marked as the north–south gradient. Overall, prevalence rates were highest in urban areas, intermediate in rural areas, and lowest for remote areas, although inconsistencies in this trend were observed in some language groups. These trends were statistically significant when the degree of geographic isolation (urban vs. rural vs. remote) was considered alone; however, this variable (assumed to be an index of acculturation) was not important in combination with other variables.

Analysis of the entire data set indicated that latitude was the most important determinant of variability in prevalence rates. However, Young et al. (1990) did not believe that strictly geographic factors (e.g., magnetism or climate) are solely responsible for these patterns. Rather, they postulated that these patterns are at least partly explained by the degree of nonnative influence on a community. In Canada, outside influences on the arctic region were most strongly felt in the western arctic, whereas Euro-Canadian influence on subarctic Algonkian populations tended to come from the south. A strong history of nonnative influences may imply shifts in lifestyle that vary in intensity along geographic gradients that correspond to the degree of nonnative influence. These shifts in lifestyle may affect diabetes prevalence more strongly than geographic differences themselves.

This study used standard epidemiological techniques to explore factors related to diabetes prevalence. However, the researchers did not simply consider the administrative units that are the traditional focus of epidemiologists; they based their group identities on *cultural* variables, such as language or cultural affinity. Cultural variables affect the preponderance of many human behaviors. Thus, studies based on such variables may capture the important factors that affect individual disease risk more effectively than studies based on politically and administratively based groups. Furthermore, Young et al. (1990) used groups that were historically linked to one another and that shared many genetic characteristics. Because of this, they were better able to make a case for how specific historical events, such as degree of contact with outside populations (which in some cases predated the existence of modern political boundaries), might affect a population's risk for diabetes.

SPECIALIZED EPIDEMIOLOGICAL METHODS

The epidemiology of human disease can be studied in several ways other than the standard techniques. Two areas that are receiving increasing attention are the methods of mathematical epidemiology, which involve the development and use of mathematical models and computer simulations to study the transmission and spread of infectious diseases, and the methods of genetic epidemiology, which centers on attempts to identify whether the underlying cause of a disease is genetic and, if so, where the genes are located within the **chromosomes**.

Mathematical Models of the Transmission of Infectious Diseases

During the past 20 years, interest in the development of mathematical models to describe the transmission of infectious diseases has skyrocketed. This interest, shared by people from several disciplines, has been fueled by two major events: the availability of relatively inexpensive, compact, and powerful computers, which allow complex models to be formulated and analyzed, and the recognition of a new, lethal infectious disease epidemic for which there is no cure: acquired immune deficiency syndrome (AIDS).

The general approach to epidemic modeling is outlined briefly in Box 7.3. Most epidemic models divide a population into groups on the basis of the disease status of people within the population. The simple example described in Box 7.3 uses three subgroups: susceptible individuals, who have not yet contracted an infection; infective individuals, who have an active case of a disease and are capable of transmitting it to susceptible people; and **recovered** individuals, who have already had the disease and are immune to further infection. Other models may add additional groups to represent other possible disease stages, such as "exposed but not yet infectious." In addition to dividing the population into disease stages, any epidemic model must describe how contact occurs between a susceptible person and an infectious person and how likely disease transmission is to follow contact. In the simplest models, the resulting probability of transmission is represented by a single parameter that gives the proportion of such contacts that result in disease transmission.

The model described in Box 7.3 is a basic model that incorporates many simplifying assumptions about both human behavior and the biology of the host–pathogen interaction. Mathematical epidemiologists adapt such simple models to reflect factors operating in actual epidemics more adequately and strive to capture the most important aspects of the host–agent–environment interaction within the structure of a model. Adaptation of the basic model to add these details may involve changes in the biology of the disease process, such as altering the number of disease stages, changing the time it takes to recover from an illness, or changing the death rate from the disease. The process may also involve changing how the model formalizes the way contact occurs between a susceptible person and an infectious person.

Two main techniques are used to analyze mathematical models of disease spread: standard mathematical analysis (a paper-and-pencil approach) and computer simulation. Most mathematical techniques are used to answer questions about the long-term behavior of the model, such as whether the epidemic will die out or be maintained in a population. Computer simulations are used primarily to gain insight into the time course of an epidemic and to estimate such parameters as the length of time it takes an epidemic to spread through a population or the number of cases in the population at any particular time.

The ultimate goal of mathematical epidemiology is to help public health–oriented epidemiologists determine how best to control outbreaks of infectious diseases. Because these diseases cause illness and perhaps death, it is difficult to engage in controlled, experimental studies of the biology of the host–parasite interaction and the spread of the infectious organism throughout a population. Mathematical models allow researchers to explore the effects of the disease itself as well as the impact on disease prevalence if particular control strategies are implemented in a population. Modeling is an ideal method to use to generate answers to questions such as,

Why did infectious diseases cause such high mortality among Native Americans at the time of European contact?

What if a cure for AIDS was available but only 50% of the population could be cured?

Who should be targeted for treatment to make sure the overall incidence of the disease is reduced as much as possible?

To gain specific information about a particular epidemic, both mathematical analysis and computer simulations require data with which to estimate **parameters** and check the validity of the models. Estimates of most biological parameters are available in books such as *Control of Communicable Diseases in Man* (Benenson 1995). Data needed to estimate parameters that reflect human behaviors are rarely

BOX 7.3   INTRODUCTION TO MODELING THE TRANSMISSION OF INFECTIOUS DISEASES

An epidemiological model is a mathematical representation of a population into which an infectious disease is introduced. In addition to describing the population involved, an epidemic model builds in factors related to a particular disease and its transmission that are of interest to the modeler. Just as one might start with blocks and strips of wood to begin building a model airplane, an epidemic modeler needs to begin with building blocks related to the population and the disease.

The initial building blocks in an epidemic model relate to the series of stages a person goes through from the time before infection until recovery and beyond. This process is represented in Figure 7.1. People who are at risk for infection by a particular pathogen are "susceptible." After infection, individuals may enter the latent period, during which they are considered "exposed." At the end of the latent period, they become capable of transmitting the pathogen and are called "infective." Eventually, infective people may recover from the disease and become immune, in which case they are referred to as "recovered." People with a disease that confers only temporary immunity will, after immunity has worn off, become susceptible again.

Different epidemic models are tailored to the specific biology of a particular disease and to the questions the modeler hopes to answer. The process of developing a mathematical model for infectious disease transmission involves first deciding what stages to include in the model and then determining the specific details of how people move from one stage to the next.

The majority of epidemic models assume that people become infectious as soon as they are infected and become permanently immune after recovery. Such models are called SIR models because they consider the three disease stages: susceptible, infective, and recovered. Some diseases confer no immunity on recovery and would be modeled using only two stages, susceptible and infective (SI models). Some diseases confer only temporary immunity, so models would consider the progression from susceptible to infective and recovered and back to susceptible (SIRS models).

SIR models are used for many common infectious diseases, including measles, influenza, hepatitis, and chicken pox. The basic model considers a population of constant size (N) and assumes that recovery from the disease brings permanent immunity. The model also assumes no births, no deaths, and no migration. The first stage in building this model is to figure out how a person passes from the susceptible class to the infective class, that is, how the pathogen is transmitted.

Transmission of pathogens involves either direct or indirect contact between an infectious person and a susceptible person. The SIR model assumes that the transmission is direct and that a susceptible person is equally likely to come into contact with any of the infective persons present. Since contacts between a susceptible person and a recovered person or between a recovered person and an infectious person do not lead to disease transmission, these contacts are ignored in the model. In a large population, where a person is equally likely to come into contact with any other person (random mixing), the number of new infections is simply $\beta SI/N$, where β is the infection rate (i.e., the proportion of contacts between susceptible and infective individuals that result in transmission of the infection), S is the number of susceptible people, and I/N is the proportion of infective people in the population.

The expression SI/N is the total number of contacts per person between susceptible and infective people in a randomly mixing population. Multiplication by ß gives the total number of new infections (which is the number of susceptibles who leave the susceptible class and enter the infective class). Because there are no births and deaths in the population, the only way the susceptible class can change in number is by loss as a consequence of infection.

A majority of epidemic models use differential equations to represent this process. The term on the left side of a differential equation represents how the variable of interest changes over time, and the term on the right side presents a mathematical function that determines the amount of change. For example, for the simple epidemic model described here, the change in the size of the susceptible class for each time unit t is described by

$$\frac{dS}{dt} = -\beta \frac{SI}{N} \tag{7.17}$$

The negative sign indicates that the number of susceptibles will decrease (because β, S, I, and N are always positive in real populations). The total decrease in this number is given by the term on the right side of the equation.

The infective class increases in number by the same term because all newly infected susceptible individuals move into the infective class. Infected people move into the recovered class at a constant rate (γ); thus, the infected class decreases at a rate γI. The differential equation describing the change in the size of the infective class is

$$\frac{dI}{dt} = \beta \frac{SI}{N} - \gamma I \tag{7.18}$$

The recovered class increases in size through recovery and is modeled as

$$\frac{dR}{dt} = \gamma I \tag{7.19}$$

These three equations constitute the simplest SIR epidemic model.

A complete solution of this model is given in Bailey (1975). Results of the analysis show that a threshold is associated with the model that is equal to γ/β. When the initial susceptible population is larger than this threshold, an epidemic is possible before the disease dies out. If the initial susceptible population is too small, then the disease never increases in incidence (i.e., no epidemic results).

Recently, as a consequence of using epidemic models to aid in making policy decisions with regard to control measures, the emphasis has shifted away from the idea of an epidemic threshold in favor of a concept called the basic reproduction ratio, or R_0. R_0 is the number of secondary infections produced by one primary infection in a totally susceptible population (Anderson and May 1992). It also can be thought of as the average number of successful offspring that a single pathogen is capable of producing. If $R_0 > 1$, then each infective person will pass the infection to more than one susceptible individual. Thus an epidemic can occur in the population. If $R_0 < 1$, then the disease will die out in the population because it is not able to reproduce at a sufficient rate.[1]

Recent work in mathematical epidemiology involves the development of models that relax some of the underlying assumptions of the SIR model and build in more realistic scenarios. These models have been developed for various diseases and have been used to study, for example, the spatial spread of epidemics, the effectiveness of different control strategies, how infectious diseases might affect the survival and reproduction of wildlife populations, and many other questions. (Interested readers are encouraged to consult Sattenspiel 1990 for a more detailed introduction to methods and applications in mathematical epidemiology.)

collected regularly. Infectious diseases can be transmitted in a very short time (sometimes in the course of a few minutes), so information is needed about the daily activities of people. Collection of such data requires extensive study of the day-to-day lives of people, is both time and money intensive, and may be perceived as an invasion of privacy if not approached properly. However, when attempts are made to collect these data in living groups or to cull the historical literature for information about such activities in historical groups, a wealth of knowledge about how human activities influence disease transmission becomes available.

A study of the spread of the 1918–9 influenza epidemic in the Keewatin District of Manitoba, Canada (Sattenspiel and Herring 1998), illustrates both the mathematical modeling approach and the use of the rich knowledge that can be derived from the historical record. The model for this study, developed by Sattenspiel and Dietz

---

[1]The basic reproduction ratio (R_0) of an infectious disease is essentially the same as Fisher's "net reproductive rate" for the parasite (Fisher 1930). Fisher's parameter resulted from an analysis of demographic models for population growth. Its interpretation in the case of human population growth is analogous to its interpretation for an infectious disease. If $R_0 = 1$, then each person is replaced by exactly one offspring, and the population size remains stable. If $R_0 < 1$, then each person produces fewer offspring than are needed for replacement, and the population will dwindle and eventually become extinct. If $R_0 > 1$, then the number of offspring produced is more than required for replacement, and the population will grow. This analysis leads to the popular conclusion that to achieve zero population growth, each person should have only one child (or, each couple should have two), so that each member of the population replaces only him- or herself.

(1995), is an adaptation of the basic SIR model (Box 7.3). It builds in patterns of travel of native peoples within the district and predicts how influenza can spread among communities as a consequence of this travel.

The Keewatin District consists of several aboriginal communities whose economic base, at the time of the influenza epidemic, depended on fur trapping. Within the region were numerous trading centers operated by the Hudson's Bay Company (HBC). HBC officials at these posts were responsible for keeping journals in which they recorded who came into the posts, what they brought, what they bought, when they left, and other information. The entries in these journals were analyzed to provide estimates for parameters in the mathematical models.

Records on births and deaths recovered from Methodist and Anglican churches in the District indicate that many of the aboriginal communities were hard hit by the 1918–9 influenza epidemic. Figure 7.8 shows the variation in mortality rates during the epidemic in six communities in the region. These and other data suggest that the main post of the region, Norway House, was more severely affected than other communities in the central subarctic (Herring 1994a, 1994b).

Simulations of the mathematical model were run to try to understand the large variation in the severity of the influenza epidemic among the affected communities. The simulations included three communities; Norway House, God's Lake, and Oxford House. Although the simulations reproduced variability in the severity of the epidemic, they also consistently predicted that there should be cases of the disease in all three communities, even though the actual influenza epidemic never reached God's

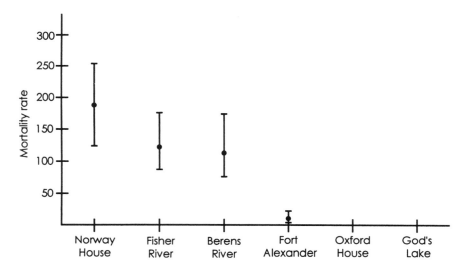

Figure 7.8   Estimated influenza mortality rates per 1000 population for aboriginal communities in central Manitoba during the 1918–19 pandemic. The disease apparently did not reach Oxford House and God's Lake. Rates are based on the 1916 population at risk. Bars show 95% confidence intervals for the mortality rate estimates.

Lake and Oxford House. This lack of fit between the model results and historical records stimulated further examination of the ethnographic sources to determine in more detail what people were doing at the time of the epidemic.

These records suggest two possibilities: (1) that tribal leaders instituted quarantine policies to prevent travel during the epidemic from Norway House to other communities or (2) that the time needed to travel from Norway House to the other communities during the winter was so long that even if an infected person escaped the quarantine, he or she would no longer be infectious (provided he or she survived both the epidemic and the trip) on arrival at Oxford House or God's Lake. Models then could be used to explore these two possibilities to see which was a more likely explanation for the observed prevalence.

These insights illustrate major advantages of combining an anthropological and historical approach with mathematical modeling. Anthropologists focus on interactions between behavior and culture and try to explain experimental (or in this case, simulation) results in light of the ways populations might behave. Mathematical models, on the other hand, are ideally suited to identifying the relative importance of the many possible variables operating in the real world. Results from models can be used by anthropologists and epidemiologists to help narrow down a search for causal factors so that the most significant ones can be identified and dealt with. Ideally, the combination of insights from people collecting the primary data—such as human biologists, anthropologists, and epidemiologists—and insights from people proficient in quantitative techniques such as mathematical modeling will become a regular and valuable addition to the search for ways to control diseases .

Genetic Epidemiology: An Introduction

It is difficult to determine the underlying cause of a disease. For many diseases, the cause is wholly or partly unknown and is likely to be a combination of genetic and environmental factors as well as a pathogenic organism, if the disease has an infectious origin. Most kinds of epidemiology focus on risk behaviors present in the host, characteristics of the disease-causing agent, and/or features in the environment that predispose the host to disease. The effects of inherited characteristics of the host, such as susceptibility or resistance to a particular disease, have not been a major focus of traditional epidemiological approaches. Genetic epidemiology, which centers on the study of the genetic factors that may be involved in determining who contracts a disease and how severely they are affected, has begun to fill this void.

Because they are interested in how **genes** influence disease, genetic epidemiologists use standard methods of human genetics to analyze human groups, particularly families linked by a pedigree. A detailed description of these methods is beyond the scope of this chapter, so I present only a brief overview of the major methods and how they were used to aid in determining whether human breast cancer is inherited. (Readers interested in additional details about the methods of genetic epidemiology are encouraged to consult Weiss 1993.)

Two primary methods are used by genetic epidemiologists: **segregation analysis** and linkage analysis. In segregation analysis, family data on the prevalence of a particular **phenotypic** trait are analyzed statistically to determine whether the trait is inherited genetically and, if so, how it is inherited. In other words, a genetic epidemiologist analyzes the pattern of phenotypes present in several families with a history of the trait and attempts to show that this pattern is consistent with a particular genetic hypothesis (e.g., inheritance of a single-gene **dominant** trait). Classical methods of segregation analysis relied on phenotypes that were either present or absent, required data on presence of the trait in families, and were based on relatively simple statistical methods. Modern computer technology has brought the development of increasingly sophisticated models and methods and has allowed geneticists to extend their analyses to traits of more complex inheritance (Elston 1981) such as those that are probably responsible for heterogeneous diseases like human breast cancer.

In most cases, segregation analysis alone is not sufficient to determine a genetic cause for human diseases; it merely indicates the likelihood that a major gene is involved in the disease and does not provide information about where in the **genome** such a gene may be found. Therefore, segregation analysis cannot be used to identify the particular gene(s) involved. For that, a genetic epidemiologist must rely on the methods of linkage analysis.

Geneticists and molecular biologists can now physically map the DNA sequence in parts of the human genome and identify specific genes (however, few genes have been identified to date). Linkage analysis is a statistical method used by genetic epidemiologists to narrow the search for a specific gene. The method depends on the fact that when two genes occur on the same chromosome, they tend to be inherited together, contrary to Mendel's principle of **independent assortment**. When **homologous** chromosomes are inherited independently, the chance that **alleles** on different chromosomes will occur together in a **gamete** is 50%. However, when the alleles are on the same chromosome (i.e., their genes are **linked**), they tend to stay together, and the probability that they won't stay together is <50%. The actual probability, or recombination fraction, is a function of the physical distance between the **loci** of the genes that code for the two alleles.

The object of linkage analysis is to estimate the value of the recombination fraction between two genes to determine whether they are linked on the same chromosome and, if so, how close the linkage is (Terwilliger and Ott 1994). When performing a linkage analysis to identify the locus involved in a particular trait, a genetic epidemiologist looks for stretches of known DNA (called **markers**) that are consistently inherited by people with a particular disease but not by unaffected family members. This pattern suggests that the actual gene involved in the disease lies on the same chromosome and is relatively close to the genetic marker. Linkage analysis allows a geneticist to narrow the search for a particular gene to a smaller area within the genome and, sometimes, to identify the gene.

In the fall of 1994, researchers announced the identification of a gene (*BRCA1*) responsible for some inherited cases of human breast cancer (Miki et al. 1994). Genetic epidemiologists and their methods—segregation analysis and linkage analysis—were instrumental in the identification of this gene. The fact that some families seemed to be predisposed to breast cancer had been known at least since Roman times (Lynch 1981). Segregation analysis on human pedigrees was used as early as

1984 and indicated that the distribution of some cases of familial breast cancer was compatible with the transmission of an **autosomal** dominant single-gene trait (Williams and Anderson 1984). (Only ~5% of all cases of human breast cancer appear to be genetically influenced, but this percentage includes ~25% of all cases diagnosed before age 30 [Claus et al. 1991].) Armed with the knowledge that some cases of breast cancer did indeed have a genetic origin, researchers used linkage analysis to narrow the search for a breast cancer gene to a region on the long arm of chromosome 17 (Hall et al. 1990). Identifying a particular region of a particular chromosome was a crucial link in the successful identification, 4 years later, of a gene that appears to be responsible for up to one-half of all cases of inherited breast cancer in humans. Identification of the gene should make possible earlier detection of breast cancer in carriers of this trait and eventually lead to a corresponding increase in their survival rates.

CHAPTER SUMMARY

Epidemiology is the study of how, when, where, and why diseases, both infectious and noninfectious, occur. Epidemiologists are especially concerned with interactions among the host, the agent (if any), and the environment. In this endeavor, they study the biology and the behavior of humans (the host) and any pathogens (agents) involved in the disease as well as characteristics of nonliving agents and environmental factors that may influence the development and manifestations of disease.

Infectious diseases can be caused by many organisms, including bacteria, rickettsiae, viruses, prions, protozoa, fungi, and metazoa. Direct spread from one human to another occurs by five primary modes of transmission: respiratory, fecal–oral, sexual, vertical, or direct physical contact. Indirect spread occurs by means of an inanimate vehicle, a biological vector, or complex cycles that involve multiple hosts and modes of transmission.

Noninfectious diseases have many causes, including nutritional deficiencies or excesses; poisoning; allergies; metabolic and hormonal disorders; and genetic, psychological, or physical factors. Often two or more causes interact with one another to produce disease.

Epidemiological studies use one of four basic designs; randomized trials, cohort studies, case–control studies, or cross-sectional studies. Most epidemiological studies are cohort or case–control studies. Associations between particular risk factors and occurrence of disease are assessed by calculating relative risk, the odds ratio, and attributable risk.

Two specialized epidemiological methods are receiving increasing attention. Mathematical epidemiology involves the development and use of mathematical models and computer simulations to study the transmission and spread of infectious diseases. The techniques of mathematical epidemiology are illustrated with a study of the spread of the 1918–9 influenza epidemic in central Canada. Genetic epidemiology centers on attempts to identify whether the underlying cause of a disease is genetic and, if so, where the genes are located within the chromosomes. The major techniques of genetic epidemiology (linkage analysis and segregation analysis) have been used to isolate a gene (*BRCA1*) that appears to be responsible for many inherited cases of breast cancer.

RECOMMENDED READINGS

Gordis L (1996) Epidemiology. Philadelphia: W. B. Saunders.

Kiple KF (ed.) (1993) The Cambridge World History of Human Disease. Cambridge: Cambridge Univ. Press.

Last JM (1995) A Dictionary of Epidemiology, 3rd edition. New York: Oxford Univ. Press.

Lilienfeld DE and Stolley PD (1994) Foundations of Epidemiology, 3rd edition. New York: Oxford Univ. Press.

Sattenspiel L (1990) Modeling the spread of infectious disease in human populations. Yearbook Phys. Anthropol. 33: 245–276.

Sattenspiel L and Herring DA (1998) Structured epidemic models and the spread of influenza in the Norway House District of Manitoba, Canada. Hum. Biol. 70: 91–115.

Szathmáry EJE (1994) Non-insulin dependent diabetes mellitus among aboriginal North Americans. Annu. Rev. Anthropol. 23: 457–482.

Weiss KM (1993) Genetic Variation and Human Disease. Cambridge: Cambridge Univ. Press.

Young TK, Szathmáry EJE, Evers S, and Wheatley B (1990) Geographical distribution of diabetes among the native population of Canada: a national survey. Soc. Sci. Med. 31: 129–139.

REFERENCES CITED

Anderson RM and May RM (1992) Infectious Diseases of Humans: Dynamics and Control. New York: Oxford Univ. Press.

Barua D (1992) History of cholera. In Barua D and WB Greenough III (eds.) Cholera. New York: Plenum, pp. 1–36.

Bailey NTJ (1975) The Mathematical Theory of Infectious Diseases and its Applications. New York: Hafner Press.

Benenson AS (ed.) (1995) Control of Communicable Diseases in Man, 16th edition. Washington, D.C.: The American Public Health Association.

Bennett PH and Knowler WC (1984) Early detection and intervention in diabetes mellitus: Is it effective? J. Chronic Dis. 37: 653–666.

Butler WJ, Ostrander LD Jr , Carman WJ, and Lamphear DE (1982) Diabetes mellitus in Tecumseh, Michigan: prevalence, incidence, and associated conditions. Am. J. Epidemiol. 116: 971–980.

Cipolla CM (1992) Miasmas and Disease. New Haven: Yale Univ. Press.

Claus EB, Risch N, and Thompson WD (1991) Genetic analysis of breast cancer in the Cancer and Steroid Hormone Study. Am. J. Hum. Genet. 48: 232–242.

Elston RC (1981) Segregation analysis. Adv. Hum. Genet. 11: 63–120.

Fisher RA (1930) The Genetical Theory of Natural Selection. Clarendon: Oxford Univ. Press.

Fox JP, Hall CE, and Elveback LR (1970) Epidemiology: Man and Disease. London: Macmillan.

Gillies E (1976) Causal criteria in African classifications of disease. In JB Loudon (ed.): Social Anthropology and Medicine. New York: Academic Press, pp. 358–395.

Gordis L (1996) Epidemiology. Philadelphia: W. B. Saunders.

Hall JM, Lee MK, Newman B, Morrow JE, Anderson LA, Huey B, and King M-C (1990) Linkage of early-onset familial breast cancer to chromosome 17q21. Science 250: 1684–1689.

Herring DA (1994a) "There were young people and old people and babies dying every week": the 1918–1919 influenza pandemic at Norway House. Ethnohistory 41: 73–105.

Herring DA (1994b) The 1918 influenza epidemic in the central Canadian subarctic. In A Herring and L Chan (eds.): Strength in Diversity: A Reader in Physical Anthropology. Toronto: Canadian Scholars' Press, pp. 365–384.

Klein R, Klein BEK, Moss SE, DeMets DL, Kaufman I, and Voss PS (1984) Prevalence of diabetes mellitus in southern Wisconsin. Am. J. Epidemiol. 119: 54–61.

Knowler WC, Pettitt DJ, Bennett PH, and Williams RC (1983) Diabetes mellitus in the Pima Indians: genetic and evolutionary considerations. Am. J. Phys. Anthropol. 62: 107–114.

Lilienfeld DE and Stolley PD (1994) Foundations of Epidemiology, 3rd edition. New York: Oxford Univ. Press.

Longmate N. 1966. *King Cholera: The Biography of a Disease.* London: Hamish Hamilton

Lynch HT (1981) Genetics and Breast Cancer. New York: Van Nostrand Reinhold.

Miki Y, Swensen J, Shattuck-Eidens D, Futreal PA, Harshman K, Tavtigian S, Liu Q, Cochran C, Bennett LM, Ding W, Bell R, Rosenthal J, Hussey C, Tran T, McClure M, Frye C, Hattier T, Phelps R, Haugen-Strano A, Katcher H, Yakumo K, Gholami Z, Shaffer D, Stone S, Bayer S, Wray C, Bogden R, Dayananth P, Ward J, Tonin P, Narod S, Bristow PK, Norris FH, Helvering L, Morrison P, Rosteck P, Lai M, Barrett JC, Lewis C, Neuhausen S, Cannon-Albright L, Goldgar D, Wiseman R, Kamb A, and Skolnick M (1994) A strong candidate for the breast and ovarian cancer susceptibility gene *BRCA1*. Science 266: 66–71.

Moore LG, Van Arsdale PW, Glittenberg JE, and Aldrich RA (1980) The Biocultural Basis of Health. Prospect Heights, Ill.: Waveland.

Mouratoff GJ, Carroll NV, and Scott EM (1969) Diabetes mellitus in Athabaskan Indians of Alaska. Diabetes 18: 29–32.

Neel JV (1962) Diabetes mellitus: a "thrifty" genotype rendered detrimental by progress? Am. J. Hum. Genet. 14: 353–362.

Risse GB (1993) History of Western medicine from Hippocrates to germ theory. In KF Kiple (ed.): The Cambridge World History of Human Disease. Cambridge: Cambridge Univ. Press, pp. 11–19.

Sattenspiel L (1990) Modeling the spread of infectious disease in human populations. Yearbook Phys. Anthropol. 33: 245–276.

Sattenspiel L and Dietz K (1995) A structured epidemic model incorporating geographic mobility among regions. Math. Biosci. 128: 71–91.

Sattenspiel L and Herring DA (1998) Structured epidemic models and the spread of influenza in the Canadian Subarctic. Hum. Biol. 70: 91–115.

Snow J. 1965. Snow on Cholera. Reprint of Two Papers. New York: Hafner

Spuhler JN (1972) Genetic, linguistic, and geographic distances in native North America. In JS Weiner and A Huizinga (eds.): The Assessment of Population Affinities in Man. Oxford, UK: Clarendon Press, pp. 72–95.

Statistics Canada (1981) Canada Health Survey. The Health of Canadians (Cat. No. 82–538). Ottawa: Statistics Canada.

Szathmáry EJE (1994) Non-insulin dependent diabetes mellitus among aboriginal North Americans. Annu. Rev. Anthropol. 23: 457–482.

Tan MH, Wornell C, and Ellis R (1983) Diabetes mellitus in Prince Edward Island. Tohoku J. Exp. Med. Suppl. 141: 301–307.

Terwilliger JD and Ott J (1994) Handbook of Human Genetic Linkage. Baltimore: The Johns Hopkins Univ. Press.

Weiss KM (1993) Genetic Variation and Human Disease. Cambridge: Cambridge Univ. Press.

Williams WR and Anderson DE (1984) Genetic epidemiology of breast cancer: segregation analysis of 200 Danish pedigrees. Genet. Epidemiol. 1: 7–20.

Wood CS (1979) Human Sickness and Health. Mountain View, Calif.: Mayfield.

Young TK, Szathmáry EJE, Evers S, and Wheatley B (1990) Geographical distribution of diabetes among the native population of Canada: a national survey. Soc. Sci. Med. 31: 129–139.

CHAPTER 8

Human Adaptations to Infectious Disease

FATIMAH L.C. JACKSON

INTRODUCTION

Infectious diseases have long been regarded as major factors influencing the direction and cadence of human evolution. In fact, the genetic composition of present-day human populations is determined largely by the interactions between the human **host** and infective agents (Fischer et al. 1998).

There has probably never been a time in our species' evolutionary history when we have not had contact with, hence been influenced by, organisms capable of causing us sickness, disease, or death. Our complex and longstanding interactions with various infectious life forms—be they **viruses, bacteria, protozoa,** or **helminths** (i.e., worms such as nematodes, cestodes, or trematodes)—have helped define us both **phenotypically** and **genotypically**. Genetic **polymorphism** in our human leukocyte antigen (**HLA**) **system**, for example, is characterized by **allelic** lineages that are extremely old. The most powerful explanation for this diversity involves our immune response to **pathogenic** microorganisms. The **coevolution** of these infectious agents with human hosts has had a major influence on human HLA polymorphism. This polymorphism is at the foundation of many current differences in human **susceptibility** and resistance to specific infections.

Infectious diseases are an ongoing threat to humanity (Garnett and Holmes 1996). According to a 1999 World Health Organization report (Kapp 1999), infectious diseases cause 48% of deaths worldwide in people under the age of 45 years. These diseases are a huge impediment to economic development in poor countries. Six illnesses cause ~90% of infectious disease deaths. Respiratory infections (mostly bacterial pneumonia and influenza) kill ~3.5 million people each year. Human immunodeficiency virus (HIV), causing acquired immune deficiency syndrome (AIDS) is responsible for ~2.3 million deaths annually. Various diarrheal

Human Biology: An Evolutionary and Biocultural Perspective, Edited by Sara Stinson, Barry Bogin, Rebecca Huss-Ashmore, and Dennis O'Rourke
ISBN 0-471-13746-4   Copyright © 2000 Wiley-Liss, Inc.

infections account for 2.2 million deaths. Tuberculosis causes 1.5 million deaths. Malaria kills 1.1 million people each year. Measles is responsible for 900,000 deaths. These figures clearly indicate the important and continuing impact of pathogenic microorganisms on human health and well-being. It is within this context that we examine the evidence for human adaptations to infectious diseases.

I have divided this chapter into three sections. In the first section, I discuss the adaptive process in general, with respect to human contact with infectious microorganisms. I consider some important biocultural perspectives on infectious diseases in contemporary and ancient human groups and briefly discuss the interrelationship between infectious disease exposures and geographical patterns of human biovariability.

In the second section, I discuss a representative cross section of human–microorganism interactions that evidence possible human adaptations. Specific case studies on malaria, onchocerciasis (river blindness), schistosomiasis (bilharzia), tuberculosis, and HIV infection/AIDS illustrate my point.

In the third section, I discuss future trends in global infectious disease control, the potential for emerging and reemerging diseases, and the possibility for the eradication of specific infectious agents.

OVERVIEW AND BACKGROUND

Patterns of Infectious Diseases in Human Prehistory and History

Anatomically and behaviorally modern *Homo sapiens* emerged and expanded their geographical range ~120,000 years before the present (years BP). It has been proposed that during this time, human–microorganism contacts were limited to species of infectious agents whose life cycles could accommodate the frequent residential shifts of migratory human gatherers and hunters. Infectious diseases were likely limited by the small group size of early modern humans and their relative genetic and behavioral homogeneity (most groups were composed of related individuals). Furthermore, the spread of infectious disease from group to group of early modern humans would have been restricted by the transitory lifestyles and broad geographical ranges annually traversed by such groups. Under these conditions, the opportunity for repetitive human contact with large numbers and a great diversity of infectious microorganisms was limited. Genetic data support an early population expansion in Africa with subsequent migrations of smaller groups out of the continent (Quintana-Murci et al. 1999). Expanding patterns of migration probably put early modern humans in contact with new infectious agents.

However, despite these contacts, many scientists hypothesize that the most dramatic shifts in human–microorganism interactions occurred within the past 20,000 years. Within this time frame, the most pronounced influences on contact with infectious agents are correlated with the human shift to an increased commitment to subsistence agriculture that necessitated a nonmigratory lifestyle. Once modern humans began the transition from gathering and hunting to agriculture, groups became more geographically stable, and populations grew. Population density also increased occupational stratification, altered land use patterns, changed the within-group exposure to pathogenic microorganisms, and increased between- and within-group heterogeneity in susceptibility or resistance to infectious agents. Interactions between groups occurred more frequently and for longer periods of time. Agriculture required the encroachment of human habitation on wilderness perimeters, which increased human contact with **vectors** of **zoonotic** viruses and bacteria.

Overall, within the past 20,000 years of human history, the potential for human contact with large numbers of a diverse array of infectious microorganisms has skyrocketed (see also Chapter 14).

Biocultural Basis of Human Adaptations

Human–microorganism interactions and the human adaptations they stimulate are often fundamentally biocultural; they embody both biological and cultural processes. Whereas adaptations may be at the behavioral, physiochemical, and/or genetic levels, these classic divisions are probably artificial.

Behavioral adaptations are rooted in the individual and group sociocultural context and are learned during the course of one's lifetime. Physiochemical adaptations are primarily stimulated as a result of direct contact with specific microorganisms during the course of one's lifetime. Such responses tend to be centered on the immune system. Adaptations also may be physiochemical with an explicit (hereditary) genetic basis that influences susceptibility or resistance to infection. Such genetic adaptations are thought to have evolved over many generations of contact and under significant selection pressure from the infective agents.

These latter kinds of adaptations are of particular interest to biological anthropologists, and their identification has been greatly facilitated by both the Human

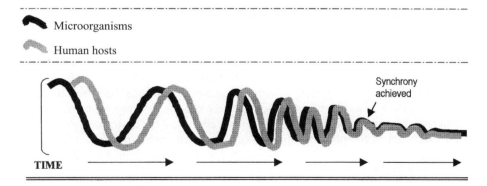

Figure 8.1 Time-related changes in the biological and behavioral synchrony of infectious microorganisms and their human hosts.

Genome Project and the **Environmental Genome Project**. Because adaptations are dynamic processes that are constantly changing in response to external and internal stimuli, successful human adaptation usually involves the synergistic effects of altered behaviors, modified physiochemistry, and genetic polymorphism. Figure 8.1 depicts the time-related changes in the human–microorganism interactions as they approach biological and behavioral synchrony (full adaptation). As specific local groups of humans and local varieties of infectious microorganisms interact, each initiates a series of actions on the other, and each in turn reacts to these stimuli. The oscillating reciprocity of these actions and reactions can be the foundation for the emergence of human–microorganism coevolution and a basis for human adaptation to the potential pathology produced by specific infectious agents.

BOX 8.1   OVERVIEW OF THE HUMAN IMMUNE SYSTEM

Human adaptations to infectious diseases are largely centered around variation in our immune system. The immune system can be distinguished by two kinds of immune responses – antibody mediated and cell mediated.

Antibodies belong to the class of proteins called immunoglobulins. Antibodies appear in serum following exposure to foreign substances such as infectious agents. These foreign substances, called **antigens**, initially stimulate a primary immune response, with a small amount of antibody produced. When the exposed human is exposed a second time to the same antigen however, a secondary immune response is produced with a rapid and elevated amount of antibody produced by the host. Antibodies can combine specifically with antigens and accelerate the destruction the antigens from the infected human host. The major serum immunoglobulin is **IgG**. IgM is a very large immunoglobulin molecule that predominates in the primary immune response. IgA is responsible for **immunity** on body surfaces. IgD is a cell-bound immunoglobulin. **IgE** mediates allergies. Immunoglobulins show significant variations in structure and functional efficacy both between individuals and within an individual.

The second kind of immune response is cell-mediated. While antibodies provide powerful protection for the human host against foreign antigens, antibodies cannot eliminate invading viruses (which grow within host cells) and some bacteria (such as those causing tuberculosis). In these conditions, a second type of immune response must come into play. Infected cells have to be destroyed by special killer cells called **lymphocytes** through cell-mediated immunity. As in antibody-mediated immunity, the goal of cell-mediated immunity is elimination of foreign material from the host.

The cells that form the first line of defense in the body are known as the myeloid system. Many bacteria or other foreign particles that enter the human body are taken up and destroyed by **neutrophils** in a process called **phagocytosis**.

Opsonins, such as antibodies and enzyme-based **complement**, promote this process. Neutrophils kill ingested bacteria using reactive oxygen metabolites and **lysosomal** enzymes. **Parasites** on the other hand are often phagocytized by **eosinophils**. Basophils are another type of granulocyte cell of the myeloid system that participates in allergic reactions.

In addition to the specialized cells of the myeloid system, other cells also participate in phagocytosis. **Macrophages** are large secretory phagocytic cells (similar to neutrophils) that produce proteins that influence inflammation, healing, and the body's response to infection. The most important of the proteins produced by macrophages are the **cytokines**, including the **interleukins** and tumor necrosis factor-α. Many of the clinical signs of infection, such as fever, are due to the action of cytokines.

Foreign antigens often must be specially processed before they can be presented to the antigen-sensitive cells of the immune system. The macrophages, dendritic cells, and **B cells** do this processing. Fragments of antigen are generated within these cells and bound to specialized receptors called MHC or major histocompatibility molecules. The MHC in humans is located on the short arm of **chromosome** 7 and contains about 3.5 megabases. There are three classes of MHC molecules, each with a specified area of immunologic action.

Class I antigen molecules are found on the surface of nucleated cells (such as white blood cells) and function to present antigen to **cytotoxic T cells**. This produces T-cell-mediated cytotoxicity. Foreign molecules synthesized within a cell (as in a virus-infected cell) are attached to MHC Class I molecules while they are being synthesized. The genes responsible for these Class I molecules fall into two categories: Class Ia genes may be highly polymorphic while Class Ib genes may be highly conserved.

Class II MHC molecules are located only on the surface of B cells, dendritic cells, and macrophages. These molecules present antigen to helper T cells. This produces T-cell-mediated help. Both Class Ia and Class II MHC molecules act as receptors that bind foreign antigens, present them to lymphocytes (either cytotoxic or helper T cells), and regulate the human immune response. Class III MHC molecules are not involved in antigen processing and are more related to the serum proteins of the complement system that protect against bacterial invasion.

In humans the MHC molecules are collectively called HLA, or human leukocyte antigen. Genetic polymorphism in HLA is characterized by allelic lineages that are extremely old. The most powerful explanation for this diversity involves our immune response to pathogenic microorganisms. The coevolution of infectious agents with human hosts has had a major influence on HLA polymorphism. This polymorphism is at the foundation of many current differences in human susceptibility and resistance to specific infections.

Adaptations to Specific Kinds of Microorganisms

Human–Virus Interactions

The most important nonimmunological adaptations to viruses include **lysozymes** (enzymes present in tears, saliva, and cells known as neutrophils), intestinal enzymes, and bile (a powerful neutralizer of viruses). Immunologically, however, the body uses several specific strategies in antiviral immunity.

Cell-mediated immunity against virus-infected cells is among the most effective mechanisms for controlling viral proliferation and disease expression. Cell-mediated immunity is facilitated through the action of two types of cells in the immune system, cytotoxic T cells and **NK (natural killer) cells**, both of which are capable of cell destruction. A second immunological strategy by which humans adapt to invading viruses is the destruction of viral envelope **proteins** by antiviral antibodies. Because viruses tend to be replicate slowly and may persist inside of the cell for extended periods, antiviral immunity can be prolonged. Although antiviral immunity may not be able to kill all of the virus, its protective value is that it can inhibit disease expression.

In response to antiviral immunity, viruses may evade immune detection directly by invading free of antigen. Viruses also may evade immune detection indirectly by evidencing regular antigenic variation in surface antigens. They also may produce immunosuppression, as is the case with AIDS due to HIV-1 or HIV-2. Third, viruses may directly interact with somatic cell **transcription** and inhibit the expression of HLA class I antigens.

Human–Bacteria Interactions

Human tissues and body fluids contain many nonimmunological protective factors that can impede bacterial invasion and proliferation. Of these (listed in Table 8.1), the most important are the lysozymes.

Some of the highest concentrations of lysozymes are in the tears and at the sites of bacterial invasion. Lysozymes kill gram-negative bacteria by attacking the bacterial cell wall. Free **fatty acid** levels also inhibit bacterial growth. The specific immune responses to bacterial infection include phagocytosis (ingestion of bacteria) by cells known as neutrophils and macrophages, lysis, neutralization of bacterial toxins by human antibodies, and destruction of bacteria with activated macrophages.

Human–Protozoa Interactions

Most protozoan parasites are fully antigenic and are capable of stimulating the immune systems of infected humans. However, many of these protozoa have developed mechanisms that allow them to undermine, evade, and elude the human immune response.

Undermining the human host's immune system usually entails inhibiting normal function. The advantage for the protozoa is that because it causes immunosuppression in the human host, the host can carry many more protozoa than would be normally permitted with an adequate immune response. This condition can lead to hyperparasitemia and can improve the chances of the protozoa being transmitted successfully from one human host to another, especially when transmission requires a third species (such as a mosquito).

Some protozoa have developed the ability to quickly and frequently change their surface antigens, thereby evading detection by the human immune system. By repeatedly altering their surface glycoproteins, these protozoa are able to impede recognition as "foreign" and avoid destruction by human defenses.

Human responses to protozoa often involve a complex and diverse array of genetically mediated physiochemical variants. Human polymorphism in red blood cell proteins, cytokines (proteins that mediate cellular interactions and regulate cell growth and secretion), the immune system, and other systems appears to play an important role in regulating human susceptibility and resistance to protozoa.

Human–Helminth Interactions

As larvae (in human tissues or the gastrointestinal tract) or as adults (in the gastrointestinal or respiratory tract), helminths are very efficient at evading the human immune system. The most significant human immunoglobulin (Ig) involved in resistance to helminths is IgE. Most humans infected with helminths also have elevated IgE levels.

Some of the most important human adaptations to helminths are behavioral. This response may because helminths tend to synchronize their reproductive cycles with the behavior and biology of their human hosts. For example, both the age and sex of the human host influence worm burdens, possibly through the interaction of behavioral and hormonal processes. Young children tend to acquire helminth infections more frequently than adults. In human groups with heavy worm levels, a few individuals appear to be predisposed to heavy infestation and harbor large numbers of worms, suggesting intergroup variability in susceptibility.

Table 8.1   Physiochemical factors protective against infectious microorganisms

Type	Name	Location	Active against
Enzyme	Lysozyme	Leukocytes and serum	Bacteria and viruses
Peptide	Leukin	Neutrophils	Bacteria
	Plakin	Platelets	Bacteria
Proteins	Transferrin	Serum	Bacteria
	Lactoferrin	Milk, Leukocytes	Bacteria
Peroxidegenerators	Myeloperoxidase	Neutrophils	Bacteria, viruses, protozoa
Interferon	β-TFN	Most cells	Viruses, protozoa in cells
Complement	Various components	Serum	Bacteria, viruses, protozoa

(*Source*: Modified from Tizard 1995)

In most infected humans, helminths tend to self-regulate their numbers via two mechanisms. The first is intraspecies competition, whereby the presence of adult worms in the intestines delays the development of larvae in the tissues. This behavior, characteristic of certain tapeworms and schistosomes, is known as concomitant immunity. The second mechanism is interspecies competition between worms for habitats in humans and access to human nutrients. Limited resources in the human host may regulate the helminth population and control the distribution of various developmental stages of the worm within a single infected human.

REPRESENTATIVE INFECTIOUS DISEASES

Human Interactions with Malaria (Infection with *Plasmodium* spp.)

Malaria is the most important infectious disease in human history, and it continues to have a profound impact on human biology and behavior. Although the disease is now concentrated in the tropics and subtropics, malaria still extends into the temperate zone in some geographical regions. No major part of the world has escaped contact with the protozoan causes of the disease.

Malaria is transmitted by several mosquito species. The malaria protozoa sexually reproduce in the abdomens of mosquitoes. When an infected female mosquito inserts her proboscis into a human, she injects infective forms of the protozoa into the human bloodstream. Once the parasite has invaded the human host, it travels to the liver and then on to the red blood cells. (The life cycle of malaria is illustrated in Figure 7.3.) While in the human red blood cell, the protozoa modifies and uses the human host's resources for its own growth and reproductive needs. In the infected human, *Plasmodium* spp. produce disease that ranges from the semibenign malariae and ovale forms of malaria (caused by *Plasmodium malariae* and *Plasmodium ovale*, respectively) to the more **virulent** vivax and falciparum forms of this infectious disease (due to *Plasmodium vivax* and *Plasmodium [Laverania] falciparum*, respectively).

We are able to reconstruct the impact of malaria on humankind through a variety of channels. By studying the history of malaria disease **epidemiology**, we can identify the location, spread, and biocultural impact of the disease on affected human groups. For example, malaria's characteristic intermittent fevers and splenomegaly (enlarged spleen) have been recognized since antiquity. In the Ebers papyrus (dated to 3500 years BP), for example, it is apparent that the ancient Egyptians of Kmt (the indigenous name for ancient Egypt) associated mosquitoes with these symptoms. They also prescribed the topical application of oil of balamite as a mosquito repellent. Hieroglyphs at Dendera, Upper Egypt, describe intermittent fevers and associate them with the annual flooding of the Nile River (when mosquito vectors would have been most abundant). Malaria appears in subsequent Greek and Roman histories as a significant definer of human settlement patterns, war strategies, and overall health status.

Another way to reconstruct the impact of malaria on humankind is to examine the geographical patterns of human biodiversity and to identify possible physiochemical and genetic adaptations to this infectious disease. Contact with malaria has modified a diverse array of **gene** families in response to the selective pressure of the disease (Weatherall 1996). At least 12 different genes affect human susceptibility or resistance to malaria (Hill 1996), including HLA variation (Hill 1998), polymorphism in α- and β-chain globulin genes, **cytoskeleton proteins**, and protein-based receptors on the surface of red blood cells (Dessein 1997).

Of the four human malaria protozoa, the best studied is *P. falciparum*. The virulence of these protozoa is thought to be due to the adherence of parasitized red blood cells to the inside of small blood vessels through several receptors. Intercellular adhesion molecule-1 (ICAM-1) seems to be a particularly important receptor for the most serious form of falciparum malaria: cerebral malaria. When a group of Africans at risk for cerebral malaria was evaluated for polymorphism at the ICAM-1 locus, a single mutation was found to predispose them to this deadly manifestation of falciparum malaria (Fernandez-Reyes et al. 1997).

As in the above study, most evidence for the involvement of genetic factors in the human response to malaria come from studies of severe clinical malaria. **Segregation analysis** was used to identify the role of major gene(s) that control blood parasite levels (Abel et al. 1992). Among nine families from southern Cameroon (Garcia et al. 1998), a locus that influences *P. falciparum* levels in malaria was located on the long arm of chromosome 5, indicating that it may be a critical region for the genetic control of malaria (as well as schistosome—see later) infections.

The genetic makeup of the host clearly has a profound influence on resistance to malaria infection. Animal studies have shown that transgenic mice that express human sickle hemoglobin (the hemoglobin responsible for sickle-cell anemia) yet have intact spleens are partially resistant to rodent malaria (Shear et al. 1993). This finding supports the many epidemiological and in vitro studies among humans that suggest that sickle hemoglobin can enhance resistance to falciparum malaria.

Malaria caused by *P. vivax* also can be limited by human genetic polymorphism. The absence of either of the **Duffy blood group** antigens Fy^a or Fy^b found on the surface of red blood cells impedes access to these cells by the vivax protozoa. In a sense, these antigens function as "doorknobs"; in their absence, it is impossible for *P. vivax* to gain entry into the cell, where it can grow, asexually proliferate, and cause disease. Outside the red blood cells, the protozoa are much more vulnerable to phagocytosis, destruction, and removal in the circulation by neutrophils and macrophages.

The Duffy-negative phenotype, Fy(a–b–), is most commonly encountered in Africa and among peoples of recent African descent elsewhere in the world. The absence of the necessary genetically determined receptor antigens on the surface of the red blood cells of these individuals make them **refractory** to *P. vivax* infection. This genetic makeup is apparently the basis for the extensive natural immunity of many peoples of African descent to vivax malaria and the virtual absence of *P. vivax* in humid tropical Africa despite the presence of appropriate mosquito vectors. Vivax malaria is found today primarily in Asia. The Duffy-negative phenotype may represent an effective genetic adaptation to past **endemic** vivax malaria.

Human Adaptations to Onchocerciasis
(Infection with *Onchocerca volvulus*)

Some 86 million people around the world are at risk of acquiring the nematode that causes human onchocerciasis, or river blindness, and another 18 million people have been infected; 600,000 people are visually impaired, one-half of whom are partially or totally blind (Basanez and Boussinesq 1999). Of all cases, 99% occur in tropical Africa, where vast stretches of the savannas are **hyperendemic** for onchocerciasis. Scattered foci exist in Latin America. Onchocerciasis is caused by the tissue nematode *Onchocerca volvulus* and is transmitted by the black fly *(Simulium* spp.).

The **pathogenesis** of onchocerciasis results from the action of the adult filarial worms (which are the least pathogenic) and/or the microfilaria. Full-blown onchocerciasis causes disfigurement and blindness in many cases. Although the disease is not fatal, it exerts a significant selective pressure on affected human groups because it is extremely debilitating and a major cause of social disruption and economic depression. Human infections result in various clinical conditions, and some evidence for an adaptive, protective immunity have been observed in certain individuals.

Initially, it was thought that adaptation to onchocerciasis was acquired through the long-term contact of individual humans with the nematode. However, current thinking is that some individuals are born genetically refractory to infection. A few individuals who are regularly exposed to *O. volvulus* transmission remain devoid of clinical signs of onchocerciasis. For example, a putatively immune population has been identified in a hyperendemic area of Ecuador. Seemingly immune individuals had lower levels of *O. volvulus*–specific immunoglobulin G (IgG), IgG subclasses, and IgE than infected individuals, and most immune persons were female (Elson et al. 1994). Molecular genetic studies of these putative immune individuals suggest that their IgG3 is able to recognize an antigen in the cuticle of adult worms and uterine microfilaria, and initiate an immune response, whereas the IgG3 of infected individuals is not able to (Stewart et al. 1997).

Recent studies in Liberia (Meyer et al. 1994), where onchocerciasis is also hyperdendemic, suggest that certain HLA variants may provide putative immunity to the disease. The *HLA-D* variants are class II genes whose function is to present antigen to helper T cells. Within the HLA-D region there are five loci called DN, DO, DP, DQ, and DR. At the DQ locus are ~21 genetically recognized alleles and 7 serologically recognized alleles. Alleles of the DQ loci were most strongly correlated with resistance to *O. volvulus* infection, specifically, the **haplotype** DQA1*0501-DQB1*0301. This HLA haplotype was significantly more frequently found among 44 resistant individuals than among the 76 nonadapted, susceptible individuals with generalized or localized onchocerciasis. The allele of most of the susceptible individuals was *DQA1*0101-DQB1*0501*. The *DQB1*0102* allele was associated with individuals with generalized onchocerciasis, whereas the *DPB1*0402* allele was most frequently found in individuals with localized onchocerciasis.

Subsequent studies in Nigeria (Murdoch et al. 1997) strongly suggest an immunogenetic basis for the spectrum of clinical presentations seen in onchocerciasis and that HLA-DQ molecules are also associated with the level of immune response to nematode antigens. Nigeria's population is highly affected by onchocerciasis, and at least two clinical forms of the disease exist (Ogunrinade et al. 1999). Protective immunity to the onchocerca nematode appears to be directed against molecules of invading larvae: Detection of parasite protein is at the level of T cells and triggers macrophage- plus eosinophil-dependent killing of *O. volvulus* larvae in vivo (Doetze et al. 1997).

Human Adaptations to Schistosomiasis (Infection with *Schistosoma* spp.)

Since ancient times, bloody urine has been a well-recognized symptom of schistosomiasis. At least 50 references to this condition have been found in surviving Egyptian papyri. In addition, calcified eggs of the flukes that cause this disease have been found in mummified tissues of ancient Egyptians from ~3200 years BP.

Today, 300 million to 600 million individuals are at risk of infection by schistosomes, and around 200,000 die each year of this disease (Marquet et al. 1996; Chiarella et al. 1998). At least three kinds of this ancient disease exist, caused by *Schistosoma mansoni, Schistosoma haematobium*, or *Schistosoma japonicum*. All three species of schistosomes are of major medical and biohistorical importance. *S. mansoni* usually is found in the portal veins that drain the large intestine. *S. haematobium* is found primarily in the veins of the urinary bladder plexus of infected humans. *S. japonicum* tends to be concentrated in the veins of the small intestine. Most of the pathology associated with schistosomiasis is due to the eggs of flukes that pass into body tissues, where they may lodge in the venules and submucosa of the sigmoid colon and rectum (*S. mansoni* and *S. japonicum*); in the bladder wall (*S. haematobium*); or in other distal tissues, such as the spleen, liver, and brain (through migration of the eggs). Lodged eggs elicit intense inflammatory reactions. Eventually, the trapped eggs become small granulate nodules. Severe clinical disease is often the consequence of heavy infection which, in endemic regions, is determined by the susceptibility or resistance of individuals. (The life cycle of schistosomiasis is illustrated in Figure 7.2.)

Human host genetic factors are probably critical in controlling schistosoma infection and disease development. Studies among Brazilians exposed to schistosomiasis indicate that a major gene, *SM1*, controls human susceptibility or resistance to infection by *S. mansoni* (Abel et al. 1991; Marquet et al. 1996). *SM1* is located on the long arm of chromosome 5 (Muller-Myhsok et al. 1997). This area, which has now been mapped at the molecular level (Abel and Dessein 1997), harbors several genes involved in the differentiation of auxillary T lymphocytes (Dessein 1997) and contains several genes that encode cytokines or cytokine receptors, which are involved in protection against schistosomes. The *SM1* gene has also been mapped (Marquet et al. 1999). It displays a **codominant** mode of inheritance, appears to account for more than 50% of the variation in infection levels (as measured by mean fecal egg counts), and does not appear to be **linked** to HLA (Chiarella et al. 1998). However, the HLA molecule may play an important role in other specific immune responses to *S. mansoni* that involve a different group of clinical traits.

Among Chinese exposed to *S. japonicum*, *HLA-DR-DQ* and *HLA-DP* alleles appear to be independently associated with susceptibility to different stages of postschistosomal liver damage (Hirayama et al. 1999). The HLA-DRB1*1101-DQA1*0501-DQB1*0301 haplotype is associated with protection against early-stage, grade I fibrosis, and HLA-DRB1*1501-DRB5*0101 is associated with susceptibility to this damage. The HLA-DPA1*0103-DPB1*0201 haplotype is associated with protection from the later development of severe liver fibrosis. These findings suggest that HLA class II molecules play a role in fibrotic liver change after deposition with schistosome eggs.

Human Adaptations to Tuberculosis (Infection with *Myobacterium tuberculosis*)

Over the course of human history, countless millions of people have died from tuberculosis, a chronic infectious disease caused by the tubercle bacillus. Tuberculosis-associated spinal caries have been found in predynastic Egyptian mummies. Both Hippocrates and Aristotle provided early descriptions of the disease.

Tuberculosis infection usually is acquired through aerosolization of virulent tubercle bacilli. When a susceptible individual inhales an infectious dried droplet nucleus that contains *Myobacterium tuberculosis* bacillus, the bacteria implants itself in the **alveolar** structure of the lung and initiates infection. The natural route of contact is airborne exposure, and the lungs are the primary organ for infection and transmission. Pulmonary tuberculosis usually involves progressive and extensive pathology in the lung. This infective disease is marked by the formation of tubercles and cavities, the development and calcification of necrotic lesions, the dissemination of infection throughout the bronchial tubes, and the scarification of parenchymal tissue (Cole et al. 1998). Tuberculosis is often associated with other medical problems, especially in aged or immunocompromised populations, which makes its diagnosis more complex (see Table 8.2).

The likelihood of genetic influences on the course of mycobacterial infections during **epidemics** and in endemic areas has long been suspected (Skamene 1994), but the precise nature of such genetic control of susceptibility and resistance is only now coming to light. For example, mutations in the interferon-gamma (**IFN-γ**) receptor 1 gene appear to increase susceptibility to infection by weakly pathogenic mycobacteria (Abel and Dessein 1997). In Cambodians, the *HLA-DQB1*0503* allele reportedly increases the susceptibility to tuberculosis significantly (Goldfeld et al. 1998). However, *NRAMP1*, a macrophage gene, appears to confer resistance to tuberculosis among Canadian Native Americans (North American Indians) (Skamene 1994) and among some Africans (Hill 1998). The *NRAMP1* gene appears to be the human analogue to the mouse *Bcg* resistant gene. In the mouse, the *Bcg* gene confers innate resistance to infection with mycobacteria (Schurr et al. 1991).

Other studies in mice suggest that interleukin-12 (IL-12) is crucial for the development of protective immunity in mice exposed to *M. tuberculosum* (Cooper 1995, 1997). **CD4 T cells** are also important in the protective immune response against tuberculosis (Caruso et al. 1999). Underlying the efficacy of both IL-12 and CD4 T cells are their abilities to stimulate IFN-γ production. Adequate IFN-γ production appears to be an adaptation that is most important in the early phases of infection and hence may crucially determine the subsequent course of tuberculosis infection.

Human Adaptations to Human Immunodeficiency Viruses (Infection with HIV/AIDS)

The primary cause of AIDS is an **RNA** retrovirus of the lentivirus family called HIV-1. HIV-1 has a worldwide distribution. A second retrovirus, HIV-2, is more localized to West Africa. Although AIDS is pandemic, the **incidence** of the disease syndrome is particularly acute in Africa and Asia, where other subclinical infections may promote invasion by HIV (e.g., coinfection with *Chlamydia trachomatis*). Additionally, subclinical infection with endemic diseases that tend to activate the immune system, such as tuberculosis and malaria, appear to accelerate the expression of AIDS.

The HIVs characteristically produce defects in the immune system of infected humans. These defects are typified by a loss of lymphocytes as CD4+ T cells are targeted for functional impairment and destruction by the virus. This loss in CD4+ T cells leads to major compromises in cell-mediated immunity, decreased cytotoxic T-cell and NK-cell responses, and a loss of T cell–mediated delayed hypersensitivity. B-cell responses may become overactivated, perhaps in a compensatory fashion. In HIV infections, B cells will spontaneously increase immunoglobulin secretion.

One of the reasons that the viruses that cause AIDS are so effective is that once they are inside a cell, they are able to use the cell's own protein-synthesizing abilities to generate new viruses. Another reason for the great success of the HIV-1 virus and its extreme pathogenicity is that HIV-1 shows extensive variability in its genetic structure. The reason for this high genetic variability is that the virus' reverse transcriptases frequently make inaccurate copies of virus RNA into proviral **DNA**. These retroviruses have no mechanism to correct their mistakes, thus leading to a rapid proliferation of new variants that have antigenically different envelope proteins from the original strain and variable pathogenicity. Host antibodies that are directed against the original envelope proteins cannot neutralize such variants. This variability helps the virus evade detection and destruction by the immune system and influences the timing and expression of AIDS-related clinical symptoms.

Table 8.2   **Biocultural factors that influence the expression of tuberculosis**

Factor	Effect
Immunosuppression	Impairment of cell-mediated immune response
Diabetes mellitus	Increased susceptibility to infectious disorders in general
Chronic alcoholism	Malnutrition, hypoalbuminemia, and anemia, increased susceptibility to infection
Psychiatric illness	Increased stress, compromised immune response
Silicosis	Impairment of macrophage function in lungs

HIV-1 requires specific **chemokine receptors** as cofactors for entry into human macrophages and T lymphocytes (Rana et al. 1997). Polymorphism in the gene for the CC chemokine receptor 5 (CCR5 Δ 32) appears to provide some protection against HIV infection (Kokkotou et al. 1998; Berger et al. 1999), particularly a delayed disease progression in HIV-infected individuals (Kantor et al. 1998; Mas et al. 1999). Individuals who are homozygous for this genetic condition often are highly resistant to infection, whereas heterozygous individuals have partial protection (Cohen et al. 1997); however, additional factors may be involved in this relationship (Howard et al. 1999). In fact, thus far, at least three genes have been identified that have variants that appear to provide resistance to infection by HIV-1 (Tashiro et al. 1999). Two of these genes encode receptors for the chemokines CCR5 and CCR2, both of which can act as entry coreceptors for HIV-1 virus, along with CXCR4 (Cammack 1999). The other gene is *SDF1* gene, a cytokine that appears to regulate the period between HIV-1 infection and the onset of AIDS (Tashiro et al. 1999).

Furthermore, an analysis of HLA alleles in 20,000 Europeans has found that HLA status may also be correlated with AIDS disease progression. HLA class I alleles *A29* and *B22* were significantly associated with rapid progression from HIV-1 infection to clinical AIDS. HLA alleles *B15* and *C8* were significantly associated with nonprogression to AIDS. HLA class I alleles *B27*, *B57*, and *C14* appear to be protective against HIV-1, whereas *C16* and *B35* appear to increase susceptibility to HIV-1 viruses. In this population, the influence of class II alleles was limited to *DR11* (Hendel et al. 1999).

FUTURE TRENDS

Biocultural Factors Influencing 21st-Century Infectious Diseases

Global change is pervasive and occurring at a dramatic rate (Sutherst 1998). Synergistic changes in human behavior, population demographics, aspects of the physical environment, climate, technology, land use patterns, and microbial adaptation favor the continued emergence of infectious diseases throughout the next century (see Wilson 1995; Sable and Mandell 1996). As we begin to understand more about the potential impact of these variables, both singly and in combination, predicting the consequences for specific infectious diseases becomes more problematic.

Increases in global temperature, estimated to rise 2°C by the year 2100, is expected to facilitate the dissemination of many serious infectious diseases (Patz et al. 1996). Vector-borne diseases such as malaria and dengue are among the infectious diseases most sensitive to climatic shifts. Over the past 20 years, dengue fever epidemics and hyperendemic transmission has been (re)established over a geographically expanding area (Monath 1994). At the root of the resurgence of dengue are changes in human demography and behavior. These changes may be the cause of other climate-sensitive infectious diseases as well (Rodier et al. 1995). Figure 8.2 illustrates seven salient biocultural factors that are likely to adversely influence future human exposure to infectious diseases.

An increase in temperature can expand the geographical ranges of disease vectors and increase the absolute number of vectors that survive. If the increase improves vector access to vulnerable human hosts, as is expected by increasing the human biting index, then genetic-based resistance in vectors and infectious agents to insecticides and therapeutic drugs may accelerate and amplify infectious disease transmission (Gubler 1998). Meterologists have predicted that an increase in climatic temperatures will raise sea surface temperatures as well as sea levels, and these events may increase the incidence of water-borne infections such as cholera. Rising sea levels also can lead to widespread human displacements, putting individuals and groups at increased risk of infections due to malnutrition (and other nutritional deficiencies) associated with climatic stress on agriculture and alterations in exposure to ultraviolet radiation.

Human displacement, too frequently due to civil war and ethnic conflicts (Kalipeni and Oppong 1998), also is associated with a loss of biodiversity that may disrupt ecological functioning and the stability of existing human–microorganism

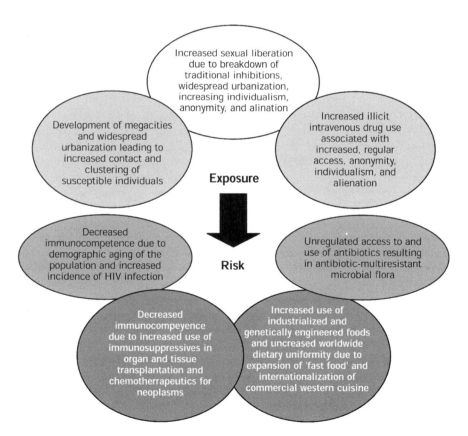

Figure 8.2 Human behavioral factors that influence infectious disease transmission and expression. (*Source*: Modified from Kumate 1997)

interactions (Jutro 1991). Infectious diseases that appeared to be declining 25 years ago may again begin to evidence dysynchrony with human hosts (see Figure 8.1) as environmental conditions shift. The epidemiological and public health literature provide many thought-provoking articles on the great potential for emerging diseases.

Emerging and Reemerging Diseases

Emerging diseases are defined as infections that are either newly recognized, often by using the new tools of molecular genetics, or whose incidence in humans has significantly increased over the past two decades (Morse 1995; Sable and Mandell 1996). Emerging diseases may include truly new diseases (resulting from new mutations); seemingly new diseases caused by old, preexisting infectious agents; and recurrent old diseases caused by mutated, previously known agents (Kilbourne 1996). For example, tuberculosis is experiencing a resurgence now because of mycobacterial mutation to antibiotic resistance and the increased number of more susceptible individuals that are immunosuppressed by HIV infection.

Technological advances hold promise for eradicating infectious diseases such as poliomyelitis and guinea worm and for the effective control of many known infectious diseases worldwide (Kumate 1997; Hinman 1998). Molecular toolkits allow scientists to create new diagnostic tests, characterize microbial virulence factors, and develop novel vaccines (Sable and Mandell 1996). However, "new" diseases such as human herpes virus 8 and Kaposi's sarcoma–associated herpesvirus (Schulz and Moore 1999) are expected, often also as a result of other technological innovations. The construction of new dams, the development of faster and more efficient transportation systems (Wilson 1995), and the use **xenobiotic** organ transplantation (usually from pig to human [Borie et al. 1998]) may carry the cost of increased infectious disease and more selective pressure on human adaptations. As new diseases emerge and old diseases reemerge, humans will be required, as throughout human history, to adapt or face extinction. Infectious disease probably will continue to be the leading cause of death worldwide in the coming century. Our challenge will be (Ewald 1998):

- to develop control measures that distinguish dangerous pathogens from those that are less virulent,
- to identify the factors contributing to the transmission of the most serious infectious agents, and
- to direct our effort and resources toward the most species-threatening infectious agents.

In the 20th century, pneumonia and influenza were the causes of the largest number of infectious disease deaths in the United States (Armstrong et al. 1999). In the early part of the century, tuberculosis caused almost as many deaths as pneumonia and influenza. Tuberculosis mortality dropped off sharply after 1945, only to return with a vengeance in the 1980s and 1990s. Infectious diseases in the United States also increased dramatically during the latter decades of the 20th century due to the emergence of AIDS, particularly in individuals 25–64 years old.

The lesson to be gained from reviewing the US pattern of fluctuation in the intensity of infectious diseases during the 20th century is that these human–microorganism interactions are dynamic and will require a multidisciplinary, biocultural approach for effective preemptive and sustainable control. Because "new" diseases will undoubtedly mandate novel behavioral, physiochemical, and genetic adaptive responses among exposed humans, biological anthropologists—particularly human biologists—have an important and integrative niche to fill in disease identification and control efforts.

CHAPTER SUMMARY

Infectious diseases have been and continue to be a major cause of human mortality. The diseases that have affected humans have changed over our evolutionary history. Probably the most dramatic shifts in human–microorganism interactions occurred within the past 20,000 years, after the adoption of agriculture. Adaptations to infectious diseases may take place at the behavioral, physiochemical, and or genetic levels; however, these divisions are probably artificial. Because adaptations are dynamic processes that are constantly changing in response to external and internal stimuli, successful human adaptation usually involves altered behaviors, modified physiochemistry, and genetic polymorphism.

The human immune system exhibits a complex set of responses to infectious agents. Evidence indicates the existence of genetic traits that decrease susceptibility to malaria, river blindness, schistosomiasis, tuberculosis, and human immunodeficiency virus (HIV), many of which involve the human leukocyte antigen (HLA) system. Changes in human behavior, population demographics, aspects of the physical environment, climate, technology, land use patterns, and microbial adaptation favor the continued emergence of infectious diseases throughout the next century.

RECOMMENDED READINGS

Ewald PM (1996) Evolution of Infectious Diseases. Oxford: Oxford Univ. Press.

Karlen A (1996) Man and Microbes: Disease and Plagues in History and Modern Times. New York: Vintage Books.

Le Guenno B (1995) Emerging viruses. Sci. Am. October:. 56–64.

Roizman B (ed.) (1995) Infectious Diseases in an Age of Change: The Impact of Human Ecology and Behavior on Disease Transmission. Washington, D.C.: National Academy of Sciences Press.

Watts S (1998) Epidemics and History: Disease, Power, and Imperialism. New Haven: Yale Univ. Press.

Williams GC and Nesse RM (1996) Why We Get Sick: The New Science of Darwinian Medicine. New York: Vintage Books.

REFERENCES CITED

Abel L, Demenais F, Prata A, Souza AE, and Dessein A (1991) Evidence for the segregation of a major gene in human susceptibility/resistance to infection by *Schistosoma mansoni*. Am. J. Hum. Genet. 48(5): 959–970.

Abel L, Cot M, Mulder L, Carnevale P, and Feingold J (1992) Segregation analysis detects a major gene controlling blood infection levels in human malaria. Am. J. Hum. Genet. 50(6): 1308–1317.

Abel L, and Dessein AJ (1997) The impact of host genetics on susceptibility to human infectious diseases. Curr. Opin. Immunol. 9(4): 509–516.

Armstrong GL, Conn LA, and Pinner RW (1999) Trends in infectious disease mortality in the United States during the 20th century. JAMA 281(1): 61–66.

Basanez MG and Bouissinesq M (1999) Population biology of human onchocerciasis. Phil. Trans. R. Soc. Lond. B, Biol. Sci. 354(1384): 809–826.

Berger EA, Murphy PM, and Farber JM (1999) Chemokine receptors as HIV-1 coreceptors: roles in viral entry, tropism, and disease. Annu. Rev. Immunol. 17: 657–700.

Borie EC, Cramer DV, Phan-Thanh L, Vaillant JC, Bequet JL, MakowkaL, and Hannoun L (1998) Microbiological hazards related to xenotransplantation of porcine organs into man. Infect. Control Hosp. Epidemiol. 19(5): 355–365.

Cammack N (1999) Human immunodeficiency virus type 1 entry and chemokine receptors: a new therapeutic target. Antivir. Chem. Chemother. 10(2): 53–62.

Caruso AM, Serbina N, Klein E, Triebold K, Bloom BR, and Flynn JL (1999) Mice deficient in CD4 T cells have only transiently diminished levels of IFN-gamma, yet succumb to tuberculosis. J. Immunol. 162(9): 5407–5416.

Chiarella JM, Goldberg AC, Abel L, Carvalho EM, Kalil J, and Dessein A (1998) Absence of linkage between MHC and a gene involved in susceptibility to human schistosomiasis. Braz. J. Med. Biol. Res. 31(5): 665–670.

Cohen OJ, Kinter A, and Fauci AS (1997) Host factors in the pathogenesis of HIV disease. Immunol. Rev. 159: 31–48.

Cole ST, Brosch R, Parkhill J, Garnier T, Churcher C, Harris D, Gordon SV, Eiglmeir K, Gas S, et al. (1998) Deciphering the biology of *Mycobacterium tuberculosis* from the complete genome sequence. Nature 393(6685): 537–544.

Cooper AM, Roberts AD, Rhoades ER, Callahan JE, Getzy DM, and Orme IM (1995) The role of interleukin-12 in acquired immunity to *Mycobacterium tuberculosis* infection. Immunology 84(3): 423–432.

Cooper AM, Magram J, Ferrante J, and Orme IM (1997) Interleukin 12 (IL-12) is crucial to the development of protective immunity in mice intravenously infested with *Mycobacterium tuberculosis*. J. Exp. Med. 186(1): 39–45.

Dessein A (1997) Genetic predisposition to infectious disease. Med. Trop. (Mars) 57(Suppl 3): 10–12.

Doetze A, Erttmann KD, Gallin MY, Fleischer B, and Hoerauf A (1997) Production of both IFN-gamma and IL-5 by *Onchocerca volvulus* S1 antigen-specific CD4+ T cells from putatively immune individuals. Int. Immunol. 9(5): 721–729.

Elson LH, Guderian RH, Araujo E, Bradley JE, Days A, and Nutman TB (1994) Immunity to onchocerciasis: identification of a putatively immune population in a hyperendemic area of Ecuador. J. Infect. Dis. 169(3): 588–594.

Ewald PW (1998) The evolution of virulence and emerging diseases. J. Urban Health 75(3): 480–491.

Fernandez-Reyes D, Craig AG, Kyes SA, Peshu N, Snow RW, Berendt AR, Marsh K, and Newbold CI (1997) A high frequency African coding polymorphism in the N-terminal domain of ICAM-1 predisposing to cerebral malaria in Kenya. Hum. Mol. Genet. 6(8): 1357–1360.

Garcia A, Marquet S, Bucheton B, Hillaire D, Cott M, Fievet N, Dessein AJ, and Abel L (1998) Linkage analysis of blood *Plasmodium falciparum* levels: interest of the 5q31-q33 chromosome region. Am. J. Trop. Med. Hyg. 58(6): 705–709.

Garnett GP and Holmes EC (1996) The ecology of emergent infectious disease. Bioscience 46(2): 127–136.

Goldfeld AE, Delgado JC, Thim S, Bozon MV, Uglialoro AM, Turbay D, Cohen C, and Yunis EJ (1998) Association of an *HLA-DQ* allele with clinical tuberculosis. JAMA 279(3): 226–228.

Gubler DJ (1998) Resurgent vector-borne diseases as a global health problem. Emerg. Infect. Dis. 4(3): 442–450.

Hendel H, Caillat-Zucman S, Lebuanec H, Carrington M, O'Brien S, Andrieu JM, Schachter F, Zagury D, Rappaport J, Winkler C, Nelson GW, and Zagury JF (1999) New class I and II HLA alleles strongly associated with opposite patterns of progression to AIDS. J. Immunol. 162(11): 6942–6946.

Hill AV (1996) Genetic susceptibility to malaria and other infectious diseases: from the MHC to the whole genome. Parasitology 112(Suppl): S75–S84.

Hill AV (1998) The immunogenetics of human infectious diseases. Annu. Rev. Immunol. 16: 593–617.

Hinman AR (1998) Global progress in infectious disease control. Vaccine 16(11–12): 1116–1121.

Hirayama K, Chen H, Kikuchi M, Yin T, Gu X, Liu J, Zhang S, and Yuan H (1999) *HLA-DR-DQ* alleles and *HLA-DP* alleles are independently associated with susceptibility to different stages of post-schistosomal hepatic fibrosis in the Chinese population. Tissue Antigens 53(3): 269–274.

Howard OM, Shirakawa AK, Turpin JA, Maynard A, Tobin GJ, Carrington M, Oppenheim JJ, and Dean M (1999) Naturally occurring CCR5 extracellular and transmembrane domain variants affect HIV-1 coreceptor and ligand binding function. J. Biol. Chem. 274(23): 16228–16234.

Jutro PR (1991) Biological diversity, ecology, and global climate change. Environ. Health Perspect. 96: 167–170.

Kalipeni E and Oppong J (1998) The refugee crisis in Africa and implications for health and disease: a political ecology approach. Soc. Sci. Med. 46(12): 1637–1653.

Kantor R, Barzilai A, Varon D, Martinowitz U, and Gershoni JM (1998) Prevalence of a CCR5 gene 32-bp deletion in an Israeli cohort of HIV-1-infected and uninfected hemophilia patients. J. Hum. Virol. 1(4): 299–301.

Kapp C (1999) WHO warns of microbial threat. Lancet 353(9171):2222.

Kilbourne ED (1996) The emergence of "emerging diseases": a lesson in holistic epidemiology. Mt. Sinai J. Med. 63 (3-4):159-166.

Kokkotou E, Philippon V, Gueye-Ndiaye A, Mboup S, Wang WK, Essex M, and Kanki P (1998) Role of the CCR5 delta 32 allele in resistance to HIV-1 infection in West Africa. J. Hum. Virol. 1(7): 469–474.

Kumate J (1997) Infectious diseases in the 21st century. Arch. Med. Res. 23(2): 155–161.

Marquet S, Abel L, Hillaire D, Dessein H, Kalil, J, Feingold, J, Weissenbach J, and Dessein AJ (1996) Genetic localization of a locus controlling the intensity of infection by *Schistosoma mansoni* on chromosome 5q31-q33. Nat. Genet. 14(2): 181–184.

Marquet S, Abel L, Hillaire D, and Dessein A (1999) Full results of the genome-wide scan which localises a locus controlling the intensity of infection by *Schistosoma mansoni* on chromosome 5q31-q33. Eur. J. Hum. Genet. 7(1): 88–97.

Mas A, Espanol T, Heredia A, Pedraza MA, Hernandez M, Caragol I, Fernando M, Bertran JM, Alcami J, and Soriano V (1999) *CCR5* genotype and HIV-1 infection in perinatally-exposed infants. J. Infect. 38(1): 9–11.

Meyer CG, Gallin M, Erttmann ED, Brattig N, Schnittger L, Gelhaus A, Tannich E, Begovich AB, Erlich HA, and Horstmann RD (1994) *HLA-D* alleles associated with generalized disease, localized disease, and putative immunity in *Onchocerca volvulus* infection. Proc. Natl. Acad. Sci. U.S.A. 91(16): 7515–7519.

Monath TP (1994) Dengue: the risk to developed and developing countries. Proc. Natl. Acad. Sci. U.S.A. 91(7): 2395–2400.

Morse SS (1995) Factors in the emergence of infectious diseases. Emerg. Infect. Dis. 1(1): 7–15.

Muller-Myhsok B, Stelma FF, Guisse-Sow F, Muntau B, Thye T, Burchard GD, Gryseels B, and Horstmann RD (1997) Further evidence suggesting the presence of a locus, on human chromosome 5q31-q33, influencing the intensity of infection with *Schistosoma mansoni*. Am. J. Hum. Genet. 61(2): 452–454.

Murdoch ME, Payton A, Abiose A, Thomson W, Panicker VK, Dyer PA, Jones BR, Maizels RM, and Ollier WE. (1997) *HLA-DQ* alleles associate with cutaneous features of onchocerciasis. The Kaduna–London–Manchester Collaboration for Research on Onchocerciasis. Hum. Immunol. 55(1): 46–52.

Ogunrinade A, Boakye D, Merriweather A, and Unnasch TR (1999) Distribution of the blinding and nonblinding strains of *Onchocerca volvulus* in Nigeria. J. Infect. Dis. 179(6): 1577–1579.

Patz JA, Epstein PR, Burke TA, and Balbus JM (1996) Global climate change and emerging infectious diseases. JAMA 275(3): 217–223.

Quintana-Murci L, Semino O, Bandelt HJ, Passarino G, McElreavey K, and Santachiara-Benerecetti AS (1999) Genetic evidence of an early exit of *Homo sapiens sapiens* from Africa through eastern Africa. Nat. Genet. 23(4):437–441.

Rana S, Besson G, Cook DG, Rucker J, Smyth RJ, Yi Y, Turner JD, Guo HH, Du JG, Peiper SC, et al. (1997) Role of CCR5 in infection of primary macrophages and lymphocytes by macrophage-tropic strains of human immunodeficiency virus: resistance to patient-derived and prototype isolates resulting from the delta ccr5 mutation. J. Virol. 71(4): 3219–3227.

Rodier GR, Parra JP, Kamil M, Chakib SO, and Cope SE. (1995) Recurrence and emergence of infectious diseases in Djibouti city. Bull. World Health Org. 73(6): 755–759.

Sable CA and Mandell GL (1996) The role of molecular techniques in the understanding of emerging infections. Mol. Med. Today 2(3): 120–128.

Schulz TF and Moore PS (1999) Kaposi's sarcoma-associated herpesvirus: a new human tumor virus, but how? Trends Microbiol. 7(5): 196–200.

Schurr E, Radzioch D, Malo D, Gros P, and Skamene E (1991) Molecular genetics of inherited susceptibility to intracellular parasites. Behring Inst. Mitt. 88: 1–12.

Shear HL, Roth EF Jr, Fabry ME, Costantini FD, Pachnis A, Hood A, and Nagel RL (1993) Transgenic mice expressing human sickle hemoglobin are partially resistant to rodent malaria. Blood 81(1): 222–226.

Skamene E (1994) The *Bcg* gene story. Immunobiology 191(4–5): 451–460.

Stewart GR, AZhu Y, Parredes W, Tree TI, Guderian R, and Bradley JE (1997) The novel cuticular collagen Ovcol-1 of *Onchocerca volvulus* is preferentially recognized by immunoglobulin G3 from putatively immune individuals. Infect. Immunol. 65(1): 164–170.

Sutherst RW (1998) Implications of global change and climate variability for vector-borne diseases: generic approaches to impact assessments. Int. J. Parasitol. 28(6): 935–945.

Tashiro K, Ikegawa M, Yabe D, and Honjo T (1999) Anti-HIV-1 genes: genetic restriction of AIDS pathogenesis by gene variants. Nippon Rinsho 57(4): 967–974.

Weatherall DJ (1996) Host genetics and infectious disease. Parasitology 112(Suppl): S23–S29.

Wilson ME (1995) Travel and the emergence of infectious diseases. Emerg. Infect. Dis. 1(2): 39–46.

CHAPTER 9

Human Nutritional Evolution

WILLIAM R. LEONARD

INTRODUCTION

The study of diet and nutrition has long been a central component of much research in human biology because it represents a critical interface between biology, culture, and the environment. Ecological variation in food availability has been an important stressor throughout our evolutionary history (see Gordon 1987; Leonard and Robertson 1992) and continues to strongly shape the biology of traditional human populations today (de Garine and Harrison 1988; Huss-Ashmore et al. 1988).

Not surprisingly, humans are broadly similar to other primate species in that they are omnivorous and have particular nutritional requirements that reflect adaptations to diets with large amounts of fruit and vegetable material (e.g., the inability to synthesize vitamin C). However, in the 5 million to 7 million years since our last common ancestor with the apes, several distinct selective pressures appear to have resulted in important nutritional and physiological differences between humans and other primates. One of the most striking differences is the diversity of diets on which modern human populations are able to subsist. This dietary plasticity evolved, in part, because of cultural and technological innovations for processing resources, and it allowed human populations to expand into newer and more marginal ecosystems.

In this chapter, we will examine the evolution of human nutritional patterns by drawing on many different sources of data. I start with a brief introduction to nutrients and nutritional requirements in humans. Next, I examine human nutrition from a comparative perspective, considering how humans conform to and deviate from general dietary and nutritional patterns seen in the mammalian and primate world. Then, I discuss major changes during human evolution to try to gain insights into origins of the distinctiveness of human dietary factors. Finally, I offer several examples of dietary variation among modern human populations, specifically considering how various ecological factors shape patterns of nutritional adaptation.

Human Biology: An Evolutionary and Biocultural Perspective, Edited by Sara Stinson, Barry Bogin, Rebecca Huss-Ashmore, and Dennis O'Rourke
ISBN 0-471-13746-4   Copyright © 2000 Wiley-Liss, Inc.

FUNDAMENTALS OF NUTRITION

Nutrients

To sustain life and maintain health, humans and other animals require many different nutrients. These nutrients are grouped into six broad classes: carbohydrates, fats, **protein**, vitamins, minerals, and water. Carbohydrates, protein, and fats, collectively known as macronutrients, are required in relatively large amounts and are critical for providing energy (measured in calories or joules) to the body. Vitamins and minerals (**micronutrients**) are required in much smaller amounts and are important for regulating many aspects of biological function.

Carbohydrates and proteins have similar energy contents; each provides ~4 **kilocalories** (kcal; ~17 kilojoules [kJ]) of metabolic energy per gram (~0.035 ounces). In contrast, fat is more calorically dense; each gram provides about 9–10 kcal (37–42 kJ). Alcohol, although not a required nutrient, also can be used as an energy source; it contributes ~7 kcal (29 kJ) per gram (Rolfes et al. 1990). Regardless of the source, excess dietary energy can be stored by the body as **glycogen** (a carbohydrate) or fat. Humans have relatively limited glycogen stores (about 375–475 grams [0.8–1.0 lb]), in the liver and muscles (McArdle et al. 1986). Fat, however, makes up ~16% of body weight in men and 27% in women (US data from Frisancho 1990).

The largest source of dietary energy for most humans (~45–55% of calories in the United States; USDA 1986, 1987) is carbohydrates. The three types of carbohydrates are **monosaccharides**, **disaccharides**, and **polysaccharides**. Monosaccharides, or simple sugars, include **glucose**, the body's primary metabolic fuel; fructose (fruit sugar); and **galactose**. Disaccharides, as the name implies, are sugars formed by a combination of two monosaccharides. Sucrose (glucose and fructose), the most common disaccharide, is found in sugar, honey, and maple syrup. **Lactose**, the sugar found in milk, is composed of glucose and galactose. Maltose (glucose and glucose), the least common of the disaccharides, is found in malt products and germinating cereals. Polysaccharides, or complex carbohydrates, are composed of three or more simple sugar molecules. Glycogen is the polysaccharide used for storing carbohydrates in animal tissues. In plants, the two most common polysaccharides are starch and cellulose. Starch is found in a wide variety of foods, such as grains, cereals, and breads, and provides an important source of dietary energy. In contrast, cellulose—the fibrous, structural parts of plant material—is not digestible to humans and passes through the gastrointestinal tract as fiber.

Fats provide the largest store of potential energy for biological work in the body. They are divided into three main groups: simple, compound, and derived. The simple or "neutral fats" consist primarily of triglycerides. A triglyceride consists of two component molecules: glycerol and **fatty acid**. Fatty acid molecules, in turn, are divided into two broad groups: saturated and unsaturated. These categories reflect the chemical bonding pattern between the carbon atoms of the fatty acid molecule. Saturated fatty acids have no double bonds between carbons, thus allowing for the maximum number of hydrogen atoms to be bound to the carbon (i.e., the carbons are "saturated" with hydrogen atoms). In contrast, unsaturated fatty acids have one (monounsaturated) or more (polyunsaturated) double bonds. Saturated fats are abundant in animal products, whereas unsaturated fats predominate in vegetable oils.

Compound fats consist of a neutral fat in combination with some other chemical substance (e.g., a sugar or a protein). Examples of compound fats include phospholipids and lipoproteins. Phospholipids are important in blood clotting and insulating nerve fibers, whereas lipoproteins are the main form of transport for fat in the bloodstream.

Derived fats are substances synthesized from simple and compound fats. The best known derived fat is **cholesterol**. Cholesterol is present in all human cells and may be derived from foods (exogenous) or synthesized by the body (endogenous). Cholesterol is necessary for normal development and function because it is critical for the synthesis of such hormones as **estradiol**, **progesterone**, and **testosterone** (Greene 1970).

Proteins, in addition to providing an energy source, are also critical for the growth and replacement of living tissues. They are composed of nitrogen-containing compounds known as **amino acids**. Of the 20 different amino acids required by the body, 9 (leucine, isoleucine, valine, lysine, threonine, methionine, phenylalanine, tryptophan, and histidine) are "essential" because they cannot be synthesized by the body and thus must be derived from food. Two others, cystine and tyrosine, are synthesized in the body from methionine and phenylalanine, respectively. The remaining amino acids are "nonessential" because they can be produced by the body and need not be derived from the diet.

Vitamins are not a source of energy; rather, they help the body use energy and carry out other metabolic activities. Vitamins are divided into two classes according to solubility. Water-soluble vitamins include the B vitamins and vitamin C; the fat-soluble vitamins include vitamins A, D, E, and K. Water-soluble vitamins generally should be consumed on a daily basis because they are not stored in the body. Excess intake of these vitamins is excreted in the urine. In contrast, fat-soluble vitamins need not be consumed every day because they can be stored in the body. High intake of these vitamins over a long time can lead to toxicity.

Minerals are inorganic elements that are key components in many biological molecules (e.g., iron in hemoglobin) and are critical for maintaining various physiological functions. Collectively, minerals make up ~4% of body weight (McArdle et al. 1986). Minerals are particularly important in regulating metabolism, because they are constituent parts of **enzymes** (e.g., zinc in carbonic anhydrase), hormones (e.g., iodine in thyroid hormones), and vitamins (e.g., selenium in vitamin E) (Gibson 1990). Minerals may appear in combination with organic compounds (e.g., calcium phosphate in bones) or in their free, ionic forms (e.g., sodium and potassium in intracellular fluids).

Water makes up about 40–60% of body weight in an adult human; about 65–75% of muscle tissue and about 25% of fat are water. The body's daily water supply, about 2.55 liters (2.7 quarts) is derived from liquid intake (~1.2 liters), food (~1 liter), and so-called "metabolic water" produced as the result of energy-yielding reactions (~0.35 liters). Water is lost from the body as urine (1–1.5 liters/day), perspiration (0.5–0.7 liters), water vapor (from expired air; 0.2–0.3 liters), and feces (~0.10 liters) (McArdle et al. 1986).

Nutritional Requirements

Much ongoing research in human nutritional science is focused on developing and refining nutrient requirements. In the United States, the most widely used standards for energy and nutrient intakes are the recommended daily allowances (RDAs) (e.g., NRC 1989). Similar recommendations have been developed in Canada (Health and Welfare Canada 1990) and in the United Kingdom (Department of Health and Social Security 1981) and by the World Health Organization (e.g., FAO/WHO/UNU 1985; FAO/WHO 1988). For some nutrients (e.g., calcium), recommended intakes differ considerably among the standards. This point is evident in Table 9.1, which compares recommended intakes of 13 selected nutrients for adult men according to these four standards. These differences, in many cases, reflect our limited knowledge about the nature of variability in human nutrient needs (see NRC 1986).

In general, the procedures for determining recommended energy intakes are different from those used for determining recommendations for specific nutrients (e.g., protein, vitamins, and minerals). For energy, recommended daily intakes (measured in kilocalories or kilojoules per day) reflect the *average* requirement for a person of a given age, sex, and body weight. In contrast, the recommendations for nutrients are set at levels sufficiently high to meet the requirements of almost all healthy individuals in a given group. Recent expert committees in the United States and Canada have set recommended nutrient intake levels at two **standard**

Table 9.1   **Recommended daily intakes for selected nutrients for adult men**

Nutrient	United States	Canada	United Kingdom	FAO/WHO
Protein (g)	56	61	72	52.5
Calcium (mg)	800	800	500	400–500
Phosphorus (mg)	800	800	—	—
Iron (mg)	10	8	10	8–23
Vitamin A (μg)	1000	1000	750	600
Vitamin D (μg)	5	2.5	2.5	2.5
Vitamin C (mg)	60	60	30	30
Folate (μg)	400	220	—	200
Vitamin E (mg)	10	9	—	—
Vitamin B-12 (μg)	3.0	2.0	—	1.0
Magnesium (mg)	350	250	—	—
Zinc (mg)	15	9	—	11
Iodine (μg)	150	160	—	—

(*Source*: Derived from Gibson 1990)

deviations above the average (mean) requirement for a given age- and sex-specific group, or about at the 97th to 98th **percentile** (NRC 1989; Health and Welfare Canada 1990). Consequently, dietary intakes at these levels should meet the nutritional needs of all but about 2–3% of the population. The different approaches for determining dietary allowances of energy and specific nutrients are illustrated in Figure 9.1.

Energy recommendations are different from those of nutrients because both underconsumption and overconsumption of energy have serious health consequences. In contrast, for most nutrients, the potential health risks of slight to moderate overconsumption are generally regarded as small compared with the risks associated with underconsumption. Consequently, nutrient requirements are purposefully set "high" to ensure that almost everyone eating the recommended level will meet his or her daily requirements. If, however, the same procedure were used for energy (i.e., recommendations at the 97th percentile), recommendations would promote obesity.

A discussion of the specific recommendations for each major nutrient is beyond the scope of this chapter. These issues are covered at length in the World Health Organization (WHO) technical reports (e.g., FAO/WHO 1988), the RDA volumes (e.g., NRC 1989), and in *Principles of Nutritional Assessment* (Gibson 1990). Instead, our discussion will focus exclusively on the recent recommendations for energy and protein intake.

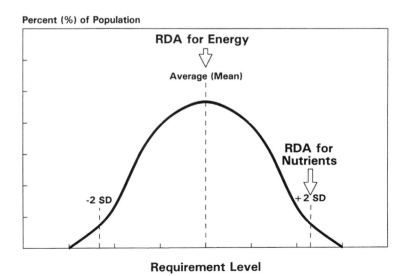

Figure 9.1   Comparison of the criteria used for establishing the recommended dietary allowances (RDAs) of energy and nutrients. Recommended intakes for energy are set at the average (mean) requirement for the reference population. In contrast, recommended intakes for nutrients (protein, vitamins, and minerals) are set well above the average requirement (by 2 standard deviations [SD]). At this level, only 2–3% of the reference population have requirements above the RDA.

Estimating Daily Energy Needs

A person's daily energy requirements are determined by numerous factors. The major components of an individual's energy budget are associated with resting or basal metabolism, activity, **thermoregulation**, **dietary thermogenesis**, growth, and reproduction (Ulijaszek 1995). **Basal metabolic rate** (BMR) represents the minimum amount of energy necessary to keep a person alive.[1] Basal metabolism is measured under controlled conditions while a subject is lying in a relaxed and fasted state (at least 12 hours after a meal; see C Schofield 1985; McLean and Tobin 1986).

In addition to basal requirements, energy is expended to perform various types of work, such as daily activities and exercise, digestion and transport of food, and regulating body temperature. The energy costs associated with food handling (i.e., the **thermic effect of food**) make up a relatively small proportion of daily energy expenditure and are influenced by amount consumed and the composition of the diet (e.g., high-protein meals elevate dietary thermogenesis). In addition, at extreme temperatures, energy must be spent to heat or cool the body. Humans (unclothed) have a thermoneutral range of 25–27 °C (77–81 °F). Within this temperature range, the minimum amount of metabolic energy is spent to maintain body temperature (Erickson and Krog 1956). Finally, during one's lifetime, additional energy is required for physical growth and for reproduction (e.g., pregnancy, lactation).

In 1985, the World Health Organization presented its most recent recommendations for assessing human energy requirements (FAO/WHO/UNU 1985). The procedure used for estimating energy needs involves first estimating BMR from body weight on the basis of predictive equations developed by WN Schofield (1985). These equations are presented in Table 9.2. After estimating BMR, the to-

Table 9.2 **Equations for predicting basal metabolic rate (BMR) based on body weight (Wt, in kg)**

	BMR (kcal/day)	
Age (years)	Males	Females
0–2.9	60.9(Wt) − 54	61.0(Wt) − 51
3.0–9.9	27.7(Wt) + 495	22.5(Wt) + 499
10.0–17.9	17.5(Wt) + 651	12.2(Wt) + 746
18.0–29.9	15.3(Wt) + 679	14.7(Wt) + 496
30.0–59.9	11.6(Wt) + 879	8.7(Wt) + 829
60+	13.5(Wt) + 487	10.5(Wt) + 596

(*Source:* FAO/WHO/UNU 1985, 71)

[1]In the human and animal energetics literature, the terms basal metabolic rate and **resting metabolic rate** (BMR and RMR, respectively) have quite distinct meanings (cf.. Kleiber 1975; WN Schofield 1985). However, for our purposes in this chapter, BMR (of humans) and RMR (of animal studies) are assumed to be functionally equivalent.

Table 9.3   Physical activity levels (PALs) associated with different types of occupational work among adults (18 years and older)

Sex	PAL			
	Minimal	Light	Moderate	Heavy
Male	1.40	1.55	1.78	2.10
Female	1.40	1.56	1.64	1.82

(*Source*: FAO/WHO/UNU 1985)

tal daily energy expenditure (TDEE) for adults (18 years old and above) is determined as a multiple of BMR, based on the individual's activity level. This multiplier, known as the **physical activity level** (PAL) index, reflects the proportion of energy *above* basal requirements that an individual spends over the course of a normal day. The PALs associated with different occupational work levels for adult men and women are presented in Table 9.3. WHO recommends that minimal daily activities such as dressing, washing, and eating are commensurate with a PAL of 1.4 for both men and women. Sedentary lifestyles (e.g., office work) require PALs of 1.55 for men and 1.56 for women. At higher work levels, however, the sex differences are greater. Moderate work is associated with a PAL of 1.78 in men and 1.64 in women, whereas heavy work levels (e.g., manual labor, traditional agriculture) necessitate PALs of 2.10 and 1.82 for men and women, respectively (see also Chapter 10).

In addition to the costs of daily activity and work, energy costs for reproduction also must be considered. WHO recommends an additional 285 kcal (1190 kJ) per day for women who are pregnant and an additional 500 kcal (2090 kJ) per day for those who are lactating.

Energy requirements for children and adolescents are estimated differently because of the extra energy costs associated with growth and because relatively less is known about variation in their activity patterns. For children and adolescents between 10 and 18 years old, WHO recommends the use of age- and sex-specific PALs. In contrast, energy requirements for children under 10 years old are determined by multiplying the child's weight by an age- and sex-specific constant. The reference values for boys and girls under 18 years old are presented in Table 9.4.

Thus, using the most recent recommendations from WHO (FAO/WHO/UNU 1985), relatively limited information is needed to effectively estimate a person's daily energy needs. (Box 9.1 shows how TDEE can be calculated for two adults by using body weight and activity level.)

Protein Requirements

Daily protein requirements are established based on controlled studies that monitor the amount of nitrogen (from protein) consumed and the amount excreted from the body (in urine, feces, and sweat). After reviewing several of these nitrogen balance

BOX 9.1   ESTIMATING DAILY ENERGY AND PROTEIN REQUIREMENTS

The most recent recommendations from the World Heath Organization (WHO; FAO/WHO/UNU 1985) provide a straightforward approach for estimating energy and protein needs. Examples of how these recommendations can be applied are presented in the following sections.

Energy Requirements

For this example, we will consider two adults who have similar sedentary office jobs. The man is 32 years old and weighs 160 lb (72.5 kg); the woman is 28 years old and weighs 120 lb (54.4 kg). From the information presented in Tables 9.2 and 9.3, the man's daily energy needs are calculated as

$$BMR = 11.6(Wt) + 879 = 11.6(72.5) + 879 = 1720 \text{ kcal (7196 kJ)}$$

$$TDEE = PAL \times BMR = 1.55 \times 1720 = 2666 \text{ kcal (11,150 kJ)}$$

where BMR is basal metabolic rate, Wt is body weight (in kg), TDEE is total daily energy expenditure, and PAL is physical activity level.

In comparison, the woman's daily energy demands would be determined as

$$BMR = 14.7(Wt) + 496 = 14.7(54.4) + 496 = 1296 \text{ kcal (5421 kJ)}$$

$$TDEE = PAL \times BMR = 1.56 \times 1296 = 2022 \text{ kcal (8460 kJ)}$$

Thus, despite having similar activity levels, the man's energy demands are 32% higher than the woman's because of the differences in weight and body composition.

Protein Requirements

Daily protein needs for the same two people may be calculated in a using the information presented in Tables 9.5 and 9.6. For the man and the woman, the base daily protein needs are calculated as

$$Protein \text{ (man)} = 0.75(Wt) = 0.75(72.5) = 54.4 \text{ grams } (\sim 2 \text{ ounces})$$

$$Protein \text{ (woman)} = 0.75(Wt) = 0.75(54.4) = 40.8 \text{ grams } (\sim 1.5 \text{ ounces})$$

Thus, if these individuals were consuming a typical US diet, the recommended protein intakes for the man and woman would be about 54 and 41 grams/day, respectively. If, however, both were consuming vegetarian diets, their protein requirements would have to be adjusted to account for the lower digestibility of the diet. Assuming that their diets were *entirely* vegetarian, the digestibility would likely be between 80 and 85. If we assume a digestibility of 82, their daily protein requirements would be determined as follows:

$$\text{Protein (man)} = (\text{Base protein}) \times (100/\text{Digestibility}) =$$
$$54.4(100/82) = 66.3 \text{ grams } (\sim 2.3 \text{ ounces})$$

$$\text{Protein (woman)} = (\text{Base protein}) \times (100/\text{Digestibility}) =$$
$$40.8(100/82) = 49.7 \text{ grams } (\sim 1.7 \text{ ounces})$$

Thus, the difference in diets results in an increase in the requirements of ~12 g/day for the man and ~9 g/day for the woman.

The influence of diet on estimated protein requirements is even more marked for infants and children, because their requirements must be adjusted for both digestibility and amino acid content (see Table 9.7). For a 4.5-year-old girl who weighs ~33 lb (15 kg), base daily protein needs would be estimated as

$$\text{Protein} = 1.10(\text{Wt}) = 1.10(15) = 16.5 \text{ grams } (\sim 0.6 \text{ ounces})$$

As with the adults, this unadjusted level of 16–17 grams/day should be sufficient to meet the child's needs if she were consuming a typical American diet. However, if the child were eating a traditional maize and beans diet typical of Central America, her daily protein needs would be

$$\text{Protein} = (\text{Base protein}) \times (100/\text{Digestibility}) \times (100/[\text{Amino acid score}]) =$$
$$(16.5)(100/82)(100/67) = 30.0 \text{ grams } (\sim 1 \text{ ounce})$$

Hence, this difference in the protein quality and digestibility results in an increase of more than 80% in estimated protein needs. It is easy to see why access to high-quality animal foods (e.g., milk) is a strong predictor of early childhood growth and nutritional status in many areas of the developing world (e.g., Dewey 1981; DeWalt 1983; Leonard et al. 1993, 1994).

studies, the most recent WHO expert committee concluded that the average protein requirement for adults was ~0.6 grams per kilogram of body weight (g/kg body wt [~0.01 ounces/pound of body weight], FAO/WHO/UNU 1985).

On the basis of these data, the WHO established a recommended intake level for adults at 0.75 g/kg body wt, approximately 2 standard deviations (SD) above the average. For children and adolescents, the recommended intake per unit weight was set substantially higher because of the additional requirements for growth and development. As shown in Table 9.5, recommended intake is highest during infancy (1.85 g/kg body wt for infants 3–6 months), when growth rates are highest. During later childhood (7–10 years old) recommendations are somewhat lower, at ~1.0 g/kg body wt; and by late adolescence (16–18 years old), recommended intake approaches adult values (0.8 and 0.9 g/kg body wt in girls and boys, respectively). Additional protein is required for women during pregnancy and lactation. WHO recommends that a woman increase her protein intake by 6 g/day during pregnancy, by 17.5 g/day during the first 6 months of lactation and by 13 g/day for the remaining period of lactation.

The recommendations presented in Table 9.5 should provide for adequate protein intakes for populations that derive most protein from animal sources, as is the case in most of North America and Europe. However, for populations that consume largely vegetarian diets, the requirements must be adjusted to compensate for the differences in digestibility and amino acid content of plant-derived proteins. **Protein digestibility** is generally scored relative to pure animal proteins (e.g., egg, meat, and cheese) as the standard. Table 9.6 presents the digestibility scores for selected food items and diets, where the digestibility of the reference proteins are set at 100. Note that the typical US diet has a digestibility of ~100 (complete digestibility), whereas that of a diet of rice and beans is only 82 (FAO/WHO/UNU 1985).

Protein derived from largely vegetarian diets also may be limiting in relative proportions of certain amino acids. This limitation is particularly important for children,

Table 9.4   Energy constants and PALs recommended for estimating daily energy requirements for individuals under the age of 18 years old

Age (years)	Males	Females
	Energy constant (kcal/kg body weight)	
<1.0	103	103
1.0–1.9	104	108
2.0–2.9	104	102
3.0–3.9	99	95
4.0–4.9	95	92
5.0–5.9	92	88
6.0–6.9	88	83
7.0–7.9	83	76
8.0–8.9	77	69
9.0–9.9	72	62
	PAL	
10.0–10.9	1.76	1.65
11.0–11.9	1.72	1.62
12.0–12.9	1.69	1.60
13.0–13.9	1.67	1.58
14.0–14.9	1.65	1.57
15.0–15.9	1.62	1.54
16.0–16.9	1.60	1.52
17.0–17.9	1.60	1.52

(*Source:* FAO/WHO/UNU 1985; James and Schofield 1990)

Table 9.5 Recommended protein intake

Age	Recommended protein intake (g/kg/day)	
	Combined sexes	
<3.0 months	2.20	
3.0–5.9 months	1.85	
6.0–8.9 months	1.65	
9.0–11.9 months	1.50	
1.0–1.9 years	1.20	
2.0–2.9 years	1.15	
3.0–4.9 years	1.10	
5.0–6.9 years	1.00	
7.0–9.9 years	1.00	
	Males	*Females*[a]
10.0–11.9 years	1.00	1.00
12.0–13.9 years	1.00	0.95
14.0–15.9 years	0.95	0.90
16.0–17.9 years	0.90	0.80
18.0+ years	0.75	0.75

[a]Pregnancy: +6.0 g/day. Lactation: <6 months, +17.5 g/day; 6+ months, +13.0 g/day.
(*Source:* Adapted from FAO/WHO/UNU 1985 and NRC 1989)

Table 9.6 Digestibility of various protein sources for humans

Protein source	Digestibility
	Selected foods
Egg, meat, milk, cheese	100
Maize	89
Rice	93
Whole wheat	90
Peanut butter	100
Beans	82
	Selected diets
Indian rice and bean	82
Brazilian mixed	82
Filipino mixed	93
Chinese mixed	98
US mixed	100

(*Source:* Adapted from FAO/WHO/UNU 1985: 119)

who have higher requirements of essential amino acids than adults do (NRC 1989). Hence, WHO recommends that where necessary, protein requirements of infants and children be adjusted based on the **amino acid score** of the diet. This score reflects the percent of the limiting amino acid in the dietary protein relative to the requirement

Table 9.7   Amino acid scores of selected diets for preschool (under 5 years old) and school-age (5–12.9 years old) children

Diet	Amino acid score[a]	
	Preschool	School-age
Indian vegetarian		
Wheat-based	76	NA
Rice-based	73	NA
Rice–legume	55	67
Tunisian		
Rural	57	75
Urban	62	82
Brazilian		
Rice/beans/maize	97	NA
Wheat/rice/beans	91	NA
Guatemalan		
Maize/beans	67	NA
Nigerian		
Rice-based	78	NA
Cassava-based	72	95

[a] For each diet, the limiting amino acid may be different. NA, not applicable; dietary protein supplies at least 100% of all essential amino acids.

(*Source:* Adapted from FAO/WHO/UNU 1985, 124)

level (FAO/WHO/UNU 1985; NRC 1989). Table 9.7 presents the amino acid scores for various largely vegetarian diets. For example, the protein of the typical maize and beans diet of rural Guatemala contains only 67% of the limiting amino acid (in this case, lysine) required for preschool children (under 5 years old) and 89% of the recommended level for school-age children (5–12 years old).

Thus, WHO advises that the protein requirements of largely vegetarian diets should be adjusted for digestibility in adolescents and adults (i.e., individuals 13 years old and older) and adjusted for both digestibility and amino acid score in infants and children (FAO/WHO/UNU 1985). An example of how to determine protein requirements for individuals of known age, sex, body weight, and diet type is given in Box 9.1.

COMPARATIVE NUTRITION

Variation in Daily Energy Needs

The basic determinants of any animal's energy requirements are the same as those outlined for humans (i.e., BMR, activity, thermoregulation, dietary thermogenesis, growth, and reproduction).

In ecology and evolutionary biology, the energy demands of an animal or a population are often divided more broadly into two components: maintenance (respiratory) and productive energy (Krebs 1972; Begon et al. 1990). Maintenance energy demands are those necessary for an animal's day-to-day survival and thus include BMR, activity, thermoregulation, and dietary thermogenesis. Productive energy costs are those associated with growth and reproduction. From an evolutionary perspective, the amount of energy that an individual or population allocates to "production" is important because it will directly influence the long-term reproductive success of the species.

Differences in BMR among different mammalian species are strongly determined by differences in body weight (see next section). Humans, on average, have BMRs similar to those of other comparably sized mammals. Similarly, the energy costs of such activities as movement (locomotion) seem to vary predictably with body weight (Taylor et al. 1970, 1982). Human bipedal locomotion is more efficient than average quadrupedal (four-legged) movement at walking speeds (Rodman and McHenry 1980; Leonard and Robertson 1995, 1997) but considerably more costly at running speeds (Margaria et al. 1963). In contrast, thermoregulatory energy costs for most human populations are probably less than for other mammalian species because of our ability to more effectively control the temperature of our immediate environments (e.g., by changing clothing and by heating and cooling our homes). Daily energy costs of growth and development also appear to be lower in human children than in the young of other species (Holliday 1986). This is because humans have a prolonged period of development relative to other mammalian species; consequently, the total cost of growing to adult size is spread over a longer period of time (Leonard and Robertson 1992).

Overall, although daily energy demands are highly variable across human populations, it appears that traditionally living human groups have TDEEs that are comparable to those of mammalian species living in the wild. This is shown in Table 9.8,

Table 9.8 **Comparison of total daily energy expenditure levels (TDEE; measured using the doubly labeled-water technique) among human and selected mammalian species**

Species	Weight (kg)	TDEE (kcal/day)	PAL (TDEE/BMR)
Field mouse (*Microtus californicus*)	0.013	9.6	3.7
Jack rabbit (*Lepus californicus*)	1.8	281	2.4
Howler monkey (*Allouatta palliata*)	6.5	552	1.7
Koala (*Phasolarctus cinercus*)	9.2	493	2.6
Mule deer (*Odocoileus hemionus*)	40.0	5584	2.4
Humans (*Homo sapiens*)			
Developing countries			
M	59.3	4227	3.0
F	50.0	2426	2.0
Western populations			
M	83.5	3401	1.8
F	72.5	2459	1.7

(*Source:* Data for humans from Schulz and Scholler 1994; data for all other mammalian species from Karasov 1992)

which compares TDEEs and PALs of selected mammalian species to those of human populations. Note that rural agricultural populations of the developing world have extremely high PALs (2.0–3.0) that are similar to those of the other mammalian species. Western populations (e.g., North America, Europe), however, have PALs comparable to "less active" mammals such as the howler monkey (see also Chapter 10).

Basal Metabolic Rate and Body Size

Since the early 1900s, researchers have studied the determinants of variation in energy requirements among mammalian species (Dubois 1927; Kleiber 1932; Benedict 1938). Results indicate that an animal's basal metabolic rate (BMR) is strongly determined by body size; however, the relationship between BMR and weight is not a simple linear function.

Rather, the work of Kleiber (1932) and others (e.g., Benedict 1938; Brody 1945) has shown that across a wide array of species, BMR increases as a function of the three-fourth power of body weight (i.e., weight$^{0.75}$).[2] On the basis of data compiled from 36 species, ranging in size from a mouse to an elephant, Kleiber (1975, 214) proposed that a mammal's basal energy needs could be best estimated by the following scaling (allometric) relationship:

$$BMR = 70 \times Wt^{0.750}$$

where BMR is measured in kilocalories per day and Wt is weight (in kilograms).

This relationship, known as **Kleiber's law**, is an example of negative **allometry** because the scaling exponent of weight is <1 (see Box 9.2 for an explanation of allometry). The functional consequence of this association is that larger mammals have *proportionally lower* metabolic rates than smaller ones. That is, across mammalian species, increases in metabolic rate do not keep pace with increases in overall size. This point is outlined in Table 9.9, which presents weight, daily BMR, and BMR per unit body weight for six mammalian species of different sizes. Note, for example, that whereas the absolute metabolic needs of a human are much greater than those of a mouse, the mass-specific requirements are more than six times higher in the mouse.

Numerous recent studies have specifically addressed whether the Kleiber relationship effectively predicts metabolic variation among primate species (Kurland and

---

[2] This explanation initially seems to contradict the previous discussion of BMR variation in humans because the predictive equations presented in Table 9.2 indicate that the relationship between BMR and weight *is* a linear function. The difference reflects the fact that the first set of equations (in Table 9.2) describe an intraspecific (within species) relationship, whereas the Kleiber equation describes an interspecific (between species) one. Intra- and interspecific equations describing the same phenomenon (e.g., BMR vs. weight, brain size vs. weight) are often different because of the factors that determine each relationship (see Peters 1983). For example, intraspecific relationships describe a much narrower range of variation that may be the result of small-scale differences in adaptations. In contrast, interspecific relationships describe a much broader range of variation that may reflect large-scale physiological and/or evolutionary processes.

BOX 9.2 ALLOMETRY

Allometry ("scaling") refers to the change in size of one biological parameter with respect to another (usually body size). The study of allometry is central to research in comparative and evolutionary biology because size and proportionality of organisms have important physiological and ecological consequences (see Peters 1983; Calder 1984; Schmidt-Nielsen 1984; Jungers 1985).

Allometric relationships are described by the following general equation, or power function:

$$Y = aX^b$$

where X and Y are two biological parameters (e.g., body weight and brain size), a is a constant, and b is a "scaling" constant (exponent).

When the $b < 1$, the dependent variable (Y) is increasing more slowly than the independent measure (X). This condition is known as negative allometry. Conversely, positive allometry occurs when $b > 1$, indicating that Y is increasing more rapidly than X. When $b = 1$, a linear relationship exists between Y and X such that Y increases as a constant multiple or fraction of X. This condition is known as **isometry**. Figure 9.2 illustrates the curves associated with the three kinds of allometric relationships.

By convention, allometric relationships are usually graphed on a double logarithmic scale:

$$\log(Y) = \log(aX^b)$$

$$\log(Y) = b\log(X) + \log(a)$$

Hence, when the log-transformed data are plotted, they appear as a linear function whose slope is equal to the scaling constant (b). Figure 9.3 presents the same relationships from Figure 9.2, plotted on a log–log scale.

Allometric studies can be divided into two broad categories: ontogenetic and static. Ontogenetic allometry, as the name implies, deals with changes in size and proportion during growth of organism(s) of a single species. Such analyses provide insights into differences in the rates of growth of different organs. Static allometry compares adult organisms of different sizes. These comparisons can be made among individuals of the same species (intraspecific allometry) or across different species (interspecific allometry).

Two examples of interspecific allometry among mammals are the relationships between brain size and body weight and between metabolic rate and body weight (i.e., Kleiber's law). Both are cases of negative allometry, with scaling constants of ~0.75 (Kleiber 1975; Martin 1989). These relationships imply that, on average, larger mammals have proportionately (i.e., relative to body mass) smaller brains and lower metabolic rates than smaller mammals.

Pearson 1986; Leonard and Robertson 1992, 1994). This question is of interest because primates have relatively larger brains than other mammalian groups. Because the brain has particularly high metabolic costs (Holliday 1986), it has been hypothesized that the high degree of **encephalization** among primates might contribute to an elevation in BMR (Mink et al. 1981). However, Figure 9.4 shows that, as a group, primates conform to the Kleiber relationship. Indeed, the allometric relationship that best explains the variation among these 31 species (11 prosimians, 19 anthropoids, and humans) is

$$BMR = 60 \times Wt^{0.81}$$

Observed BMR in humans deviates by only ~1% from that predicted by the Kleiber equation. The similarity in the BMR-versus-weight relationship between primates and other mammals implies that primates spend a larger share of their daily energy budget on brain metabolism. Human appear to expend some 20–25% of their BMR on brain metabolism, whereas other primate species expend 8–10% and most other mammals only 3–5% (Mink et al. 1981; Holliday 1986; Leonard and Robertson 1992).

As with any general relationship, however, some individual species deviate markedly from the overall pattern. Several prosimian species are "hypometabolic"; that is, they have BMRs substantially lower than those predicted for their size. In fact, prosimians as a group have significantly lower BMRs than predicted for their size (BMR = $44 \times Wt^{0.64}$), whereas anthropoids or "higher primates" (including humans) have a BMR-versus-weight relationship that is almost identical to the Kleiber equation (BMR = $69 \times Wt^{0.78}$).

In sum, energy requirements across mammalian species vary predictably with respect to body weight. Primates in general, and humans in particular, appear to conform to the mammalian pattern of BMR increasing as a function of the three-fourth power of body weight. The primary consequence of this relationship is that smaller mammals (including primates) have higher mass-specific energy needs than larger animals do. Moreover, the fact that primates conform to this relation-

TABLE 9.9 Comparison of body weights and BMRs for selected mammalian species

Species	Weight (kg)	BMR (kcal/day)	BMR/kg body wt (kcal/kg)
Mouse (*Mus musculus*)	0.021	3.6	171.4
Pygmy marmoset (*Cebuella pygmaea*)	0.11	10.1	91.8
Holwer monkey (*Aloutta palliata*)	4.67	231.9	49.7
Dog (*Canis familiaris*)	14.10	534.0	37.9
Chimpanzee (*Pan troglodytes*)	18.30	581.9	31.8
Human (*Homo sapiens*)	54.00	1438.5	26.6
Cow (*Bos taurus*)	300.00	4221.0	14.1

(*Source:* Data for primate species (pygmy marmoset, howler monkey, chimpanzee, and human) from Leonard and Robertson 1992; data for other species from Kleiber 1975)

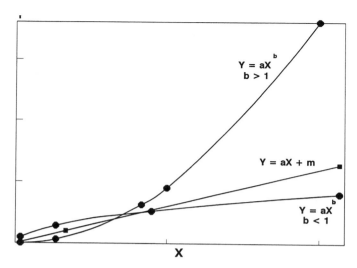

Figure 9.2   Curves depicting three kinds of scaling relationships. With positive allometry (scaling exponent, b > 1), the dependent variable (Y) increases more rapidly than the independent variable (X), producing an exponential-type curve. In a negative allometric relationship (b < 1), Y increases more slowly than X. An isometric relationship (b = 1) is depicted by a straight line, because Y increases as a constant proportion of X.

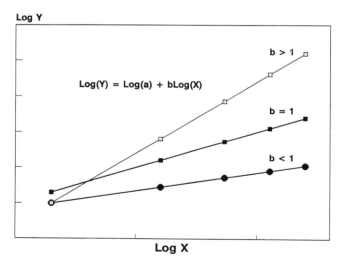

Figure 9.3   Logarithmic transformations of the curves presented in Figure 9.2. When both the X and Y axes are log-transformed, all three relationships appear as straight lines with different slopes equal to the scaling exponent (b). Thus, the log-transformed positive allometric relationship has a slope >1, whereas the slopes of the isometric and negative allometric relationships are 1 and <1, respectively.

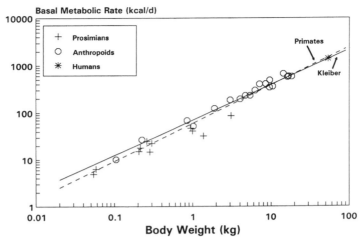

Figure 9.4   Body weight plotted as a function of basal metabolic rate (BMR) for 31 primate species (logarithmic scale). The solid line represents the allometric relationship predicted by the Kleiber equation, and the broken line represents the linear regression best fitting the data. (*Source*: Adapted from Leonard and Robertson 1994)

ship implies that they allocate a larger share of their daily energy needs to brain metabolism than most other mammals. The dietary consequences of the Kleiber relationship are explored later.

Diet and Body Size

Primates, as a group, are eclectic feeders and subsist on a wide variety of different foods. The food choices of most primates can be divided into five main categories: faunivory (animal foods, including insects and other invertebrates), gumnivory (saps and gums), **frugivory** (fruit and reproductive plant parts), gramnivory (seeds and nuts), and **folivory** (leaves and structural plant parts).

The diets of most primate species contain foods from at least two of these categories as a means of meeting both energy and protein demands (Richard 1985). Small primates, such as the pygmy marmoset (*Cebuella pygmaea*), subsist mostly on insects (a good protein source) and gums (high in carbohydrate energy). In contrast, large-bodied species, such as orangs (*Pongo pygmaeus*) and gorillas (*Gorilla gorilla*) consume mostly leaves (with structural protein) and fruits (good carbohydrate sources).

An important nutritional consequence of these different feeding strategies is that the diets of smaller primates tend to be much denser in nutrients and energy than those of larger primates. Animal material, for example, provides 100–200 kcal (400–850 kJ) per 100 grams (3.5 ounces) of edible portion, whereas fruit provides 50–100 kcal/100 g (200–400 kJ) and foliage only 10–20 kcal/100 g (40–80 kJ) (Wu Lueng 1968; Eaton and Konner 1985). Gorillas, which weigh ~90 kg (~200 pounds), derive a full 86% of their diet from leaves and barks, whereas tiny pygmy marmosets

consume no foliage and derive one-third of their food from insects and other invertebrates (Richard 1985).

The negative relationship between dietary quality and body weight (on a logarithmic scale) is shown for 72 nonhuman primate species in Figure 9.5. Diet quality was measured using an index developed by Sailer et al. (1985), whereby the
nutrient density is ranked on a scale ranging from 100 (100% foliage) to 350
(100% animal material). The **correlation** between diet quality and weight is robust
(correlation coefficient [r] = –0.661; p < 0.001) and implies that diet quality declines with increased body size. This association between size and diet quality,
sometimes referred to as the Jarman–Bell relationship (Gaulin 1979), is seen
among mammals in general and is thought to result from the metabolic consequences of the Kleiber relationship (Bell 1971; Jarman 1974). Because basal metabolism scales to the three-fourths power of body weight, smaller animals have
much higher metabolic costs per unit of body weight and thus need to consume
foods of high caloric density. In contrast, large animals have higher total energy
needs but relatively low requirements per unit weight. Larger animals thus tend to
fulfill their nutritional needs by consuming large amounts of widely available, low-
quality foods such as leaves and bark.

Looking again at Figure 9.5, we find that humans depart substantially from the
primate diet quality–body weight relationship. The human data are for five modern
hunting and gathering populations; these groups were used because the foraging
(food-collecting) subsistence strategy represents the feeding strategy used by early
Homo sapiens.

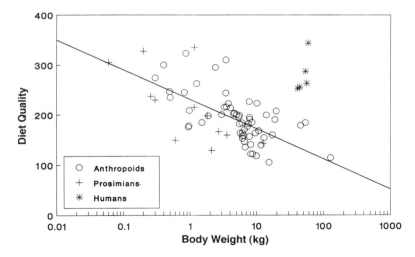

Figure 9.5   Plot of diet quality versus log-body weight for 72 nonhuman primate species and
5 human hunting and gathering groups. Among the nonhuman primate sample, diet quality is
strongly negatively correlated with weight (r = –0.661; p < 0.001). In contrast, humans have
substantially higher diet quality values than predicted for their size. (*Source*: Adapted from
Leonard and Robertson 1994)

Among these contemporary foragers, animal foods contribute an average of 59% of energy intake, with the range spanning one-third of daily calories in African groups such as the !Kung (Lee 1968) to 95% in Eskimos (Gaulin and Konner 1977). By comparison, chimpanzees, the most predatory of the large-bodied apes, derive only about 5–7% of their daily energy intake from animal material (estimated from Wu Leung 1968; McGrew 1974; Teleki 1981). Thus, the diets of the five hunter-gatherer groups are of much higher quality than expected for primates of their size. This generalization holds for most all subsistence-level human populations, because even agricultural groups that consume cereal-based diets have estimated diet qualities higher than those of other primates of comparable size.

If we assume that the last common ancestor of apes and humans had a diet similar to that of a modern ape and that the earliest members of our species (*Homo sapiens sapiens*) consumed a diet whose composition fell within the range encompassed by modern hunter-gatherers, then it appears that **hominid** evolution was characterized by increasing diet quality. I will explore the significance of this relatively higher quality diet in later sections.

Brain Size, Gut Size, and Dietary Quality

One primary explanation for our species' high-quality diet is that it reflects an adaptation to the high metabolic costs of our brain (Martin 1989). As noted previously, humans expend a large proportion of their daily energy budget on brain metabolism. Figure 9.6 shows the plot of brain size vs. BMR for the same 31 primate species presented in Figure 9.2. Brain size scales isometrically with body weight ($b = 0.98$) across the primate order, and humans are the outliers, expending about two to three times more energy for brain metabolism than other primates and five times more than the average mammal.

Humans are also distinct from most other primates in their gut (gastrointestinal) morphology, having very reduced large intestines (colons), as displayed in Figure 9.7. Most large-bodied primates display a greatly expanded large intestine, which is regarded as an adaptation to maximize the surface area over which nutrients from low-quality foods can be absorbed (Milton 1987, 1993). In contrast, human gut morphology is more similar to that of a carnivore (Sussman 1987), reflecting an adaptation to an easily digested, nutrient-rich diet. Hence, it has been argued that the adoption of an easy-to-digest, nutrient-rich diet likely shaped the evolution of both our large brains and reduced guts.

The results of two studies support this hypothesis. Leonard and Robertson (1994) demonstrated that among a sample of 25 primate species (including humans), relative diet quality (i.e., whether the quality of a species' diet was higher or lower than predicted for its size) was associated with relative brain size (i.e., deviations from brain size predicted from body weight). That is, species with better quality diets had larger brain sizes relative to body weight.

Similarly, Aiello and Wheeler (1995) showed that among a sample of 18 primates (including humans), large gut size was associated with relatively smaller brain size. These results imply that changes in diet quality during hominid evolution were

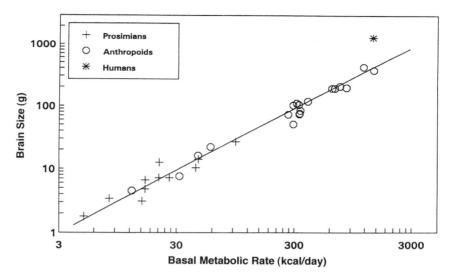

Figure 9.6   Brain size plotted as a function of BMR for 31 primate species (logarithmic scale). The best-fit regression shows that brain size scales isometrically with BMR. Humans depart substantially from the regression, having brains that require two to three times more energy to maintain than most primates. (*Source*: Adapted from Leonard and Robertson 1994)

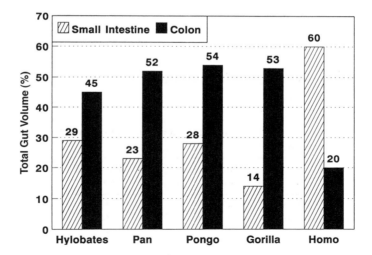

Figure 9.7   Relative proportions of the small intestine and large intestine (colon) in humans and four species of apes. Expansion of the colon reflects an adaptation to a low-quality, fibrous diet. Colon volume of humans is smaller than that of apes, indicative of a high-quality diet that includes large amounts of animal foods. (*Source*: Data from Milton 1987)

linked with the evolution of brain size and gut morphology. The shift to a more calorically dense diet was probably needed to support the increased energy demands of the hominid brain (Leonard and Robertson, 1992, 1994). Thus, shifting dietary strategies likely contributed to a unique co-evolution of brain and gut size in the human lineage. Although nutritional factors alone are not sufficient to explain the evolution of our large brains, certain dietary changes were probably necessary for substantial brain evolution to take place.

EVOLUTIONARY CHANGES IN BRAIN AND BODY SIZE IN HOMINIDS: DIETARY IMPLICATIONS

The comparative data presented thus far indicate that the distinctive aspects of human dietary patterns and metabolism appear to have been shaped by our relatively large brains. Some tentative insights into when these distinctive patterns emerged can be gained from looking at the prehistoric record. In the past 4 million years of hominid evolution, brain and body size appear to have increased dramatically (McHenry 1992); However, the rates of evolutionary change over this period were not constant but highly variable.

Table 9.10 presents data on geologic age, mean cranial capacity (brain size), and estimated body weight (kg) for 6 fossil hominid species. We see that with the evolution of *Homo erectus* in Africa at 1.7 million years ago, brain and body size appear to have grown substantially.

Figure 9.8a is a plot of brain size versus body weight for 31 extant primate species (including humans) along with the 6 fossil hominid species in Table 9.10. The solid line is the **regression** derived from the extant primate sample. Figure 9.8b plots the **residuals** of individual species from the best-fit regression. These values indicate the degree to which observed brain size deviates from the brain size predicted from the regression. Humans are the extreme outliers of the group. Examining the fossil species, we see that the four australopithecine species deviate from their predicted

Table 9.10   Geologic age, cranial capacity, and estimated body weight for six fossil hominid species

Species	Geologic age (Mybp)	Cranial capacity (cm³)	Body weight (kg)
Australopithecus afarensis	4.0–2.9	404	37.0
Australopithecus africanus	3.0–2.4	442	35.5
Australopithecus robustus	1.8–1.6	530	36.1
Australopithecus boisei	2.0–1.3	515	41.3
Homo habilis	2.4–1.6	632	41.6
Homo erectus (Africa)	1.7–0.7	871	57.7

Mybp, million years BP.

(*Source*: Data from McHenry 1994)

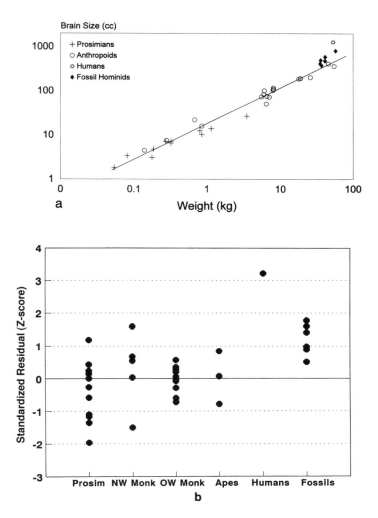

Figure 9.8   (a) Plot of brain size versus body weight for 31 primates species and 6 fossil hominid species (logarithmic scale). (b) Residuals of individual species from the best-fit regression line of brain size versus body weight. Modern humans depart from the regression by more than 3 standard deviation (SD) units, whereas the australopithecines deviate by an average of +0.95 and early members of the genus Homo depart by +1.69 SD.

brain size by about one-third as much as modern humans, whereas the two early *Homo* species (*H. habilis* and *H. erectus*) deviate by about one-half the amount of humans. Indeed, the brain size deviations with both early *Homo* species are greater than those observed for any of the extant nonhuman primates species. Thus, with the evolution of early *Homo*, the brain–body weight relationship has moved beyond the range for nonhuman primates and is now approaching that of modern humans.

It is intriguing that the clear departure from the general primate brain size regression occurs with the emergence of species of our own genus, because other important anatomical and behavioral changes also appear at this time. Archeological and skeletal evidence indicate that these early members of the genus *Homo* incorporated greater amounts of animal material in their diet than did the australopithecines (Bunn 1981). With early *Homo*, we also observe fundamental changes in the way resources are used and transported (Potts 1988, 1993). Hence, both the higher quality and greater stability of the diet probably supported the rapid expansion of brain size in *H. habilis* and *H. erectus*. Evidence from the fossil record indicates that such a dietary shift may have appeared with the emergence of the genus *Homo*. The evolution of early *Homo* was associated with rapid rates of brain evolution and important changes in material culture (Gowlett 1984; Falk 1987; Potts 1988). Moreover, this period appears to have coincided with a marked drying trend in eastern and southern Africa (Behrensmeyer and Cooke 1985; Vrba 1988).

Vrba (1988, 1993) demonstrated that at ~2.5 million years ago, a major period of global cooling had begun, which led to a dramatic increase in the amount of open grassland environments in Africa. Such ecological change would have directly influenced both the density and distribution of high-quality plant foods and likely would have made animals an increasingly attractive food resource to early hominids (Vrba 1988). Different hominid species of that time appear to have adapted to the environmental drying in different ways. Early members of the genus *Homo* (*H. habilis* and *H. erectus*) appear to have included larger amounts of meat in their diet, whereas those of the robust australopithecine lineage (e.g., *Australopithicus robustus*, *Australopithicus boisei*) continued to subsist largely on fibrous low-quality plant foods. Thus, these ecologically initiated changes in foraging behavior and diet likely provided the fuel for sustaining rapid brain evolution in early members of the genus *Homo*. Among the later australopithecines, however, the continued reliance on a poorer quality diet may have been responsible for their relative stasis in brain size.

Humans, thus, have a much higher quality diet than expected for their size or their basal metabolic needs. Although hardly carnivores, humans do consume substantially more meat than any other primate of equal size. Contemporary human foraging groups obtain at least 30% of dietary energy from animal foods, compared with ≤5–7% in chimpanzees. Adaptation to this calorically dense, easily digested diet is evident in our gut morphology, because the digestive tract of humans is reduced compared with those of most other primates (Chivers and Hladik 1980; Milton 1987; Sussman 1987).

Our distinct diet appears to be linked to the high metabolic costs of our brain. In general, primate brain size varies as a direct (linear) function of body metabolism. This means that the proportion of metabolic energy spent on the brain is relatively constant across primates of all size (about 8–9% of BMR). Species that spend a larger proportion of BMR on their brain have a higher quality diet than expected for their body size. Conversely, small brains relative to metabolic turnover are associated with low-quality diets. Humans represent the positive extreme, having both a very high quality diet and a brain that accounts for 20–25% of basal metabolic energy.

INTERPOPULATIONAL VARIATION IN NUTRITION AND METABOLISM

Patterns of nutritional and metabolic variation among human populations provide additional insight into the influence of environmental and cultural factors on the evolution of nutrition in our species. The evolution of distinct cultural practices (adaptations) have been important in expanding the range of foods on which human populations can subsist. Traditional techniques used for processing corn throughout the Americas, potatoes in the Andean region of South America, and soybeans in Asia all appear to have been critical for establishing these crops as staple foods. Additionally, certain aspects of human biological variation, such as interpopulational differences in BMR or the ability to digest milk sugar (lactose), appear to reflect adaptations to specific dietary/nutritional selective pressures.

In many cases, however, metabolic adaptations that were once adaptive under traditional conditions often are rendered maladaptive under conditions of modernization and acculturation. **Hypertension** among African Americans, the high rates of adult-onset diabetes among native North American populations, and increasing rates of obesity among human populations worldwide are all examples of diseases that have resulted from interactions between genes and the environment. These issues are examined below in detail.

Food Processing and Human Dietary Diversity

Traditional techniques for processing foods have been widely studied and provide clear evidence of how cultural and behavioral strategies directly influence human nutrition and health (Johns 1990). Katz (1987) proposed a lock-and-key model for understanding how cultural knowledge about food processing can enhance the quality of particular food resources. He argues that processing techniques likely develop first in areas where a particular crop is initially domesticated. Because the information about processing is distinct to the particular culture, it is less likely to spread than the crop itself. Consequently, this model predicts that utilization of the crop will be less effective outside of the "core" area until the appropriate cultural knowledge has been transmitted.

Maize (corn) processing among Amerind populations is one good example of this lock-and-key biocultural interaction. Although corn is relatively high in protein, it is deficient in the amino acids lysine and tryptophan and the B vitamin niacin. Hence, corn-based diets tend to be marginal in niacin and low in protein quality (i.e., digestibility and amino acid scores), placing young children at particular risk for malnutrition. Katz and colleagues (1974, 1975), however, have shown that the use of alkali products (e.g., ash, lime, and lye) by Native American populations in cooking corn substantially increases the concentrations of niacin and the limiting amino acids, thus improving its nutritional quality. They also demonstrated that among 51 traditional societies from throughout the New World, both the levels of cultivation and consumption of maize are strongly associated with the use of alkali processing. That is, the more important maize is in the diet, the more likely the population is to use alkali to enhance its protein quality. In contrast, in parts of the world such as Africa, where maize is not processed with alkali, pellagra (a niacin deficiency disease) is often seen among populations with high levels of maize consumption (Katz 1987).

Similar work has been done in the Andes, where the traditional methods of potato processing have been examined. The problem is somewhat different because the potential food items contain toxins that are hazardous to human health. The hearty potato varieties that will grow at high elevations (3000–4500 meters [~10,000–14,500 feet] above sea level) in the Andes tend to have high concentrations of bitter substances known as glycoalkaloids (Johns and Keen 1986). Because these bitter potatoes are one of the few crops that will grow in such a harsh climate, the indigenous Quechua and Aymara populations of Peru and Bolivia have developed various techniques for reducing the potatoes' glycoalkaloid content. One involves first freezing the potatoes (by leaving them outside in the cold overnight) and then placing them in a shallow well or stream for several weeks (Werge 1979). This process allows for the toxins to be leached from the potatoes, resulting in a final product (known as *chuño blanco* or *moraya*) that has only ~3% of its original alkaloid content (Johns 1990). The other approach involves consuming these bitter potatoes fresh along with clays and limestone pastes (in soups and sauces). Johns (1990) has shown that the minerals in the clays bind to the toxins, thus preventing them from being absorbed by the body. These cultural innovations have been critical to the survival of Andean populations, because potatoes and other tubers contribute >40% of the calories in their diets (Leonard and Thomas 1988).

Another crop that requires processing before it can be consumed is soybeans. Like the bitter potatoes of the Andes, soybeans contain a potentially toxic substance known as antitrypsin factor (ATF), which inhibits the digestion of protein. Katz (1987) showed that the traditional processing techniques for deactivating ATF (e.g., roasting, fermenting, or producing bean curd) were initially developed in China and later spread throughout Asia. The use of calcium or magnesium salts to separate the soybean protein into a curd (e.g., tofu) appears to be the most effective processing technique. Today, various types of soybean curd account for ~90% of the soybean consumption in Asia (Katz 1987).

These examples highlight how aspects of recent human nutritional evolution have been a biocultural process in which learned cultural practices are critical to effectively use of particular food resources. Katz and others have demonstrated that whereas domestication and the emergence of agriculture were critical to increasing the level of productivity of many plant species, these crops often could not be used for widespread human consumption until processing techniques necessary for unlocking their full nutritional benefits were developed.

Interpopulational Variation in Metabolic Rate

Although the relationship between BMR and body mass is strong, both within our species and across mammalian species in general, significant interpopulational variation does exist. In *Climate and Human Variability*, Roberts (1978) demonstrated a strong negative correlation between BMR and mean annual temperature, suggesting that adaptations to local or regional climatic stressors, in part, help to explain variation in BMR.

More recently, Henry and Rees (1991) and Soares et al. (1993) demonstrated that the WHO's standard equations for predicting BMR from body weight (FAO/WHO/UNU 1985; WN Schofield 1985) overestimate BMRs for several tropical populations. In contrast, many studies suggest that standard predictive equations substantially underestimate BMR in indigenous northern populations (see Rode and Shephard 1995; Shephard and Rode 1996; Galloway et al. 2000).

Whereas population differences in BMR have been demonstrated, several important questions remain unresolved. First, it is unclear whether these population differences are largely genetically determined or whether they reflect shorter term functional **acclimatization**. Some evidence indicates that migration of individuals from temperate environs to either tropical or arctic climes results in substantial changes in BMR. Second, it is difficult to determine which ecological stressor(s) promotes differences in BMR. Among tropical populations, for example, heat stress, chronic undernutrition (and its implications for reduced muscle mass and organ weights), and a largely carbohydrate diet have been offered as contributors to reduced BMRs (Henry and Rees 1991; Henry 1996). Similarly, among arctic groups, severe cold stress, marked seasonal changes in day length (photoperiod), and a traditional diet high in protein and fat have been linked to elevations in metabolic costs (Shephard and Rode 1996).

In addition to being of theoretical importance to human biologists, interpopulational differences in BMR also have important nutritional and public health implications. Recent work among the Pima indians indicates that low BMRs are strong contributors to high rates of obesity and adult-onset diabetes among the Pima (Knowler et al. 1983, 1991; Howard et al. 1991). Whether low BMRs are a primary contributor to the high rates of obesity in the United States and other western countries remains a major point of debate and ongoing research in human nutrition (cf. DeBoer et al. 1987; Ravussin et al. 1988; Westrate et al. 1990; Armellini et al. 1992; Welle et al. 1992).

Lactose Intolerance

Interpopulational variation in the ability to digest lactose in adulthood is a well-studied example of nutritional adaptation (Simoons 1970; McCracken 1971; Kretchmer 1972). (See Chapter 6 for a discussion of how the ability to digest lactose is related to skin color.)

Lactose is broken down into its component sugars—glucose and galactose—by the enzyme **lactase** in the villi of the small intestine. The monosaccharides are readily absorbed and rapidly appear in the bloodstream (Alpers 1981). Consequently, one method of assessing lactose tolerance is by measuring blood glucose levels after the ingestion of a dose of lactose. Alternative approaches involve measuring the hydrogen gas (H_2) content of expired breath samples (Solomons et al. 1980). Undigested lactose in the gut undergoes fermentation, which results in the production of H_2. Hence, increases in breath H_2 levels after the consumption of lactose indicate reduced intestinal lactase activity.

In most mammals, lactase activity in the gut substantially declines after weaning, resulting in the loss of the ability to digest milk sugar in childhood and adulthood. Most human groups follow this general mammalian pattern, with lactase deficiency being well established by 3–7 years of age (Scrimshaw and Murray 1988). However, in a few populations, such as those of northern Europe and some herding populations of Africa, high intestinal lactase activity is retained into adulthood. This interpopulational variation in lactose tolerance is shown in Figure 9.9.

The most widely accepted model for explaining variation in lactose digestion is the "cultural historical" hypothesis (Simoons 1970), which posits that high frequency of adult lactose tolerance is found in populations with a history of animal husbandry and milk consumption. According to this model, adult lactase deficiency was the primitive condition for our species; however, with the evolution of animal domestication, **natural selection** favored the retention of lactase activity in those populations that relied heavily on dairy products for nutrition. This model thus implies that the evolution of adult lactose tolerance in our species is a relatively recent event, occurring within the last 10,000–12,000 years.

Adult lactase persistence is generally thought to be controlled by an **autosomal dominant** gene (Scrimshaw and Murray 1988). This simple mode of inheritance has been questioned by some who argue that milk "tolerance" and the ability to digest

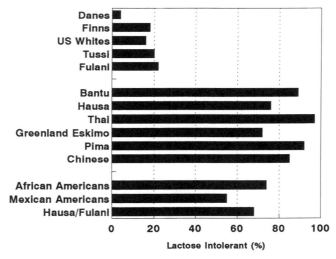

Figure 9.9   Variation in lactose tolerance among selected human populations. Populations in the uppermost group (Danes through Fulani) are from dairying societies that have largely retained the ability to digest lactose. The middle group (Bantu through Chinese) are populations without a history of herding and milk consumption; these populations are largely lactose intolerant. The bottom group (African Americans through Hausa/Fulani) are admixed populations that show somewhat greater tolerance levels than those in the middle group. (*Source*: Data from Kretchmer 1972, 1981)

lactose are not the same (Bolin and Davis 1970). That is, some "malabsorbers" are able to tolerate milk, whereas a small proportion of "absorbers" are unable to tolerate milk (Johnson et al. 1981; Johnson et al. 1993). The ability of certain malabsorbers to tolerate *some* milk in adulthood does seem to be attributable to dietary habits during development. Some researchers have argued that this pattern suggests that lactase is an inducible enzyme (e.g., Bolin and Davis 1970; Kretchmer 1972; Kotler 1982); however, the majority of the evidence indicates that this is not the case (e.g., Gilat et al. 1972; Lerebours et al. 1989; Johnson et al. 1993). Johnson et al. (1993) found that among a sample of 22 lactose-intolerant African Americans, gradual introduction of increasing amounts of lactose in milk resulted in the eventual tolerance of at least 1 glass of milk per day (12 g lactose) in 17 (77%) of the subjects. This increased tolerance, however, occurred with without evidence of greater lactase activity. Consequently, these authors argue that adaptation to higher milk intakes among lactose malabsorbers is the result of increased tolerance of the colon to lactose fermentation products.

Hence, although the adult lactase activity is under strong genetic control, it does not entirely determine the variation in milk utilization among human populations. Rather, tolerance to at least some milk (~1 cup/day) can be acquired even among individuals with little or no intestinal lactase activity (Scrimshaw and Murray 1988). As a result, a great deal of plasticity in milk use reflects a more complex biocultural interaction than would be implied by a single-**locus** genetic trait.

Hypertension in African Americans

High blood pressure is a health risk, especially among African American populations. In 1991, the rates of hypertension (defined as **systolic blood pressures** of at least 140 millimeters of mercury [mmHg] or **diastolic pressures** of at least 90 mmHg) among African American men and women (20–74 years old) were 37% and 31%, respectively, considerably higher than those observed among European Americans (25% in men; 18% in women) (NCHS 1996). Over the past two decades, many different evolutionary and biosocial models have been proposed to explain these high rates hypertension in African Americans. The two most prominent evolutionary models suggest that the problem is a consequence of a genetic adaptation for efficient sodium (Na^+) storage under tropical conditions.

Gleibermann (1973) and Denton (1982) argued that under such environments, efficient Na^+ metabolism would have been selected for because of its limited availability and because it is readily lost in perspiration. As support for this model, they note that for many traditional tropical societies, salt (NaCl; the principal source of sodium in preindustrial diets) was such a limited resource that is often was used as a form of currency (Denton 1982). Additionally, early physiological research among African Bantu and Bushman populations demonstrated that under conditions of heat stress, these groups had lower sweat rates and lower sodium concentrations in their sweat than did the control subjects of European ancestry (Wyndham 1966). Denton and Gleibermann argue that although such a physiological trait would have been adaptive under traditional conditions, when sodium was in limited supply, it becomes maladaptive in the modern Western society, where sodium is a ubiquitous food additive. Clinical studies have shown that African American subjects retain intravenous sodium loads longer (Luft et al. 1979) and are better able to normalize elevations in blood pressure with salt-excreting diuretics than their European American counterparts (Freis et al. 1988). Thus, this genetic predisposition, combined with the environmental "trigger" (high rates of sodium in the modern diet), results in elevated blood pressure among African Americans.

More recently, Wilson and Grim (1991) further refined this model, arguing that the problem of hypertension is particularly acute among African Americans because of the legacy of slavery. After reviewing historical records of the slave trade, these authors concluded that salt-depleting dehydration (due to limited water, heat stress, vomiting, and diarrheal diseases) was a major contributor to mortality on the slave ships. Consequently, they maintain that the slave trade functioned as a strong form of selection, favoring those individuals who were best able to retain sodium in the face of the severe stressors of dehydration. However, this hypothesis has been criticized by Curtin (1992), who argues that deaths on the slave ships were not generally caused by salt-depleting diseases. Additionally, he notes that most of Wilson and Grim's evidence was drawn from the West Indies and hence is not directly relevant to populations in the United States.

In contrast to the above models, Dressler (1991) argued that hypertension among African Americans is largely a product of social stress rather than an underlying genetic predisposition. He demonstrates that in two "color-conscious" societies (Brazil and the southern United States), skin color and social class interact to influence blood pressure such that dark-skinned individuals of *higher* social class have higher blood pressure than those of lower classes. Dressler maintains that this relationship reflects the fact that dark-skinned individuals of upper social classes experience more discrimination and psychosocial stress than those of the lower classes because they have achieved a lifestyle *above* what their society had prescribed for them. Hence, for Dressler, skin color is a predictor of blood pressure only through its interaction with social processes.

A physiological mechanism for Dressler's model has been outlined by Anderson et al. (1991). The model, illustrated in Figure 9.10, proposes that chronic stress in African Americans leads to increased **sympathetic nervous** activity, which results in the release of **norepinephrine** and **adrenocorticotropic hormone** (ACTH). Increased ACTH and norepinephrine levels elevate blood pressure by increasing sodium retention, expanding blood volume, and enhancing the **vasoconstriction** of peripheral blood vessels (Denton 1982).

Overall, considerable debate continues regarding the extent to which genetic versus socioeconomic factors contribute to elevated blood pressure in African Americans. Whereas it seems clear that social stress, poverty, and the legacy of slavery are important determinants of hypertension and other chronic diseases in the African American population, the role of genetics cannot be discounted. Indeed, even among studies such as Dressler's, which find a significant social com-

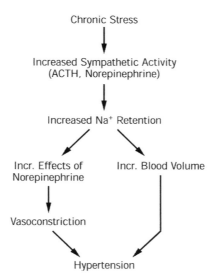

Figure 9.10   Proposed model for how chronic stress may contribute to increased blood pressure among African Americans. Chronic stress results in increased sympathetic nervous activity and the release of norepinepherine and adrenocorticotropic hormone (ACTH). These actions, in turn, result in increased Na+ retention, which contributes to increased blood pressure by increasing the vasoconstrictive effects of norepinepherine and increasing blood volume. (*Source*: Developed from Anderson et al. 1991)

ponent, the proportion of variation in blood pressure that is explained by social factors is relatively small.

Additionally, work by Frisancho and colleagues (1996, 1999) has found evidence of elevated blood pressure among Afro-Bolivians, a population of African ancestry without a long history racial oppression. In the study community from the Yungas region of Bolivia, this Afro-Bolivian sample had significantly higher blood pressure than their indigenous (Aymara) sample (Frisancho 1999) despite having similar social and economic status (Woodill 1996). The higher blood pressure in the Afro-Bolivian sample appears to be the result of higher sodium retention (Frisancho 1996). As such, the evidence to date suggests that high blood pressure among African Americans is influenced by both social and genetic factors.

Non-Insulin-Dependent Diabetes

Like hypertension, **non-insulin-dependent (adult-onset) diabetes mellitus** (NIDDM) is a disease thought to be the product of an interaction between genes and the environment. NIDDM is distinct from **insulin-dependent** (juvenile-onset) **diabetes** in being the result of reduced tissue sensitivity to **insulin**, rather than the inability of the pancreatic β **cells** to produce insulin (Nowak and Handford 1994).

Lifestyle factors play an important role in the etiology of NIDDM, with the disease being most common among the obese and generally occurring in people more than 40 years old. Obesity appears to directly contribute to the disease process by reducing the number of insulin receptor sites in the cells of the target tissues (i.e., muscle and adipose) (Nowak and Handford 1994).

Population variation in the **prevalence** of NIDDM has been a focus of considerable research over the past 30 years. As discussed in Chapters 7 and 13, Neel (1962) initially postulated that NIDDM was the product of a genetic adaptation being "rendered detrimental" by dietary and lifestyle changes. Specifically, he argued that among our hunting and gathering ancestors, who likely were faced with seasonal and year-to-year fluctuations in food availability, selection would have favored a "**thrifty genotype**" that would have allowed for quick release of insulin to enhance glucose storage during times of plenty. With the shift to a modern lifestyle, with an abundant and less variable food supply that contains higher levels of fats and simple carbohydrates, this once-adaptive genotype would have led to the development of diabetes by promoting the overproduction of insulin and desensitization of the target cell for glucose uptake. Neel (1982) later refined his model in light of advances in our understanding of diabetes, offering alternative metabolic pathways through which the "thrifty genes" might operate.

Weiss et al. (1984) proposed a similar model to explain the high prevalence of diabetes among Amerind populations. Diabetes was largely unknown among Native American populations during the first half of the 20th century; however, since the 1950s, prevalences have increased dramatically (West 1974; Weiss et al. 1984; Young 1993). As shown in Figure 9.11, indigenous North American groups have much higher prevalences of NIDDM than populations of European ancestry, whereas His-

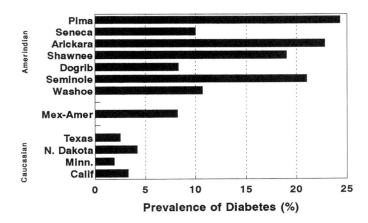

Figure 9.11 Variation in the prevalence of non-insulin-dependent diabetes mellitus (NIDDM) among North American populations. Prevalences are highest among Amerindian groups, lowest among Caucasians, and intermediate among Hispanics. (*Source*: Derived from Weiss et al. 1984)

panic populations of mixed ancestry have intermediate prevalences. Weiss and colleagues (1984) hypothesized that the small migrant populations that initially peopled the New World would have been exposed to very marginal food supplies. Hence, according to this model, the high frequency of thrifty genes among indigenous North and South American populations can be attributed to the influence of both selection and founder effect. This genetic predisposition, in combination with the lifestyle changes that have occurred among these populations over the past 50 years (e.g., less activity, greater food availability, and more fats and simple sugars) are thought to explain the high rates of a diabetes as well as several other metabolic disorders (e.g., obesity, hyperlipidemia, cholesterol gallstones, and gallbladder cancer). Thus, in this model, diabetes is seen as one part of a larger New World syndrome of diseases.

Szathmáry (1986, 1990) questioned the utility of the thrifty gene model for explaining the high rates of diabetes in native North Americans, particularly those who live in arctic and subarctic environs. She notes that because the diets of the founding New World populations were likely high in protein and fats and low in carbohydrates, there would not have been intense selection for rapid glucose uptake. Instead, she argues that among these founding prehistoric populations, enhanced gluconeogenesis—the process of converting noncarbohydrate sources into glucose—would have been selected for. Such an adaptation may have made individuals of these populations less able to uptake the high glucose loads associated with modern diets that are high in simple carbohydrates.

An additional problem with the Weiss model is that it implies that the genetic basis for diabetes is unique to New World (and perhaps Central Asian) populations. Yet, data from other parts of the world suggest that similar transitions from a traditional subsistence lifestyle are frequently associated with increases in the prevalence of diabetes. O'Dea (1991), for example, found marked increases in diabetes among Australian aborigines undergoing acculturation. Comparable trends have been shown among Pacific Island populations (Zimmet et al. 1977, 1982; Crews and MacKeen 1982).

Barker (1994) proposed a nongenetic "thrifty **phenotype**" model to explain the etiology of NIDDM. This model posits that prenatal and early postnatal malnutrition results in decreased β-cell mass of the pancreas and increased resistance to insulin in the muscle tissue (Hales and Barker 1992). These conditions, in the presence of adult overnutrition, are thought to lead to the development of diabetes. As support for this model, Barker and colleagues have shown that low birth weight is a significant predictor of diabetes and impaired glucose tolerance in adulthood, even after other determinants of diabetes (e.g., adult body weight) have been controlled (Phipps et al. 1993; Phillips et al. 1994).

Despite this recent emphasis on functional and developmental determinants in NIDDM, much ongoing research is investigating potential **candidate genes** for the disease (De Fronzo et al. 1992; Ferrell and Iyengar 1993). At this point considerable debate continues over the roles of genetic versus environmental factors in contributing to diabetes. Ferrell and Iyengar (1993) suggested that the difficulty in identifying a genetic basis for NIDDM likely reflects the fact that it is influenced by the interaction of multiple **alleles** at different loci. Hence, the single-locus approach of most candidate gene studies may not be powerful enough to detect the influence of individual alleles. Therefore, it would seem that, like many diseases of complex etiology (e.g., obesity, hypertension), diabetes may be a condition that arises via a variety of genetic–environmental pathways among different populations.

Obesity

Obesity has increasingly become a major public health problem in the United States and throughout much of the world. Over the past two decades, body weights have steadily increased in the United States such that currently, ~34% of American men and women are considered overweight or obese (Kuczmarski et al. 1994). It is also clear that obesity is not just a problem of adulthood, because body weights of children and adolescents have dramatically increased in recent years as well (Gortmaker and Dietz 1990; Must et al. 1991). In addition to being a direct health risk, obesity also increases one's risk of other chronic diseases, such as hypertension, adult-onset diabetes, and high blood cholesterol (hyperlipidemia).

Obesity refers to an excess of body fat and often is defined in terms of weight relative to height or by various measures of fatness. Reference tables, such as those compiled by the Metropolitan Life Insurance Company (1983), are frequently used to assess excess weight for height. In these cases, obesity is often defined as being >20% above the reference standard (Frisancho 1993). An alternative way of measuring "relative weight" is through the use of the **body mass index** (BMI). The BMI is calculated as weight (in kilograms) divided by height (in meters) squared. Thus, for example, a 72.5-kg (160-lb), 183-cm (6-foot) man would have a BMI of

$$BMI = (Weight)/(Height)^2 = (72.5)/(1.83)^2 = 21.7 \text{ kg/m}^2$$

Epidemiological research suggests that the lowest mortality risks are associated with BMIs between 20 and 25 kg/m^2 in adult men and women (Gray and Bray 1991). Risks of chronic health problems appear to increase substantially with BMIs above 27 (Gray and Bray 1991; Must et al. 1991).

Although weight-for-height tables and the BMI are widely used because of their relative ease of use, both measurements are problematic because they do not measure body composition (i.e., the relative proportions of muscle and fat). Several different approaches are available for assessing body composition, including **underwater weighing**, **bioelectrical impedance**, and **skinfold** (fat-fold) measures (see Shephard 1991). The most accurate of these techniques is underwater (hydrostatic) weighing. This technique involves determining body density based on the difference between weight measured in air versus that measured while submerged under water (McArdle et al. 1986). In contrast, skinfold measurements are the simplest and most widely used approach for assessing body fatness. Special calipers are used to measure fatness at specific sites on the body such as over the triceps muscle on the back of the upper arm or below the shoulder blade (scapula). These measures can then be compared to age- and sex-specific reference standards (e.g., Frisancho 1990), with obesity being defined as skinfolds greater than the 85th percentile.

Obesity results from the consumption of more energy than is expended, thus resulting in the accumulation of stored energy as fat. Whereas this root cause is clear, considerable debate continues regarding the relative importance of genetic and environmental (e.g., lifestyle) factors in contributing to obesity. That is, does obesity result largely from limited activity, or does evidence suggest that obese individuals are genetically more "metabolically efficient" (i.e., requiring less energy to perform the same task)? These aspects are discussed in the next sections.

Genetic Factors

Over the past decade, several different studies have attempted to quantify the **heritability** of body weight and body fatness (see Chapter 5). Heritability refers to the proportion of variation in a physical trait (the phenotype—in this case, body weight or fatness) that is attributable to variation in genetics (i.e., the **genotype**). Heritabilities (h^2) range between 0 (no genetic contribution) and 1 (entirely genetic), with values for most traits falling midway between the extremes (i.e., 0.4–0.6). Studies of heritability examine variation among related individuals, and studies of identical (monozygotic) twins are among the most powerful.

Stunkard et al. (1990) compared the BMIs of samples of identical Swedish twin pairs who had been raised together versus those who had been raised apart. They found high correlations in BMIs between twin pairs of both groups, indicating a high heritability for the BMI ($h^2 = 0.6$–0.7) and relatively limited influence of shared environment. A subsequent study on a large sample of related individuals (including parents and offspring, siblings, and twins) from Norway (Tambs et al. 1991) found a lower h^2 for BMI of ~0.4. Taken together, these results have been interpreted as strong evidence for large genetic component to obesity (e.g., Stunkard et al. 1990). However, these studies are limited by the fact that they have evaluated only overall body mass rather than composition. Because the BMI is highly correlated with measures of skeletal size and muscle mass as well as fatness, it is possible that these studies have overestimated the genetic basis of obesity (Garn et al. 1986).

Studies that examine the heritability of measures of fatness have produced variable results (Bouchard 1991; Bouchard and Pérusse 1993). However, one study (Bouchard et al. 1988) is particularly notable in that it used underwater weighing to assess body composition in many related individuals (1698 individuals from 409 families) from Quebec, Canada.. These researchers found the genetic component of body fatness to be 25%, as compared to 30% for fat-free mass. These results suggest that the heritability of body fatness is more modest than suggested by the some of the BMI studies.

Other research has focused on the genetic basis for variation in metabolic rates, thus investigating differences in the potential risk of obesity. Twin studies by Bouchard et al. (1989) and Fontaine et al. (1985) estimated the heritability of BMR to be between 0.4 and 0.8 after controlling for sex, body size, and body composition. Heritabilities reported for the thermic effect of food are of similar magnitude, 0.4–0.6 (Bouchard et al. 1989). Not surprisingly, studies of the heritability of physical activity levels suggest a much weaker genetic contribution (Pérusse et al. 1988).

Genetic differences in the tendency to gain weight have been directly explored in experimental studies of overfeeding (Poehlman et al. 1986; Bouchard et al. 1990). One long-term overfeeding study (Bouchard et al. 1990) examined 12 identical twin pairs (young adult men) for 84 days under conditions of restricted physical activity. During an initial 2-week baseline period, body composition and daily energy needs were established. Each subject was then overfed by 1000 kcal (4184 kJ) per day for the entire 84 days of the study. Considerable variation in weight gain was observed, ranging from a low of 4.3 kg (9.5 lb) to a high of 14.3 kg (29 lb). However, as shown in Figure 9.12, individual twin pairs tended to respond similarly to the overfeeding. That is, certain pairs of twins were more likely to gain weight than others. In addition, both the amount and the distribution of fat gain significantly differed among the twin pairs. These findings suggest a genetic component to weight and fat gain as well as patterns of fat distribution in response to overfeeding.

Research is also identifying specific genetic loci that are associated with body composition and obesity. As of 1998, the number of loci that had been associated or linked with human obesity was almost 200 (Pérusse et al. 1999). Many of these loci are referred to as QTLs (**quantitative trait loci** [see Chapter 5]), because they are believed to contain genes that, in combination with other genes and the environment, contribute to variation in a complex, continuously distributed trait (e.g., body fatness, body mass). Results recently obtained from the San Antonio Family Heart Study, for example, indicate that QTLs on **chromosomes** 2 and 8 are important in

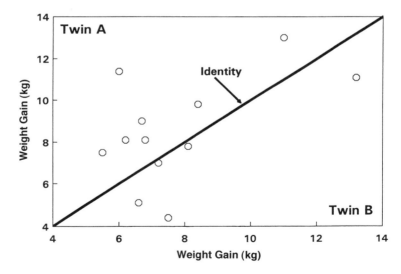

Figure 9.12 Changes in body weight in response to overfeeding among 12 pairs of male twins. Each point represents the weight gains for a pair of twins (denoted A and B). The line of identity demarcates equal weight gains for twin pairs. The degree of similarity in weight gains within twin pairs is high, as indicated by the significant correlation between weight gains among the pairs ($r = 0.55$; $p < 0.02$). (*Source*: Adapted from Bouchard et al. 1990)

shaping variation in body fatness among Mexican Americans (Comuzzie et al. 1997; Rogers et al 1999). Similarly, Bouchard and colleagues' (1997) recent work on the Quebec Family Study has linked variation in resting metabolism and body fatness to a locus on chromosome 11.

Overall, despite considerable variability among studies, a significant heritable (genetic) component to variation in body mass and body fatness clearly exists. Some of the differences in heritability estimates reflect differences in the nature of the samples and the statistical approaches employed. Additionally, the number of genetic loci that have been linked to human obesity has increased dramatically over the last few years, and continues to grow with new research. What remains less clear, however, are the mechanisms through which these genes directly, and in combination with environmental factors, contribute to obesity.

Lifestyle Factors

Although the importance of genetics to variation in weight and obesity cannot be discounted, lifestyle factors also play an important role in promoting obesity. In particular, changes associated with the transition from a traditional subsistence way of life to a more modern lifestyle have been shown to increase rates of obesity and other chronic diseases in populations throughout the world (e.g., Baker et al. 1986; Friedlaender 1987; McGarvey 1991; Shephard and Rode 1996). As discussed earlier, traditional food-producing populations of the developing world have much higher physical activity levels than populations of the industrialized (Western) world (see Table 9.8). These high levels of energy expenditure, combined with periods of limited food availability, contribute to the low levels of obesity observed in nonindustrial societies.

One of the most extensive studies of the changes associated with this lifestyle transition was conducted by Shephard and Rode (1996) among the Inuit of Igloolik, Northwest Territories (Canada). This research documented sharp increases in obesity and declines in fitness and general health associated with the transition from the Inuit's traditional hunting lifestyle. When first studied in the late 1960s and early 1970s, the Inuit displayed high levels of daily energy expenditure, **aerobic** fitness, strength, and modest levels of body fatness (Shephard 1974). By 1989–90, however, the Inuit's strength and aerobic capacity had declined to levels comparable to those of sedentary urban populations, whereas their level of body fatness had increased dramatically (Rode and Shephard 1994). In Table 9.11, the BMIs and averages of three skinfold measures are compared between Inuit adults measured in 1970 and in 1990. Although the increases in BMI are modest in some of the age groups, the skinfold measures, on average, more than doubled over this 20-year period.

The rapid increase in rates of obesity seen with acculturating populations such as the Inuit likely reflects the fact that only small to moderate changes in daily activity patterns are necessary to produce substantial changes in daily caloric requirements. For example, as outlined in Table 9.12, the difference in daily energy needs for a 70-kg (155-lb) man with a "moderate" versus a "light" daily activity budget is only ~400 kcal. The shift from a "heavy" to a "light" activity budget for the same man would result in almost a 1000-kcal decline in TDEE. Thus, changes in activity, com-

Table 9.11   Comparison of the body mass index (BMI) and average skinfold thickness of Inuit adults in 1970 and 1990

Age group (years)	BMI (kg/m^2)		Skinfold thickness (mm)	
	1970	1990	1970	1990
Males				
20–29	24.4	24.1	5.5	10.4
30–39	24.9	24.3	6.3	10.9
40–49	25.3	28.1	5.4	15.6
50–59	25.8	27.1	7.9	15.7
Females				
20–29	23.2	23.3	8.5	15.1
30–39	23.9	25.0	9.2	21.4
40–49	23.7	29.0	7.0	28.5
50–59	27.5	31.7	19.0	34.8

Skinfold thickness is the average of measures from the triceps, subscapular, and suprailiac sites.
(*Source:* Data from Shephard and Rode 1996)

Table 9.12   Energetic consequences of changes in daily activity level of a 70-kg man

Activity level	Weight (kg)	BMR (kcal/day)	TDEE (kcal/day)
Heavy	70	1750	3675
Moderate	70	1750	3115
Light	70	1750	2712

BMR and TDEE estimates were calculated based on the information presented in Tables 9.2 and 9.3.

bined with an increase in food availability, likely account for the much of the change in body size observed with acculturating populations.

It appears that variation in body weight and fatness is promoted through various genetic and environmental interactions. Although significant genetic components to variation in the utilization of excess energy clearly exist, largely nongenetic factors such as daily activity patterns also exert a strong influence on one's energy balance. As such, obesity is like many other chronic diseases in having multiple etiologies. Hence, it is critical for future research to identify and better understand the alternative pathways that promote obesity in different subpopulations.

CHAPTER SUMMARY

In examining human nutrition from an evolutionary perspective, many important themes emerge. First, it is clear that we share many aspects of our nutritional biology with our primate kin. Humans are eclectic, omnivorous feeders whose metabolic rates are about what would be predicted for a comparably sized primate or mammal. Yet despite these similarities, humans depart substantially from other primates in having a much higher quality, more nutrient-dense diet than expected. Adaptation to this dietary regime is reflected in our gut morphology, which in some aspects more closely resembles a carnivore than a folivore or frugivore. It appears that this dietary adaptation is a product of evolutionary trends for increased encephalization in the hominid lineage over the past 2 million to 2.5 million years. With a relatively larger brain size, the proportion of daily energy requirements used by the brain increases and necessitates a more easily digestible and energy-rich diet.

Beyond these broad species-level similarities and differences between humans and other primates, humans display remarkable intraspecific dietary and metabolic diversity. Ethnic and interpopulational variation in lactose tolerance and diabetes, for example, suggest that distinct dietary and metabolic adaptations (both functional or physiological and genetic) can emerge quite rapidly over evolutionary time. Hence, whereas attempts to broadly define the "natural" (i.e., evolutionarily prescribed) diet for the human species may have some heuristic value (e.g., Eaton and Konner 1985; Eaton et al. 1988), they are problematic in that they obscure important nutritional differences among human groups.

Moreover, such models ignore the distinct biocultural interactions that produce variation in nutritional and metabolic diseases, such as hypertension and obesity. A better understanding of this interpopulational variation is necessary if we are to deal with the ongoing global changes in morbidity and mortality patterns that are emerging as the result of changes in diet and lifestyle.

RECOMMENDED READINGS

Bouchard C and Pérusse L (1993) Genetics of obesity. Annu. Rev. Nutr. 13: 337–354.

Bouchard C, Tremblay A, Després JP, Nadeau A, Lupien PJ, Thériault G, Dussault J, Moorjani S, Pinault S, and Fournier G (1990) The response to long-term overfeeding in identical twins. N. Engl. J. Med. 322: 1477–1482.

Denton D (1982) The Hunger for Salt: An Anthropological, Physiological, and Medical Analysis. New York: Springer-Verlag.

Eaton SB and Konner M (1985) Paleolithic nutrition: a consideration of its nature and current implications. N. Engl. J. Med. 312: 283–289.

Frisancho AR (1993) Human Adaptation and Accommodation. Ann Arbor: Univ. of Michigan Press.

Gibson R (1990) Principles of Nutritional Assessment. Oxford: Oxford Univ. Press.

Harding RSO and Telecki G (eds.) (1981) Omnivorous Primates: Gathering and Hunting in Human Evolution. New York: Columbia Univ. Press.

Harris M and Ross EB (eds.) (1987) Food and Evolution: Toward a Theory of Human Food Habits. Philadelphia, Pa.: Temple Univ. Press.

Holliday MA (1986) Body composition and energy needs during growth. In F Falkner and JM Tanner (eds.): Human Growth: A Comprehensive Treatise, 2nd edition. New York: Plenum, Vol. 2, pp. 101–117.

James WPT and Schofield EC (1990) Human Energy Requirements. New York: Oxford Univ. Press.

Johns T (1990) The Origins of Human Diet and Medicine. Tucson: Univ. of Arizona Press.

Johnston FE (ed.) (1987) Nutritional Anthropology. New York: Alan R. Liss.

Kleiber M (1975) The Fire of Life: An Introduction to Animal Energetics, 2nd edition. Huntington, N.Y.: Krieger.

Leonard WR and Robertson ML (1992) Nutritional requirements and human evolution: a bioenergetics model. Am. J. Hum. Biol. 4: 179–195.

Leonard WR and Robertson ML (1994) Evolutionary perspectives on human nutrition: the influence of brain and body size on diet and metabolism. Am. J. Hum. Biol. 6: 77–88.

McArdle WD, Katch FI, and Katch VL (1986) Exercise Physiology: Energy, Nutrition, and Human Performance, 2nd edition. Philadelphia, Pa.: Lea and Febiger.

McHenry HM (1992) How big were early hominids? Evol. Anthropol. 1: 15–20.

Milton K (1993) Diet and primate evolution. Sci. Am. 269(2): 86–93.

Rogers J, Mahaney MC, Almasy L, Comuzzie AG, and Blangero J (1999) Quantitative trait linkage mapping in Anthropology. Yrbk. Phys. Anthropol. 42:127-151.

Scrimshaw NS and Murray EB (1988) The acceptability of milk and milk products in populations with a high prevalence of lactose intolerance Am. J. Clin. Nutr. 48(Suppl. 1): 1083–1159.

Shephard RJ and Rode A (1996) Health Consequences of "Modernization": Evidence from Circumpolar Populations. Cambridge: Cambridge Univ. Press.

Stinson S (1992) Nutritional adaptation. Annu. Rev. Anthropol. 21: 143–170.

Ulijaszek SJ (1995) Human Energetics in Biological Anthropology. Cambridge: Cambridge Univ. Press.

Ulijaszek SJ and Strickland SS (1993) Nutritional Anthropology: Prospects and Perspectives. London: Smith-Gordon.

Weiss KM, Ferrell RE, and Hanis CL (1984) A New World Syndrome of metabolic diseases with a genetic and evolutionary basis. Yearbook Phys. Anthropol. 27: 153–178.

Whiten A and Widdowson EM (eds.) (1991) Foraging Strategies and Natural Diet of Monkeys, Apes, and Humans. Oxford: Oxford Univ. Press.

REFERENCES CITED

Aiello LC and Wheeler P (1995) The expensive-tissue hypothesis: the brain and digestive system in human and primate evolution. Curr. Anthropol. 36: 199–221.

Alpers DH (1981) Carbohydrate digestion: effects of monosaccharide inhibition and enzyme degradation on lactase activity. In DM Paige and TM Bayliss (eds.): Lactose Digestion: Clinical and Nutritional Implications. Baltimore, Md.: Johns Hopkins Univ. Press, pp. 58–68.

Anderson NB, McNeilly M, and Myers H (1991) Autonomic reactivity and hypertension in blacks: a review and proposed model. Ethnicity Dis. 1: 154–185.

Armellini F, Robbi R, Zamboni M, Todesco T, Castelli S, and Bosello O (1992) Resting metabolic rate, body fat distribution, and visceral fat in obese women. Am. J. Clin. Nutr. 56: 981–987.

Baker PT, Hanna JT, and Baker TS (eds.) (1986) The Changing Samoans: Behavior and Health in Transition. Oxford: Oxford Univ. Press.

Barker DJP (1994) Mothers, Babies, and Disease in Later Life. London: BMJ Publishing Group.

Begon M, Harper JL, and Townsend CR (1990) Ecology: Individuals, Populations and Communities, 2nd edition. New York: Blackwell Scientific.

Behrensmeyer AK and Cooke HBS (1985) Paleoenvironments, stratigraphy, and taphonomy in the African Pliocene and early Pleistocene. In E Delson (ed.): Ancestors: The Hard Evidence. New York: Alan R. Liss, pp. 60–62.

Bell RH (1971) A grazing ecosystem in the Serengeti. Sci. Am. 225(1): 86–93.

Benedict FG (1938) Vital Energetics (Publication no. 503). Washington, D.C.: Carnegie Institution.

Bogin B (1988) Patterns of Human Growth. Cambridge: Cambridge Univ. Press.

Bolin TD and Davis AE (1970) Primary lactase deficiency: genetic or acquired? Am. J. Dig. Dis. 15: 679–692.

Bouchard C (1991) Genetic aspects of anthropometric dimensions relevant to the assessment of nutritional status. In JH Himes (ed.): Anthropometric Assessment of Nutritional Status. New York: Wiley-Liss, pp. 213–231.

Bouchard C and Pérusse L (1993) Genetics of Obesity. Annu. Rev. Nutr. 13: 337–354.

Bouchard C, Pérusse L, LeBlanc C, Tremblay A, and Thériault G (1988) Inheritance of the amount and distribution of human body fat. Int. J. Obes. 12: 205–215.

Bouchard C, Tremblay A, Nadeau A, Després JP, and Thériault G (1989) Genetic effects in resting and exercising metabolic rates. Metabolism 38: 364–370.

Bouchard C, Tremblay A, Després JP, Nadeau A, Lupien PJ, Thériault G, Dussault J, Moorjani S, Pinault S, and Fournier G (1990) The response to long-term overfeeding in identical twins. N. Engl. J. Med. 322: 1477–1482.

Bouchard C, Pérusse L, Chagnon YC, Warden C, and Ricquier D (1997) Linkage between markers in the vicinity of the uncoupling protein 2 gene and resting metabolic rate in humans. Hum Mol Genet. 6:1887–1889.

Brody S (1945) Bioenergetics and Growth. New York: Reinhold.

Bunn HT (1981) Archeological evidence for meat eating by Plio-Pleistocene hominids from Koobi-Fora and Olduvai Gorge. Nature 291: 574–577.

Calder WA (1984) Size, Function, and Life History. Cambridge, Mass.: Harvard Univ. Press.

Chivers DJ and Hladik CH (1980) Morphology of the gastrointestinal tract in primates: comparisons with other mammals in relation to diet. J. Morphol. 166: 337–386.

Comuzzie AG, Hixson JE, Almasy L, Mitchell B, Mahaney MC, Dyer TD, Stern MP, MacCluer JW, and Blangero J (1997) A major quantitative trait locus determining serum leptin levels and fat mass is located on chromosome 2. Nature Genet. 15:273–275.

Crews DE and MacKeen P (1982) Mortality related to cardiovascular disease and diabetes mellitus in a modernizing population. Soc. Sci. Med. 16: 175–181.

Curtin PD (1992) The slavery hypothesis for hypertension among African Americans: the historical evidence. Am. J. Public Health 82: 1681–1686.

De Boer JO, Van ES, Van Raaij JMA, and Hautvast JGAJ (1987) Energy requirements and energy expenditure of lean and obese women measured by indirect calorimetry. Am. J. Clin. Nutr. 46: 13–21.

De Fronzo RA, Bondonna RC, and Ferrannini E (1992) Pathogenesis of NIDDM – a balanced overview. Diabetes Care 15: 318–368.

De Garine I and Harrison GA (eds.) (1988) Coping with Uncertainty in Food Supply. Oxford: Oxford Univ. Press.

Denton D (1982) The Hunger for Salt: An Anthropological, Physiological, and Medical Analysis. New York: Springer-Verlag.

Department of Health and Social Security (1981) Recommended Daily Amounts of Food Energy and Nutrients for Groups of People in the United Kingdom. Second Impression (Report on Health and Social Subjects No. 15). Committee on Medical Aspects of Food Policy, London: Her Majesty's Stationary Office.

DeWalt KM (1983) Nutritional Strategies and Agricultural Change in a Mexican Community. Ann Arbor: UMI Press.

Dewey KG (1981) Nutritional consequences of the transformation from subsistence to commercial agriculture in Tabasco, Mexico. Hum. Ecol. 9: 157–181.

Dressler WW (1991) Social class, skin color, and arterial blood pressure in two societies. Ethnicity Dis. 1: 60–77.

Dubois EF (1927) Basal Metabolism in Health and Disease. Philadephia: Lea & Febiger.

Eaton SB and Konner M (1985) Paleolithic nutrition: a consideration of its nature and current implications. N. Engl. J. Med. 312: 283–289.

Eaton SB, Shostak M, and Konner M (1988) The Paleolithic Prescription: A Program of Diet and Exercise and a Design for Living. New York: Harper & Row.

Erickson H and Krog J (1956) Critical temperature for naked man. Acta Physiol. Scand. 37: 35–39.

Falk D (1987) Hominid paleoneurology. Annu. Rev. Anthropol. 16:13–30.

Ferrell RE and Iyengar S (1993) Molecular studies of the genetics of non-insulin-dependent diabetes mellitus. Am. J. Hum. Biol. 5: 415–424.

Fontaine E, Savard R, Tremblay A, Després JP, and Poehlman ET (1985) Resting metabolic rate in monozygotic and dizygotic twins. Genet. Med. Gemellol. 34: 41–47.

Food and Agriculture Organization and World Health Organization (FAO/WHO) (1988) Requirements of Vitamin A, Iron, Folate, and Vitamin B-12 (Report of the joint FAO/WHO Expert Consultation). Rome: Food and Agriculture Organization.

Food and Agriculture Organization, World Health Organization, and United Nations Univ. (FAO/WHO/UNU) (1985) Energy and Protein Requirements (Report of joint FAO/WHO/UNU expert consultation; WHO technical report series no. 724). Geneva: World Health Organization.

Freis, ED, Reda DJ, and Materson BJ (1988) Volume (weight) loss and blood pressure following thiazide diuretics. Hypertension 12: 244–250.

Friedlaender JS (ed.) (1987) The Solomon Islands Project: A Long-Term Study of Health, Heredity, and Modernization. Oxford, UK: Clarendon Press.

Frisancho AR (1990) Anthropometric Standards for the Assessment of Growth and Nutritional Status. Ann Arbor: Univ. of Michigan Press.

Frisancho AR (1993) Human Adaptation and Accommodation. Ann Arbor: Univ. of Michigan Press.

Frisancho AR, Farrow S, Lockette W, Frisancho H, Friedenzohn I, Johnson T Kapp B, Miranda C, Perez M, Rauchle I, Sanchez N, Swaninger K, Wheatcroft G, Woodill L, Ayllon I, Soria R,

Rodriguez A, and Machicao J (1996) Biological and environmental components of variability of blood pressure among Bolivian blacks. Am. J. Hum. Biol. 8: 115 (abstract).

Frisancho AR, Farrow S, Lockette W, Friedenzohn I, Johnson T, Kapp B, Miranda C, Perez M, Rauchle I, Sanchez N, Wheatcroft G, Woodill L, Ayllon I, Bellido D, Rodriguez A, Machicao J, Villena M, and Vargas E (1999) Role of genetic and environmental factors in the increased blood pressures of Bolivian blacks. Am. J. Hum. Biol. 11: 489-498.

Galloway VA, Leonard WR, and Ivakine E (2000) Basal metabolic adaptation of the Evenki reindeer herders of Central Siberia. Am. J. Hum. Biol. 12:75-87.

Garn SM, Leonard WR, and Hawthorne VW (1986) Three limitations of the Body Mass Index. Am. J. Clin. Nutr. 44: 996–997.

Gaulin SJC (1979) A Jarman/Bell model of primate feeding niches. Hum. Ecol. 7: 1–20.

Gaulin SJC and Konner M (1977) On the natural diets of primates, including humans. In R Wurtman and J Wurtman (eds.): Nutrition and the Brain. New York: Raven Press, Vol. 1, pp. 1–86.

Gibson R (1990) Principles of Nutritional Assessment. Oxford: Oxford Univ. Press.

Gilat T, Russo S, Gelman-Malachi E, and Aldor TAM (1972) Lactase in man: a nonadaptable enzyme. Gastroenterology 62: 1125–1127.

Gleibermann L (1973) Blood pressure and dietary salt in human populations. Ecol. Food Nutr. 2: 142–156.

Gordon KD (1987) Evolutionary perspectives on the human diet. In FE Johnston (ed.): Nutritional Anthropology. New York: Alan R. Liss, pp. 3–39.

Gortmaker SL and Dietz WH (1990) Secular trend in body mass in the United States, 1960–1980. Am. J. Epidemiol. 132:194–197.

Gowlett JAJ (1984) Mental abilities of early man: a look at some hard evidence. In R Foley (ed.): Hominid Evolution and Community Ecology. New York: Academic Press, pp. 167–192.

Greene R (1970) Human Hormones. New York: McGraw-Hill.

Gray DS and Bray GA (1991) Anthropometric assessment in an adult obesity clinic. In JH Himes (ed.): Anthropometric Assessment of Nutritional Status. New York: Wiley-Liss, pp. 383-398.

Hales CN and Barker DJP (1992) Type 2 (non-insulin dependent) diabetes mellitus: the thrifty phenotype hypothesis. Diabetologia 35: 595–601.

Health and Welfare Canada (1990) Nutrition Recommendations: The Report of the Scientific Review Committee. Ottawa, Ontario: Canadian Government Publishing Centre.

Henry CJK (1996) Early environment and later nutritional needs. In CJK Henry and SJ Ulijaszek (eds.): Long Term Consequences of Early Environment: Growth, Development, and the Lifespan Developmental Perspective. Cambridge: Cambridge Univ. Press, pp. 124–138.

Henry CJK and Rees DG (1991) New predictive equations for the estimation of basal metabolic rate in tropical peoples. Eur. J. Clin. Nutr. 45: 177–185.

Holliday MA (1986) Body composition and energy needs during growth. In F Falkner and JM Tanner (eds.): Human Growth: A Comprehensive Treatise, 2nd edition. New York: Plenum, Vol. 2, pp. 101–117.

Howard BV, Bogardus C, Ravussin E, Foley JE, Lillioja S, Mott DM, Bennett PH, and Knowler WC (1991) Studies of the etiology of obesity in Pima Indians. Am. J. Clin. Nutr. 53: 1577S–1585S.

Huss-Ashmore R, Curry JJ, and Hitchcock RK (eds.) (1988) Coping with Seasonal Constraints. Philadelphia, Pa.: MASCA Research Papers in Science and Archeology.

James WPT and Schofield EC (1990) Human Energy Requirements. New York: Oxford Univ. Press.

Jarman PJ (1974) The social organization of antelope in relation to their ecology. Behaviour 58: 215–267.

Johns T (1990) The Origins of Human Diet and Medicine. Tucson, Ariz.: Univ. of Arizona Press.

Johns T and Keen SL (1986) Taste evaluation of potato glycoalkaloids by the Aymara: a case study in human chemical ecology. Hum. Ecol. 14: 437–452.

Johnson AO, Semenya JG, Buchowski MS, Enwonwu CO, and Scrimshaw NS (1993) Adaptation of lactose maldigesters to continued milk intake. Am. J. Clin. Nutr. 58: 879–881.

Johnson RC, Cole RE, and Ahern FM (1981) Genetic interpretation of racial/ethnic differences in lactose absorption and tolerance: a review. Hum. Biol. 53: 1–13.

Jungers WL (ed.) (1985) Size and Scaling in Primate Biology. New York: Plenum.

Karasov WH (1992) Daily energy expenditure and the cost of activity in mammals. Am. Zool. 32: 238–248.

Katz SH (1987) Food and biocultural evolution: a model for the investigation of modern nutritional problems. In FE Johnston (ed.): Nutritional Anthropology. New York: Alan R. Liss, pp. 41–63

Katz SH, Hediger ML, and Valleroy LA (1974) Traditional maize processing techniques in the New World. Science 184: 765–773.

Katz SH, Hediger ML, and Valleroy LA (1975) The anthropological and nutritional significance of traditional maize processing techniques in the New World. In ES Watts, FE Johnston, and GW Lasker (eds.): Biosocial Interrelations in Population Adaptation. Chicago: Aldine, pp. 195–231.

Kleiber M (1932) Body size and metabolism. Hilgardia 6: 315–353.

Kleiber M (1975) The Fire of Life: An Introduction to Animal Energetics, 2nd edition. Huntington, N.Y.: Krieger.

Knowler WC, Pettitt D, Bennett PH, and Williams RC (1983) Diabetes mellitus in the Pima indians: genetic and evolutionary considerations. Am. J. Phys. Anthropol. 62: 107–114.

Knowler WC, Pettitt DJ, Saad MF, Charles MA, Nelson RG, Howard BV, Bogardus C, and Bennett PH (1991) Obesity in the Pima indians: its magnitude and relationship with diabetes. Am. J. Clin. Nutr. 53: 1543S–1551S.

Kotler DP (1982) Lactose absorption in man is adaptable. Gastroenterology 82: 1105.

Krebs CJ (1972) Ecology: The Experimental Analysis of Distribution and Abundance. New York: Harper & Row.

Kretchmer N (1972) Lactose and lactase. Sci. Am. 227(4): 71–78.

Kretchmer N (1981) The significance of lactose intolerance. In DM Paige and TM Bayliss (eds.): Lactose Digestion: Clinical and Nutritional Implications. Baltimore, Md.: Johns Hopkins Univ. Press, pp. 3–7.

Kuczmarski RJ, Flegal KM, and Campbell K (1994) Increasing prevalence of overweight among US adults. The National Health and Nutrition Examination Surveys, 1960–1991. JAMA 272: 205–211.

Kurland JA and Pearson JD (1986) Ecological significance of hypometabolism in nonhuman primates: allometry, adaptation, and deviant diets. Am. J. Phys. Anthropol. 71: 445–457.

Lee RB (1968) What hunters do for a living, or how to make out on scarce resources. In RB Lee and I Devore (eds.): Man the Hunter. Chicago: Aldine, pp. 30–48.

Leonard WR and Robertson ML (1992) Nutritional requirements and human evolution: a bioenergetics model. Am. J. Hum. Biol. 4: 179–195.

Leonard WR and Robertson ML (1994) Evolutionary perspectives on human nutrition: the influence of brain and body size on diet and metabolism. Am. J. Hum. Biol. 6: 77–88.

Leonard WR and Robertson ML (1995) Energetic efficiency of human bipedality. Am. J. Phys. Anthropol. 97: 335–338.

Leonard WR and Robertson ML (1997) Rethinking the energetics of bipedality. Curr. Anthropol. 38: 304–309.

Leonard WR and Thomas RB (1988) Changing dietary patterns in the Peruvian Andes. Ecol. Food Nutr. 21: 245–263.

Leonard WR, DeWalt KM, Uquillas JE, and DeWalt BR (1993) Ecological correlates of dietary consumption and nutritional status in highland and coastal Ecuador. Ecol. Food Nutr. 31: 67–85.

Leonard WR, DeWalt KM, Uquillas JE, and DeWalt BR (1994) Diet and nutritional status among cassava producing agriculturalists of coastal Ecuador. Ecol. Food Nutr. 32: 113–127.

Lerebours E, N'Djitoyap Ndam C, Lavoine A, Hellot MF, Antoine JM, and Colin R (1989) Yogurt and fermented-then-pasteurized milk: effects of short-term and long-term ingestion on lactose absorption and mucosal lactase activity in lactase-deficient subjects. Am. J. Clin. Nutr. 49: 823–827.

Luft FC, Rankin LI, Bloch R, Weymen AE, Willis LR, Murray RH, Grim CE, and Weinberger MH (1979) Cardiovascular and humoral responses to extremes of sodium intake in normal white and black men. Circulation 60: 697–706.

Margaria R, Cerretelli P, Aghemo P, and Sassi G (1963) Energy cost of running. J. Appl. Physiol. 18: 367–370.

Martin RD (1989) Evolution of the brain in early hominids. OSSA 14: 49–62.

McArdle WD, Katch FI, and Katch VL (1986) Exercise Physiology: Energy, Nutrition, and Human Performance, 2nd edition. Philadelphia, Pa.: Lea and Febiger.

McCracken RD (1971) Lactase deficiency: an example of dietary evolution. Curr. Anthropol. 12: 479–517.

McGarvey ST (1991) Obesity in Samoans and a perspective on its etiology in Polynesians. Am. J. Clin. Nutr. 53: 1586S–1594S.

McGrew WC (1974) Tool use by chimpanzees in feeding upon driver ants. J. Hum. Evol. 3: 501–508.

McHenry HM (1992) Body size and proportions in early hominids. Am. J. Phys. Anthropol. 87: 407–431.

McHenry HM (1994) Behavioral ecological implications of early hominid body size. J. Hum. Evol 27: 77–87.

McLean JA and Tobin G (1986) Animal and Human Calorimetry. Cambridge: Cambridge Univ. Press.

Metropolitan Life Insurance Company (1983) Metropolitan height and weight tables. Stat Bull 64:2-9.

Milton K (1987) Primate diets and gut morphology: implications for hominid evolution. In M Harris and EB Ross (eds.): Food and Evolution: Toward a Theory of Human Food Habits. Philadelphia, Pa.: Temple Univ. Press, pp. 93–115.

Milton K (1993) Diet and primate evolution. Sci. Am. 269(2): 86–93.

Mink JW, Blumenschine RJ, and Adams DB (1981) Ratio of central nervous system to body metabolism in vertebrates: its constancy and functional basis. Am. J. Physiol. 241: R203–R212.

Must A, Dallal GE, and Dietz WH (1991) Reference data for obesity: 85th and 95th percentiles of body mass (wt/ht^2) and triceps skinfold thickness. Am. J. Clin. Nutr. 53: 839–846.

National Center for Health Statistics (NCHS) (1996) Health, United States, 1995. Hyattsville, Md.: Public Health Service.

National Research Council (NRC) (1986) Nutrient Adequacy: Assessment Using Food Consumption Surveys. Washington, D.C.: National Academy Press.

National Research Council (NRC) (1989) Recommended Dietary Allowances, 10th edition. Washington D.C.: National Academy Press.

Neel JV (1962) Diabetes mellitus: a "thrifty" genotype rendered detrimental by "progress"? Am J Hum Genet 14:353-362.

Neel JV (1982) The thrifty genotype revisited. In J Kobberling and J Tattersall (eds.): The Genetics of Diabetes Mellitus. New York: Academic Press, pp. 283–293.

Nowak TJ and Handford AG (1994) Essentials of Pathophysiology. Dubuque, Iowa: WC Brown Publishers.

O'Dea, K (1991) Traditional diet and food preferences of Australian aboriginal hunter-gatherers. In A Whiten and EM Widdowson (eds.): Foraging Strategies and Natural Diet of Monkeys, Apes, and Humans. Oxford: Oxford Univ. Press, pp. 73–81.

Pérusse L, LeBlanc C, and Bouchard C (1988) Inter-generation transmission of physical fitness in a Canadian population. Can. J. Sports Sci. 13: 8–14.

Pérusse L , Chagnon YC, Weisnagel J, and Bouchard C (1999) The Human Obesity Gene Map: the 1998 update. Obes. Res. 7:111–129.

Peters RH (1983) The Ecological Implications of Body Size. Cambridge: Cambridge Univ. Press.

Phillips DIW, Barker DJP, Hales CN, Hirst S, and Osmond C (1994) Thinness at birth and insulin resistance in adult life. Diabetologia 37: 150–154.

Phipps K, Barker DJP, Hales CN, Fall CHD, Osmond C, and Clark PMS (1993) Fetal growth and impaired glucose tolerance in men and women. Diabetologia 36: 225–228.

Poehlman ET, Tremblay A, Després JP, Fontaine E, Pérusse L, Thériault G, Bouchard C (1986) Genotype-controlled changes in body composition and fat morphology following overfeeding in twins. Am. J. Clin. Nutr. 43: 723–731.

Potts R (1988) Early Hominid Activities at Olduvai. New York: Aldine.

Potts R (1993) Archeological interpretations of early hominid behavior and ecology. In DT Rasmusssen (ed.): The Origin and Evolution of Human and Humanness. Boston: Jones and Bartlett, pp. 49–74.

Ravussin E, Lillijoa S, and Knowler WC (1988) Reduced rate of energy expenditure as a risk factor for body-weight gain. N. Engl. J. Med. 318: 467–472.

Richard AF (1985) Primates in Nature. New York: W.H. Freeman.

Roberts DF (1978) Climate and Human Variability, 2nd edition. Menlo Park, CA: Cummings.

Rode A and Shephard RJ (1994) Growth and fitness of Canadian Inuit: secular trends, 1970–1990. Am. J. Hum. Biol. 6: 525–541.

Rode A and Shephard RJ (1995) Basal metabolic rate of Inuit. Am. J. Hum. Biol. 7: 723–729.

Rodman PS and McHenry HM (1980) Bioenergetics and the origin of hominid bipedalism. Am. J. Phys. Anthropol. 52: 103–106.

Rolfes SR, De Bruyne LK, and Whitney EN (1990) Life Span Nutrition: Conception through Life. New York: West Publishing.

Sailer LD, Gaulin SJC, Boster JS, and Kurland JA (1985) Measuring the relationship between dietary quality and body size in primates. Primates 26: 14–27.

Schmidt-Nielsen K (1984) Scaling: Why Is Animal Size So Important? Cambridge: Cambridge Univ. Press.

Schofield C (1985) An annotated bibliography of source material for basal metabolic rate data. Hum. Nutr. (Clin. Nutr.) 39C(Suppl, 1): 42–91.

Schofield WN (1985) Predicting basal metabolic rate: new standards and a review of previous work. Hum. Nutr. (Clin. Nutr.) 39C(Suppl. 1): 5–41.

Schulz LO and Schoeller DA (1994) A compilation of total daily energy expenditures and body weights in healthy adults. Am. J. Clin. Nutr. 60: 676–681.

Scrimshaw NS and Murray EB (1988) The acceptability of milk and milk products in populations with a high prevalence of lactose intolerance. Am. J. Clin. Nutr. 48(Suppl. 1): 1083–1159.

Shephard RJ (1974) Work physiology and activity patterns of circumpolar Eskimo and Ainu: A synthesis of IBP data. Hum Biol 46: 263–294.

Shephard RJ (1991) Body composition in Biological Anthropology. Cambridge: Cambridge Univ. Press.

Shephard RJ and Rode A (1996) Health consequences of 'modernization': evidence from circumpolar populations. Cambridge: Cambridge Univ. Press.

Simoons FJ (1970) Primary adult lactose intolerance and the milking habit: a problem of biologic and cultural interrelations II. The culture historical hypothesis. Am. J. Dig. Dis. 15: 695–710.

Soares MJ, Francis DG, and Shetty PS (1993) Predictive equations for basal metabolic rates of Indian males. Eur. J. Clin. Nutr. 47: 389–394.

Solomons NW, García-Ibañez R, and Viteri FE (1980) Hydrogen breath test of lactose absorption in adults: the application of physiological doses and whole cow milk sources. Am. J. Clin. Nutr. 33: 545–554.

Stunkard AJ, Harris JR, Pedersen NL, and McClearn GE (1990) The body-mass index of twins who have been reared apart. N. Engl. J. Med. 322: 1483–1487.

Sussman RW (1987) Species-specific dietary patterns in primates and human dietary adaptations. In WG Kinzey (ed.): The Evolution of Human Behavior: Primate Models. Albany, N.Y.: SUNY Press, pp. 131–179.

Szathmáry EJE (1986) Diabetes in arctic and subarctic populations undergoing acculturation. Coll. Anthropol. 10: 145–158.

Szathmáry EJE (1990) Diabetes in Amerindian populations: the Dogrib studies. In AC Swedlund and GJ Armelagos (eds.): Disease in Populations in Transition. New York: Bergin and Garvey, pp. 75–103.

Tambs K, Moum T, Eaves L, Neale M, Midthjell K, Lund-Larsen PG, Næss S, and Holmen J (1991) Genetic and environmental contributions to the variance of the body mass index in a Norwegian sample of first- and second-degree relatives. Am. J. Hum. Biol. 3: 257–267.

Taylor CR, Schmidt-Nielsen K, and Raab JL (1970) Scaling of energetic cost to body size in mammals. Am. J. Physiol. 219: 1104–1107.

Taylor CR, Heglund NC, and Maloiy GMO (1982) Energetics and mechanics of terrestrial locomotion. I. Metabolic energy consumption as a function of speed and body size in birds and mammals. J. Exp. Biol. 97: 1–21.

Teleki G (1981) The omnivorous diet and eclectic feeding habits of chimpanzees in Gombe National Park, Tanzania. In RSO Harding and G Teleki (eds.): Omnivorous Primates: Gathering and Hunting in Human Evolution. New York: Columbia Univ. Press, pp. 303–343.

Ulijaszek SJ (1995) Human Energetics in Biological Anthropology. Cambridge: Cambridge Univ. Press.

US Department of Agriculture (USDA) (1986) Nationwide Food Consumption Survey: Continuing Survey of Food Intakes by Individuals. Men 19–50 years, 1 Day, 1985 (Report No. 85-3). Hyattsville, Md.: US Department of Agriculture.

US Department of Agriculture (USDA) (1987) Nationwide Food Consumption Survey: Continuing Survey of Food Intakes by Individuals. Women 19–50 years and their Children 1–5 years, 4 Days, 1985 (Report No. 85-4). Hyattsville, Md.: US Department of Agriculture (USDA).

Vrba ES (1988) Late Pliocene climatic events and hominid evolution. In FE Grine (ed.): Evolutionary History of the "Robust" Australopithecines. Hawthorne, N.Y.: Aldine, pp. 405–426.

Vrba ES (1993) The pulse that produced us. Nat. Hist. 102(5): 47–51.

Weiss KM, Ferrell RE, and Hanis CL (1984) A New World Syndrome of metabolic diseases with a genetic and evolutionary basis. Yearbook Phys. Anthropol. 27: 153–178.

Welle S, Forbes GB, Statt M, Barnard RR, and Amatruda JM (1992) Energy expenditure under free-living conditions in normal-weight and overweight women. Am. J. Clin. Nutr. 55: 14–21.

Werge RW (1979) Potato processing in the central highlands of Peru. Ecol. Food Nutr. 7: 229–234.

West KM (1974) Diabetes in American Indians and other native populations of the New World. Diabetes 23:841-855.

Westrate JA, Dekker J, Stoel M, Begheijn L, Deurenberg P, and Hautvast JGAJ (1990) Resting energy expenditure in women: impact of obesity and body-fat distribution. Metabolism 39: 11–17.

Wilson TW and Grim CE (1991) Biohistory of slavery and blood pressure differences in blacks today: a hypothesis. Hypertension 17 (Suppl. I): I122–I128.

Woodill LM (1996) Growth variation in a rural community of lowland Bolivia (MSc Thesis). Guelph, Ontario: Univ. of Guelph.

Wu Leung W-T (1968) Food composition table for use in Africa. Bethesda, Md.: US Department of Health, Education, and Welfare.

Wyndham CH (1966) Southern African ethnic adaptation to temperature and exercise. In PT Baker and JS Weiner (eds.): The Biology of Human Adaptability. Oxford: Clarendon Press.

Young TK (1993) Diabetes mellitus among native Americans in Canada and the United States: an epidemiological review. Am. J. Hum. Biol. 5: 399–413.

Zimmet P, Taft P, Guinea A, Guthrie W, and Thoma K (1977) The high prevalence of diabetes mellitus on a central Pacific island. Diabetologia 13: 111–115.

Zimmet P, Kirk R, Serjeantson S, Whitehouse S, and Taylor R (1982) Diabetes in Pacific populations: genetic and environmental interactions. In JS Melish, J Hanna, and S Baba (eds.): Genetic Environmental Interactions in Diabetes Mellitus. Amsterdam: Excerpta Medica, pp. 9–17.

CHAPTER 10

Work and Energetics

STANLEY J. ULIJASZEK

INTRODUCTION

The term "work" is commonly used to describe all economically productive activities. However, in traditional societies, physical work is the way in which populations service their need for food—often transforming their environment in the process. For the purposes of this chapter, we define work as physical exertion with the goal of generating subsistence. It excludes sedentary categories of work, such as office work. Although work can be estimated and quantified in many ways, it is fundamentally energetic. Work is measured as the amount of energy (in **kilocalories** [kcal] or **megajoules** [MJ; 1 MJ $\approx$ 240 kcal]) expended in the course of performing a task or a series of tasks.

In this chapter, we describe the principles of energetics and the human biological components of energy expenditure at different stages of the life cycle. We consider the factors that influence variations in energy expenditure within and between populations and discuss the ways in which energetics are associated with subsistence food production. We also give some attention to the use of energetics in the study of evolutionary ecology.

Patterns of energy production and use in a community or population are a function of the type and extent of local energy acquisition, exchange relationships with other groups, and other activities that are important for the maintenance of group function. These activities include individual biological as well as group economic maintenance and reproduction (Nydon and Thomas 1989) and are summarized in Figure 10.1.

Estimates of energy intake, expenditure, cost of activity, and the balance between intake and expenditure (in respect to gain and/or loss of body mass) have been used in attempts to understand human subsistence within the adaptation and adaptability framework (see Chapter 1). Research has focused on

- how the need for dietary energy and the ways in which it is obtained affect different aspects of human population biology or ecology (Thomas et al. 1989);

Human Biology: An Evolutionary and Biocultural Perspective, Edited by Sara Stinson, Barry Bogin, Rebecca Huss-Ashmore, and Dennis O'Rourke
ISBN 0-471-13746-4 Copyright © 2000 Wiley-Liss, Inc.

- the implications of different subsistence and foraging strategies for fertility and biological **fitness** (Hill and Kaplan 1988);
- human responses and adaptations to seasonal energetic stresses (Dugdale and Payne 1986, Ferro-Luzzi et al. 1990);
- the energetics of different physiological states, including undernutrition, obesity, pregnancy, and lactation (Shetty 1993; Durnin 1988);
- the ecological correlates and functional consequences of small body size (Spurr 1988) (see Chapter 12);
- ovarian function and fecundability (Ellison 1991) (see Chapter 15);
- levels of work effort and output in different types of subsistence system (Pasquet and Koppert 1993).

CONCEPTS IN ENERGETICS

In any community, dietary energy generated through various processes involving energy expenditure can be divided and consumed in a variety of ways. Division and consumption, in turn, influence the pattern of energy expenditure, storage, and mobilization of bodily energy stores. The relationship among intake, expenditure, and storage within a group or community is usually complex but can be studied by using appropriate energy balance and energy accounting methods (Ulijaszek 1992). Furthermore, bodily maintenance, locomotion, work and physical activity, pregnancy, lactation, and physical growth can be measured by using energetics methods. Underpinning these methods is thermodynamic theory.

Thermodynamics

Of the laws of thermodynamics, the law of energy conservation is the one that underpins bioenergetics as a discipline. The law states that energy can be neither gained nor lost but may be transformed in nature. In living systems, transformation includes the transfer of chemical energy into heat, work, and electrical energy. During growth and development, chemical energy in the form of food is used to support the creation of morphological structures. The energy balance equation is simply written

$$\text{Energy balance} = \text{Energy in} - \text{Energy out}$$

If output exceeds input, energy balance is negative, and body mass will be mobilized to meet energy needs. If prolonged, negative energy balance will result in weight loss, the composition of which will depend on body composition at the onset of imbalance: someone with greater body fatness will lose more fat than muscle, whereas someone with lower fatness will lose more muscle than fat. Figure 10.2 illustrates the relationship between the proportion of body weight lost as muscle at dif-

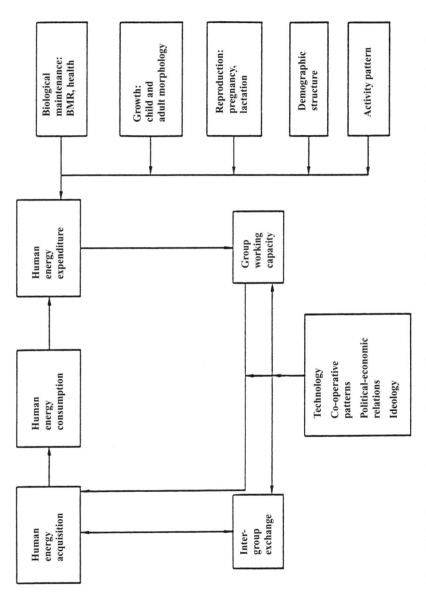

Figure 10.1 Patterns of energy production and use at community and individual levels. Balance between intake and expenditure depends on many linked factors, including population size, distribution and pyramid, organization of work, and extent to which the group trades with others. (*Source:* Nydon and Thomas 1989)

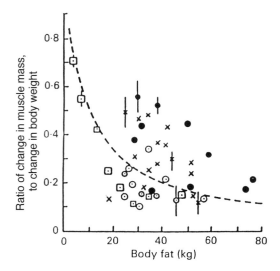

Figure 10.2 Change in muscle mass during weight loss as a function of total body fat. Underfeeding experiments were at least 4 weeks long. Individuals with higher body fat lose more fat than muscle during starvation. (*Source*: Prentice et al. 1991, after Forbes 1989)

ferent levels of fat stores (Forbes 1989; Prentice and Prentice, 1995). In general, individuals in industrialized countries lose a greater proportion of body weight as fat than do subjects in the rural developing countries because they start with a higher ratio of body fat to body weight.

Metabolic Interchangeability of Macronutrients

The existence of intermediary metabolism enables humans and other animal species to use carbohydrate, fat, and protein interchangeably as dietary energy (Figure 10.3). This biochemical system allows common chemical intermediates to be formed in the body. It can be used for building up bodily tissues and structures; creating the chemical energy needed to drive this process, which also can be used for muscular work; or depositing fat and **glycogen** as energy stores.

Human diets vary in their relative composition of carbohydrate, fat, and **protein**. In general, however, carbohydrate forms the largest component of intake in all societies, except in some groups of strict hunter-gatherers. If carbohydrate intake exceeds requirements for immediate energy expenditure, then the surplus is stored as fat. This is also the case for protein intake if intake of protein and energy is in excess of maintenance requirements. The **tricarboxylic cycle** (or Krebs cycle) links pathways of

- **glycolysis**, which involves the breakdown of glucose, the common intermediary of carbohydrate metabolism;

- the breakdown and synthesis of **fatty acids**, which are the common intermediaries of fat metabolism; and
- the interconversion of many of the **amino acids**, which are the constituent units of protein. This process includes making the amino acids available for use as energy when excess protein has been eaten.

In total, intermediary metabolism allows the energetic interchangeability of carbohydrate, fat, and protein and consequently, dietary flexibility.

ENERGY EXPENDITURE

Maintenance Metabolism

Human energy expenditure can be broken down into several components, which include energy cost of maintenance, **thermic effect of food**, physical activity, **thermoregulation**, growth, and reproduction. These components vary between individuals and populations. Bodily maintenance includes all functions that preserve bodily integrity, including cardiac activity, and the maintenance of cell function and body temperature (Ulijaszek 1992). An approximation of this function can be obtained by measuring **basal metabolic rate** (BMR), or the **resting metabolic rate** (RMR). The energy cost of digestion, absorption, transport, metabolism, and storage of ingested food is called **dietary thermogenesis** or the thermic effect of food (TEF) and accounts for approximately 10% of energy intake (Horton 1983). Energy is expended in physical activity by all individuals at all times, but particularly for growth in infants and children, and for pregnancy and childbirth in most women during adult-

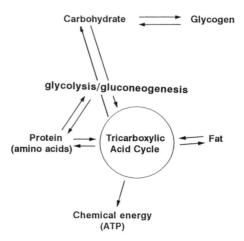

Figure 10.3   Intermediary metabolism of macronutrients. The biochemical interconvertibility of carbohydrate, fat, and protein is shown. ATP, adenosine triphosphate.

hood. The metabolic rate is known to fluctuate in women as a function of the **menstrual cycle** (Solomon et al. 1982, Meijer et al. 1992); for example, the sleeping metabolic rate during the post-**ovulation** phase is, on average, 7–8% greater than before ovulation.

Growth

Like all physiological processes, growth and development has an energetic cost. However, it is high only in the earliest stages of postnatal life. At 1 month of age, the energy cost of growth is about one-third of total energy expenditure, declining to about 4% by 1 year, and by 2 years, the energy cost of growth is a mere 1% of total energy expenditure (Bergmann and Bergmann 1986).

In the age groups during which the growth of children living under disadvantaged circumstances begins to slow down relative to the growth of children living in advantaged circumstances (i.e., 6–12 months), the energy available for physical activity is many times greater than that available for growth. The incremental energetic cost of sustaining a level of growth similar to that recommended by the U.S. National Center for Health Statistics (1977) references is a mere 2–3% of total energy expenditure in addition to that already committed to growth. Therefore, simple energy deficiency does not explain **growth faltering** among children in the developing world.

One possible explanation is that a shortage of dietary factors other than energy causes growth faltering. Deficiencies of protein, zinc (Golden 1988; Prentice and Bates 1994), calcium (Prentice and Bates 1994), and sulfur (Golden 1994) have been suggested as contributing factors to growth faltering in children at ages where their requirements for those nutrients are high. Another possibility is that repeated infections reduce appetite and absorption of nutrients in the gut, thus creating a shortage of dietary energy and other nutrients secondary to infection. Some infections actually increase the requirement for dietary energy, especially those that involve fever.

Reproduction

Weight loss has been associated with changes in reproductive function in women living in seasonal environments in Zaire (Bailey et al. 1992) and Nepal (Panter-Brick et al. 1993) (see Chapter 15). This occurrence is similar to the loss of menstrual function observed in dieters and may be independent of plane of nutrition. The mechanism that facilitates this change in function is still uncertain.

However, Ulijaszek (1996) suggested a possible mechanism that links two series of observations: the effects of reduced dietary energy intake on the **sympathetic nervous system** (SNS), which regulates basal energy expenditure; and the effects of reduced activity on the regulatory system for ovarian function, the hypothalamic **gonadotrophin release hormone** (GnRH) pulse generator. Reduced **norephinephrine** output as a consequence of depressed SNS activity may indirectly reduce the activity of the GnRH pulse generator (Ulijaszek 1996). Associated with reduced GnRH pulsatility are reduced output of **luteinizing hormone** and **follicle-stimulating hormone** and a concomitant reduction in ovulatory function (Rosetta 1990).

Work

Physical working capacity is a fundamental determinant of human survival in traditional societies (Shephard 1978). Sustained physical activity implies the steady **aerobic** replacement of the high-energy compounds **adenosine triphosphate** and **creatine phosphate**, which are found in skeletal muscle, because they are used in muscular contraction. This process usually is limited by the rate of oxygen transport from the atmosphere to the working tissues (Shephard 1974). The most important determinant of endurance fitness is the **maximum oxygen uptake** or **aerobic power,** the amount of oxygen the body can extract. **Physical work capacity** (PWC), or maximum oxygen consumption (VO_2**max**), is a measure of this (see Box 10.1). The upper limit of work that can be tolerated over an 8-hour period is equivalent to an intensity of 35–40% of an individual's VO_2max (Barac-Nieto 1987).

Many studies have been carried out under a variety of conditions to observe the variation in PWC under different ecological conditions. Between-population differences in PWC cannot be adequately explained by genetic factors (Shephard 1980). Although the absolute work performance of undernourished adults is lower than that of their better nourished peers, this difference can be almost totally accounted for by differences in body size and muscle mass.

VARIATION IN ENERGY EXPENDITURE

Cross-Cultural

Total daily energy expenditure (TDEE) of adults varies greatly between populations. Table 10.1 shows selected mean total daily energy expenditures of adult males in developing and industrialized countries. The ratio of highest to lowest mean values is 1.55 for hunter-gatherers and hunter-horticulturalists, 1.7 for agriculturalists, and 2.2 for groups engaged in paid labor. The lowest mean values for all three categories is similar; the greater ratio reflects a trend toward greater total daily energy expenditure across the categories.

The ratio of highest to lowest expenditure in industrialized countries is 3.5, much higher than for any of the developing country categories. However, when the highest energy expenditure categories of military and polar expedition members are removed, as well as measurements made prior to 1970, the ratio comes down to 2. This value is very close to the occupational category differences of developing countries. It illustrates the extent to which energy expenditures have declined in the industrialized nations as a result of the automation of physical tasks at work and the widespread availability of motorized transport.

Mean TDEEs also are slightly lower in industrialized nations than in developing ones. These differences are largely attributable to differences in activity levels. Differences in body size could mask the true extent of these differences, because industrialized populations with larger mean body mass also have greater maintenance energy expenditures. Within populations, the range of TDEEs is as large as between-population variation. It can be attributed to differences in body size, activity levels, and BMR.

Table 10.1   **Mean daily energy expenditures of some groups of adult males**

A. Traditional subsistence

Group or district	Country	Total energy expenditure (MJ/day)	Reference
Hunter-gatherers, fishers, and hunter-horticulturalists			
Eskimos	Canada	15.36	Godin and Shephard 1973
Machiguenga	Peru	13.40	Montgomery and Johnson 1974
Tukanoans	Colombia	12.05	Dufour 1992
Keto	Russia	11.42	Katzmarzyk et al. 1994
Yassa	Cameroon	11.41	Pasquet and Koppert 1993
Mvae	Cameroon	9.90	Pasquet and Koppert 1993
Pastoralists and agropastoralists			
Evenk	Russia	11.91	Katzmarzyk et al. 1994
Turkana	Kenya	9.04	Galvin 1985
Nuñoa	Peru	8.36	Leonard 1988
Nuñoa	Peru	8.31	Thomas 1976
Agriculturalists			
	Guatemala	15.50	Viteri et al. 1971
	Philippines	13.80	de Guzman et al. 1974a
Varanin	Iran	13.57[a]	Brun et al. 1981
Gurung	Nepal	12.80[b]	Strickland et al. 1997
	Myanmar	12.55[b]	Tin-May-Than and Ba-Aye 1985
	Burkina Faso	12.30[c]	Brun et al. 1981
Savaii	Western Samoa	12.16	Pearson 1990
non-Gurung	Nepal	11.30	Strickland et al. 1997
Lufa	Papua New Guinea	10.75	Norgan et al. 1974
Tamil Nadu	India	10.60[a]	McNeill et al. 1988
Genieri	Gambia	9.83[c]	Fox 1953
Kaul	Papua New Guinea	9.82	Norgan et al. 1974
Pari	Papua New Guinea	9.33	Hipsley and Kirk 1962
Kaporaka	Papua New Guinea	9.11	Hipsley and Kirk 1962

B. Paid labor

Country	Occupation	Total energy expenditure (MJ/day)	Reference
India	Rickshaw pullers	20.42	Bannerjee 1962
	Cottonmill workers	12.76	Bannerjee et al. 1959b
	Stone cutters	12.66	Ramanamurthy et al. 1962
	Laborers	12.59	Devadas et al. 1975
	Laboratory technicians	8.66	Bannerjee et al. 1959a
Philippines	Shoemakers	11.30	de Guzman et al. 1974b
	Jeepney drivers	10.40	de Guzman et al. 1974c
	Textile mill work	10.40	de Guzman et al. 1979
	Clerk-typists	9.20	de Guzman et al. 1978

(continued)

Table 10.1   *(Continued)*

C. Industrialized nations

Country	Group or Occupation	Total energy expenditure (MJ/day)	Reference
Germany	Forestry workers	27.41	Kaminsky 1953
Sweden	Forestry workers	23.84	Lundgren 1946
Holland	Forestry workers	23.00	Streef et al. 1959
UK	Polar expedition	20.08	Masterton et al. 1957
UK	Polar expedition	16.00	Kinloch 1959
UK	Polar expedition	15.77	Hampton 1960
UK	Polar expedition	15.48	Rogers et al. 1964
Japan	Forestry workers	15.48	Kusunoki 1956
UK	Forestry workers	15.36	Durnin and Passmore 1967
UK	Coal miners	15.31	Garry et al. 1955
UK	Farm workers	14.85	Durnin and Passmore 1967
Switzerland	Elderly peasants	14.77	Durnin and Passmore 1967
UK	Students	14.64	Passmore et al. 1952
UK		14.62	Livingstone et al. 1990
UK	Military cadets	14.39	Edholm et al. 1955
UK	Military cadets	14.31	Widdowson et al. 1954
Italy	Shipyard workers:		Norgan and Ferro-Luzzi, 1978
	Heavy workload	13.84	
	Moderate workload	13.13	
	Sedentary	11.95	
UK	Steel workers	13.72	Durnin and Passmore 1967
	Building workers	12.55	
	Students	12.26	
	Laboratory technicians	11.88	
	Colliery clerks	11.72	
UK	Students	11.80	Norgan and Durnin 1980
UK	Office clerks	11.72	Garry et al. 1955
	Office workers	10.54	Durnin and Passmore 1967
UK	Elderly	9.75	Durnin and Passmore 1967
USA	Young	9.01	Vaughan et al. 1991
	Elderly	7.81	Vaughan et al. 1991

[a] Average of four seasons. [b] Average of three seasons. [c] Average of two seasons

Genetics

The extent to which genetic factors contribute to variation in energy expenditure within populations is not clear. The evidence for genetic adaptation in energy expenditure is limited, and consists of parent-offspring relationships in energy expenditure and twin studies of BMR and other components of total energy expenditure.

On the basis of twin studies, Bouchard et al. (1991) concluded that the **genotype** accounts for a significant fraction of the individual differences in resting metabolic rate, TEF, and energy cost of exercise—even after age, sex, body mass, and composition are taken into account. They also concluded that body energy gains during overfeeding vary among individuals, and variations in response are probably genotype-dependent.

Sex and Age

Table 10.2 compares mean daily energy expenditure of males and females as multiples of BMR. Expressing data in this way controls for differences in body size and, where BMR has been measured as opposed to estimated, for metabolic variation. This comparison, often called the **physical activity level** (PAL) (James et al. 1988), is a reasonably standardized measure of daily exertion.

Population values vary enormously, and no evidence indicates that women work harder than men in the developing countries or that agriculturalists work harder than hunter-gatherers or hunter-horticulturalists. Indeed, of 15 populations where male–female comparisons of PAL have been made, women work harder than men in only 2: the Mvae millet cultivators of Cameroon and one group of New Guinea highlanders. Men work harder than women in 6 of the 15 populations, and no real difference was seen in 7 populations.

However, time budget data show male–female differences in the division of labor in many populations, including agriculturalists in Africa (Bleiberg et al. 1980; Brun et al. 1981; Pasquet and Koppert 1993), even where the PAL is not significantly different. For example, in the rainy season in Burkina Faso, both men and women work 7 hours per day. Men spend all of this time farming, whereas women spend just more than 2 of the 7 hours doing household tasks. Among sago-gatherers in Papua New Guinea, women work nearly 10 times longer than men in processing the palm sago (Ulijaszek and Poraituk 1993).

In industrialized societies, energy expenditure often declines with increasing age. Although the evidence is limited, the lack of hard-and-fast rules about retirement from economic activity in the rural developing world probably lead to a more limited reduction in physical activity, more closely related to changes in physical ability and physique in later life.

Seasonality

Seasonality of climate and subsistence productivity can lead to seasonal variation in energy balance and influence reproductive performance, child growth, nutritional status, and within-household food allocation. In most communities in which climatic seasonality might affect human biology in some stressful way, economic or sociocultural strategies are used to reduce the impact of seasonality. In Northern Nigeria, for example, different types of subsistence crops are intercropped to buffer against the uncertain onset of the rainy season, which may affect crop yields (Watts 1988). Strategies for coping with seasonal food shortages may be complex and graded, such that short-term dietary stress is dealt with by adopting so-called famine foods, borrowing grain from kin, or migrating short-term and selling labor power. Long-term stress may involve the sale of domestic assets, pledging or sale of farmland (Watts

Table 10.2   Physical activity levels for adults engaged in a variety of subsistence practices

Group or district	Country	Subsistence type	PAL (TDEE/BMR)	Reference
Males				
	Guatemala	Maize cultivation	2.32	Viteri et al. 1971
	Philippines	Rice cultivation	2.25	de Guzman et al. 1974a
Varanin	Iran	Wheat cultivation	2.10[a]	Brun et al. 1979
Machiguenga	Peru	Hunter- horticulturalist	2.09[b]	Montgomery and Johnson 1974
Aché	Paraguay	Hunter-gatherer	2.08[c]	Hill et al. 1985
Gurung	Nepal	Rice, maize and millet cultivation	2.05[d]	Strickland et al. 1997
	Myanmar	Rice cultivation	2.02[b]	Tin-May-Than and Ba-Aye 1985
	Gambia	Rice and peanut cultivation	2.02[b]	Fox 1953
Tamil Nadu	India	Rice cultivation	1.96[a]	McNeill et al. 1988
Sundanese	Indonesia	Rice cultivation	1.96[c]	Suzuki 1988
Non-Gurung	Nepal	Rice, maize, and millet cultivation	1.91[d]	Strickland et al. 1997
	Burkina Faso	Millet cultivation	1.89[b]	Brun et al. 1981
Igloolik	Canada	Hunters	1.82	Shephard 1974
Ivenk	Russia	Pastoralism	1.78[c]	Katzmarzyk et al. 1994
Keto	Russia	Fishing	1.72[c]	Katzmarzyk et al. 1994
Yassa	Cameroon	Millet cultivation, fishing	1.71	Pasquet and Koppert 1993
	Ivory Coast	Mixed cultivation	1.68[c]	Dasgupta 1977
Wopkaimin	Papua New Guinea	Hunter horticulturalist	1.65[b,e]	Ulijaszek and Brown, unpublished results
Lufa	Papua New Guinea	Sweet potato cultivation	1.64[c]	Norgan et al. 1974
	Uganda	Maize and plantain cultivation	1.63[c]	Cleave 1970
Savaii	Western Samoa	Cultivation	1.60[c]	Pearson 1990
Mvae	Cameroon	Millet cultivation, hunting	1.60	Pasquet and Koppert 1993
	India	Rice cultivation	1.56	Edmundson and Edmundson 1988
Kaul	Papua New Guinea	Taro and plantain cultivation	1.52[c]	Norgan et al. 1974

(continued)

Table 10.2 *(Continued)*

Group or district	Country	Subsistence type	PAL (TDEE/BMR)	Reference
Pari	Papua New Guinea	Hunter-horticulturalist	1.42	Hipsley and Kirk 1962
Nuñoa	Peru	Agro-pastoralist	1.32	Thomas 1976
Turkana	Kenya	Pastoralists	1.29	Galvin 1985
Females				
Tamang	Nepal	Rice, maize, millet, and wheat cultivation	1.82[a]	Panter-Brick 1993a
Lufa	Papua New Guinea	Sweet potato cultivation	1.82	Norgan et al. 1974
	Burkina Faso	Millet cultivation	1.80	Brun et al. 1981
Igloolik	Canada	Hunting	1.79	Shephard 1974
Machiguenga	Peru	Hunter-horticulturalist	1.72	Montgomery and Johnson 1974
Mvae	Cameroon	Millet cultivation, hunting	1.72	Pasquet and Koppert 1993
Tukanoa	Colombia	Horticulture	1.71	Dufour 1984
	India	Rice cultivation	1.69[a]	McNeill et al. 1988
Yassa	Cameroon	Millet cultivation, fishing	1.67	Pasquet and Koppert 1993
Gurung	Nepal	Rice, maize and millet cultivation	1.67[d]	Strickland et al. 1997
Iven	Russia	Pastoralism	1.62[c]	Katzmarzyk et al. 1994
	Guatemala	Maize cultivation	1.62	Schutz et al. 1980
Savaii	Western Samoa	Cultivation	1.60[c]	Pearson 1990
Kaul	Papua New Guinea	Taro and plantain cultivation	1.57	Norgan et al. 1974
Non Gurung	Nepal	Rice, maize, and millet cultivation	1.56[d]	Strickland et al. 1997
Keto	Russia	Fishing	1.55[c]	Katzmarzyk et al. 1994
	India	Rice cultivation	1.53	Edmundson and Edmundson 1988
	Ethiopia	Cultivation	1.47	Ferro-Luzzi et al. 1990
Turkana	Kenya	Pastoralists	1.37	Galvin 1985
Swazi	Swaziland	Cultivation	1.35	Huss-Ashmore et al. 1989

[a]From estimates of BMR from body weight, and total energy expenditure from activity diariesm. [b]Average of two seasons. [c]Estimate of total energy expenditure from activity diaries. [d]Average of four seasons. [e]Average of three seasons

1988). However, where the potential for economic or sociocultural strategies is limited, physiological and behavioral strategies may be brought into play.

Among the most common biological phenomenon observed is seasonality of weight change. Ferro-Luzzi and Branca (1993) summarized data on seasonal weight loss in several developing country populations and concluded that loss rarely exceeds 5% of the seasonal high in adults; women usually lose less weight than men, possibly because men are subjected to greater seasonal stress than women.

Furthermore, less well nourished people resist mobilization of body energy stores when they are low, so as to safeguard against the depletion of muscle mass. The physiological mechanisms that underly this phenomenon must include the down-regulation of basal metabolism and increased efficiency of physical work of people who are less well nourished. These mechanisms result in lower energy expenditure of bodily maintenance and of physical activity, respectively.

However, weight loss is not necessarily greatest in areas of highest climatic seasonality (Ulijaszek and Strickland 1993). This fact underlines the importance of cultural and behavioral factors in buffering against seasonality. The lower seasonal weight loss in women than in men may be related to the greater importance of women to reproductive success. Reduced body mass may carry less functional significance in males than females, especially if most subsistence tasks are well within the productive capacity of males. However, weight loss in females carries with it the possibility of reduced reproductive potential, through **anovulation** (Ellison 1991).

It is likely that small reductions in body weight are sufficient to induce metabolic responses to negative energy balance, whereas large changes may also affect work capacity and performance. Even if small, seasonal imbalances between intake and expenditure may be important. For example, between the end of April and the end of September, male Gambian farmers lose on average 2.4 kilograms (kg) (5.3 pounds [lb]) in body weight (Fox 1953). This loss is equivalent to a mean weight loss of 16 grams/day (0.5 ounces/day). At a lean-to-fat mass-change ratio of 1, this loss equates to a negative energy balance of 435 kJ/day (~100 kcal/day), which is ~5% of daily energy intake. Therefore, such weight loss could be avoided by spending only 20 fewer minutes per day (5% of a 6-hour working day) performing hard physical work. That the farmers do not make such a small adjustment in reducing their physical activity to accommodate the negative energy balance suggests that they are unable to, presumably because doing so might threaten overall agricultural productivity.

Across several studies, weight loss—even if small—is the only universally observed response to seasonal energy imbalance. Changes in BMR appear to follow intake changes more closely than changes in weight. BMR often declines simultaneously with, or even before, weight loss. This observation ties in with what is known about thyroid hormone responses to underfeeding and their lowering effects on basal metabolism (Dauncey 1990). Physical activity, potentially the most powerful mechanism of saving energy, is usually maintained at the expense of weight loss. This mechanism contradicts a suggestion by Ferro-Luzzi (1990): behaviorally mediated reduction in physical activity should minimize the extent of weight loss in third-world communities in the face of restricted food intake. Economically important activities are likely to be curtailed only as a last resort, after all other responses have failed to accommodate the energy imbalance.

DIVISION OF LABOR

Work organization varies with the type of subsistence practice adopted. In turn, it affects both long-term and short-term work group organization. Long-term organization includes the formation of a community, band, or village in which the sharing of dietary energy between households is in long-term, but not necessarily short-term, balance. Short-term organization includes the formation of work groups to perform specific tasks that may vary from day to day, or may be seasonal in nature.

Social relations are also important when considering group work performance. For example, kin-group fission and fusion as well as cycles of exchange, ritual, and production may be central to group function. They may involve cooperative behavior that

---

BOX 10.1   MEASURING PHYSICAL WORK CAPACITY

The measurement of oxygen consumption associated with work has been extensively and critically reviewed (Astrand and Rodahl 1986; Barac-Neito 1987). It is assumed that heart rate has a well-defined relationship to work and oxygen consumption. The aim of any test in which maximum aerobic power (VO_2max) is measured is to determine the capacity of an individual's body to perform aerobic work. This measurement can be done either

- directly, by exercising a subject to exhaustion by a stepwise protocol involving the use of a bicycle ergometer, treadmill, or standard step test and measuring oxygen consumption when close to exhaustion or
- indirectly, by predicting maximal oxygen consumption by extrapolating the relationship between heart rate and oxygen to an hypothetical, or population maximal heart rate.

Although it is preferable to obtain a direct measure of physical work capacity (PWC, or VO_2max), it is rarely obtained among rural subsistence-oriented populations. Tests to exhaustion may alienate subjects from taking part in further study, or subjects may be unwilling to push themselves to the limit. Furthermore, tests to exhaustion are potentially dangerous—particularly for individuals with undiagnosed heart conditions and for older adults.

When matched with the energy cost of performing important subsistence tasks, PWC can be used to determine the extent to which individuals within a group might be physically stressed by hard work. Such quantification has allowed human biologists to study subsistence ecology in new and different ways (Spurr 1988). **Performance effort** is the proportion of maximum aerobic power expended in performing any given subsistence task. The greater the proportion of an individual's VO_2max, the greater the performance effort required. When used in association with activity diaries across the year, this measure provides a quantitative way of determining whether work in subsistence is more stressful at one time of year than another. It can also be used to determine whether women work harder than men.

BOX 10.2 MEASURING ENERGY INTAKE

Numerous techniques available for measuring energy intake (Margetts and Nelson 1997). Regardless of method used, data must give estimates of weights of different foods eaten, either in households or by individuals. These weights must then be converted into energy values by using food composition tables, acknowledging the problems attendant with their use. More detailed examination, beyond the level of the group or the household, requires energy intake data to be disaggregated by sex and age group. Furthermore, intakes may be compared with international dietary references to the presence or absence of notable dietary energy deficiency, from either observations of clinical pathologies or **anthropometric** measures (e.g., **weight for height** or age, **skinfold** thickness) (Jelliffe and Jelliffe 1989; Gibson 1990).

Although the corroboration of food intake with other lines of evidence for nutritional stress might seem appropriate, this approach presents significant problems: estimates of energy intake are associated with much greater error than measures of energy expenditure and may be prone to underestimation (Ulijaszek 1992), and no ideal reference for nutritional assessment from dietary data exists.

In any study, a balance must be sought between the desires to maximize accuracy and to keep the duration of study to a minimum. Operating within these constraints, researchers should use equations available for the estimation of the reliability of energy intake (Cole 1991) to arrive at their own estimates of reliability, having decided on the number of days of observation per subject, and the sample size possible after collecting the data.

In estimating the energy intake of a group of people, it is important to use a sampling period that is representative of long-term habitual intake or of intake in a particular season. The weighed dietary record has in the past been regarded as the reference method of dietary methodology (Bingham 1991). In this method, the weight of food consumed is determined by weighing each food before it is eaten and subtracting from it the weight of any uneaten food. Only recently has the validity of the weighed dietary record method for assessing the habitual diet of free-living individuals been questioned (Ulijaszek 1992; Huss-Ashmore 1996).

---

runs along lines of kinship or alliance. In addition, subsistence production may require cooperative behavior that, at its simplest, involves a husband and wife team—sometimes with the support of children, according to their abilities. More complicated cooperative behavior includes the formation of male hunting parties and female field labor groups. Cooperative work may permit greater efficiency in the quest for subsistence. In cooperative hunting, larger animals may be stalked or the unpredictability of hunting reduced for the individuals taking part; in cooperative labor, tasks which may have a limited window of opportunity may be completed more quickly.

Panter-Brick (1993) considered the characteristics of different labor strategies: the process of decision making between different labor strategies in Nepal. Of five categories of labor identified (household, free assistance, festive, reciprocal exchange, and hired), all except hired labor are based on kinship or alliances. In general, family labor is preferred above all others (Chibnick and de Jong 1989) because it ensures that the work done is of high quality.

In many traditional societies, a clear division separates tasks that are carried out by women and those that are carried out by men. For example, among the Machiguenga in Peru, men spend a significantly greater proportion of their day engaged in garden labor and collecting wild foods than do women, whereas women spend much more time in food preparation and child rearing (Johnson 1975). Similar divisions are seen in rural Mexico, where men spend much more time in economic activities and women spend more time in household activities (Erasmus 1955). Furthermore, child participation in economic activities has been demonstrated for several societies (Strickland 1990), and where significant, child labor has been argued to promote high fertility.

MODELING

Even though models are simplifications of reality, they have been useful in the study of complex subsistence work and output relationships. Most models use optimization procedures in some sort of cost–benefit analysis. Such procedures have been used in the study of hunter-gatherer societies according to the postulates of synthetic evolutionary theory. Pianka (1988) stated that "natural selection and competition are inevitable outgrowths of heritable reproduction in a finite environment." Thus, direct competition for resources within and between groups gives advantages to subgroups or groups that have efficient techniques of acquiring energy and nutrients. These advantages can be turned into offspring or used to avoid predators. In this way, the more efficient individuals pass on more copies of their **genes**, and relative fitness increases for the group as a whole. Energy has become the nutrient of choice in such studies, because the drive for dietary energy is the primary one in most societies; the meeting of human energy needs is the most fundamental in nutritional terms.

Many foraging strategy **models** are based on geometric representations of the relationships of foragers to resources, and solutions are found by usings graphs or differential calculus (Cody 1974). After the currency has been chosen, an appropriate cost–benefit function must be adopted and the function solved for an optimum. It is usually assumed that fitness varies directly with the rate of net energy capture that can be achieved while foraging (Pyke et al. 1977). Although many factors can intervene to influence this relationship (Schoener 1971), in general, it holds true.

Such optimization modeling has been useful in the study of foraging strategies from an evolutionary perspective. For example, Smith (1991) examined the foraging strategies of Inuit hunters in Canada and found that traditional hunting took advantage of proven methods that give the best returns. However, in a modernizing world, Smith's modeling also showed foraging to be less profitable than alternative sources of livelihood.

Different models allow different aspects of foraging strategy to be examined, and Table 10.3 lists the optimality assumptions possible for a range of models that could be energy limited. For example, application of the patch resource choice model

Table 10.3 Optimal foraging models; cost–benefit criteria, and major constraining variables

Category	Strategic goal	Domain of choice	Cost–benefit criteria	Major constraints
Diet breadth	Optimal set of resource types to harvest	Which types to harvest	Return per unit handling time for each type; overall foraging time (including search time)	Search-and-pursuit abilities of forager, encounter rates with high-ranked types
Patch choice	Optimal array of habitats to exploit	Which set of patches to visit	Average rate of return with patch types and average overall patches (including travel time between patches)	Efficiency ranking of patch types, habitat richness, travel time between patches.
Time allocation	Optimal pattern of time allocated to alternatives	Time spent foraging in each alternative	Marginal return rate for each alternative, average return rate for entire set	Resource richness, depletion rates for each alternative
Foraging group size	Formation of optimal-sized groups for foraging	Size of groups to join for foraging under specified conditions	Average per capita return rate at each group size, marginal cost and gain to joiner or group members	Return-rate curves for each group size under each condition possibilities for group formation, rules governing division of harvest
Settlement pattern	Optimal location of home base for foraging efficiency	Settlement location of each foraging unit (individual)	Mean travel costs and/or search costs per unit harvest	Spatiotemporal dispersion and predictability of effects of cooperation and competition

(*Source:* Modified from Smith 1983)

allows examination of the efficiency ranking of various resource patches used, whereas application of the time allocation model allows the use of these patches to be related to their depletion rates. Thus, what may appear to be illogical by the application of one model may appear perfectly logical when the other is applied. If several bands or groups are compared in this way, it is possible to examine the different rationales behind the various foraging strategies seen in real life.

In the contemporary world, energy economic relationships can reveal aspects of exchange that could be of adaptive significance in communities that rely minimally on cash exchange or in parts of the system that cash cannot represent. For example, a taller and heavier man may be physically fitter, work harder, and produce more food for his family than a man who is shorter and lighter. Consequently, the taller man's family may be bigger and suffer less disease than the smaller man's family. Furthermore, the wife of the taller man may have larger babies, bear more viable offspring, and breastfeed more successfully. Such relationships could not be revealed if money were the currency of exchange being used (Ulijaszek 1995a).

The most common type of modeling has been descriptive, a convenient way of summarizing energy relationships in a community and determining the likelihood of energetic stress. Most aspects of human activity can be included in a descriptive model, as long as an energy value can be obtained for it. The relative importance of different components will vary from group to group, and some components prominent in one society may be completely absent in another. For example, the Tsembaga Maring of New Guinea rely heavily on the mechanical energy of human labor and work an average of 9.5 hours per week in food-producing activities. The greatest work input is planting and weeding sweet potato, taro, and yam gardens and herding pigs. Overall, the Tsembaga Maring generate a small surplus of food energy with relatively small work inputs (Bayliss-Smith 1982). Descriptive modeling has shown that crop production subsidizes pig keeping and that hunting and gathering as practiced in the 1960s could not sustain this population.

In contrast, the Quechua Indians who live at high altitudes in the Peruvian Andes practice agro-pastoralism. The vast majority of their human subsistence work is herding animals. However, herding and agriculture are related, because the impoverished soils of the Peruvian Altiplano require plentiful fertilizer for adequate and sustained agriculture. This fertilizer comes in the form of dung from the sheep, alpaca, and llamas. Thus, herded animals consume scrubland vegetation on marginal land and excrete the fertilizer that allows sustained cultivation.

Trade is of fundamental importance to this group of Quechua. Although it might seem that crop production should sustain this group even in the absence of other subsistence activities, it would only supply ~70% of possible energy returns if herding were also to be practiced. This return would be low even though the energetic returns of herding are already extremely low. However, animal products command a respectable price on the open market, and for every kilocalorie or megajoule of meat, wool, or animal skin sold, about five kilocalories or megajoules of wheat or maize flour can be bought. Without trade, the existing subsistence system could not sustain the population.

Further examination of the energetics of the Quechua subsistence system shows that the time spent herding exceeds the possible time budgets of adults alone and that the size of animal herds used by the Quechua to obtain animal products for trade could be maintained only with the extensive use of child labor. Indeed, the Quechua practice a high-fertility strategy in which large numbers of children are involved in the subsistence economy. Although many children die before reaching adulthood, on becoming adults, many of them migrate in search of work in the towns and cities at lower altitudes. Thus, the children provide a source of labor that does not increase the local population, because most of the surplus adults that result from a high-fertility strategy leave the system permanently before reaching reproductive age (Thomas et al. 1982). However, the problem of population expansion among the Quechua is exported to the towns.

NUTRITION AND HEALTH

Prolonged undernutrition usually results in reduced physical work capacity as a function of reduced muscle mass. However, the energy cost of physical activities also varies with body mass. In the Gambia, the energy cost of daily activities is lower in heavier individuals than in lighter individuals, when expressed per kilogram body weight (Lawrence 1988). Because the energy cost of activity appears to be more closely related to BMR than to body weight (Lawrence 1988), decreased body weight or a down-regulation of BMR results in a lower energy cost of common daily activities.

The impact of iron-deficiency anemia on work performance can be considerable. Anemic subjects have lower PWC (see Box 10.1) than nonanemic ones (Gardner et al. 1977). With reduced hemoglobin concentrations, the capacity of the bloodstream to supply oxygen for aerobic metabolism is limited. In addition, iron deficiency in tissues also may limit work capacity by reducing the total activity of oxidative enzymes needed to support muscular work (Dallman 1980). The body's greater ability to synthesize red blood cells at high altitudes than at low altitudes may be important in sustaining the high levels of PWC seen in populations at high altitudes (Haas et al. 1988).

Although schistosomiasis (*Schistosoma mansoni*) (see Chapters 7 and 8) infection has been associated with low PWC (Awad el Karim et al. 1981), little direct evidence demonstrates the deleterious effects of schistosomiasis on self-paced work in the natural environment (Collins et al. 1988). However, it is difficult to assess the true biological effect of this infection by considering PWC and self-paced work alone. For example, in Sudan, women infected with *Shistosoma mansoni* have spent less time in agricultural activities other than the central economic one (cotton picking), all claiming to be too tired to work in the afternoons (Parker 1993). Thus, unmeasured economic deficits may result from not being able to work during all the daylight hours.

Another aspect of infection in relation to work is absenteeism. Less well nourished adults are likely to lose more time from work due to illness than are better-nourished individuals. For example, chronically energy-deficient Indian and Ethiopian adults spend less time working than their better nourished counterparts (Ferro-Luzzi et al. 1992).

Much of the literature on metabolic adaptation to low energy intake in adults (e.g., Shetty 1984; Ferro-Luzzi 1985; Waterlow 1986, 1990) is concerned with the maintenance of energy balance at low levels of intake. Shetty (1993) summarized research in this field over the past 30 years and concluded that

- the BMR of undernourished (in energy balance at low levels of intake) subjects, on an absolute basis (MJ/day), is significantly lower than in well-nourished (in energy balance at higher levels of intake) subjects, and this discrepancy is largely accounted for by the low body weights of these individuals.
- Differences in BMR per unit of lean body mass are unlikely to explain these changes, because the composition of lean body and organ mass in chronically energy-deficient subjects is significantly different.
- No evidence suggests that TEF is lower in chronically energy-deficient subjects.
- Most studies show that energy-deficient adults are able to perform physical work at a lower energy cost than are better nourished adults.

However, it is important to distinguish the state of being in energy balance at low levels of intake from the process of responding physiologically to a decline in energy availability. Under these conditions, often regardless of initial plane of nutrition, adults can demonstrate a decline in body weight, a reduction in the amount and energy cost of physical activity, and a reduced BMR (Waterlow 1990). In addition, mental states such as apathy due to dietary restriction may be appropriate to conserving energy. Some of the earliest data that show such changes come from the Minnesota starvation experiment of 1944–45 (Keys et al. 1950).

In the Minnesota starvation study, 32 adult male conscientious objectors were partially starved, on a dietary intake equal to one-half their habitual levels, for 24 weeks.

Table 10.4   Mean body weight and energy expenditure of 32 subjects semi-starved for 24 weeks: Minnesota starvation experiment, 1944–5

Parameter measured	Initial	After 24 weeks of semi-starvation	24-week value as a proportion of the initial value (%)
Body weight (kg)	69.4	52.3	75
Total energy expenditure (MJ/day)	14.5	7.8	54
BMR (MJ/day)	6.7	4.2	63
Activity[a]	7.8	3.6	46

[a] Difference between total energy expenditure and basal metabolic rate (BMR)

(*Source:* Keys et al. 1950)

Mean body weights and daily energy expenditures at the beginning and end of the study are given in Table 10.4. After 24 weeks, mean weight was 75% of its initial value and the large negative energy balance at the start of the study—in excess of 7 MJ per day (~1,700 kcal)—was reduced to a very small value at 24 weeks. The subjects did not, on average, achieve energy balance at their lower level of intake by this time (Ulijaszek 1995b). Total energy expenditure fell from ~15 MJ/day (3,400 kcal) to ~7 MJ/day over the 24 weeks. Most of this decline was due to loss of body mass, particularly metabolically active tissue. At the end of semistarvation, BMR was 63% of the initial value and energy expended in physical activity was 46% of the starting value. The decline in BMR, in excess of the decline in body mass, has been taken as evidence for down-regulation of BMR, perhaps as a result of increased metabolic efficiency at the cellular level (Waterlow 1986). The apathy associated with low thyroid hormone status is neurological and poorly understood. The thyroid hormone–related effects on energy metabolic responses to low food availability have been unified in a model that incorporates the effects on physical activity and basal metabolism, showing the link between energetic adaptability in physical activity and maintenance metabolism (Ulijaszek 1996).

Body Size

Various studies have shown relationships between body size and physical work capacity (Ferro-Luzzi 1985) and work between body size and productivity (Spurr 1988; Satyanarayana, Naidu, Chaterjee and Rao 1977). In all such studies, these

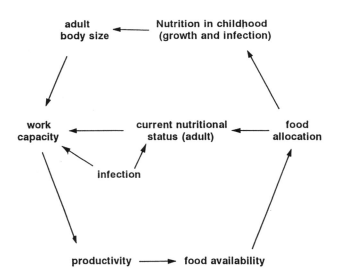

Figure 10.4   Relationships among growth, body size, work capacity, and subsistence productivity. This diagram shows how the small body size of an adult can influence the nutritional status of both adults and children in the same household through the adult's physical work capacity and associated work productivity and food production.

relationships were examined in situations where adult males were engaged in paid labor. However, it is not clear whether similar relationships exist in populations occupied primarily in traditional subsistence economies (Strickland 1990).

Understanding such relationships in traditional subsistence systems is difficult because of great variation within and between communities in regard to

- intensity and duration of work,
- day-to-day work patterns,
- patterns of group work according to task and season, and
- direct seasonal and climatic influences on work performed.

Although complex, a general model in which the relationships among growth, body size, work capacity, and subsistence productivity might operate in such contexts is summarized in Figure 10.4.

This model is an elaboration of an earlier model proposed by Martorell and Arroyave (1988), in which households, families, or units of reproduction can be considered "energy trapped" if their rates of energy flow and productivity are so low that some aspects of biological function—including reproductive success—are impaired.

ENERGETICS THROUGHOUT THE LIFE CYCLE

Pregnancy

Although the total energy requirement for tissue gain and increased metabolism during pregnancy is currently estimated to be 356 MJ (~85,000 kcal; FAO/WHO/UNU 1985), evidence suggests that the energy cost of pregnancy is considerably less. Estimates vary from 22% to 58% of this value in the Gambia and the Philippines, respectively (Durnin 1987). In many less developed countries, mean daily energy intakes do not increase during pregnancy, and various energy-saving mechanisms have been invoked to account for the disparity between the increased energy cost of pregnancy and the lack of increased food intake. These mechanisms include downregulation of basal metabolism and reduced energy cost of performing physical activities (Durnin 1987).

A more obvious way of saving energy during pregnancy is to reduce the total amount of physical activity performed. In the Gambia, women become progressively less active during pregnancy (Roberts et al. 1982). They reduce the amount of walking within the village and the duration of household work and tend to go to the fields less often (Lawrence and Whitehead 1988). Such reduction may be seasonal.

In Nepal, Tamang women rarely modify their work schedules during pregnancy and lactation during the monsoon season, when they cannot afford to interrupt or significantly modify their subsistence work. However, during the less strenuous late winter season, pregnant and lactating women spend less time in outdoor subsistence activities (Panter-Brick 1993).

Lactation

The energy cost of lactation is equal to the energy content of the milk secreted plus the energy required to produce it. To maintain long-term energy balance, child-bearing women build up fat stores during pregnancy that will provide the energy required to sustain lactation. Table 10.5 shows that body fat gain during pregnancy varies widely across populations. If we assume that energy intake during lactation does not increase above that of the nonpregnant, nonlactating state, then weight gains of 1–3.5 kg (2–8 lb) represent low increases in fat stores compared with those recommended by international agencies (FAO/WHO/UNU 1985).

Data indicate that child-bearing women do not have high energy stores for breast milk synthesis. In turn, these data suggest considerable potential for energy depletion in women in traditional societies, where prolonged breastfeeding is the norm. However, the lactational performance of women in developing countries is adequate compared with Western women (Prentice and Prentice 1995). At low levels of energy intake, women may reduce their levels of physical activity to conserve energy. Energy savings of ~15% due to reduced activity have been reported in pregnant and lactating women in India (McNeill and Payne 1985) and in the Gambia (Lawrence and Whitehead 1988). As in pregnancy, the energy-sparing mechanisms may be inadequate, and a negative energy balance in lactation may have long-term implications for the mother.

Age

As people get older, physical ability may decrease as a result of reduced physical activity due to impaired motor skills or judgment, or accidents that may result from such situations. It also come from the reduced capacity of **mitochondria** to supply energy for muscular work (Bittles and Sambuy 1986) and from the reduced muscle mass that results from reduced levels of physical activity. Decreased muscle strength with increasing age also increases the likelihood of falling. Levels of bone mineralization determine whether such falls result in bone fractures. In turn, bone mineral mass and density are influenced by muscle strength.

Of the many factors that can influence the capacity for physical work (Figure 10.5), sociocultural systems that determine the microenvironment, health, nutrition,

Table 10.5   Fat gain during pregnancy in various studies

Country	Fat gain (kg)	Reference
England	3.4	Hytten and Leitch 1971
Scotland	2.3	Durnin 1988
Thailand	1.4	Durnin 1988
The Netherlands	1.3	Durnin 1988
Philippines	1.3	Durnin 1988
Gambia	0.6	Durnin 1988

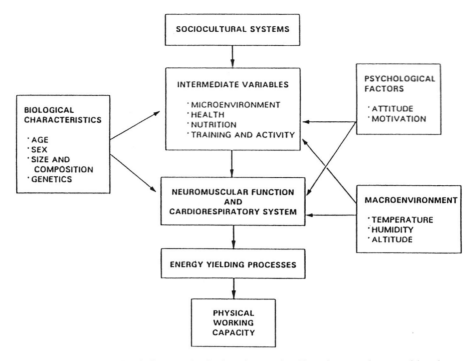

Figure 10.5   Factors that influence physical work capacity. Size, shape, and composition do not impinge directly on work capacity but act through other sets of variables. (*Source*: Norgan 1994)

and levels of habitual activity of older people affect neuromuscular function as well as the skeletal system. These factors, in turn, influence physical work capacity in a potentially downward spiral. Women lose bone density earlier and more quickly than men (Stini et al. 1994) and are more likely to lose activity due to skeletal injury at an earlier age than men.

CHAPTER SUMMARY

In this chapter, the principles of energetics were described as were the factors that influence variations in energy expenditure within and between populations, the human biological components of energy expenditure at different stages of the life cycle, and models that use optimization procedures in a cost–benefit analysis.

Human energy expenditure can be broken down into several components: energy cost of maintenance, thermic effect of food, physical activity, thermoregulation, growth, and reproduction. The total daily energy expenditure of adults varies greatly between populations, and mean total daily energy expenditure is slightly lower in industrialized nations than in developing ones. Physical activity level (i.e., energy expenditure, expressed as units of BMR) also varies widely between populations.

The law of conservation of energy states that energy can be neither gained nor lost. If energy output exceeds input, then energy balance is negative, and body mass will be mobilized to meet energy needs. Seasonality of climate and subsistence productivity can lead to seasonal variation in energy balance and seasonal weight loss. Reductions in physical activity can play some role in balancing decreased energy intake or increased energy demand (e.g., in pregnancy and lactation) but are not always possible.

SUGGESTED READINGS

Bouchard C, Despres JP, and Tremblay A (1991) Genetics of obesity and human energy metabolism. Proc. Nutr. Soc. 50: 139–147.

Cole TJ (1991) Sampling, study size, and power. In BM Margetts and M Nelson (eds.): Design Concepts in Nutritional Epidemiology. Oxford: Oxford University Press, pp. 53–78.

Ferro-Luzzi A, Sette S, Franklin M, and James WPT (1992) A simplified approach of assessing adult chronic energy deficiency. Eur. J. Clin. Nutr. 46: 173–186.

Gibson RS (1990) Principles of Nutritional Assessment. Oxford: Oxford University Press.

Haas JD, Tufts DA, Beard JL, Roach RC, and Spielvogel H (1988) Defining anaemia and its effect on physical work capacity at high altitudes in the Bolivian Andes. In KJ Collins and DF Roberts (eds.): Capacity for Work in the Tropics. Cambridge: Cambridge University Press, pp. 85–106.

Nydon J and Thomas RB (1989) Methodological procedures for analysing energy expenditure. In GH Pelto, PJ Pelto, and E Messer (eds.): Research Methods in Nutritional Anthropology. Tokyo: United Nations University, pp. 57–81.

Shetty PS (1993) Chronic undernutrition and metabolic adaptation. Proc. Nutr. Soc. 52: 267–284.

Smith EA (1983) Optimal foraging theory and hunter-gatherer societies. Current Anthropol. 24: 625–651.

Spurr GB (1988) Body size, physical work capacity, and productivity in hard work: is bigger better? In JC Waterlow (ed.): Linear Growth Retardation in Less Developed Countries. New York: Raven Press.

Ulijaszek SJ (1995) Human Energetics in Biological Anthropology. Cambridge: Cambridge University Press.

REFERENCES CITED

Astrand PO and Rodahl K (1986) Textbook of Work Physiology, 3rd edition. New York: McGraw-Hill.

Awad el Karim MA, Collins KJ, Sukkar MY, Omer AHS, Amin MA, and Dore C (1981) An assessment of anti-schistosome treatment on physical work capacity. J. Trop. Med. Hyg. 85: 65–70.

Bailey RC, Jenike MR, Ellison PT, Bentley GR, Harrigan AM, and Peacock NR (1992) The ecology of birth seasonality among agriculturalists in central Africa. J. Biosoc. Sc. 24: 393–412.

Bannerjee S (1962) Studies in energy metabolism. Special Report Series of the Indian Council for Medical Research No. 43, New Delhi.

Bannerjee S, Acharya KN, Chattopadhyay D, and Sen RN (1959a) Studies on energy metabolism of labourers in a spinning mill. Indian J. Med. Res. 47: 657–662.

Bannerjee S, Sen RN, and Acharaya KN (1959b) Energy metabolism in laboratory workers. J. Appl. Physiol. 14: 625–628.

Barac-Nieto M (1987) Physical work determinants and undernutrition. World Review of Nutrition and Dietetics 49: 22–65.

Bayliss-Smith (1982) Ecology of Agricultural Systems. Cambridge: Cambridge University Press.

Bergmann RL and Bergmann KE (1986). Nutrition and growth in infancy. In F Falkner and JM Tanner (eds.): Human Growth: A Comprehensive Treatise. New York: Plenum Press, pp. 389–413.

Bingham, SA (1991). Assessment of food consumption and nutrient intake. Current intake. In BM Margetts and M Nelson (eds.): Design Concepts in Nutritional Epidemiology. Oxford: Oxford University Press, pp. 154–167.

Bittles AH and Sambuy Y (1986) Human cell culture systems in the study of ageing. In AH Bittles and KJ Collins (eds.): The Biology of Human Ageing. Cambridge: Cambridge University Press, pp. 49–66.

Bleiberg FM, Brun TA, and Goihman S (1980) Duration of activities and energy expenditure of female farmers in dry and rainy season in Upper Volta. Br. J. Nutr. 43: 71–82.

Bouchard C, Despres JP, and Tremblay A (1991) Genetics of obesity and human energy metabolism. Proc. Nutr. Soc. 50: 139–147.

Brun T, Bleiberg F, and Goihman S (1981) Energy expenditure of male farmers in dry and rainy seasons in Upper Volta. Br. J. Nutr. 45: 67–75.

Brun TA, Geissler CA, Mirbagheri I, Hormozdiary H, Bastani J, and Heydayat H (1979) The energy expenditure of Iranian agricultural workers. Am. J. Clin. Nutr. 32: 2154–2161.

Chibnick M and de Jong W (1989) Agricultural labor organization in Ribereno communities of the Peruvian Andes. Ethnology 28: 75–95.

Cleave JH (1970) Labour in the development of African agriculture: the evidence of farm surveys. Stanford University: PhD Thesis.

Cody ML (1974) Optimisation in ecology. Science 183: 1156–1164.

Cole TJ (1991) Sampling, study size, and power. In BM Margetts and M Nelson (eds.): Design Concepts in Nutritional Epidemiology. Oxford: Oxford University Press, pp. 53–78.

Collins KJ, Abdel-Rahaman TA, and Awad el Karim MA (1988) Schistosomiasis: Field studies of energy expenditure in agricultural workers in the Sudan. In KJ Collins and DF Roberts (eds.): Capacity for Work in the Tropics. Cambridge: Cambridge University Press, pp. 235–247.

Dallman PR (1980) Iron deficiency: distinguishing the effects of anemia from muscle iron deficiency on work performance. In P Saltman and J Hegenauer (eds.): The Biochemistry and Physiology of Iron. Elsevier: North Holland, pp. 509–523.

Dasgupta B (1977) Village Society and Labour Use. Delhi: Oxford University Press.

Dauncey MJ (1990) Thyroid hormones and thermogenesis. Proc. Nutr. Soc. 49, 203–215.

Devadas RP, Anuradha V, and Rani AJ (1975) Energy intake and expenditure of selected manual labourers. Indian J. Nutr. Diet. 12: 279–284.

Dufour DL (1984) The time and energy expenditure of indigenous women horticulturalists in the northwest Amazon. Am. J. Phys. Anthropol. 65: 37–46.

Dufour D (1992) Nutritional ecology in the tropical rainforests of Amazonia. Am. J. Hum. Biol. 4: 197–207.

Dugdale AE, and Payne PR (1986) Modelling seasonal changes in energy balance. In TG Taylor and NK Jenkins (eds.): Proceedings of the XIII International Congress of Nutrition. London: John Libbey, pp. 141–144.

Durnin JVGA (1987) Energy requirements of pregnancy: an integration of the longitudinal data from the five country study. Lancet 2: 1131–1134.

Durnin JVGA (1988) The energy requirements of pregnancy and lactation. In B Schurch and NS Scrimshaw (eds.): Chronic Energy Deficiency: Consequences and Related Issues. Lausanne: Nestle Foundation, pp. 135–152.

Durnin JVGA and Passmore R (1967) Energy, Work and Leisure. London: Heinemann.

Edholm OG, Fletcher JG, Widdowson EM, and McCance RA (1955) The energy expenditure and food intake of individual men. Br. J. Nutr. 9: 286–300.

Edmundson WC and Edmundson SA (1988) Food intake and work allocation of male and female farmers in an impoverished Indian village. Br. J. Nutr. 60: 433–439.

Ellison PT (1991) Reproductive ecology and human fertility. In GW Lasker and CGN Mascie-Taylor (eds.): Applications of Biological Anthropology to Human Affairs. Cambridge: Cambridge University Press, pp. 14–54.

Erasmus CJ (1955) Culture, structure, and process: the occurrence and disappearance of reciprocal farm labor. Southwestern J. Anthropol. 12: 444–469.

Food and Agriculture Organization, World Health Organization, and United Nations University (FAO/WHO/UNU) (1985) Energy and Protein Requirements. Technical Report Series No. 724. Geneva: World Health Organization.

Ferro-Luzzi A (1985) Work capacity and productivity in long-term adaptation to low energy intakes. In K Blaxter and JC Waterlow (eds.): Nutritional Adaptation in Man. London: John Libbey, pp. 61–69.

Ferro-Luzzi A (1990) Seasonal energy stress in marginally nourished rural women: interpretation and integrated conclusions of a multicentre study in three developing countries. Eur. J. Clin. Nutr. 44 (Suppl. 1): 41–46.

Ferro-Luzzi A and Branca F (1993) Nutritional seasonality: the dimensions of the problem. In SJ Ulijaszek and SS Strickland (eds.): Seasonality and Human Ecology. Cambridge: Cambridge University Press, pp. 149–165.

Ferro-Luzzi A, Sette S, Franklin M, and James WPT (1992) A simplified approach of assessing adult chronic energy deficiency. Eur. J. Clin. Nutr. 46: 173–186.

Ferro-Luzzi A, Scaccini C, Taffese S, Aberra B, and Demeke T (1990) Seasonal energy deficiency in Ethiopian rural women. Eur. J. Clin. Nutr. 44 (Suppl 1): 7–18.

Forbes GB (1989) Changes in body composition. In: Report of the 98th Ross Conference on Pediatric Research. Columbus, Ohio: Ross Laboratories, pp. 112–118.

Fox RH (1953) A study of the energy expenditure of Africans engaged in various activities, with special reference to some environmental and physiological factors which may influence the efficiency of their work (PhD thesis). London: University of London.

Galvin KA (1985) Food procurement, diet, activities, and nutrition of Ngisonyoka, Turkana pastoralists in an ecological and social context (doctoral dissertation). Binghamton: State University of New York.

Gardner GW, Edgerton VR, Senewiratne V, Barnard RJ, and Ohira Y (1977) Physical work capacity and metabolic stress in subjects with iron-deficiency anemia. Am. J. Clin. Nutr. 30: 910–918.

Garry RC, Passmore R, Warnock GM, Durnin JVGA (1955) Expenditure of energy and consumption of food by miners and clerks, Fife, Scotland 1955. Medical Research Council Special Report Series No. 289. London: Her Magesty's Stationery Office.

Gibson RS (1990) Principles of Nutritional Assessment. Oxford: Oxford University Press.

Godin G and Shephard RJ (1973) Activity patterns of the Canadian Eskimo. In OG Edholm and EKE Gunderson (eds.): Polar Human Biology. Chichester, UK: Heinemann Books, pp. 193–215.

Golden MHN (1988) The role of individual nutrient deficiencies in growth retardation of children as exemplified by zinc and protein. In JC Waterlow (ed.): Linear Growth Retardation in Less Developed Countries. New York: Raven Press, pp. 143–160.

Golden MHN (1994) Is complete catch-up possible for stunted malnourished children? Eur. J. Clin. Nutr. 48 (Suppl. 1): S58–S70.

de Guzman PE, Dominguez SR, Kalaw JM, Basconcillo RO, and Santos VF (1974a) A study of the energy expenditure, dietary intake, and pattern of daily activity among various occupational groups. I. Laguna rice farmers. Philippine J. Sc. 103: 53–65.

de Guzman PE, Dominguez SR, Kalaw JM, Buning MN, Basconcillo RO, and Santos VF (1974b) A study of the energy expenditure, dietary intake, and pattern of daily activity among various occupational groups. II. Marikina shoemakers and housewives. Philippine J. Nutr. 27: 21–30.

de Guzman PE, Kalaw JM, Tan RH, Recto RC, Basconcillo RO, Ferrer VT, Tombokon MS, Yuchingtat GP, and Gaurano AL (1974c) A study of the energy expenditure, dietary intake, and pattern of daily activity among various occupational groups. III. Urban jeepney drivers. Philippine J. Nutr. 27: 182–188.

de Guzman PE, Cabrera JP, Basconcillo RO, Gaurano AL, Yuchingtat GP, Tan RM, Kalaw JM, and Recto RC (1978) A study of the energy expenditure, dietary intake, and pattern of daily activity among various occupational groups. V. Clerk-typists. Philippine J. Nutr. 31: 147–156.

de Guzman PE, Recto RC, Cabrera JP, Basconcillo RO, Gaurano AL, Yuchingtat GP, Abanto ZU, and Math BS (1979) A study of the energy expenditure, dietary intake and pattern of daily activity among various occupational groups. VI. Textile mill workers. Philippine J. Nutr. 32: 134–148.

Haas JD, Tufts DA, Beard JL, Roach RC, and Spielvogel H (1988) Defining anaemia and its effect on physical work capacity at high altitudes in the Bolivian Andes. In KJ Collins and DF Roberts (eds.): Capacity for Work in the Tropics. Cambridge: Cambridge University Press, pp. 85–106.

Hampton IFG (1960) Energy expenditure and calorie intake of an undergraduate expedition. Nutrition 14: 174–183.

Hill K and Kaplan H (1988) Tradeoffs in male and female reproductive strategies among the Ache: part 1. In L Betzig, M Borgerhoff Mulder, and P Turke (eds.): Human Reproductive Behaviour. Cambridge: Cambridge University Press, pp. 277–290.

Hill K, Kaplan H, Hawkes K, and Hurtado AM (1985) Men's time allocation to subsistence work among the Ache of Eastern Paraguay. Hum. Ecol. 13: 29–47.

Hipsley EH and Kirk NE (1962) Studies of dietary intake and the expenditure of energy by New Guineans (Technical Paper No. 147). South Pacific Commission: Nouméa, New Caledonia.

Horton ES (1983) Introduction: an overview of the assessment and regulation of energy balance in human. Am. J. Clin. Nutr. 38: 972–977.

Huss-Ashmore RA (1996) Issues in the measurement of energy intake for free-living human populations. Am. J. Hum. Biol. 8: 159–167.

Huss-Ashmore RA, Goodman JI, Sibiya TE, and Stein TP (1989) Energy expenditure of young Swazi women as measured by the doubly-labelled water method. Eur. J. Clin. Nutr. 43: 737–748.

Hytten FE and Leitch I (1971) The Physiology of Human Pregnancy, 2nd edition. Oxford: Blackwell Scientific Publications.

James WPT, Ferro-Luzzi A, and Waterlow JC (1988). Definition of chronic energy deficiency in adults. Eur. J. Clin. Nutr. 42: 969–981.

Jelliffe DB and Jelliffe EP (1989) Assessment of Nutritional Status of the Community. Oxford: Oxford University Press.

Johnson A (1975) Time allocation in a Machiguenga community. Ethnology 14: 301–310.

Kaminsky G (1953) Untersuchungen beim Holztransport mit Schlitten im Winterlichen Hochgebirge. Arbeitsphysiologie 15: 47–56.

Katzmarzyk PT, Leonard WR, Crawford MH, and Sukernik RI (1994) Resting metabolic rate and daily energy expenditure among two indigenous Siberian populations. Am J. Hum. Biol 6: 719–730.

Keys A, Brozek J, Henschel A, Michelson O, and Taylor HL (1950) The Biology of Human Starvation. Minneapolis: University of Minnesota Press.

Kinloch JD (1959) The dietary intake and activities of an Alaskan mountaineering expedition. Br. J. Nutr. 13: 85–99.

Kusunoki T (1956) Forest labourers in Japan. Rep. Inst. Sci. Labour 49: 23–25.

Lawrence M (1988) Predicting energy requirements: is energy expenditure proportional to the BMR or to body weight? Eur. J. Clin. Nutr. 42: 917–919.

Lawrence M and Whitehead RG (1988) Physical activity and total energy expenditure of child-bearing Gambian women. Eur. J. Clin. Nutr. 42: 145–160.

Leonard WR (1988) The impact of seasonality on caloric requirements of human populations. Hum. Ecol. 16: 343–346.

Livingstone MBE, Stain JJ, McKenna PG, Nevin GB, Barker ME, Hickey RJ, Prentice AM, Coward WA, Ceesay SM, and Whitehead RG (1990) Simultaneous measurement of free living energy expenditure by the doublelabelled water $(H_2 O)^{18}$ method and heart rate monitoring. Am. J. Clin. Nutr. 52: 59–65.

Lundgren NPV (1946) Physiological effects of time schedule work on lumber workers. Acta Physiol. Scand. 13 (Suppl. 41): XX–XX.

Margetts BM and Nelson M (1997). Design Concepts in Nutritional Epidemiology, 2nd edition. Oxford: Oxford University Press.

Martorell R and Arroyave G (1988) Malnutrition, work output, and energy needs. In KJ Collins and DF Roberts (eds.): Capacity for Work in the Tropics. Cambridge: Cambridge University Press, pp. 57–75.

Masterton JP, Lewis HE, and Widdowson EM (1957) Food intakes, energy expenditures, and faecal excretions of men on a polar expedition. Br. J. Nutr. 11: 346–358.

McNeill G and Payne PR (l985) Energy expenditure of pregnant and lactating women. Lancet 1: 1237–l238.

McNeill G, Payne PR, Rivers JPW Enos AMT, de Britto JJ, and Mukarji DS (1988) Socio-economic and seasonal patterns of adult energy nutrition in a South Indian village. Ecol. Food Nutr. 22:85–95.

Meijer GAL, Westerterp KR, Saris WHM, and ten Hoor F (1992) Sleeping metabolic rate in relation to body composition and the menstrual cycle. Am. J. Clin. Nutr. 55: 637–640.

Montgomery E and Johnson A (1974) Machiguenga energy expenditure. Ecol. Food Nutr. 6: 97–105.

National Center for Health Statistics (NCHS) (1977) NCHS Growth Curves for Children (PHS 78-1650). Hyattsville, Md.: US Department of Health, Education, and Welfare.

Norgan NG (1994) Anthropometry and physical performance. In SJ Ulijaszek and CGN Mascie-Taylor (eds.): Anthropometry: the Individual and the Population. Cambridge: Cambridge University Press, pp. 141–159.

Norgan NG and Durnin JVGA (1980) The effects of six weeks of overfeeding on the body weight, body composition, and energy metabolism of young men. Am. J. Clin. Nutr. 33: 978–988.

Norgan NG and Ferro-Luzzi A (1978) Nutrition, physical activity, and physical fitness in contrasting environments. In J Parizkova and VA Rogozkin (eds.): Nutrition, Physical Fitness, and Health. Baltimore: University Park Press, pp. 167–193.

Norgan NG, Ferro-Luzzi A, and Durnin JVGA (1974) The energy and nutrient intake and the energy expenditure of 204 New Guinean adults. Phil. Trans. R. Soc. Lond. B. 268: 309–348.

Nydon J and Thomas RB (1989) Methodological procedures for analysing energy expenditure. In GH Pelto, PJ Pelto, and E Messer (eds.): Research Methods in Nutritional Anthropology. Tokyo: United Nations University, pp. 57–81.

Panter-Brick C (1993) Seasonal organisation of work patterns. In SJ Ulijaszek and SS Strickland (eds.): Seasonality and Human Ecology. Cambridge: Cambridge University Press, pp. 220–234.

Panter-Brick C, Lotstein DS, and Ellison PT (1993) Seasonality of reproductive function and weight loss in rural Nepali women. Hum. Reprod. 8: 684–690.

Parker M (1993) Bilharzia and the boys: questioning common assumptions. Soc. Sci. Med. 37: 481–492.

Pasquet P and Koppert GJA (1993) Activity patterns and energy expenditure in Cameroonian tropical forest populations. In CM Hladik, A Hladik, OF Linares, H Pagezy, A Semple, and M Hadley (eds.): Tropical Forests, People, and Food. Paris: United Nations Educational, Scientific, and Cultural Organization (UNESCO), pp. 311–320.

Passmore R, Thomson JG, and Warnock GM (1952) Balance sheet of the estimation of energy intake and energy expenditure as measured by indirect calorimetry. Br. J. Nutr. 6: 253–264.

Pearson JD (1990) Estimation of energy expenditure in Western Samoa, American Samoa, and Honolulu by recall interviews and direct observation. Am. J. Hum. Biol. 2:313–326.

Pianka ER (1988) Evolutionary Ecology, 4th edition. New York: Harper and Row.

Prentice AM, Goldberg GR, Jebb SA, Black AE, and Murgatroyd PR (1991) Physiological responses to slimming. Proc. Nutr. Soc. 50: 441–458.

Prentice A and Bates CJ (1994) Adequacy of dietary mineral supply for human bone growth and mineralisation. Eur. J. Clin. Nutr. 48 (Suppl. 1): S161–S176.

Prentice AM and Prentice A (1995) Evolutionary and environmental influences on human lactation. Proc. Nutr. Soc. 54: 391–400.

Pyke GH, Pulliam HR, and Charnov EL (1977) Optimal foraging: a selective review of theory and tests. Q. Rev. Biol. 52: 137–154.

Ramanamurthy PSV and Dakshayani R (1962) Energy intake and expenditure in stone cutters. Indian J. Med. Res. 50: 804–809.

Roberts SB, Paul AA, Cole TJ, and Whitehead RG (1982) Seasonal changes in activity, birth weight, and lactational performance in rural Gambian women. Trans. R. Soc. Trop. Med. Hyg. 76: 668–678.

Rogers TA, Setliff JA, and Klopping JC (1964) Energy cost, fluid and electrolyte balance in subarctic survival situations. J. Appl. Physiol. 19: 1–8.

Rosetta L (1990) Biological aspects of fertility among third world populations. In V Reynolds and J Landers (eds.): Fertility and Resources. Cambridge: Cambridge University Press, pp. 18–34.

Satyanarayana K, Naidu NA, Chaterjee B, and Narasinga Rao BS (1977) Body size and work output. Am. J. Clin. Nutr. 30: 322–325.

Schoener TW (1971) Theory of feeding strategies. Annu. Rev. Ecol. Systematics 2: 369–404.

Schutz Y, Lechtig A, and Bradfield RB (1980) Energy expenditures and food intakes of lactating women in Guatemala. Am. J. Clin. Nutr. 33: 892–903.

Shephard RJ (1974) Work physiology and activity patterns of circumpolar Eskimos and Ann. Hum. Biol. 46: 263–294.

Shephard RJ (1978) Human Physiological Work Capacity. Cambridge: Cambridge University Press.

Shephard RJ (1980) Population aspects of physical work capacity. A review. Ann. Hum. Biol. 7: 1–28.

Shetty PS (1984) Adaptive changes in basal metabolic rate and lean body mass in chronic undernutrition. Hum. Nutr. Clin. Nutr. 38C: 443–452.

Shetty PS (1993) Chronic undernutrition and metabolic adaptation. Proc. Nutr. Soc. 52: 267–284.

Smith EA (1983) Optimal foraging theory and hunter-gatherer societies. Curr. Anthropol. 24: 625–651.

Smith EA (1991) Inujjuamiut Foraging Strategies. New York: Aldine de Gruyter.

Solomon SJ, Kurzer MS, and Calloway DH (1982) Menstrual cycle and basal metabolic rate in women. Am. J. Clin. Nutr. 36: 611–616.

Spurr GB (1988) Body size, physical work capacity, and productivity in hard work: is bigger better? In JC Waterlow (ed.): Linear Growth Retardation in Less Developed Countries. New York: Raven Press, pp. 215–239.

Stini WA, Chen Z, and Stein P (1994) Aging, bone loss, and the body mass index in Arizona retirees. Am. J. Hum. Biol. 6: 43–50.

Streef GM, Gerritsen AG, and Bol M (1959) Arbeidsfysiologisch Onderzoek bij Vellingswerk in de Bosbouw. Med. Landb. Logesch. Wageningen 59, 14: 1–39.

Strickland SS (1990) Traditional economies and patterns of nutritional disease. In GA Harrison and JC Waterlow (eds.): Diet and Disease. Cambridge: Cambridge University Press, pp. 209–239.

Strickland SS, Tuffrey VT, and Gurung G (1997) Form and Function. London: Smith-Gordon and Company.

Suzuki S (1988) Villagers' daily life and the environment. In S Suzuki (ed.): Health Ecology in Indonesia. Tokyo: Gyosei Corporation, pp. 13–22.

Thomas RB (1976) Energy flow at high altitude. In PT Baker and MA Little (eds.): Man in the Andes. Stroudsburg, Pa.: Dowden, Hutchinson, and Ross, pp. 379–404.

Thomas RB, McRae SD, and Baker PT (1982) The use of models in anticipating effects of change on human populations. In GA Harrison (ed.): Energy and Effort. London: Taylor and Francis, pp. 243–281.

Thomas RB, Gage TB, and Little MA (1989) Reflections on adaptive and ecological models. In MA Little and JD Haas (eds.): Human Population Biology. Oxford: Oxford University Press, pp. 296–319.

Tin-May-Than and Ba-Aye (1985) Energy intake and energy output of Burmese farmers at different seasons. Hum. Nutr. Clin. Nutr. 39C: 7–15.

Ulijaszek SJ (1992) Human energetics methods in biological anthropology. Yearbook Phys. Anthropol. 35: 215–242.

Ulijaszek SJ (1995a) Human Energetics in Biological Anthropology. Cambridge: Cambridge University Press.

Ulijaszek SJ (1995b) Plasticity, growth, and energy balance. In CGN Mascie-Taylor and B Bogin (eds.): Human Variability and Plasticity. Cambridge: Cambridge University Press.

Ulijaszek SJ (1996) Energetics, adaptation, and adaptability. Am. J. Hum. Biol. 8: 169–182.

Ulijaszek SJ and Poraituk SP (1993) Making sago in Papua New Guinea: is it worth the effort? In CM Hladik, A Hladik, OF Linares, H Pagezy, A Semple, and M Hadley, (eds.): Tropical Forests, People, and Food. Biocultural Interactions and Applications to Development. Paris: United Nations Educational, Scientific, and Cultural Organization (UNESCO), pp. 271–280.

Ulijaszek SJ and Strickland SS (1993) Nutritional Anthropology. Prospects and Perspectives. London: Smith-Gordon.

Vaughan L, Zurlo F, and Ravussin E (1991) Aging and energy expenditure. Am. J. Clin. Nutr. 53: 821–825.

Viteri FE, Torun B, Garcia JC, and Herrera E (1971) Determining energy costs of agricultural activities by respirometer and energy balance techniques. Am. J. Clin. Nutr. 24: 1418–1430.

Waterlow JC (1986) Notes on the new estimates of energy requirements. Proc. Nutr. Soc. 45: 351–360.

Waterlow JC (1990) Mechanisms of adaptation to low energy intakes. In GA Harrison and JC Waterlow (eds.): Diet and Disease. Cambridge: Cambridge University Press, pp. 5–23.

Watts M (1988) Coping with the market: uncertainty and food security among Hausa peasants. In I de Garine and GA Harrison (eds.): Coping with Uncertainty in Food Supply. Oxford: Oxford University Press, pp. 260–289.

Widdowson EM, Edholm OG, and McCance RA (1954) The food intake and energy expenditure of cadets in training. Br. J. Nutr. 8: 147–155.

CHAPTER 11

Evolution of the Human Life Cycle

BARRY BOGIN and B. HOLLY SMITH

INTRODUCTION

The study of human growth has been a part of anthropology and human biology since the founding of these disciplines. European "anthropology" of the early to mid-19th century was basically a combination of anatomy and **anthropometry,** the science of human body measurements. American anthropology of the late 19th century incorporated anthropometry into its foundation, and the early practitioners, especially Franz Boas, are known as much for their studies of human growth as for anything else (see Chapter 2).

An interest in human growth is natural for anthropologists and human biologists because the way a human being grows is the product of an interaction among the biology of our species; the physical environment in which we live; and the social, economic, and political environment that every human culture creates. The basic pattern of human growth is shared by all people and is the outcome of the 4-million-year evolutionary history of the **hominids,** living human beings and our fossil ancestors. Thus, human growth and development reflect the biocultural nature and evolutionary history of our species.

GROWTH AND EVOLUTION

If there is a "secret" to life, it is hidden in the process that converts a single cell, with its complement of deoxyribonucleic acid (**DNA**) into a multicellular organism composed of hundreds of different tissues, organs, behavioral capabilities, and emotions. That process is no less wondrous when it occurs in an earthworm, a whale, or a human being. Because this book is about human biology, in this chapter we focus on the process of human growth and development; however, the reader must be aware that much of what we know about human growth is derived from research on nonhu-

Human Biology: An Evolutionary and Biocultural Perspective, Edited by Sara Stinson, Barry Bogin, Rebecca Huss-Ashmore, and Dennis O'Rourke
ISBN 0-471-13746-4   Copyright © 2000 Wiley-Liss, Inc.

man animals. The two reasons for this are ethical limits on the kind of experimental research that may be performed on human beings and the evolutionary history that connects all living organisms.

Many growth processes that occur in humans are identical to those in other species and attest to a common evolutionary origin. Powerful evidence for the common evolutionary origin of the eye came in 1995 with the discovery of a "master-control gene" for eye growth and development. This gene is common to species as diverse as marine worms, squid, fruit flies, mice, and humans. Other organs, and the mechanisms that control their growth and development, also are shared among many diverse species. Some events in the human life cycle may be unique, such as the **adolescent growth** spurt in height and **menopause,** and they attest to the ongoing evolution of our species.

Biological evolution is the continuous process of genetic adaptation of organisms to their environments. **Natural selection** determines the direction of evolutionary change and operates by **differential mortality** between individual organisms prior to reproductive maturation and by **differential fertility** of mature organisms. Thus, genetic adaptations that enhance the survival of individuals to reproductive age and that increase the production of similarly successful offspring will increase in frequency in the population. The unique stages and events of human growth and development evolved because they conferred reproductive advantages to our species.

Before we explore the evolution of the human life cycle, let us review some of the basic principles of human growth and development.

BASIC PRINCIPLES OF HUMAN GROWTH AND DEVELOPMENT

Human beings, like all animals, begin life as a single cell, the fertilized ovum. Guided by the interaction of the genetic information provided by each parent and the environment, this cell divides and grows. It differentiates and develops into the embryo, fetus, child, and adult.

Although growth and development may occur simultaneously, they are distinct biological processes. **Growth** may be defined as a quantitative increase in size or mass. Measurements of height or weight indicate how much growth has taken place in a child. Additionally, the growth of a body organ, such as the liver or the brain, also may be described by measuring the number, weight, or size of cells present. **Development** is defined as a progression of changes, either quantitative or qualitative, that lead from an undifferentiated or immature state to a highly organized, specialized, and mature state. **Maturity** is measured by functional capacity (e.g., the development of motor skills in a child) because these skills are related to development of the skeletal and muscle systems. Even though these definitions are broad, they allow us to consider the growth, development, and maturation of organs (e.g., kidneys), systems (e.g., the reproductive system), and the person.

The process of growth and development from fertilized ovum to human newborn is counter-intuitive to our expectations, which are formed on the basis of our experience with child growth after birth. In fact, throughout much of human history, scholars and physicians did not know or believe that it occurred. Preformation (Figure 11.1) was a popular belief about the nature of human growth. In 1651, William Harvey, a physician, presented some evidence that the embryo is not a preformed adult. Incontrovertible evidence against preformation was presented in 1799, when S. T. Sommerring published drawings of the human embryo and fetus from the fourth week after **fertilization** to the fifth month. These drawings clearly showed that the embryo is not a miniature human being; rather, during development, each new life passes through a series of embryological stages that are distinct in appearance from the form visible just before and after birth (Bogin 1999 provides greater historical detail and references).

Stages in the Life Cycle

Many of the basic principles of human growth, development, and maturation are best presented in terms of the events that take place during the life cycle. One of the many possible orderings of events is given in Table 11.1, in which growth periods are divided into developmentally functional stages. This is only one possible ordering, because declaring that one moment (e.g., fertilization) is the beginning of life is arbitrary in a continuous cycle that passes through fixed stages in each individual person and in generation after generation.

Prenatal Life

The course of pregnancy may be divided into three periods, or **trimesters.** During the first trimester, one of the major events is the multiplication of a single cell, the fertilized ovum, into tens of thousands of new cells. At first, cell division may produce ex-

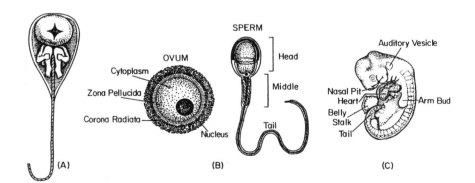

Figure 11.1   (A) Preformationist rendering of a human spermatozoon, from a drawing published by Hartsoeker in 1694 (from Singer, 1959). (B) Diagram of the actual appearance of human ovum and spermatozoon (not to scale). (C) Diagram of human embryo 32 days after fertilization. The union of sperm with ovum initiates the process of growth by hyperplasia and hypertrophy. Development also begins soon after fertilization as groups of cells differentiate, or specialize, and become functional cell types and tissues.

act copies of the original parent cell. However, within hours of the first division, distinct groups of cells begin to form. The rate of cell division in the separate groups is unequal; these cells have begun to differentiate and will eventually form different kinds of tissue (the "germ layers" of endoderm, mesoderm, and ectoderm) that will constitute the growing embryo. Growth, an increase in cell number, and development (in this case, cellular differentiation) begin almost simultaneously with conception.

Although the human body is composed of dozens of kinds of tissues and organs, their generation and growth during prenatal life, and postnatal life as well, takes place through a few ubiquitous processes. Goss (1964, 1978) described two types of

Table 11.1 Stages in the human life cycle

Stage		Growth events/duration (approximate or average)
Prenatal Life		
Fertilization		
	First trimester	Fertilization to 12th week: embryogenesis
	Second trimester	Fourth through sixth lunar month: rapid growth in length
	Third trimester	Seventh lunar month to birth: rapid growth in weight and organ maturation
Birth		
Postnatal Life		
	Neonatal period	Birth to 28 days: extrauterine adaptation, most rapid of postnatal growth and maturation
	Infancy	Second month to end of lactation, usually by 36 months: rapid growth velocity, but with steep deceleration in growth rate, feeding by lactation, deciduous tooth eruption, many developmental milestones in physiology, behavior, and cognition
	Childhood	Years 3 to 7: Moderate growth rate, dependency on older people for care and feeding, midgrowth spurt, eruption of first permanent molar and incisor, cessation of brain growth by end of stage
	Juvenile	Years 7–10 for girls, 7–12 for boys: slower growth rate, capable of self-feeding, cognitive transition leading to learning of economic and social skills

(continued)

Table 11.1 *(Continued)*

Stage		Growth events/duration (approximate or average)
	Puberty	Occurs at end of juvenile stage and is an event of short duration (days or a few weeks): reactivation of central nervous system mechanism for sexual development, dramatic increase in secretion of sex hormones
	Adolescence	The stage of development that lasts for 5–10 years after the onset of puberty: growth spurt in height and weight; permanent tooth eruption almost complete; development of secondary sexual characteristics; sociosexual maturation; intensification of interest in and practice of adult social, economic, and sexual activities
Adulthood		
	Prime and transition	From 20 years old to end of child-bearing years: homeostasis in physiology, behavior, and cognition; menopause for women by age 50
	Old age and senescence	From end of child-bearing years to death: decline in the function of many body tissues or systems

cellular growth, **hyperplasia** and **hypertrophy**. Hyperplasia involves cell division by **mitosis**. For instance, **epidermal** cells of the skin form by the mitotic division of germinative cells in the deep layers of the skin. Hypertrophic growth involves the enlargement of already existing cells, as in the case of adipose cells growing by incorporating more lipid (fat) within their cell membranes.

Goss also describes three strategies of growth used by different tissues: renewal, expansion, and stasis. **Renewing tissues** include blood cells, **gametes** (sperm and egg cells), and the epidermis. Mature cells of renewable tissue are incapable of mitosis and have relatively short lives; for instance, red blood cells (erythrocytes) survive in circulation for ~6 months. New erythrocytes are produced from a reserve of undifferentiated cells located in the bone marrow and lymphatic organs. The rate of replacement of renewing tissues is carefully balanced against the rate of death of the mature cells. **Expanding tissues** include the liver, kidney, and the endocrine glands, the cells of which retain their mitotic potential even in the differentiated state. The liver, for example, has no special germinative layer or compartment, and most liver cells are capable of hyperplasia to replace damaged cells. **Static tissues,** such as nerve cells and striated muscle, give up the ability to divide by mitosis early in their lives. Once formed, a static tissue can grow in size by hypertrophy but not by hyperplasia. Thus, when a person exercises a muscle, its size may increase by enlarging the individual muscle cells already present, not by adding new cells. Also, because the reserve of germ cells is limited and is usually depleted early in life, the pool of static tissues cannot be renewed if damaged or destroyed. The destruction of brain tissue, as a result of accident or stroke, often means the permanent loss of the damaged cells and the functions they once performed. However, unlike renewable tissues, which have short lives, static tissues usually live as long as the person survives.

The biological substrate of the individual is not permanent. From embryonic life through adulthood, the human body is in a constant state of decomposition and reorganization. Young adult men renew about 2–3% of their muscle mass each day. In infancy, when new muscle tissue is forming by hyperplasia, the rate of protein renewal is about 6–9% per day. The magnitude of this metabolic renewal may be appreciated by the fact that much of the **basal metabolic rate** of the body (which may be measured by the heat that the body produces when at complete rest) is due to protein turnover. A similar turnover of cellular material occurs in other static tissues, such as nerve cells in the body and in the brain, and in expanding tissues. Tanner (1978) wrote,

> This dynamic state enables us to adapt to a continuously changing environment, which presents now an excess of one type of food, now an excess of another; which demands different levels of activity at different times; and which is apt to damage the organism. But we pay in terms of the energy we must take in to keep the turnover running. Enough food must be taken in to provide this energy, or the organism begins to beak up.

During the years and decades of life, the turnover and renewal of the molecular constituents of a human being's cells must take place often enough to recreate the entire body many times over.

The metabolic dynamic of the human organism is most active during the first trimester of prenatal life. The multiplication of millions of cells from the fertilized ovum—and the differentiation of these cells into hundreds of different body parts—makes this earliest period of life highly susceptible to growth pathology caused by either the inheritance of genetic **mutations** or exposure to harmful environmental agents that disrupt the normal course of development (e.g., certain drugs, smoking, malnutrition, disease, and psychological trauma that the mother may experience). Due to these causes and others, it is estimated that ~10% of human fertilizations fail to implant in the wall of the uterus, and of those that do, ~50% are spontaneously aborted. It is consoling to know, perhaps, that most of these spontaneous abortions occur so early in pregnancy that the mother and father are not aware that a conception took place.

After the initial embryonic tissues are formed, the first trimester is taken up with organogenesis, the formation of organs and physiological systems of the body. By the eighth week, the embryo is recognizably human. By the start of the second trimester of pregnancy, the differentiation of cells into tissues and organs is complete, and the embryo has become a fetus. During the next 12 or so weeks, most of the growth that takes place in the fetus is in length. At eight weeks after conception, the embryo is ~30 millimeters (mm) long (1 inch). By the fourth month, the length from the crown of the head to the bottom of the buttocks (crown–rump length) is ~205 mm (8 inches), and by the sixth month, between 356 and 381 mm (14 inches), which is ~70% of average birth length.

Increases in weight during this same period are much less rapid. Eight weeks after conception, the embryo weighs 2.0–2.7 grams (g; 0.08 ounces [oz.]), and by the sixth month, the fetus weighs only 700 g (1.5 pounds [lb]), which is ~20% of birth weight. It is during the third trimester of pregnancy that growth in weight takes place at a relatively faster rate. During the third trimester, several physiological systems (e.g., circulatory, respiratory, and digestive) also develop and mature, preparing the fetus for the transition to extrauterine life after birth.

Birth

Birth is a critical transition between life in utero and life independent of the support systems provided by the uterine environment. The **neonate** moves from a fluid to a gaseous environment, from a nearly constant external temperature to one with potentially great volatility. The newborn is also removed from a supply of oxygen and nutrients (provided by the mother's blood and passed through the placenta, which also handles the elimination of fetal waste products) to a reliance on his or her own systems for digestion, respiration, and elimination.

The difficulty of the birth transition is illustrated in the data presented in Figure 11.2. Nearly one-half of all neonatal deaths occur during the first 24 hours after birth. Of course, most of these deaths are not attributable to the birth process itself; rather, the leading factor associated with neonatal death is **low birth weight** (defined as a weight <2500 g [5.5 lb] for a full-term birth). Low birth weight is the result of growth retardation during fetal life. The cause of this growth retardation may be **congenital** (hereditary or inborn) problems with the fetus; placental insufficiency; maternal undernutrition, disease, smoking, or alcohol consumption; or other causes.

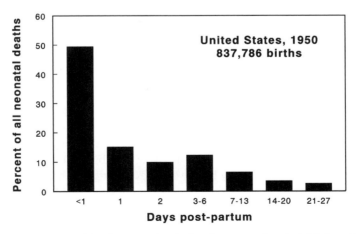

Figure 11.2 Percent of deaths occurring during the neonatal period (birth to day 28). Data from the United States, all registered births for 1950. These data are presented in lieu of more recent data because the technology for extraordinary neonatal medical care in existence today reduces neonatal deaths.

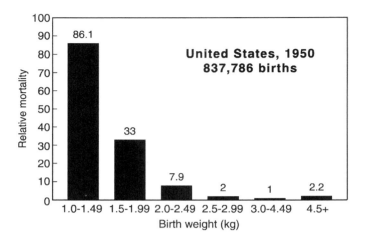

Figure 11.3 Index of relative mortality by birth weight during the neonatal period. Data from the United States, all registered births for 1950. Relative mortality is calculated as the risk of death for an infant born at a given weight. Infants of normal birth weight, 3.0–4.5 kg (6.5–10 lb), have a relative mortality of 1.0. Infants born at 2.5–3.0 kg (5.5–6.5 lb) have twice the risk of death as infants with normal birth weights, and so on.

An index of relative **mortality** during the neonatal period by birth weight is given in Figure 11.3. Relative mortality is defined as the percentage of deaths in excess of the number that occur for infants within the normal birth weight range of 3.0–4.5 kilograms (6.5–10 lb). These data are for infants at all gestational ages. **Prematurity,** defined as birth before 37 weeks gestation, may cause additional complications that increase the chances of neonatal death. However, an infant small for gestational age—that is, of low birth weight—is usually at greater risk of death than a premature child of the expected weight for gestational age.

There also is a strong relationship between low birth weight, with or without prematurity, and **socioeconomic status** (SES). SES is a concept devised by the social sciences to measure some aspects of education, occupation, and social prestige of a person or social group. The incidence of low birth weight in the wealthy, high-SES, developed nations is 5.9% of all live births; in the poor, low-SES, developing nations, the incidence is 23.6%. Even in the United States, the socioeconomic relationship with birth weight is strong. When educational attainment is used to estimate general SES, researchers found that 10.1% of births to women with less than 12 years of schooling had low birth weights, compared with 6.8% of births to women with 12 years and 5.5% to those with 13 or more years of formal education. African Americans and other minority groups show consistent lower average birth weights compared with Whites, and part of this difference is due to the lower SES of the minority groups. However, when White and Black women are matched for socioeconomic status, Black women still give birth to a higher percentage of low-birth-weight infants. This suggests that ethnic or genetic factors also determine birth weight. However, recent evidence shows that the lower birth weight of the

Black infants is likely due to several generations of SES differences and discrimination. Emanuel and colleagues (1992) try to explain the reason for the persistence of lower birth weights for infants born to Black mothers by the "intergenerational effect hypothesis." By this, the researchers mean that the SES matching is valid only for the current generation of adult women. The mothers and grandmothers of these Black and White women (from the United States and Britain) were less likely to be equally matched for SES. Given the social history of the United States and Britain, previous generations of Black women were likely to be of lower SES than their White counterparts. The intergenerational effect hypothesis predicts that the poor growth and development of women from older generations will have a lasting effect on the current generation.

Several lines of research estimate the **variance** in birth weight due to fetal genotype to be 10%, the variance due to maternal genotype to be 24%, and the variance due to nongenetic maternal and environmental factors to be 66% of the total variance. Because of the predominance of nongenetic factors, public health workers often use birth weight statistics as one indication of the well-being of a population.

Weight at birth is only one measurement that is commonly taken to indicate the amount of growth that took place during prenatal life. **Recumbent length** (length of the body when lying down); circumference of the head, arm, and chest; and **skinfolds** are other measurements. Circumferences measure the contribution made by various tissues to the size of different body parts. For example, head circumference measures the maximum girth of the skull, which includes bone and, more importantly, the size of the brain. Some representative data on size at birth and at 18 years of age for several measures are given in Table 11.2. At birth, **sexual dimorphism** (a difference in appearance or behavior between males and females) in size is biologically insignificant. At 18 years of age, however, there is considerable dimorphism between the sexes in average stature, body weight, and fatness as measured by **triceps** and **subscapular skinfolds**.

Table 11.2   Size at birth and at age 18 years for children born in the United States. Data from various sources of nationally representative statistics

	Birth		18 years	
	Boys	Girls	Boys	Girls
Recumbent length/stature (cm)	49.9	49.3	176.6	163.1
Weight (kg)	3.4	3.3	71.4	58.3
Head circumference (cm)	34.8	34.1	55.9	54.9
Triceps skinfold (mm)	3.8	4.1	8.5	17.5
Subscapular skinfold (mm)	3.5	3.8	10.0	12.0
Arm muscle area (mm$_2$)	20.4[a]	19.6[a]	75.0	57.2

[a]Measured at 1 year old.

(*Source*: Bogin 1988; arm muscle data from Frisancho 1990)

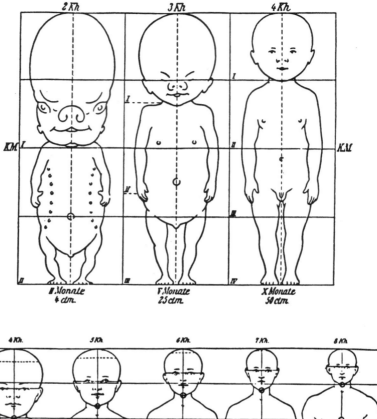

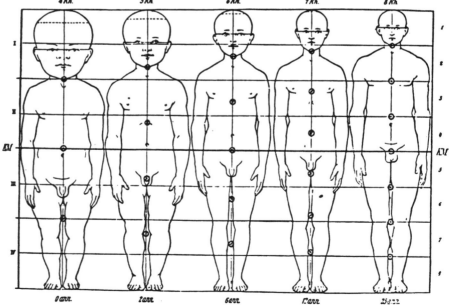

Figure 11.4 Diagrams illustrating the changes in body proportions of human beings that occur during prenatal and postnatal growth. Human body segments (head, trunk, arms, and legs) grow at different rates and generally mature from head to foot. (*Source*: Stratz 1909)

From birth to adulthood, humans experience many changes in body size, shape, composition (e.g., fat, muscle, and water content), proportions, and skeletal formation (Figures 11.4 and 11.5). The importance of studying these contrasts between early and later life is twofold. First, they allow clinicians and researchers to assess a child's stage of biological maturation for different organs, tissues, or the body as a whole independent of chronological age. Biological maturation is used to help determine whether a child is developing too slowly or too quickly, either of which may indicate the presence of some disorder. Second, the contrasts between early and later life are also conceptually important. They show that the infant may take one of several different paths for growth, maturation, and functional development. Adult human morphology, physiology, and behavior are plastic and in no way rigidly predetermined.

Plasticity refers to the ability of an organism to modify its biology or behavior to respond to changes in the environment, particularly stressful changes. Of course, people cannot sprout wings or breathe under water, but the sizes, shapes, colors, emotions, and intellectual abilities of people can be significantly altered by environmental stress, training, and experience. When the biology and behavior of people are considered together (i.e., in a biocultural perspective), it seems that human beings are perhaps the most plastic of all species, hence the most variable and adaptable.

Postnatal Life

Life after birth can be divided into distinct periods in many ways. Bogin (1999) proposes a five-stage model of human postnatal growth and development: infant, child, juvenile, adolescent, and adult. The rationale for this model begins with an analysis of the amount and rate of growth from birth to adulthood.

To visualize the amount and rate of growth that takes place during each of these stages, the growth in height (or length) for normal boys and girls is depicted in Figure 11.6; growth in weight follows very similar curves. The stages of growth are also outlined in Table 11.1. In Figure 11.6, the **distance curve** of growth, that is, the amount of growth achieved from year to year, is labeled on the right Y-axis. The **velocity curve,** which represents the rate of growth during any one year, is labeled on the left Y-axis. Below the velocity curve are symbols that indicate the average duration of each stage of development. Clearly, changes in growth rate are associated with each stage of development. Each stage also may be defined by characteristics of the dentition, changes related to methods of feeding, physical and mental competencies, or maturation of the reproductive system and sexual behavior.

Infancy is characterized by the most rapid velocity of growth of any of the postnatal stages. The infant's rate of growth is also characterized by a steep decline in velocity, a deceleration. The infant's curve of growth, rapid velocity and deceleration, is a continuation of the fetal pattern; the rate of growth in length reaches a peak in the second trimester and then begins a deceleration that lasts until childhood (Figure 11.7). As for all mammals, human infancy is the period when the mother provides all or some nourishment to her offspring via lactation or some culturally derived imitation of lactation. During infancy, the deciduous dentition (the so-called milk teeth) erupt through the gums. Human infancy ends when the child is weaned from the breast, which in preindustrialized societies occurs between 24 and 36 months of age. By this age, all the deciduous teeth have erupted, even for very late-maturing infants.

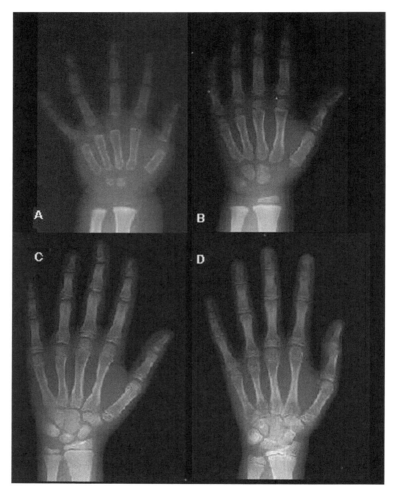

Figure 11.5 Radiographs of the hand and wrist at different skeletal ages, illustrating the sequence of bone maturation events. All radiographs from healthy girls. (A) At 6 months, most wrist bones and the growing ends of the finger bones (called epiphyses) are formed of cartilage. At certain X-ray exposures, this cartilage is "invisible." The image shows few **centers of ossification** (i.e., places where bone is present) in the wrist and few visible epiphyses. (B) At 3 years old, some wrist ossification centers are present, epiphysis of radius is present, and most epiphyses of the hand are calcified (i.e., forming bone). (C) At 8 years old, all ossification centers are calcified. (D) At 13 years old, all bones have assumed final shape, but growth in size remains to be completed.

Motor skills (i.e., what a baby can do physically) develop rapidly during infancy. At birth, states of wakefulness and sleep are not sharply differentiated, and motor coordination is variable and transient. By 1 month, the infant can lift its chin when prone, and by 2 months, lift its chest by pushing up with the hands and arms. The infant can sit with support by 4 months, sit without support by 7 months, crawl by 8 months, and walk with support by 12 months. By 2 years of age, the infant can walk well and turn the pages of a book, one at a time. By 3 years of age, the end of the infancy stage, the youngster can run smoothly, pour water from a pitcher, and manipulate small objects, such as blocks, well enough to control them. There is a similar progression of changes in the problem solving, or cognitive, abilities of the infant.

The development of the skeleton, musculature, and the nervous system account for all of these motor and cognitive advancements. The rapid growth of the brain, in particular, is important. Later in this chapter, we explain how the evolution of the relatively large human brain has influenced the total pattern of human growth and development. We point out that the human brain grows rapidly during infancy, much more rapidly than almost any other tissue or organ of the body (Figure 11.8). All parts of the brain seem to take part in this fast pace of growth and maturation, including the structures that control the reproductive system.

The **hypothalamus,** a center of neurological and endocrine control, is one of these brain structures. During fetal life and early infancy, the hypothalamus produces relatively high levels of **gonadotropin-releasing hormone** (GnRH). This

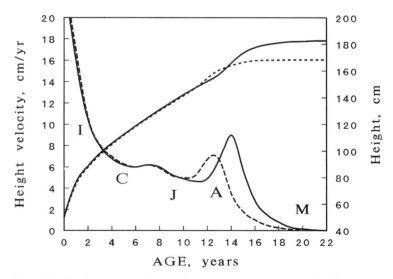

Figure 11.6   Idealized mean velocity and distance curves of growth in height for healthy girls (dashed lines) and boys (solid lines) showing the postnatal stages of the pattern of human growth. Note the spurts in growth rate at midchildhood and adolescence for both girls and boys. The postnatal stages: I, infancy; C, childhood; J, juvenile; A, adolescence; M, mature adult. (*Source*: Data from Prader 1984 and Bock and Thissen 1980)

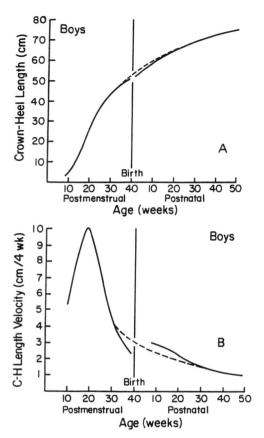

Figure 11.7 Distance (A) and velocity (B) curves for growth in body length during human pre-natal and postnatal life. The graphs are diagrammatic, because they are based on several sources of data. Dashed lines depict the predicted curve of growth if no uterine restriction takes place. In fact, such restriction does take place toward the end of pregnancy and may impede the flow of oxygen or nutrients to the fetus. Growth rate slows but rebounds after birth and returns the infant to the size she or he would be without any restriction. C-H, crown–heel. (*Source*: Tanner 1990)

hormone causes the release of **luteinizing hormone** (LH) and **follicle-stimulating hormone** (FSH) from the **pituitary** gland. LH and FSH travel in the bloodstream to the gonads (**ovaries** or **testes**), where they stimulate the production and release of **estrogen** or **androgen** hormones (Figure 11.9). These gonadal hormones are, in part, responsible for the rapid rate of growth during early infancy. By late infancy, however, the hypothalamus is inhibited for reasons that are not completely known. GnRH secretion almost stops, and the levels of the sex hormones fall, which sus-pends reproductive maturation (see Figure 11.8, "Reproductive" curve). The hypo-thalamus is reactivated just before **puberty**, the event of development that marks the onset of sexual maturation.

The **childhood** stage follows infancy, encompassing the ages of about 3–7 years. Childhood may be defined by its own pattern of growth, feeding behavior, and motor and cognitive development. The growth deceleration of infancy ends at the beginning of childhood, and the growth rate levels off at ~5 centimeters (cm) per year. This leveling-off in growth rate is unusual for mammals, because almost all other species continue a pattern of deceleration after infancy (look ahead to Figures 11.13 and 11.14).

This slow and steady rate of human growth maintains a relatively small-sized body during the childhood years. In terms of feeding, children are weaned from the breast or bottle but still depend on older people for food and protection. Most mammalian species move from infancy and its association with dependence on nursing to a stage of independent feeding. Postweaning dependency is found in several species of social mammals, especially carnivores (such as lions, wild dogs, and hyenas) and in some species of primates. Lion cubs, for example, are weaned at about 6–8 months old but remain dependent on their mothers until ~24 months old. During that time, the cubs must learn how to hunt for themselves. For many species of primates, learning to hunt for high-quality foods, such as insects, and learning how to open fruits and seeds with tough skins also requires a period of postweaning dependence on the mother and, sometimes, the father (as for marmosets and tamarins).

Postweaning dependency is, by itself, not a sufficient criteria to define human childhood. Human children do, of course, learn how to find and prepare food, but a suite of features define the childhood stage. Not all of these features are found for the

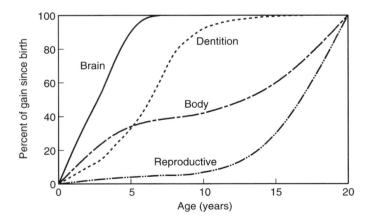

Figure 11.8 Growth curves for different body tissues. The "Brain" curve is for total weight of the brain (Cabana et al. 1993). The "Dentition" curve is the median maturity score for girls based on the seven left mandibular teeth (I1, I2, C, PM1, PM2, M1, M2) (Demirjian (1986). The "Body" curve represents growth in stature or total body weight, and the "Reproductive" curve represents the weight of the gonads and primary reproductive organs (Scammon 1930).

social carnivores and nonhuman primates. Human children require specially prepared foods because of the immaturity of their dentition, the small size of their stom-
achs and intestines, and the rapid growth of their brain (Figure 11.8). Again, we
emphasize that the human brain is especially important. The newborn uses 87% of its
resting metabolic rate (RMR) for brain growth and function. By the age of 5 years,

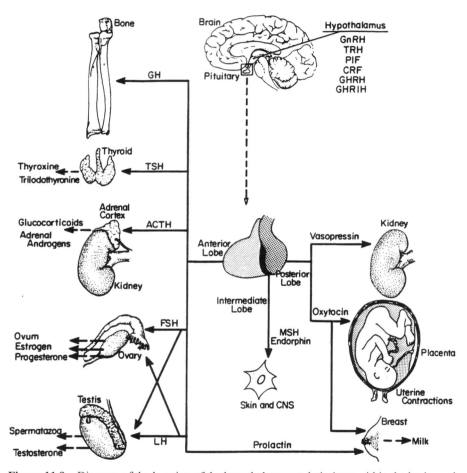

Figure 11.9   Diagram of the location of the hypothalamus and pituitary within the brain, and
a schematic illustration of the target organs and tissues of the pituitary hormones. GnRH, gonadotropin-releasing hormone; TRH, thyrotropin-releasing hormone; PIF, **prolactin**-release inhibiting factor; CRF, adrenocorticotropin-releasing factor; GHRH, growth hormone-releasing
hormone; GHRIH, growth hormone release-inhibiting hormone; GH, growth hormone; TSH,
thyroid stimulating hormone; ACTH, **adrenocorticotropic hormone**; FSH, follicle-stimulating hormone; LH, luteinizing hormone; MSH, melanocyte-stimulating hormone. (*Source*:
Schally et al. 1977)

the percent RMR usage is still high at 44%, whereas in the adult human, the figure is between 20 and 25% of RMR. At comparable stages of development, the RMR values for the chimpanzee are about 45, 20, and 9% respectively.

The human constraints of immature dentition and small digestive system necessitate a childhood diet that is easy to chew and swallow and low in total volume. The child's relatively large and active brain, almost twice the size of an adult chimpanzee's brain, requires that the low-volume diet be dense in energy, lipids, and proteins. Children do not yet have the motor and cognitive skills to prepare such a diet for themselves. Children are also especially vulnerable to predation and disease and thus require protection. Children will not survive in *any* society if deprived of the care provided by older individuals. So-called wolf children and even street children, who are sometimes alleged to have lived on their own, are either myths or not children at all. A search of the literature finds no case of a child (i.e., a youngster under the age of 6 years) living alone, either in the wild or on urban streets.

Two of the important physical developmental milestones of childhood are the replacement of the deciduous teeth with the eruption of the first permanent teeth, and completion of brain growth (in weight). First molar eruption takes place, on average, between the ages of 5.5 and 6.5 years in most human populations. Eruption of the central incisor quickly follows, or sometimes precedes, the eruption of the first molar. By the end of childhood, usually at the age of 7 years, most children have erupted the four first molars, and permanent incisors have begun to replace "milk" incisors. Along with growth in size and strength of the jaws and the muscles for chewing, these new teeth provide sufficient capabilities to eat a diet similar to that of adults.

A close association between human dental development and other aspects of growth and maturation was noted many years ago by anatomists and anthropologists. More recently, Smith (1991) analyzed data from humans and 20 other primates species and found that age of eruption of the first molar is highly associated with brain weight (**correlation** coefficient $[r] = 0.98$, 1.00 is a perfect correlation) and a host of other growth and maturation variables. Other research, based on direct measurements of victims of accidents and disease, shows that human brain growth in weight is complete at a mean age of 7 years. Thus, at this stage of development, not only is the child capable dentally of processing an adult-type diet; the nutrient requirements for brain growth also diminish. Moreover, cognitive and emotional capacities mature to new levels of self-sufficiency. Language and symbolic thinking skills mature rapidly, social interaction in play and learning become common, and the 7-year-old individual can perform many basic tasks, including food preparation with little or no supervision.

Another feature of the childhood phase of growth associated with these physical and mental changes is the modest acceleration in growth velocity at about 6–8 years, called the **midgrowth spurt** (shown in Figure 11.6). Some studies note the presence of the midgrowth spurt in the velocity curve of boys but not girls. Others find that up to two-thirds of boys and girls have midgrowth spurts. The midgrowth spurt is linked with an endocrine event called **adrenarche,** the progressive increase in the secretion of adrenal androgen hormones. Adrenal androgens produce the midgrowth spurt in height, a transient acceleration of bone maturation, and the appearance of axillary and pubic hair. They also seem to regulate the development of body fatness and fat distribution. The mechanism controlling adrenarche is not understood because no known hormone appears to cause it. Adrenarche is found only in chimpanzees and humans, and the midgrowth spurt is apparently unique to human beings.

In the 1920s, there was speculation that adrenarche was associated with sexual maturation. However, today it is known that there is no connection between the occurrence or timing of adrenarche and **gonadarche,** the reactivation of production and secretion of gonadal hormones at the start of puberty. Perhaps the evolution of adrenarche and the human midgrowth spurt may be explained as mechanisms that serve as biocultural markers of physical and mental maturation. We mean that the physical changes induced by adrenarche are accompanied by a change in cognitive function, called the "5-to 7-year-old shift" by some psychologists, or the shift from the preoperational to concrete operational stage, using the terminology of Piaget. This shift leads to new learning and work capabilities in the juvenile. Adrenarche and the human midgrowth spurt may function as a **life history** event, marking the transition from the childhood to the juvenile growth stage.

In summary, human childhood is defined by the following traits:

- slow and steady rate of growth and relatively small body size;
- a large, fast-growing brain;
- higher RMRs than any other mammalian species;
- immature dentition;
- motor immaturity;
- cognitive immaturity; and
- both adrenarche and the midgrowth spurt.

No other mammalian species has this entire suite of features.

Juveniles may be defined as "prepubertal individuals that are no longer dependent on their mothers (parents) for survival" (Pereira and Altmann 1985, 236). This definition is derived from ethological research with social mammals, especially nonhuman primates, but applies to the human species as well. In contrast to infants and human children, juvenile primates can survive the death of their adult caretaker. Ethnographic research shows that juvenile humans have the physical and cognitive abilities to provide much of their own food and to protect themselves from predation and disease. The so-called street children mentioned above are in fact "street juveniles"!

The human juvenile stage begins at about 7 years old. In girls, the juvenile period ends, on average, at about the age of 10, 2 years before it usually ends in boys, the difference reflecting the earlier onset of adolescence in girls. The juvenile stage is characterized by the slowest rate of growth since birth. Studies of juvenile primates and human juveniles in many cultures indicate that much social learning takes place during this stage. Human boys and girls learn a great deal about important adult activities, including the production of food and methods of infant and child care. The completion of growth in weight of the brain and the onset of new cognitive competencies allow for this increased intensity of learning. Because juveniles are prepubertal, they can attend to this kind of social learning without the distractions caused by sexual maturation. As an aside, the start of the juvenile stage coincides with entry into traditional formal schooling in the industrialized nations. The connection is hardly a coincidence, because the juvenile stage allows for the kinds of learning and socialization found in school environments.

Human **adolescence** is the stage of life when social and sexual maturation takes place. Adolescence begins with puberty, or more technically with gonadarche, which is an event of the neuroendocrine system. The current understanding of the control of gonadarche is that one, or perhaps a few, centers of the brain change their pattern of neurological activity and their influence on the hypothalamus. The hypothalamus, which has been basically inactive in terms of sexual development since about age 3 years, is again stimulated to produce GnRH. It is not known exactly how this change takes place. As stated above, the production of GnRH by the hypothalamus becomes inhibited by the age of about 3 years. The "inhibitor" has not been identified but likely is located in the brain and certainly not in the gonads. Human children born without gonads as well as rhesus monkeys and other primates whose gonads have been surgically removed still undergo both hypothalamus inhibition in infancy and hypothalamus reactivation at puberty (Figure 11.10a). The transition from juvenile to adolescent stages requires not only the renewed production of GnRH but also its secretion from the hypothalamus in pulses (Figure 11.10b). Gonadarche is triggered when the pulsatile secretion reaches the necessary frequency (i.e., number of pulses in a given time period) and amplitude (i.e., peak amount of secretion during each pulse).

None of these hormonal changes can be seen without sophisticated technology, but the effects of gonadarche can be noted easily as visible and audible signs of sexual maturation. One such sign is a sudden increase in the density of pubic hair (indeed, the term "puberty" is derived from the Latin *pubescere,* "to grow hairy"). In boys, the deepening of the voice is another sign of puberty. In girls, a visible sign is the development of the breast bud, the first stage of breast development. The pubescent boy or girl, his or her parents, and relatives, friends, and sometimes everyone else in the social group can observe these signs of early adolescence.

The adolescent stage also includes development of **secondary sexual characteristics,** such as development of the external genitalia, sexual dimorphism in body size and composition (Figure 11.11), and the onset of greater interest and practice of adult patterns of sociosexual and economic behavior. These physical and behavioral changes occur with puberty in many species of social mammals. What makes human adolescence unusual amongst the primates are two important differences. The first is the length of time between age at puberty and age at first birth. Humans take, on average, at least 10 years for this transition. The average ages for girls are puberty at 9 and first birth at 19 years old; for boys, puberty takes place at 11 and fatherhood at 21-25 years old. The reasons for delay between puberty and first birth or fatherhood are discussed below. The point to make here is that monkeys and apes take less than 3 years to make the transition from puberty to parenthood.

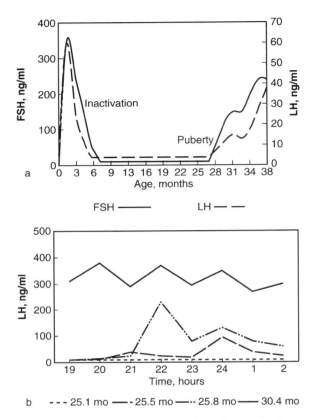

Figure 11.10 (a) Pattern of secretion of FSH and LH in a male rhesus monkey (genus **Macaca**). The testes of the monkey were removed surgically at birth. The curves for FSH and LH indicate the production and release of GnRH from the hypothalamus. After age 3 months (i.e., during infancy), the hypothalamus is inactivated. Puberty takes place at ~27 months, and the hypothalamus is reactivated. (b) Development of hypothalamic release of GnRH during puberty in a male rhesus monkey with testes surgically removed. At 25.1 months (mo) of age, the hypothalamus remains inactivated. At 25.5 and 25.8 mo, modest hypothalamic activity is observed, indicating the onset of puberty. By 30.4 mo, the adult pattern of LH release is nearly achieved. This pattern shows increases in both the number of pulses of release and the amplitude of release. In human beings, a very similar pattern of infant inactivation and late juvenile reactivation of the hypothalamus takes place. (*Source*: Adapted, with some simplification, from Plant 1994)

The second human difference is that during this life stage, both boys and girls experience a rapid acceleration in the growth velocity of almost all skeletal tissue— the adolescent growth spurt (see Figure 11.6, velocity curve). The magnitude of this acceleration in growth was calculated for a sample of healthy Swiss boys and girls measured annually between the ages of 4 and 18 years. At the peak of their adolescent growth spurt, the average velocity of growth in height was +9.0 cm/year (3.5

inches/year) for boys and +7.1 cm/year (2.8 inches/year) for girls. Similar average values are found for adolescents in all human populations, making the human adolescent growth spurt a species-specific characteristic.

Other primate species may show a rapid acceleration in soft tissue growth, such as muscle mass in many male monkeys and apes. However, unlike humans, other primate species either have no acceleration in skeletal growth or a very small increase in growth rate. Watts and Gavan (1982) found that for chimpanzees, the increase in the velocity of growth at the time of sexual maturation of individual long bones is "usually less than a centimeter" and often <5.0 mm/year (0.2 inches/year). In contrast, the velocity of human long bone growth may be five times as rapid as that of the ape. An analysis of the growth of individual limb segments of British boys by Cameron and colleagues (1982) found that peak adolescent velocities ranged from 1.34 cm/year (0.5 inches/year) for the forearm to 2.44 cm/year (1 inch/year) for the tibia.

Another important ape–human difference in growth is that by the time a chimpanzee begins its modest acceleration in long bone growth, the animal has already completed 88% of its skeletal growth. At the onset of the human adolescent growth spurt, boys and girls have completed only 81% of their skeletal growth.

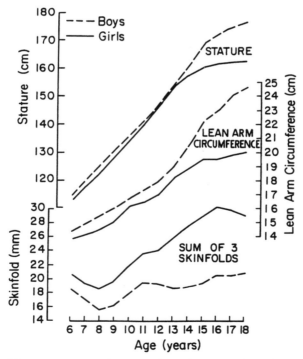

Figure 11.11   Mean stature, mean lean arm circumference, and median of the sum of three skinfolds for Montreal boys and girls. Notice that sexual dimorphism increases markedly after puberty (~12–13 years old). (*Source*: Baughn et al. 1980)

That 9% difference in **skeletal maturation** will disappear only after the human adolescent completes his or her adolescent growth spurt. Clearly, the human pattern of growth following gonadarche is quantitatively different in terms of amount, rate, and duration of growth from the pattern for other primates. The human skeletal growth spurt is unequaled by other species, and when viewed graphically, the growth spurt almost defines human adolescence (Figure 11.6). We say almost, because human adolescence is also defined by several other changes in behavior and cognition that are found only in our species. We discuss these changes in later sections.

Adolescence ends and early **adulthood** begins with the completion of the growth spurt, the attainment of adult stature, and the achievement of full reproductive maturity, meaning both physical and psychosocial maturity. Height growth stops when the long bones of the skeleton (e.g., femur, tibia, etc.) lose their ability to increase in length. Usually this occurs when the **epiphysis,** the growing end of the bone, fuses with the **diaphysis,** the shaft of the bone. As shown in Figure 11.5, the process of epiphyseal union can be observed from radiographs of the skeleton. The amount of growth that occurs late in adolescence is a function of skeletal maturation; late maturers grow more than average or early maturers. This fact has been known for many years, and an estimate of skeletal maturation, often called **skeletal age,** is incorporated into equations used to predict the adult height of children. The fusion of epiphysis and diaphysis (Figure 11.12) is stimulated by the gonadal hormones, the androgens and estrogens. However, it is not merely the fusion of epiphysis and diaphysis that stops growth, because children without gonads or whose gonads are not functional never have epiphyseal fusion, even though they stop growing. Rather, it is a change in the sensitivity to growth stimuli of cartilage and bone tissue in the **growth plate region** (Figure 11.12) that causes these cells to lose their hyperplastic growth potential.

Reproductive maturity is another hallmark of adulthood. The production of viable spermatozoa in boys and viable **oocytes** in girls is achieved during adolescence, but these events mark only the early stages, not the completion, of reproductive maturation. Socioeconomic, and psychobehavioral maturation must accompany physiological development. All of these developments coincide, on average, by about age 19 in women and 21–25 years of age in men. We discuss possible reasons for the human sequence of biological, social, and psychological development toward adulthood, in terms of the evolutionary background of human reproductive development, later in this chapter.

The transition to adulthood is marked by several events, including the cessation of height growth and achievement of full reproductive maturity. In contrast, the course of growth and development during the prime reproductive years of adulthood are relatively uneventful. Most tissues of the body lose the ability to grow by hyperplasia (cell division), but many may grow by hypertrophy (enlargement of existing cells). Exercise training can increase the size of skeletal muscles, and caloric oversufficiency certainly will increase the size of adipose tissue. However, the most striking feature of the prime adult stage of life is its stability, or **homeostasis,** and its resistance to pathological influences, such as disease-promoting organisms and psycho-

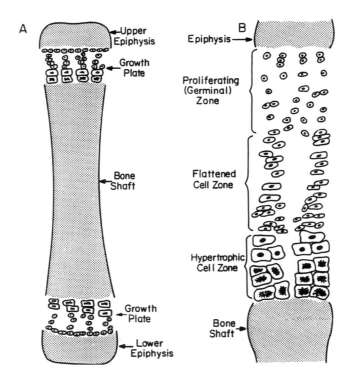

Figure 11.12   (A) Diagram of a limb bone with its upper and lower epiphyses. (B) Diagrammatic enlargement of the growth plate region: New cells are formed in the proliferation zone and pass to the hypertrophic zone to add to the bone cells accumulating on top of the bone shaft. (*Source*: Tanner 1990)

logical stress. It contrasts with the preceding stages of life, which were characterized by change and susceptibility to pathology.

Old age and **senescence** follow the prime years of adulthood. The aging period is one of gradual or sometimes rapid decline in the ability to adapt to environmental stress. The pattern of decline varies greatly between individuals. Although specific molecular, cellular, and organismic changes can be measured and described, not all changes occur in all people, and rarely do they follow a well-established sequence. Unlike the biological self-regulation of growth prior to adulthood, the aging process appears to follow no biological or genetic plan. Menopause may be the only event of the later adult years that is experienced universally by women who live past 50 years of age; men have no similar event. We discuss the biology and possible value of menopause later in this chapter.

There are many theories about the aging process and about why we must age at all. Chapter 13 discusses some of them and shows that aging is a multicausal process. We may state simply here that the inability of all cell types—renewing, expanding, and static—to use nutrients and repair damage leads to aging and death.

HUMAN POSTNATAL GROWTH IN COMPARATIVE PERSPECTIVE

In contrast to the process of aging and death, growth and development follow a predictable pattern from conception to adulthood. During the evolutionary history of our species and of those species ancestral to ours, selective pressures operated to shape our pattern of growth. Thus, to understand why we grow the way we do, we must examine some of the events that occurred during human evolution. To do this, we first examine human life history and then discuss the evidence for the evolution of the pattern of human growth.

Life History and Stages of the Life Cycle

Life history might be defined as the strategy an organism uses to allocate its energy toward growth, maintenance, reproduction, raising offspring to independence, and avoiding death. For a mammal, it is the strategy of when to be born, when to be weaned, how many and what type of prereproductive stages of development to pass through, when to reproduce, and when to die. Living things on earth have greatly different life history strategies, and understanding what shapes these histories is one of the most active areas of research in whole-organism biology.

Anthropologists have become increasingly interested in explaining the significance of human life history. This interest is due to the discovery that the human life cycle stands in sharp contrast to other species of social mammals, even other primates. Anthropological theory needs to explain how humans successfully combined a vastly extended period of offspring dependency and delayed reproduction with helpless newborns, a short duration of breast feeding, an adolescent growth spurt, and menopause. A central question is, did these characteristics evolve as a package

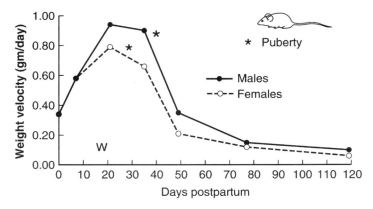

Fig. 11.13   Velocity curves for weight growth in the mouse. In both sexes puberty (*; vaginal opening for females or spermatocytes in testes of males) occurs just after weaning (W) and maximal growth rate. Weaning takes place between days 15 and 20. Sexual maturity follows weaning by a matter of days.

BOX 11.1   COLLECTION AND ANALYSIS OF GROWTH AND LIFE HISTORY DATA FROM MAMMALIAN SPECIES: A CLASS EXERCISE

The concept of life history and its importance in the study of human growth and development are easier to understand if we compare several features of human growth and the human life cycle with those of other species. In this exercise, we confine such comparisons to the mammals.

Each student in the class should select a species of mammal for study. The mammals chosen should include both large and small species and some that are terrestrial, aquatic, and capable of flight. Try to choose species that live in different habitats, such as tropical, temperate, desert, and arctic zones.

Complete the following data list for each chosen mammal:

Species: _____

Habitat: _____

Adult body size: _____ kilograms

Adult brain size: _____ kilograms (or cubic centimeters)

Length of gestation: _____ days

Age at weaning: _____ months

Age at reproductive maturity: _____ months

Number of offspring per pregnancy: _____ (mean litter size)

Interval between births: _____ (mean number of months)

Body temperature: _____ (mean adult value)

Mean life span: _____ months

Maximum life span: _____ months

This information can be found in encyclopedias such as *Walker's Mammals of the World* (Nowak and Paradiso 1983) and in books devoted to the various families and genera of mammals.

After you have collected the data, perform several analyses. As a start, enter all of the data into a database and use any statistical software package to compute correlations between each of the variables. Discuss the meaning of the pattern of correlations. Consult review articles on life history (e.g., Harvey et al. 1987) to help interpret the meaning of the data analysis. Use the correlation analysis to frame new questions about the pattern of life history variables. Then, devise a way to analyze the data to answer your questions.

or a mosaic? The present evidence suggests that human life history evolved as a mosaic and may have taken form over more than a million years.

Understanding the human condition requires a comparative approach, and we restrict such comparisons to the mammals. Recent work in mammalian life history and

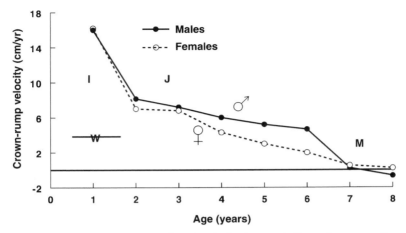

Figure 11.14   Baboon crown–rump length velocity. Letters indicate the stages of growth as in Figure 11.6. Weaning (W) may take place anytime between 6 and 18 months. (*Source:* Adapted from Coelho 1985)

its evolution focuses on the period of the life cycle from birth to adulthood. We can use some of the same criteria that we used for human development to define the stages of the life cycle for nonhuman mammals, especially changes in the rate of growth and the onset of reproductive maturation. The majority of mammals progress from infancy to adulthood seamlessly, without any intervening stages, and puberty occurs after the peak velocity of their postnatal growth. This pattern of postnatal growth is illustrated in Figure 11.13 using data for the mouse. Highly social mammals such as wolves, wild dogs, lions, elephants, and the primates (e.g., the baboon, Figure 11.14) postpone adulthood by inserting a period of juvenile growth and behavior between infancy and adulthood.

Some debate ensues as to the function of the juvenile stage for social mammals. The traditional "learning hypothesis" is that the juvenile period allows for the extended period of brain growth and learning necessary for success in the species of group-living social mammals: the social hierarchy, feeding skills such as hunting, and reproductive skills. A recent corollary to the learning hypothesis is the "starvation aversion hypothesis" proposed by Janson and Van Schaik (1993). Juveniles must forage for their own food, a skill that must be practiced until mature levels of success are achieved. Much of the foraging of juveniles is in competition with adults. This competition becomes clear during times of food scarcity, when juvenile primates die in greater numbers than infants or adults. The point is that a slow-growing juvenile requires less food input than a fast-growing infant, and the juvenile may practice feeding skills with less risk of starvation during the learning period.

Another possible explanation for a juvenile stage for social mammals may be called the "dominance hypothesis." Research with wild and captive primates shows that high-ranking individuals in the social hierarchy can suppress and inhibit the reproductive maturation of low-ranking individuals. The inhibition may be due to the stress of social intimidation acting directly on the endocrine system, or the suppression may be secondary to inadequate nutrition due to feeding competition. Juveniles are almost always low-ranking members of primate social systems. In the past, individuals with slow growth and delayed reproductive maturation after infancy may have survived to adulthood more often than individuals with rapid growth and maturation, and thus the juvenile stage may have evolved. Whatever the cause, Alexander (1990) points out that in broad perspective, "juvenile life has two main functions: to get to the adult stage without dying and to become the best possible adult." Adding a juvenile stage must have served this purpose well.

During human evolution, childhood and adolescence were added as new life stages between birth and adulthood. These new stages presumably add additional security and value to the whole of life history. Universally, human females who live long enough experience menopause, another new stage for primates and an event that may mark the passage from one phase of adult life to another. In the sections that follow, we offer some current ideas as to why and when these additional stages of the human life cycle evolved.

EVOLUTION OF THE HUMAN LIFE CYCLE

Why Do New Life Stages Evolve?

In *Size and Cycle,* Bonner (1965) develops the idea that the stages of the life cycle of an individual organism, a colony, or a society are "the basic unit of natural selection." Bonner's focus on life-cycle stages follows in the tradition of many of the 19th-century embryologists who proposed that speciation is often achieved by altering rates of growth of existing life stages and by adding or deleting stages. Bonner shows that the presence of a stage and its duration in the life cycle relate to such basic adaptations as locomotion, reproductive rates, and food acquisition. From this theoretical perspective, it is profitable to view the evolution of human childhood, adolescence, and perhaps menopause as adaptations for both feeding and reproduction.

Why Childhood?

Consider the data shown in Figure 11.15, which depicts several **hominoid** developmental landmarks. Compared with living apes, human beings experience developmental delays in eruption of the first permanent molar, age at **menarche,** and age at first birth. However, humans have a shorter infancy and a shorter birth interval, which in apes and traditional human societies almost coincide.

We discussed earlier that dental development is an excellent marker for life history in the primates. There is a very strong correlation between age at eruption of the first molar (M1 eruption) and cessation of brain growth. In general, primates wean infants about the time M1 erupts. This timing makes sense, because the mother must nurse her current infant until it can process and consume an adult diet, which requires at least some of the permanent dentition.

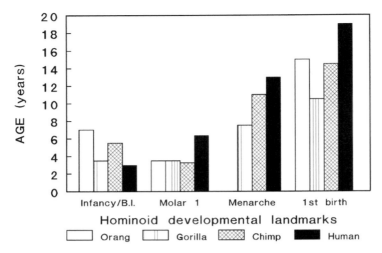

Figure 11.15 Hominoid developmental landmarks. Data based on observations of wild-living individuals or, for humans, healthy individuals from various cultures. Infancy/B.I. is the period of dependency on mother for survival, usually coincident with mean age at weaning and/or a new birth (where B.I. is birth interval); Molar 1 is mean age at eruption of first permanent molar; Menarche is mean age at first estrus or menstrual bleeding; 1st birth is mean age of females at first offspring delivery. Orang, *Pongo pygmaeus*; gorilla, *Gorilla gorilla*; chimp, *Pan troglodytes*; human, *Homo sapiens*. (*Source*: Bogin and Smith 1996)

Earlier in this chapter, we discussed primate and social carnivore species that provide food to their young during the postweaning period. In these species, the molar and premolar teeth may erupt before the postweaning dependency period ends. These primates and carnivores are similar in that both are predatory (the primates hunt insects) and the young need time to learn and practice hunting skills. The effect of postweaning dependency on the mother is delayed reproduction. In a review of reproductive behavior of the carnivores, Ewer (1973) found that female lions, hyenas, the sea otter (a fisher-hunter), and many bears wait two or more years between pregnancies. The same delay holds for the more carnivorous primates, such as marmosets, tamarins, and chimpanzees. From these data, we may conclude that reproduction in the social and predatory mammals occurs either at or well after the age at which the first permanent teeth erupt.

The human species is a striking exception to this relationship between permanent tooth eruption and birth interval. Women in traditional societies wait, on average, three years between births, not the six years expected on the basis of M1 eruption. The short birth interval gives humans a distinct advantage over other apes, because we can produce and rear two offspring through infancy in the time it takes chimpanzees or orangutans to produce and rear one offspring. By reducing the length of the infancy stage of life (i.e., lactation) and by developing the special features of the human childhood stage, humans have the potential for greater lifetime fertility than any ape.

Selection for increased reproductive success is the force that drives much of biological evolution. The evolution of the human childhood stage gave our species this reproductive advantage, because children no longer are fed by nursing. Children are still dependent on older individuals for feeding and protection. The child must be given foods that are specially chosen and prepared, which may be provided by older juveniles, adolescents, or adults. The mother does not have to provide 100% of offspring nutrition and care directly. Indeed, traditional societies deal with the problem of child care by spreading the responsibility among many individuals.

In Hadza society, for example (African hunters and gatherers), grandmothers and great aunts supply a significant amount of food and care to children. In Agta society (Philippine hunter-gatherers), women hunt large game animals but still retain primary responsibility for child care. They accomplish this dual task by living in extended family groups—two or three brothers and sisters, their spouses, children, and parents—and sharing the child care. Among the Maya of Guatemala (horticulturists and agriculturists), many people live together in extended family compounds. Women of all ages work together in food preparation, clothing manufacture, and child care. In some societies, fathers provide significant child care, including the Agta and the Aka pygmies, hunter-gatherers of central Africa. Summarizing the data from many human societies, Lancaster and Lancaster (1983) call this kind of child care and feeding "the hominid adaptation," because no other primate or mammal does all of this.

Childhood also may be viewed as a mechanism that allows for more precise "tracking" of ecological conditions by allowing more time for developmental plasticity. The **fitness** of a given **phenotype** (i.e., the physical features and behavior of an individual) varies across the range of variation of an environment. When phenotypes are fixed early in development, such as in mammals that mature sexually soon after weaning (e.g., rodents), environmental change and high mortality are positively correlated. Social mammals (carnivores, elephants, primates) prolong the developmental period by adding a juvenile stage between infancy and adulthood. Adult phenotypes develop more slowly in these mammals because the juvenile stage lasts for years. These social mammals experience a wider range of environmental variation, such as seasonal variation in temperature and rainfall. They also experience years of food abundance and food shortage as well as changes in the number of predators and in types of diseases. The result on the phenotype is a better conformation between the individual and the environment. Fitness is increased in that more offspring survive to reproductive age than in mammalian species without a juvenile stage. For example, ~4% of infant Norway rats (no juvenile stage) born in the wild survive to adulthood versus 14–16% of lions, which have a juvenile stage.

Monkeys and apes have juvenile periods that last as long or longer than those of social carnivores. Consequently, these primates rear between 12 and 36% of their offspring to adulthood. In addition to the primate juvenile stage, humans have a childhood stage of life. The human childhood stage adds another four years of relatively slow physical growth and allows for behavioral experience that further enhances developmental plasticity. The combined result is that humans in traditional societies (hunting and gathering, or horticultural groups) rear ~50% of their off-spring to adulthood. In the technologically advanced nations today, survival to adult-hood is appreciably higher. For the United States in the year 1990, an estimated 98.2% of infants born in 1970 survived to age 20 years.

The bottom line, in a biological sense, is that the evolution of human childhood decreases the interbirth interval and increases reproductive fitness. The relatively slow rate of body growth and small body size of children reduces competition with adults for food resources. Slow-growing small children require less food per day than larger juveniles, adolescents, and adults. Thus, although provisioning children is time-consuming, it is not as onerous a task of investment as it would be, for instance, if both brain and body growth were rapid simultaneously. Finally, the additional time for development added by childhood increases the quality of the young and its chance of surviving to adulthood.

Why Adolescence?

An adolescent stage of human growth may have evolved to provide the time to prac-tice complex social skills required for effective parenting. The evolution of childhood afforded hominid females the opportunity to give birth at shorter intervals, but pro-ducing offspring is only a small part of reproductive fitness. Rearing the young to their own reproductive maturity is a more sure indicator of success.

Studies of yellow baboons, toque macaques, and chimpanzees show that between 50 and 60% of first-born offspring die in infancy (see Bogin 1999 for references). By contrast, in hunter-gatherer human societies, between 39% (the Hadza of eastern Africa) and 44% (the !Kung of southern Africa) of offspring die in infancy. Studies of wild baboons by Altmann (1980) show that whereas the infant mortality rate for the first-born is 50%, mortality for second-born drops to 38%, and for third- and fourth-born reaches only 25%. The difference in infant survival is, in part, due to ex-perience and knowledge gained by the mother with each subsequent birth.

Such maternal information is internalized by human females during their juvenile and adolescent stages, giving the adult women a reproductive edge. The initial hu-man advantage may seem small, but it means that up to 21 more people than baboons or chimpanzees survive out of every 100 first-born infants—more than enough over the vast course of evolutionary time to make the evolution of human adolescence an overwhelmingly beneficial adaptation.

In human societies, juvenile girls often are expected to provide significant amounts of child care for their younger siblings, whereas in most other social mam-mal groups, the juveniles are often segregated from adults and infants. Thus, human girls enter adolescence with considerable knowledge of the needs of young children. Adolescent girls gain knowledge of sexuality and reproduction because they look mature sexually, and are treated as such, several years before they actually become fertile. The adolescent growth spurt serves as a signal of maturation. Early in the spurt, before peak height velocity is reached, girls develop pubic hair and fat deposits on breasts, buttocks, and thighs. They appear to be maturing sexually. About a year after peak height velocity, girls experience menarche, an unambiguous external signal of internal reproductive system development. However, most girls experience one to three years of **anovulatory menstrual cycles** after menarche. Nevertheless, the dramatic changes of adolescence stimulate both the girls and the adults around them to participate in adult social, sexual, and economic behavior. For the postmenarchial adolescent girl, this participation is "risk free" in terms of pregnancy for one or more years.

It is noteworthy that female chimpanzees and bonobos, like human girls, also experience up to three years of postmenarchial infertility, so this time of life may be a shared hominoid trait. Like human adolescents, the postmenarchial but infertile chimpanzees and bonobos participate in a great deal of adult social and sexual behavior. Primate researchers observing these apes point out that this participation, without pregnancy, allows for practicing many key behaviors that are needed to successfully rear an infant.

Although ape and human females may share a year or more of adolescent sterility, apes reach adulthood sooner than humans. Full reproductive maturation in human women is not achieved until about 5 years after menarche. The average age at menarche in the United States is 12.4 years, which means that the average age at full sexual maturation occurs between the ages of 17 and 18 years. Although adolescents younger than these ages can have babies, both the teenage mothers and the infants are at risk because of the reproductive immaturity of the mother. Risks include a low-birth-weight infant, premature birth, and high blood pressure in the mother. The likelihood of these risks declines and the chance of successful pregnancy and birth increases markedly after age 18.

Another feature of human growth not found in the African apes is that female fertility tracks the growth of the pelvis. Ellison (1982) and Worthman (1993) found that age at menarche is best predicted by biiliac width, the distance between the iliac crests of the pelvis. A median width of 24 cm (9.4 inches) is needed for menarche in American girls living in Berkeley, California, Kikuyu girls of East Africa, and Bundi girls of highland New Guinea. The pelvic width constant occurs at different ages in these three cultures, about 13, 16, and 17 years old, respectively. The later ages for menarche are due to chronic malnutrition and disease in Kenya and Bundi.

Moerman (1982) also reported a special human relationship between growth in pelvic size and reproductive maturation. She found that the crucial variable for successful first birth is size of the **pelvic inlet,** the bony opening of the birth canal. Moerman measured pelvic X-rays from a sample of healthy, well-nourished American girls who achieved menarche between 12 and 13 years. These girls did not attain adult pelvic inlet size until 17–18 years of age. Quite unexpectedly, the adolescent growth spurt, which occurs before menarche, does not influence the size of the pelvis in the same way as the rest of the skeleton. Rather, the female pelvis has its own slow pattern of growth, which continues for several years after adult stature is achieved.

Cross-cultural studies of reproductive behavior shows that human societies acknowledge (consciously or not) this special pattern of pelvic growth. The age at marriage and first childbirth clusters around 19 years for women from such diverse cultures as the Kikuyu of Kenya, Mayans of Guatemala, Copper Eskimos of Canada, and both the colonial and contemporary United States. Why the pelvis follows this unusual pattern of growth is not clearly understood. Perhaps another human attribute, bipedal walking, is a factor. Bipedalism is known to have changed the shape of the human pelvis from the basic ape-like shape. Apes have a cylindrical-shaped pelvis, but humans have a bowl-shaped-pelvis. The human shape is more efficient for bipedal locomotion but less efficient for reproduction because it restricts the size of the birth canal. Whatever the cause, this special human pattern of pelvic growth helps explain the delay from menarche to full reproductive maturity. That time of waiting provides the adolescent girls with many opportunities to practice and learn important adult behaviors that lead to increased reproductive fitness in later life.

Why Do Boys Have Adolescence?

The adolescent development of boys is quite different from that of girls. Boys become fertile well before they assume the size and the physical characteristics of men. Analysis of urine samples from boys 11–16 years old show that they begin producing sperm at a median age of 13.4 years. Yet cross-cultural evidence indicates that few boys successfully father children until they are into their third decade of life. In the United States, for example, only 3.09% of live-born infants in 1990 were fathered by men under 20 years of age. Among the traditional Kikuyu of East Africa, men do not marry and become fathers until about age 25 years, although they become sexually active after their circumcision rite at around age 18.

The explanation for the lag between sperm production and fatherhood is not likely to be a simple one of sperm performance, such as not having the endurance to swim to an egg cell in the woman's fallopian tubes. More likely is the fact that the average boy of 13.4 years is only beginning his adolescent growth spurt (Figure 11.6). Growth researchers have documented that in terms of physical appearance, physiological status, psychosocial development, and economic productivity, the 13-year-old boy is still more a juvenile than an adult. Anthropologists working in many diverse cultural settings report that few women (and more important from a cross-cultural perspective, few prospective in-laws) view the teenage boy as a biologically, economically, and socially viable husband and father.

The delay between sperm production and reproductive maturity is not wasted time in either a biological or social sense. The obvious and the subtle psychophysiological effects of testosterone and other androgen hormones that are released after gonadal maturation may "prime" boys to be receptive to their future roles as men. Alternatively, it is possible that physical changes provoked by the endocrines provide a social stimulus toward adult behaviors. Whatever the case, early in adolescence, sociosexual feelings including guilt, anxiety, pleasure, and pride intensify. At the same time, adolescent boys become more interested in adult activities, adjust their attitude to parental figures, and think and act more independently. In short, they begin to behave like men.

However—and this is where the survival advantage may lie—they still look like boys. One might say that a healthy, well-nourished 13.5-year-old human male, at a median height of 160 cm (62 inches) "pretends" to be more childlike than he really is. Because their adolescent growth spurt occurs late in sexual development, young males can practice behaving like adults before they are actually perceived as adults. The sociosexual antics of young adolescent boys are often considered to be more humorous than serious. Yet, they provide the experience to fine-tune their sexual and social roles before their lives or those of their offspring depend on them. For example, competition between men for women favors the older, more experienced man. Because such competition may be fatal, the childlike appearance of the immature but hormonally and socially primed adolescent male may be life-saving as well as educational.

When Did Childhood and Adolescence Evolve?

The stages of the life cycle may be studied directly only for living species. However, we can postulate on the on the life cycle of extinct species. Such inferences for the hominids are, of course, hypotheses based on comparative anatomy, comparative physiology, comparative ethology, and archaeology. Examples of such methods are found in the work of Martin (1983) and Harvey et al. (1987) on patterns of brain and body growth in apes, humans, and their ancestors.

Apes have a pattern of brain growth that is rapid before birth and relatively slower after birth. In contrast, humans have rapid brain growth both before and after birth (Figure 11.16). This difference may be illustrated by comparing ratios of brain weight divided by total body weight. The data are given in Table 11.3. At birth, this ratio averages 0.09 for the great apes and 0.12 for humans, showing that in proportion to body size, humans are born with brains that average 1.33 times larger than those of the apes. At adulthood the ratio averages 0.008 for the great apes and 0.028 for humans, meaning that the difference between apes and humans in the brain-to-body size proportion has increased to 3.5 times. It is the faster rate of human brain growth after birth that accounts for most of the difference. Indeed, the rate of human brain growth exceeds that of most other tissues of the body during the first few years after birth (Figure 11.8).

Martin's analysis of ape and human trajectories of growth indicate that a "human-like" pattern of brain and body growth becomes necessary after adult hominid brain size reaches ~850 cubic centimeters (cc). This biological marker is based on an analysis of the size of the head of the fetus and the size of the pelvic inlet (birth canal) of the mother across a wide range of social mammals, including the living primates and fossil hominids. Given the mean rate of postnatal brain growth for living apes, an 850-cc adult brain size may be achieved by all hominoids, including extinct hominids, by lengthening the fetal stage of growth. At brain sizes >850 cc, the size of the pelvic inlet of the fossil hominids and living humans does not allow for sufficient fetal growth. Thus, a period of rapid postnatal brain growth and slow body growth—the human pattern—is needed to reach adult brain size.

From this analysis, we can see clearly why so much of human postnatal growth and development is intimately associated with brain size. We presented earlier the figures on the percent of RMR due to brain growth and activity. The relation of human life history to our large and active brain can be looked at as an energetic problem. Large brains are costly investments; recall that the adult human brain uses 20%

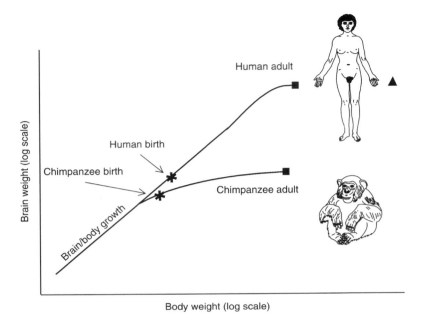

Figure 11.16 Brain and body growth curve for humans compared with chimpanzees. The length of the human fetal phase, in which brain and body grow at the same rate for both species, is extended for humans. In chimpanzees, brain growth slows after birth, but in humans, the high rate of brain growth is maintained during the postnatal phase. In contrast, the rate of human body growth slows after birth. If human brain and body growth rates were equal to those of chimpanzees, then adult humans would weigh 454 kilograms (998.8 pounds) and stand nearly 3.1 meters tall (9.9 feet). That body size is indicated by the "▲" symbol.

of RMR whereas the chimpanzee uses only 9% and an average marsupial only 2%. Moreover, larger brains have lower tolerances for temperature extremes, blood pressure, and oxygenation. The large human brain may increase obstetric risks (birth defects and maternal death). The costs are potentially high, but what is the payoff? The explanation we favor here is that a large brain is an investment that pays off on a long time scale. An organism recoups its energetic "investment" in a large brain through complex behavior, which is itself a combined product of large brains, slow development, extended care by older individuals, enhanced learning, and phenotypic plasticity, among other influences. The benefits of large brains probably accrue slowly over a long life. For primates in general and humans in particular, much of life history may support a substantial investment in brains.

We are, perhaps, fortunate that brains are so important, because after teeth and jaws, skulls are one of the more common pieces of fossil evidence preserved in the record of primate evolution. Having skulls, or at least sufficient skull parts to reconstruct the whole, allows paleontologists to estimate brain size. Having teeth and jaws in relative abundance is also fortuitous because of the strong correlation between tooth formation and eruption and so many life history events.

Given this background, Figure 11.17 is an attempt to summarize the evolution of the human pattern of growth and development. This figure must be considered as "a work in progress," because only the data for *Pan* and *Homo sapiens* are known with some certainty. Known ages for eruption of the M1 are given for *Pan* and *H. sapiens.* Estimated ages for M1 eruption in other species were calculated by Smith and Tompkins (1995). Age of eruption of M1 is an important life history event that correlates very highly with other life history events. Known or estimated adult brain sizes are given at the top of each bar; the estimates are averages based on reports in several textbooks of human evolution. Brain size is another crucial influence on life history evolution (further details and references for this and the following discussion of Figure 11.17 may be found in Bogin 1999).

Australopithecus afarensis appears in the fossil record about 3.9 MYA and is one of the oldest hominid fossil species. *A. afarensis* shares many anatomical features with nonhominid pongid (ape) species including an adult brain size of about 400 cc and a pattern of dental development indistinguishable from extant apes. Therefore, the chimpanzee and *A. afarensis* are depicted as sharing the typical tripartite stages of postnatal growth of social mammals—infant, juvenile, adult. Following the definitions used throughout this chapter, infancy represents the period of feeding by lactation, the juvenile stage represents a period of feeding independence prior to sexual maturation, and the adult stage begins following puberty and sexual maturation. The duration of each stage and the age at which each stage ends are based on empirical data for the chimpanzee. A probable descendent of *A. afarensis* is the fossil species *A. africanus,* dating from about 3.0 MYA. To achieve the larger adult brain size of *A. africanus* (average of 442 cc) may have required an addition to the length of the fetal and/or infancy periods. Figure 11.17 indicates an extension to infancy of one year.

The first permanent molar (M1) of the chimpanzee erupts at 3.1 years, but chimpanzees remain in infancy until about age five years. Until that age the young chimpanzee is dependent on its mother, and will not survive if the mother dies or is otherwise not able to provide care and feeding. After erupting M1 the young

Table 11.3   Neonatal and adult brain weight and total body weight for the great apes and human beings

Species	Neonatal weight (grams)		Adult weight (grams)	
	Brain	Body	Brain	Body
Pongo (orangutan)	170.3	1,728.0	413.3	5.300
Pan (chimpanzee)	128.0	1,756.0	410.3	3,635
Gorilla	227.0	2,110.0	505.9	12,650
Homo sapiens	384.0	3,300.0	1,250.0	4,400

Adult body weight is the average of male and female weight.

(*Source:* Data from Harvey et al. 1987).

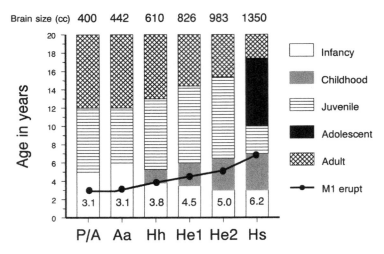

Figure 11.17 The evolution of hominid life history during the first 20 years of life. Abbreviated nomenclature as follows: P/A—Pan and *Australopithecus afarensis*, Aa—*Australopithecus africanus*, Hh—*Homo habilis*, He1—early *Homo erectus*, He2—late *Homo erectus*, Hs—*Homo sapiens*. Mean brain sizes are given at the top of each histogram. Mean age at eruption of the first permanent molar (M1) is graphed across the histograms, and given below the graph.

chimpanzee may be able to eat adult-type foods, but still must learn how to find and process foods. Learning to successfully open fruits that are protected by hard shells and to extract insects from nests (such as ants and termites) requires more than one year of observation and imitation by the infant of the mother. For these reasons, chimpanzees extend infancy for more than one year past the eruption of M1. Based on brain size details of dental anatomy the mean age of M1 eruption, and eruption of M2 and M3, for *A. afarensis,* and *A. africanus* is estimated to be identical to the chimpanzee. It is likely that these early hominids followed a pattern of growth and development very similar to chimpanzees. Behavioral capacities of these fossil hominids also seem to be similar to living chimpanzees. For these reasons, the early hominids also may have extended infancy for at least one year beyond the age of M1 eruption.

About 2.2 MYA fossils with several more human-like traits including larger cranial capacities and greater manual dexterity appear. Also dated to about this time are stone tools of the Oldowan tradition. Given the biological and cultural developments associated with these fossils they are considered by most paleontologists to be members of the genus *Homo* (designated as *H. habilis, H. rudolfensis,* or early *H. erectus*—refered to collectively here as *H. habilis*). The rapid expansion of adult brain size during the time of *H. habilis* (650 to 800 cc) might have been achieved with further expansion of both the fetal and infancy periods, as Martin's "cerebral Rubicon" was not surpassed. However, the insertion of a brief childhood stage into hominid life history may have occurred. Tardieu (1998) shows that *H. habilis* has a pattern of growth of the femur that is distinct from that of the australopithecines, but consistent with that of later hominids. The distinctive femur shape of the more recent hominids is due to the addition of a prolonged childhood stage of growth. *H. habilis,* then, may have had a short childhood stage of growth.

A childhood stage of growth for the earliest members of the genus *Homo* is also supported by a comparison of human and ape reproductive strategies. There are limits to the amount of delay possible between birth and sexual maturity, and between successful births, that any species can tolerate. The great apes are examples of this limit. Chimpanzee females in the wild reach menarche at 11 to 12 years of age and have their first births at an average age of 14 years. The average period between successful births in the wild is 5.6 years, as infant chimpanzees are dependent on their mothers for about five years. Actuarial data collected on wild-living animals indicate that between 35 % and 38 % of all live-born chimpanzees survive to their mid-twenties. Although this is a significantly greater percentage of survival than for most other species of animals, the chimpanzee is at a reproductive threshold. Goodall (1983) reports that for the period 1965 to 1980 there were 51 births and 49 deaths in one community of wild chimpanzees at the Gombe Stream National Park, Tanzania. During a 10 year period at the Mahale Mountains National Park, Tanzania Nishida et al. (1990) observed, "...74 births, 74 deaths, 14 immigrations and 13 emigrations..." in one community. Chimpanzee population size in these two communities is, by these data, effectively in equilibrium. Any additional delay in age of females at first birth or the time between successful births would likely result in a decline in population size.

The great apes, and fossil hominids such as *Australopithecus,* may have reached this demographic dilemma by extending the length of the infancy stage and forcing a demand on nursing to its limit. Early *Homo* may have overcome this reproductive limit by reducing the length of infancy and inserting childhood between the end of infancy and the juvenile period. Free from the demands of nursing and the physiological brake that frequent nursing places on **ovulation,** mothers could reproduce soon after their infants became children. This certainly occurs among modern humans. An often cited example, the !Kung, are a traditional hunting and gathering society of southern Africa. A !Kung woman's age at her first birth averages 19 years and subsequent births follow about every 3.6 years, resulting in an average fertility rate of 4.7 children per woman. Women in another hunter-gather society, the Hadza, have even shorter intervals between successful births, stop nursing about one year earlier, and average 6.15 births per woman.

For these reasons, a brief childhood stage for *H. habilis* is indicated in Figure 11.17. This stage begins after the eruption of M1 and lasts for about one year. That year of childhood would still provide the time needed to learn about finding and processing adult-type foods. During this learning phase, *H. habilis* children would need to be supplied with special weaning-foods. There is archaeological evidence for just such a scenario. *H. habilis* seems to have intensified its dependence on stone tools. There are both more stone tools, more carefully manufactured tools, and a greater diversity of stone tool types associated with *H. habilis*. There is considerable evidence that some of these tools were used to scavenge animal carcasses, especially to break open long bones and extract bone marrow. This behavior may be interpreted as a strategy to feed children. Such scavenging may have been needed to provide the essential **amino acids,** some of the minerals, and, especially the fat (dense source of energy) that children require for growth of the brain and body.

Further brain size increase occurred during *H. erectus* times, which begin about 1.6 MYA. The earliest adult specimens have mean brain sizes of 826 cc, but many individual adults had brain sizes between 850 to 900cc. This places *H. erectus* at or above Martin's "cerebral Rubicon" and seems to justify insertion and/or expansion of the childhood period to provide the biological time needed for the rapid, human-like, pattern of brain growth. It should be noted from Figure 11.17 that the model of human evolution proposed here predicts that from the *Australopithecus* to the *H. erectus* stage the infancy period shrinks as the childhood stage expands. Perhaps by early *H. erectus* times the transition from infancy to childhood took place before M1 eruption. Of course, it is not possible to know if this was the case or to state the cause of such a life history change with any certainty. Maybe the evolution of ever larger brains led to a delay in M1 eruption, which in turn led to both the need for a childhood stage and the expansion of the childhood stage as brains continued to enlarge. Alternatively, a delay in dental maturation may have precipitated the need for childhood, and in turn the biocultural ecology of childhood and its effects on hominid social learning and behavior selected for ever larger brains. No matter what the cause of childhood may be, if an expansion of childhood led to a shrinking of the infancy stage, then *H. erectus* would have enjoyed a greater reproductive advantage than any previous hominid. This seems to the case, as *H. erectus* populations certainly did increase in size and begin to spread throughout Africa and into other regions of the world.

Later *H. erectus*, with average adult brain sizes of 983cc, are depicted with further expansion of the childhood stage. In addition to bigger brains (some individuals had brains as large as 1100cc), the archaeological record for later *H. erectus* shows increased complexity of technology (tools, fire, and shelter) and social organization. These techno-social advances, and the increased reliance on learning that occur with these advances, may well be correlates of changes in biology and behavior associated with further development of the childhood stage of life (Bogin and Smith, 1996). The evolutionary transition to archaic, and finally, modern *H. sapiens* expands the childhood stage to its current dimension. Note that M1 eruption becomes one of the events that coincides with the end of childhood. Perhaps no further extension of childhood beyond M1 eruption is possible, given the significant biological, cognitive, behavioral, and social changes that are also linked with dental maturation and the end of childhood. With the appearance of *H. sapiens* comes evidence for the full gamut of human cultural capacities and behaviors. The technological, social, and ideological requisites of culture necessitate a more intensive investment in learning than at any other grade of hominid evolution. The learning hypothesis for childhood, while not sufficient to account for its origins, certainly plays a significant role in the later stages of its evolution.

The *H. sapiens* grade of evolution also sees the addition of an adolescent stage to post-natal development. The single most important feature defining human adolescence is the skeletal growth spurt that is experienced by virtually all boys and girls. There is no evidence for a human-like adolescent growth spurt in any living ape. There is no evidence for adolescence for any species of *Australopithecus*. There is some tentative evidence that early *Homo*, dating from 1.8 MYA, may have a derived pattern of growth that is leading toward the addition of an adolescent stage of development. This evidence is based on an analysis of shape change during growth of the femur (Tardieu 1998). Modern humans have a highly diagnostic shape to the femur, a shape that is absent in fossils ascribed to *Australopithecus*, but present in fossils ascribed to *Homo habilis, Homo rudolfensis,* or early African *Homo erectus.* The human shape is produced by growth changes during both the prolonged childhood stage and the adolescent stage. The more human-like femur shape of the early *Homo* fossils could be due to the insertion of the childhood stage alone, or to the combination of childhood and adolescent stages. Due to the lack of fossils of appropriate age at death, the lack of dental and skeletal material from the same individuals, and the lack of sufficient skeletal material from other parts of the body it is not possible to draw any more definitive conclusions.

A remarkable fossil of early *Homo erectus* is both of the right age at death and complete enough to allow for an analysis of possible adolescent growth. The fossil specimen is catalogued formally by the name KMN-WT 15000, but is called informally the "Turkana boy" as it was discovered along the western shores of Lake Turkana in 1984. This fossil is 1.6 million year old, making it an early variety of *Homo erectus.* The skeletal remains are almost complete, missing the hands and feet and a few other minor bones. Smith (1993) analyzed the skeleton and dentition of the Turkana fossil and ascertained that, indeed, it is most likely the remains of an immature male. The youth's deciduous upper canines were still in place at the time of death, and he died not long after erupting second permanent molars. These dental features place him firmly in the juvenile stage by comparison with any hominoid. The boy was 160 cm (63 inches) tall at the time of death, which makes him one of the tallest fossil youths or adults ever found.

Part of Smith's analysis focused on patterns of growth and development, especially the question "Did early *H. erectus* have an adolescent growth spurt?" Based on her analysis the answer to that question is no. Judged according to modern human standards, the Turkana boy's dental age of 11 years is in some conflict with his bone age (skeletal maturation) of 13 years and his stature age of 15 years. If the Turkana boy grew along a modern human trajectory, then dental, skeletal and stature ages should be about equivalent. By chimpanzee growth standards, however, the boy's dental and bone ages are in perfect agreement, both at seven years of age. As *Homo erectus* is no chimpanzee, the Turkana boy's true age at death was probably between seven and 11 years. What is clear is that the Turkana boy followed a pattern of growth that is neither that of a modern human nor that of a chimpanzee. Based on Smith's analysis the boy's large stature becomes more explicable. The reason for his relatively large stature-for-age is that the distinct human pattern of moderate to slow growth prior to puberty followed by an adolescent growth spurt had not yet evolved in early *Homo erectus.* Rather, the Turkana boy followed a more ape-like pattern of growth in stature making him appear to be tall in comparison with a modern human boy at the same age. At the time of puberty, the chimpanzee has usually achieved 88 percent of stature growth, while humans have achieved only 81 percent. Smith and Tompkins state that the human pattern of growth suppression up to puberty followed by a growth spurt after puberty had not evolved by early *H. erectus* times. "Because of this, any early *H. erectus* youth would seem to us to be too large" (1995, 273).

Unfortunately there are no appropriate fossil materials of later *H. erectus* available to analyze for an adolescent growth spurt. There are several fossils of a species called *Homo antecessor,* found in Spain (Castro de Bermudez et al. 1999). This species is also called early *H. sapiens* by many paleontologists. These fossils date from about 800,000 BP. Based on an analysis of tooth formation, this species seems to have a pattern of dental maturation equal to that of living people. Perhaps, they also had a pattern of skeletal growth, including an adolescent growth spurt, like that of living people as well.

One possible descendant of *Homo antecessor* are the Neandertals. There is one fossil of a Neandertal in which the associated dental and skeletal remains needed to assess adolescent growth are preserved. The specimen is a juvenile, most likely a male. It is called Le Moustier 1 and was found in 1908 in Western France. The specimen is dated at between 42,000 and 37,000 years BP (before present). Thompson and Nelson (1998) use information on crown and root formation of the molar teeth to estimate a dental age of 15.5 +/- 1.25 years. Compared with modern human standards for length of the long bones of the skeleton, Thompson and Nelson estimate that Le Moustier 1 has a stature age of about 11 years, based on the length of his femur. The dental age of 15.5 years and the stature age of 11 years are in very poor agreement, and indicate that like the Turkana boy, Le Moustier 1 may not have followed a human pattern of adolescent growth. The dental age indicates that by human standards Le Moustier 1 was in late adolescence at the time of death, but the stature age indicates he was still a juvenile or had just entered adolescence. Quite unexpectedly, these differences in dental and skeletal maturity are exactly the opposite of those for the Turkana boy.

What is clear is that adolescent growth, at least the pattern of adolescent growth found in modern humans, does not seem to be present in either the Turkana boy or in the Le Moustier 1 fossil. In modern humans, certain diseases, prolonged undernutrition, and unusual individual variations in growth may produce the skeletal and dental features of the fossils. While it is possible that these two fossil specimens fall into one of these categories of unusual growth, the most parsimonious conclusion that one may draw from these findings is that the adolescent growth spurt in skeletal growth evolved only in the *Homo sapiens* line. Quite likely this would be no earlier than the appearance of archaic *H. sapiens* in Africa at about 125,000 years ago, or in *H. antesessor* 800,000 years ago. If Neandertals are direct ancestors to modern humans, as some scientists believe, then the adolescent skeletal growth spurt may be less than 37,000 years old.

The Valuable Grandmother, or Could Menopause Evolve?

In addition to childhood and adolescence, human life history has another unusual aspect: menopause. One generally accepted definition of menopause is "the sudden or gradual cessation of the menstrual cycle subsequent to the loss of ovarian function" (Timiras 1972). The process of menopause is closely associated with but distinct from the adult female postreproductive stage of life. Reproduction usually ends before menopause. In traditional societies, such as the !Kung, the Dogon of Mali, and the rural-living Maya of Guatemala, women rarely give birth after age 40 and almost never give birth after age 44. Even in the United States from 1960 to 1990, with modern health care, good nutrition, and low levels of hard physical labor, women rarely gave birth after age 45. This is true even among social groups attempting to maximize lifetime fertility, such as the Old Order Amish.

As among the !Kung, Dogon, and Maya, menopause occurs well after this fertility decline, at a mean age of 49 years for living women in the United States. After age 50, births are so rare that they are not reported in the data of the US National Center for Health Statistics or by the Amish. (However, they are sensationalized in the tabloids sold at supermarket checkouts.) We report these ages for the onset of human female postreproductive life versus the ages for menopause because some scholars incorrectly equate menopause with the beginning of the postreproductive stage, so one must read the literature carefully to interpret in what sense the term "menopause" is used.

Menopause, and a significant period of life after menopause, appear to be uniquely human female characteristics. Wild-living nonhuman primate females do not share the universality of human menopause, and human males have no comparable life history event. In a review of the data for all mammals, Austad (1994) finds that no wild-living species, except possibly pilot whales, "are known to commonly exhibit reproductive cessation." Female primates studied in captivity, including langurs, baboons, rhesus macaques, pigtailed macaques, and chimpanzees, usually continue estrus cycling until death, although fertility declines with age. These declines are best interpreted as a normal part of aging. Only one captive bonobo more than 40 years old and one captive pigtail macaque older than 20 are known to have ceased estrus cycling. In contrast to the senescent decline in fertility of other female primates, the human female reproductive system is abruptly "shut down" well before other systems of the body, which usually experience a gradually decline toward senescence.

In Chapter 13, some of the hypotheses for the evolution of menopause and a postreproductive life stage are reviewed. In terms of basic biology, it appears that a 50-year age barrier exists to female mammalian fertility because by that age, all oocytes are depleted. By that age, the females of most mammalian species are dead. The few exceptions are elephants and whales, the largest mammalian species, and humans. Much ethnographic evidence shows that significant numbers of women in almost every human society, traditional and industrial, live for many years after oocyte depletion (menopause). The only reproductive strategy open to postmenopausal females is to provide increasing amounts of aid to their offspring and their grandoffspring. Elephants do this, because the leader of the social group is usually an old matriarch. Little is known about the social lives of whales in terms of grandmother behavior. The ethnographic evidence shows that human grandmothers and other postreproductive women are beneficial to the survival of children in many human societies. Grandmothers provide food, child care, and a repertoire of knowledge and life experiences that assist in the education of their grandchildren. In sum, the inevitabilities of mammalian biology, combined with the creativity of human culture, allow women (and men?) of our species to develop biocultural strategies to take greatest advantage of a postreproductive life stage. Viewed in this context, human grandmotherhood may be added to human childhood and adolescence as distinctive stages of the human life cycle.

BIOCULTURAL MODELS OF HUMAN GROWTH

Biosocial Risks for Children, Adolescents, and Grandmothers

The evolution of any new structure, function, or stage of development may bring about many biosocial benefits; however, it also incurs risks. Bipedalism, the method of human locomotion unique among the primates, is one example of evolutionary benefits and risks. Often considered to be one of the crucial feeding and reproductive adaptations of our species, bipedalism also brings about many physical ailments, including lower back pain, fallen arches, and inguinal hernias. Similarly, the benefits of childhood need to be tempered against the hazards of this developmental stage. Dependency on older individuals for food and protection, small body size, slow rate of growth, and delayed reproductive maturation convey liabilities to the child. The "charms" of children and childhood do not provide for total security.

To illustrate this point, one can examine traditional societies of both historic and prehistoric eras. In such societies, including hunter-gatherers and horticulturists, ~35% of live-born humans died by age 7—that is, by the end of childhood. Even if two-thirds of these deaths occurred during infancy, the childhood period still had an appreciable risk of death. In hunter-gatherer societies, starvation, accidents, and predation accounted for most childhood deaths.

Today, hunter-gatherer and traditional horticultural societies account for <1% of human cultures, so it may be more instructive to examine current risks to children in postcolonial and industrialized societies. Mild-to-moderate energy undernutrition is, perhaps, the most common risk. Worldwide estimates are that between 60 and 75% of all children are undernourished. Malnutrition is a serious threat to adequate growth, physical and cognitive development, health, and performance at school or work (see Chapter 12). Undernutrition is equally likely to be due to food shortages as to work loads and infectious disease loads placed on children that compromise their energy balance. Viewed historically, unreasonable work loads for children and many childhood diseases are products of the agricultural and industrial revolutions. Agriculture, for example, increased social stratification and reduced the variety and quality of foods consumed by people of the lower social classes. Industrialization further increased social disparities and resulted in the forced labor of children in mines and factories.

Despite programs of legal protection and welfare for children in the 20th century, many risks remain for children in the contemporary world. Abuse and neglect are two. One estimate of the worldwide mortality from abuse and neglect is between 13 and 20 infants and children per 1000 live births. The incidence of all suffering from abuse and neglect is probably higher but very difficult to estimate because data are not reported by most nations. Some industrialized nations do maintain statistics for nonfatal abuse and neglect of children. In the United States, for example, "in 1991

2.7 million abused or neglected children were reported to child protection agencies" (Kliegman 1995). These numbers constitute a rate of 38.6 children per 1000.

Some cases of child abuse and neglect may result from a severance between the biology of childhood and the rapid pace of technological, social, and ideological change relating to families and their children. It is now technologically possible to nourish infants without breast feeding, and this development allows parents (mothers) an opportunity to pursue economic activities or have another baby. Among the poor populations of the developing countries, short birth intervals (<23 months) compromise the health of both the infant and the mother. A major negative effect on the infant is low birth weight, which impairs both physical growth and cognitive development during childhood and later life stages.

In the populations of the more developed nations, such as among the US middle class, ~20% of infants are breast fed through 6 months of age. The weaning

---

BOX 11.2   COLLECTING AND COMPARING ETHNOGRAPHIC DATA ON THE TREATMENT OF CHILDREN, ADOLESCENTS, AND GRANDMOTHERS IN SEVERAL CULTURES

To better understand the concept of a biocultural model of human growth presented in this chapter, we recommend the following exercise. Seek out official government agency reports, or analyses that reference such reports, concerning the treatment of children, adolescents, and the elderly—especially grandmothers. We emphasize these life history stages because they are highly developed and important to the human species, at least among the primates. Look for data regarding health, the most common diseases and conditions, any incidence of abuse and neglect, and economic and political status. Students concentrating in subject areas relating to children, adolescents, or the elderly may think of other topics for research.

The data for this exercise may be found in various places. Many large university libraries are repositories for official government reports. Social service agencies and charitable organizations devoted to children or the elderly may have these reports in addition to their own publications. There are many electronic sources for these data, including the databases and publications on the Internet (for example the Centers for Disease Control at www.cdc.gov).

When searching for data, each student or group may chose a life history stage, a particular country in the world, or an ethnic or social group found in one or more countries. Devise a method to record the data from each study so that you can compare findings. Use the data collection form from Box 11.1 as an example to assist in this task. Some of the data will be statistical in nature. You may enter these values into a database for statistical analysis, as you did for the life history data. Other information may be in the form of narratives, which will require a nonstatistical analysis.

Be creative when thinking about ways to make sense of all the data. Remember, the goal of this exercise is to see how human biology and culture interact and influence each other in terms of growth, development, and life history stages.

process—from bottles and formula—may begin by 3 months of age, severely curtailing infancy. These "premature children" present a problem for care, because they are still biologically in the infancy stage of development. The problem is often "solved" by relegating these young to restraining devices such as high chairs, playpens, and cribs or segregating them from the family by placement in daycare centers or preschools. When the employees of these centers and preschools are trained properly, these arrangements are suitable. However, if not well trained, and especially if the infants react poorly to these arrangements, the frustrated parents or caregivers may respond with abusive or neglectful behavior.

The influence of culture change can introduce considerable discord into the normally harmonious relationship between human biology and behavior during development. Two brief examples may be offered here. First, Kenyan mothers with more formal (European style) education believe that sibling care responsibilities teach juveniles to be passive and that domestic work, including child care, is menial. Second, forced settlement of the Inuit (hunters of North America) resulted in loss of their nomadic hunting lifestyle. This change required them to acculturate to settled life, the economics of wage labor—parents at work and children at school—and the social values of television. Such shifts in values and behavior may present significant changes for future generations of Kenyans and Inuit. For example, the average juvenile may not learn about human growth and development until after the birth of his or her first infant. The consequences of these changes in social learning for the health, nutritional status, physical growth, and development of the next generation are less clear.

The evolution of human menopause is associated with several risks for older women. The hormonal changes and bone loss that occur with the cessation of ovarian function (reviewed in detail in Chapter 13) are one kind of risk. These biological changes may bring about several degenerative diseases, such as **osteoporosis.** Postmenopausal women often must assume new social and economic roles for which they need adequate training and social support. But grandmothers, like their adolescent grandchildren, may no longer receive appropriate training for postreproductive sexual, social, and economic expectations in some "modern" societies.

The elderly also may be denied a productive social role and even be segregated away from productive society—in "retirement communities" for those who can afford it and "old age homes" for those of limited means. The social isolation that these sequestered elderly people may experience exacerbates the normal degenerative process of aging. Moreover, research shows that children living in households with little or no contact with grandparents suffer more abuse and neglect than children in multigenerational households—another testament to the value of grandparents.

CHAPTER SUMMARY

In this chapter, we have taken a life history approach to the study of human growth and development. We reviewed several of the basic principles of human growth and development and set these basic principles in their evolutionary context. Social mammals have three basic stages of postnatal development: infant, juvenile, and adult.

Some species also have a brief female post-reproductive stage. The human life cycle, however, is best described by five stages: infant, child, juvenile, adolescent, and adult. It is hypothesized that the new life stages of the human life cycle represent feeding and reproductive specializations of the genus *Homo*. Human women, in both traditional and industrial societies, may also have a long post-reproductive stage. Analyses of bones and teeth of early hominids who died as subadults suggest that the evolution of the new life stages of childhood and adolescence are not of ancient origin. It is now fairly certain that the outline of human life history took shape with the evolution of *Homo* and not before. Indeed, australopithecines probably lived at a pace nearly twice as fast as modern humans. The evidence from fossil remains of *Homo habilis, Homo erectus*, and early *Homo sapiens* suggests that the elements of human life history evolved as a mosaic, not as a package, over more than 1 million years.

We have tried to take a biocultural perspective of human development, a perspective that focuses on the constant interaction taking place during all phases of human development, both between genes and hormones within the body and with the sociocultural environment that surrounds the body. Research from social anthropology, developmental psychology, endocrinology, primate ethology, and physical anthropology shows how the biocultural perspective enhances our understanding of human development.

We also discussed the risks of the new stages of human development, especially when these stages impact with culture change. Malnutrition, child abuse, and neglect of both infants, children, and the elderly are some of the risks. With this knowledge, we hope that some readers of this chapter will conduct new biocultural research on human growth and development. We also hope that the findings of your research may help bring about peaceful improvement in the social, economic, and political conditions of life and will lead to good growth for all human beings.

RECOMMENDED READINGS

Bogin B (1993) Why must I be a teenager at all? New Sci. 137(Mar. 6): 34–38.

Bogin B (1995) Growth and development: recent evolutionary and biocultural research. In NT Boaz and LD Wolfe (eds.): Biological Anthropology: The State of The Science. Bend, Oreg.: International Institute for Human Evolutionary Research, pp. 49–70.

Bogin B (1997) The evolution of human nutrition. In L Romanucci-Ross, D Moerman, and LR Tancredi (eds.): The Anthropology of Medicine, 3rd edition. South Hadley, Mass.: Bergen and Garvey, pp. 96–142.

Bogin B (1999) Patterns of Human Growth, 2nd edition. Cambridge: Cambridge Univ. Press.

Condon RG (1990) The rise of adolescence: social change and life stage dilemmas in the Central Canadian Arctic. Hum. Org. 49: 266–279.

DeRousseau CJ (ed.) (1990) Primate Life History and Evolution. New York: Wiley-Liss.

Smith BH (1992) Life history and the evolution of human maturation. Evol. Anthropol. 1: 134–142.

Weisner TS (1987) Socialization for parenthood in sibling caretaking societies. In JB Lancaster, J Altmann, AS Rossi, and LR Sherrod (eds.): Parenting Across the Life Span: Biosocial Dimensions. New York: Aldine de Gruyter, pp. 237–270.

REFERENCES CITED

Alexander RD (1990) How Did Humans Evolve? Reflections on the Uniquely Unique Species (Special Publication No. 1). Ann Arbor: Univ. of Michigan Museum of Zoology.

Altmann, J (1980) Baboon Mothers and Infants. Cambridge: Harvard Univ. Press.

Austad SN (1994) Menopause: An evolutionary perspective. Exp. Gerontol. 29: 255–263

Baughn B, Brault-Dubuc M, Demirjian A, and Gagnon G (1980) Sexual dimorphism in body composition changes during the pubertal period: as shown by French-Canadian children. Am. J. Phys. Anthropol. 52: 85–94.

Bermudez de Castro, JM, Rosas, A, Carbonell, E, Nicolas, ME, Rodroguez, J, and Arsuaga, JL (1999) A modern human pattern of dental development in Lower Pleistocene hominids from Atapuerca-TD6 (Spain). Proc. Natl. Acad. Sci. 96: 4210–4213.

Bock RD and Thissen, D (1980). Statistical problems of fitting individual growth curves. In FE Johnston, AF Roche, and C Susanne (eds). Human Physical Growth and Maturation, Methodologies and Factors. New York: Plenum, pp. 265–90.

Bogin B (1999) Patterns of Human Growth, 2nd edition. Cambridge: Cambridge Univ. Press.

Bogin B and Smith BH (1996) Evolution of the human life cycle. Am. J. Hum. Biol. 8: 703–716.

Bonner JT (1965) Size and Cycle. Princeton, N.J.: Princeton Univ. Press.

Cabana T, Jolicoeur P,and Michaud J (1993) Prenatal and postnatal growth and allometry of stature, head circumference, and brain weight in Québec children. Amer. J. Hum. Biol. 5: 93–99.

Cameron N, Tanner JM, and Whitehouse RH (1982) A longitudinal analysis of the growth of limb segments in adolescence. Ann. Hum. Biol. 9: 211–220.

Coelho AM (1985) Baboon dimorphism: growth in weight, length, and adiposity from birth to 8 years of age. In ES Watts (ed.): Nonhuman Primate Models for Human Growth. New York: Alan R. Liss., pp. 125–159.

Demirjian A (1986) Dentition. In F. Falkner and J. M. Tanner (eds.): Human Growth, Volume 2, Postnatal Growth. New York: Plenum, pp. 269–298.

Ellison PT (1982) Skeletal growth, fatness, and menarcheal age: a comparison of two hypotheses. Hum. Biol. 54:269–281.

Emanuel I, Filakti H, Alberman E, Evans SJW (1992) Intergenerational studies of human birthweight from the 1958 birth cohort. 1. Evidence for a multigenerational effect. Brit. J. Obst. Gynecol, 99. 67–74.

Ewer RF (1973) The Carnivores. Ithaca, N.Y.: Cornell Univ. Press.

Frisancho AR (1990) Anthropometric Standards for the Assessment of Growth and Nutritional Status. Ann Arbor: Univ. of Michigan Press.

Goodall, J (1983) Population dynamics during a 15-year period in one community of free-living chimpanzees in the Gombe National Park, Tanzania. Zietschrift fur Tierpsychologie, 61: 1–60.

Goss R (1964) Adaptive Growth. New York: Academic Press.

Goss R (1978) The Physiology of Growth. New York: Academic Press.

Harvey P, Martin RD, and Cluton-Brock TH (1987) Life histories in comparative perspective. In B Smuts , DL Cheney, RM Seyfarth, RW Wrangham, and TT Struhsaker (eds.): Primate Societies Chicago, Ill.: Univ. of Chicago Press, pp. 181–196.

Janson CH and Van Schaik, C P (1993) Ecological risk aversion in juvenile primates: slow and steady wins the race. In ME Perieira and LA Fairbanks (eds.): Juvenile Primates: Life History, Development, and Behavior. New York: Oxford Univ. Press, pp. 57–74.

Kliegman RM (1995) Neonatal technology, perinatal survival, social consequences, and the perinatal paradox. Am. J. Public Health 85: 909–913.

Lancaster JB and Lancaster CS (1983) Parental investment: the hominid adaptation. In DJ Ortner (ed.): How Humans Adapt. Washington, D.C.: Smithsonian Institution Press, pp. 33–65.

Martin RD (1983) Human Brain Evolution in an Ecological Context (Fifty-second James Arthur Lecture). New York: American Museum of Natural History.

Moerman ML (1982) Growth of the birth canal in adolescent girls. Am. J. Obstet. Gynecol. 143: 528–532.

Nishida T, Takasaki H, and Takahata, Y (1990) Demography and reproductive profiles. In T. Nishida (ed.): The Chimpanzees of the Mahale Mountains: Sexual and Life History Strategies. Tokyo: Univ. of Tokyo Press, pp. 63–97.

Nowak RM and Paradiso JL (1983) Walker's Mammals of the World, 4th edition. Baltimore, Md.: John Hopkins Univ. Press.

Pereira ME and Altmann J (1985) Development of social behavior in free-living nonhuman primates. In ES Watts (ed.): Nonhuman Primate Models for Human Growth and Development. New York: Alan R. Liss, pp. 217–309.

Plant TM (1994) Puberty in primates. In E Knobil and JD Neill (eds.): The Physiology of Reproduction, 2nd edition. New York: Raven, pp. 453–485.

Prader A (1984). Biomedical and endocrinological aspects of normal growth and development. In J Borms, RR Hauspie, A Sand, C Susanne, and M Hebbelinck (eds.): Human Growth and Development. New York: Plenum, pp. 1–22.

Scammon RE (1930). The measurement of the body in childhood. In JA Harris et al. (eds.): The Measurement of Man. Minneapolis: University of Minnesota Press, pp. 173–215.

Schally AV, Kastin AJ, and Arimura A (1977) Hypothalamic hormones: the link between brain and body. Am. Sci. 65: 712–719.

Singer C (1959). A Short History of Scientific Ideas to 1900. London: Oxford Univ. Press.

Smith BH (1991) Dental development and the evolution of life history in Hominidae. Am. J. Phys. Anthropol. 86: 157–174.

Smith BH (1993) Physiological age of KMN-WT 15000 and its significance for growth and development of early *Homo*. In AC Walker and RF Leakey (eds.): The Nariokotome *Homo erectus* Skeleton. Cambridge, Mass.: Belknap Press, pp. 195–220.

Smith BH and Tompkins RL (1995) Toward a life history of the hominidae. Ann. Rev. Anthropol., 25:257–79.

Stratz CH (1909) Wachstum und Proportionen des Menschen vor und nach der Geburt. Arch. Anthropologie 8: 287–297.

Tanner JM (1978) Fetus into Man. Cambridge, Mass.: Harvard Univ. Press.

Tanner JM (1990) Fetus into Man, 2nd edition. Cambridge, Mass.: Harvard Univ. Press.

Tardieu C (1998) Short adolescence in early hominids: infantile and adolescent growth of the human femur. Amer. J. Phys. Anthropol. 197, 163–178.

Thompson JL and Nelson AJ (1997) Relative postcranial development of Neandertals. J. Hum. Evol., 32: A23-A24.

Timiras PS (1972) Developmental Physiology and Aging. New York: MacMillan Publishing Co.

Watts ES and Gavan JA (1982) Postnatal growth of nonhuman primates: the problem of the adolescent spurt. Hum. Biol. 54: 53–70.

Worthman CM (1993) Biocultural interactions in human development. In ME Perieira and LA Fairbanks (eds.) Juvenile Primates: Life History, Development, and Behavior. New York: Oxford Univ. Press, pp. 339–357.

CHAPTER 12

Growth Variation:
Biological and Cultural Factors

SARA STINSON

INTRODUCTION

Body size and shape are among the most obvious ways in which humans vary, both within and among populations. This variation is the result of complex interaction among many factors. Growth has a substantial genetic component, but we cannot yet identify most of the specific **genes** that affect body size and shape. Thus, children tend to resemble their parents in body size and shape, and some of the differences in body size and shape among populations have a genetic basis as well.

But growth is also strongly influenced by environmental factors, so two individuals who have identical genes but were raised under different environmental circumstances would not be identical in body size and shape. In previous chapters, we learned how temperature, high altitude, and physical activity can affect body size and shape. Other environmental factors that affect body size and shape are nutritional status and disease. Both genetic and environmental influences on growth are modified by cultural practices. Beliefs about who appropriate marriage partners are, how heavily clothed a child should be before going out in the cold, what suitable foods for growing children are, and when a child should be given medical treatment can all ultimately influence variation in body size and shape.

Why is growth variation such a frequently studied aspect of human biology? Partly because body size and shape are conspicuous components of appearance, partly because measurements such as height, weight, body circumferences, arm and leg lengths, and amount of **subcutaneous fat** are relatively easy to measure with portable instruments and therefore can be studied in a wide variety of populations throughout the world. Body size and shape also are important indicators of adaptation. Because growth is so influenced by nutritional status and disease, it can provide important information about the health of children.

Human Biology: An Evolutionary and Biocultural Perspective, Edited by Sara Stinson, Barry Bogin, Rebecca Huss-Ashmore, and Dennis O'Rourke
ISBN 0-471-13746-4   Copyright © 2000 Wiley-Liss, Inc.

VARIATION AMONG AND WITHIN POPULATIONS

Average body size and shape differ between human populations as well as among individuals within every population. Before discussing the overall causes of this variation, I present a few examples to illustrate the amount of variation that exists.

Figure 12.1 shows the variation in average adult height among living human populations. For both males and females, the range in height between the average for the tallest and shortest populations is >30 centimeters (cm; >1 foot). The shortest of the populations shown in Figure 12.1 is Efe pygmies, who live in the tropical forests of central Africa. The term pygmy refers to many separate populations with small body size, most frequently defined as an average adult male height <150 cm (<59 inches). African pygmies such as the Efe live by foraging and exchanging forest products for foods cultivated by neighboring farmers.

The Efe are small from birth, with average birth weights of 2.7 kilograms (kg; a little less than 6 pounds [lb]). As shown in Figure 12.2, by age 5, individuals are >18 cm (>7 inches) shorter than the US average (Bailey 1991). The cause of the extremely small body size of African pygmies such as the Efe has been researched a great deal. Although poor nutrition and health certainly contribute to their size, the most recent research indicates that hormonal factors also play a major role.

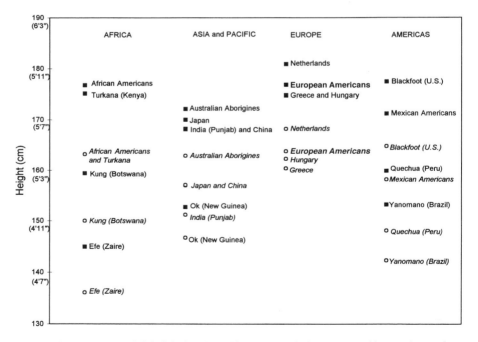

Figure 12.1   Average adult height in selected human populations, grouped by continent of ancestry and differentiated as males (■) and females (○). (*Source:* Data from Eveleth and Tanner 1976, 1990; Little et al. 1983; and Najjar and Kuczmarski 1989; Lin et al. 1992)

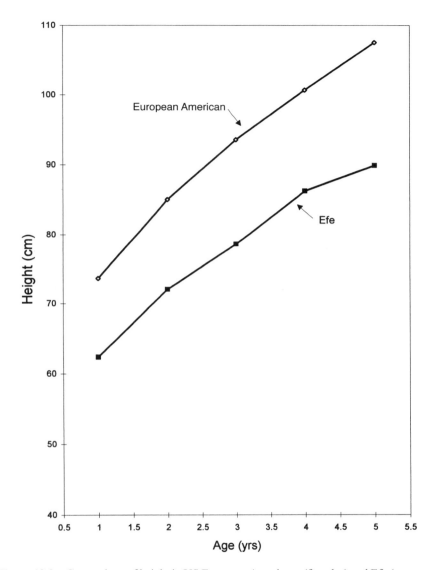

Figure 12.2   Comparison of height in US European Americans (females) and Efe (sexes combined). (*Source:* Data from Bailey 1991; Eveleth and Tanner 1990)

Efe appear to be resistant to the growth-promoting effects of the **hormone insulin-like growth factor I** (IGF-I), apparently because they have a reduced number of sites for this hormone to attach to cells (Geffner et al. 1995). This hormonal feature is probably genetic, but the specific gene or genes responsible have not been identified.

Why might small body size have been favored by **natural selection** in pygmy populations? Small body size could be an advantage in a hot, humid environment because it reduces the amount of heat the body produces. Short stature might enhance one's ability to traverse dense forest vegetation or to withstand starvation under conditions of limited food availability (Diamond 1991). At present, these explanations remain speculative because evidence is limited regarding the extent to which any of these factors really constitutes an advantage. We do not even know for certain that the small body size of pygmies is the result of natural selection.

In terms of *average* height, Efe are among the shortest populations in the world. Yet, this does not mean that every Efe is shorter than every individual in other populations. Within every human population, we find considerable variation in body size. As shown in Figure 12.3, US male and female heights vary >30 cm (>1 foot). Notice that this range is about equal to the range in average height among the populations shown in Figure 12.1. So, looking at the range of heights among several populations, we would observe substantial overlap.

Males and females differ in *average* body size and shape. In the United States, the average adult height of females is ~163 cm (64 inches), whereas that of males is 177 cm (70 inches). Although males are on average taller than females, the height distri-

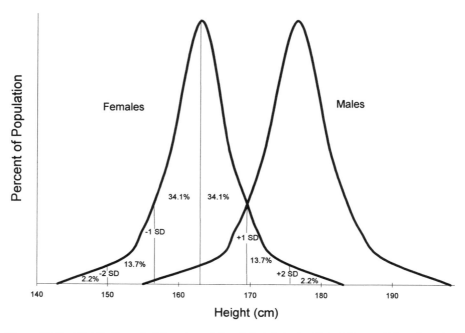

Figure 12.3   Distribution of adult male and female height in the United States. Although males have a greater mean height than females, distributions overlap considerably. Vertical lines on the female distribution indicate standard deviations (SD) above and below the mean, and percent values indicate the percentages of population falling between standard deviations (see Box 12.2).

butions of the two sexes overlap (see Figure 12.3); more than 15% of US females are taller than males. Males and females also differ in amount and distribution of body fat. As shown in Figure 12.4, females on average have more subcutaneous fat than males do. Although males have more fat on the trunk (**subscapular skinfold**) than on the arms (**triceps skinfold**), in most populations, females have more fat on the arms. As for body size, the amount of body fat also varies considerably within populations, so not all males are leaner than all females.

Sexual dimorphism in body size, body fat, and fat distribution exist in every human population. These differences certainly have a strong genetic component (however, again, the specific genes have not been identified). Although hormonal differences between males and females at **puberty** are a major cause of sex differences in **adolescent** growth, they are not the only (and probably not even the main) cause of adult sexual dimorphism (Prader 1984). Females have more subcutaneous fat than males throughout life, and a major contributor to sex differences in overall body size is that males grow for longer than females do. Genes on the Y chromosome—other than those that result in the secretion of **androgen** hormones—might lead to greater height in males. This suggestion arises from studies of individuals who have both an X and a Y **chromosome** but develop as females because they do not respond to androgen hormones such as **testosterone**. These XY females are taller than XX females.

Although the differences in body size and shape between males and females certainly have a genetic basis, they also are affected by cultural factors. In some societies, male children are given preference in terms of feeding and health care (Chen et al. 1981), and in many societies, fatness is a valued characteristic in females (Brown and Konner 1987). Both of these cultural preferences would increase differentiation between the sexes.

ENVIRONMENTAL EFFECTS ON GROWTH

The small body size of pygmies and dimorphism in body size emphasize the genetic component of variation in growth. Environmental causes, however, are at least as important as genes in causing growth variation. In this section, I discuss some environmental effects on growth.

Socioeconomic Status

Socioeconomic status (SES) is usually measured by looking at household characteristics such as income, education, occupation, and household possessions. In almost every study that has investigated the existence of height differences between children of different SES, high-SES children have been taller than low-SES children. These results hold true for children living in both affluent countries and poor countries. As shown in Figure 12.5, the height differences are greater between the well-off and the poor in developing countries (e.g., Guatemala, Brazil, Kenya, and China) than in industrialized countries, reflecting the wide range of SES where the conditions of poverty are much more severe.

One notable exception to the finding that height and SES are positively related comes from studies of Swedish urban schoolchildren born in the 1950s (Lindgren 1976; Lindgren and Cernerud 1992). No height differences were observed between children of different socioeconomic groups. This finding indicates that conditions for all groups were adequate to support the same level of growth, almost certainly the result of the social welfare policies in place in Sweden at that time. It is noteworthy that height differences among socioeconomic groups reappeared for Swedish children born 10 years later (Figure 12.5).

Height differentials related to SES result from differences in nutritional status, disease rate, and medical care between socioeconomic groups. However, these environmental factors may not be the only explanation for these differences. In most societies, tallness is valued, particularly in males. One indicator from the United States is that almost all US presidents have been taller than the average for their period. Whereas US presidential stature may partly be a result of their privileged backgrounds, it also reflects the fact that we equate tallness with power (Cassidy 1991).

As a result of the value placed on height, tall individuals may be more likely to improve their SES and be upwardly mobile. If so, this ability would contribute to the greater stature of high-SES individuals. Among investigations indicating that height

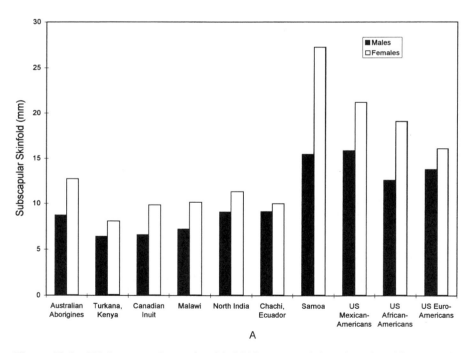

Figure 12.4 (**A**) Average subscapular skinfold in young adult males (closed bars) and females (hatched bars) in selected populations. Amounts of subcutaneous fat are higher in females than in males and higher in more affluent populations than in more traditional populations.

influences social mobility, a study in Germany found that young adults of a higher social class than their parents were taller than their siblings who had remained in the same social class (Schumacher and Knussmann 1979). One problem with this and other studies of the relationship between height and social mobility is that it is also possible that the upwardly mobile and the socially stationary have been exposed to different environments during growth. Taller individuals may have benefited from better circumstances that simultaneously allowed for better growth and improved educational and career opportunities (Bielicki and Waliszko 1992). This complication makes it difficult to determine the extent to which social mobility of tall individuals influences socioeconomic differences in height.

SES is also associated with amount of body fat. Children of higher SES generally have more subcutaneous fat (see Figure 12.6). Although this difference sometimes reflects undernutrition among low-SES groups, in others it is the result of overnutrition and higher levels of obesity in high-SES groups. In poorer countries, high-SES individuals also have greater amounts of fat in adulthood. In affluent countries and increasingly in more affluent developing countries, the differences in

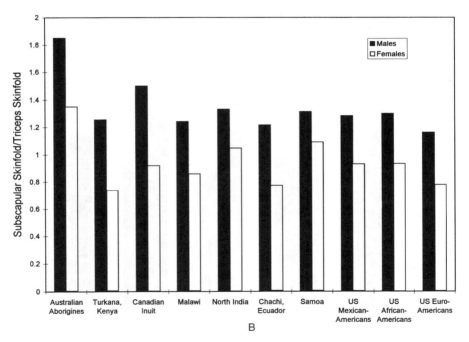

Figure 12.4   (**B**) The ratio of subscapular skinfold to triceps skinfold in the same populations (males, closed bars; females, hatched bars) as shown at left. Ratios >1.0 indicate greater subscapular than triceps skinfolds; ratios <1.0 indicate greater triceps than subscapular skinfolds. This ratio varies between populations and is greater in males than in females. (*Source:* Data from Eveleth and Tanner 1976; Little et al. 1983; Pawson 1986; Najjar and Rowland 1987; Najjar and Kuczmarski 1989; Stinson 1989; Pelletier et al. 1991; Sangeev et al. 1991)

fatness between socioeconomic groups sometimes are reversed in adulthood, especially in females (see Figure 12.6). The exact reasons for this finding are not entirely clear (Brown and Konner 1987). Diets high in calories, sedentary jobs that do not provide leisure time for additional exercise, and the high cost of fresh fruits and vegetables are among the possible causes. In addition, low-SES women are less likely to place a high value on thinness than are high-SES women.

Body fat distribution also differs between SES groups. Several studies have found that low-SES individuals have more of their fat on the trunk and less of their fat on the extremities than high-SES individuals (e.g., Georges et al. 1991). Potential health consequences of a centralized distribution of body fat (i.e., on the trunk rather than on the limbs) include a higher risk for diabetes and heart disease. The central distribution of body fat in low-SES groups may be a result of hormonal changes that occur because of increased exposure to stress.

Urban–Rural Comparisons

Growth differences between children living in urban and rural areas mirror differences in living conditions between the two regions (Bogin 1988). In the 19th century and the early part of 20th century, rural children in industrialized countries were generally

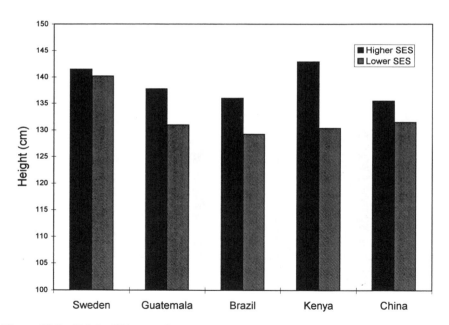

Figure 12.5   Height differences between 10- and 11-year-old boys with different socioeconomic status (SES). Higher SES (closed bars) is associated with taller stature and lower SES (hatched bars) with shorter stature. (Source: Data from Johnston et al. 1973; Bogin and MacVean 1978; Kulin et al. 1982; Instituto Nacional de Alimentação e Nutrição 1990; Lin et al. 1992; Lindgren and Cernerud 1992)

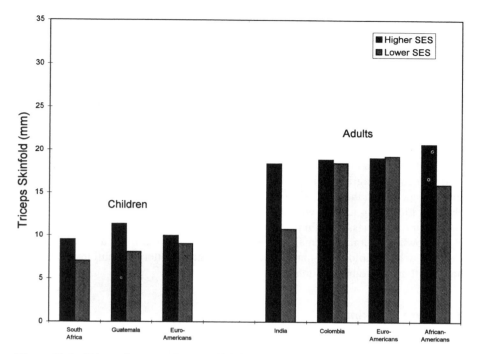

Figure 12.6   Triceps fatness at 7 years old (Children) and in adult females of different SES. In childhood, high SES (closed bars) is associated with greater fatness and low SES (hatched bars) with less fatness, but this is not always the case in adulthood. (*Source:* Data from Garn and Clark 1975; Bogin and MacVean 1981; Flegal et al. 1988; Sanjeev et al. 1991; Cameron et al. 1992; Dufour et al. 1994)

taller and heavier than urban children. Cities were not particularly healthy places because of the lack of sanitation, crowded living conditions, child labor, and industrial pollution. As public health measures were instituted and wealth became more concentrated in urban areas, conditions in cities improved.

By about 1930, urban children were larger than rural children. This difference still exists in most countries; however, in the most affluent industrialized countries (e.g., the United States, the Netherlands, and Australia), growth differences between urban and rural children are small or nonexistent. In these countries, conditions affecting growth do not differ greatly between urban and rural areas.

Although on average, urban children in developing countries are taller and heavier than rural children, this broad comparison conceals the variation that exists within urban environments. Cities in developing countries contain both the elite and the extremely poor, the latter frequently living in slums without access to basic municipal services. Not surprisingly, researchers in developing countries often find that the poorest urban children have heights and weights similar to those of rural children; sometimes, they are even shorter and lighter than rural children.

Family Size

In societies where income comes primarily from wages earned by adults, children from small families tend to be larger than children from large families. This result is true even after controlling for large family size, which in many cases is associated with lower SES. Growth differences related to family size usually are explained by the fact that large families have fewer resources available per person and that children in large families are exposed to more disease.

We might expect that the relationship between family size and growth would be different in agricultural societies, where children make an economic contribution to the family. Whereas increasing family size could be an asset, because more workers lead to increased agricultural production, the amount that children are able to contribute depends on their age. Older children may provide substantial assistance, but children too young to help with work are an economic drain on the household. In agricultural societies, both family size and the age structure of the household affect child growth. In rural Guatemala (Russell 1976) and in rural Bolivia (Stinson 1980) children in households with many older children grow better than children in households with many young children.

Exposure to Toxins

Exposure to a variety of environmental pollutants has been associated with reduced growth. One of the most investigated toxins is cigarette smoke. Women who smoke during pregnancy give birth to babies weighing ~200 grams (~0.4 lb) less than the infants of nonsmokers (Garn 1985). Lead is another toxin that affects growth. Children with high blood lead levels have slower prenatal and postnatal growth. Evidence also indicates that exposure to high levels of noise or toxic chemicals may lead to reduced growth. In most cases, the effects of toxins are found after controlling for SES, so the reduced growth associated with exposure to toxins is not the result of the fact that such exposure is likely to be greater in low SES groups. Rather, it appears that pollutants directly interfere with growth, but the exact mechanisms for the action of various toxins are not yet fully known (Schell 1991).

Secular Trends

One of the best examples of the effect of the environment on growth is the increases in body size that have taken place during the last century and a half in affluent industrialized countries. These changes are referred to as **secular trends**, or changes that take place over time. In most populations, secular trends include increases in height and weight, which are called positive secular trends. As discussed below, however, in some populations, body size has decreased over time—a negative secular trend.

The increases in the height of Swedish females between 1883 and 1982 are shown in Figure 12.7. During this period, average height at age 14 increased by 12 cm (4.75 inches). Increases in the height of Swedish males during the same time were even greater, 19 cm (7.5 inches) at age 14. The magnitude of the positive secular trend in

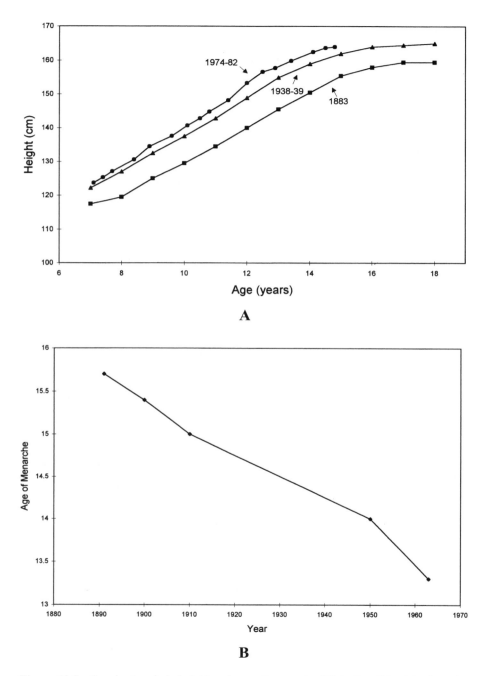

Figure 12.7   Secular trends in height and age of menarche (B) in Swedish girls since the 1880s. Height has increased, but age of first menstruation has decreased. (*Source:* Data from Ljung et al. 1974; Lindgren and Hauspie 1989; Tanner 1990)

Sweden is similar to that in other industrialized countries. Summarizing positive secular trends in several countries from the late 1800s to the 1970s, Meredith (1976) found that height in late childhood had increased at a rate of ~1.3 cm (0.5 inches) per decade, height in adolescence had increased at a rate of ~1.9 cm (0.75 inches) per decade, and adult stature had increased at a rate of ~0.6 cm (0.25 inches) per decade.

Secular increases in height are greater during adolescence than during childhood or adulthood because faster growth rates have been accompanied by earlier maturation. As a result, the **adolescent growth spurt** occurs earlier and the period of growth is shorter. In industrialized countries today, a 14-year-old girl is closer to her adult height than a 14-year-old was in the 19th century. A striking illustration of the trend toward earlier maturation is that the age of first menstruation (**menarche**) has decreased markedly over the past 150 years, from between 15 and 16 years in the middle 19th century to between 12 and 13 years in most industrialized countries today (see Figure 12.7).

Positive secular trends do not affect all parts of the body equally. Secular increases in leg length have generally been greater than increases in trunk length. Thus, body proportions change as a result of positive secular trends, with the legs becoming an increasing percentage of total body height. As shown in Figure 12.8, body proportions differ among human populations. Populations of European and Asian ancestry have shorter legs relative to their heights than do populations of African or Australian aboriginal ancestry. We might expect leg length to be relatively longer in populations that live under more advantaged circumstances, because positive secular trends will be greater in those circumstances. However, Australian Aborigines and populations of African ancestry have longer legs, even though they generally live under less advantaged conditions than populations of European ancestry. This discrepancy probably reflects genetic differences among these groups. Because positive secular trends are ongoing in many of these populations, we do not know how body proportions will differ when and if secular trends are completed.

What has caused the positive secular trends toward faster growth and earlier maturation? Most research suggests that secular trends are the result of environmental changes (Roche 1979). Because so many changes have occurred during the past 150 years, it is difficult to identify any individual cause. Multiple factors probably are involved, and specific causes may vary from location to location. The improved health status that results from reducing the spread of **infectious disease** is certainly one important cause, and improvements in overall nutrition also contribute to positive secular trends.

Secular trends have occurred too rapidly to be the result of genetic changes brought about by natural selection for larger body size. However, there may be some genetic influence on secular trends. During the past 150 years, **gene flow** has increased among populations because of increased population mobility. The increased **heterozygosity** that results from outbreeding may contribute to secular trends. Although studies indicate that offspring of **exogamous** matings are taller than offspring of **endogamous** matings, they have not always controlled for factors known to affect growth, such as SES. In any case, the effect of increased heterozygosity on height is relatively small and could account for only a small part of the observed positive secular trends.

In affluent countries, the increases in body size have slowed or stopped in recent decades. (As shown in Figure 12.7, secular trends in Sweden were greater between the 1880s and the 1930s than between the 1930s and the 1980s). One possible explanation is that the environmental factors that affect growth and maturation are no longer improving. Another interpretation is that positive secular trends are coming to a halt because individuals have grown as tall as they can.

Positive secular trends are associated with improvements in environmental condition. It is not unexpected that in populations that have not experienced improved conditions there is very little evidence for positive secular trends. One such example is from the state of Oaxaca in southern Mexico (Malina et al. 1983). A comparison of adult heights measured in the 1970s with heights measured in the previous 80 years indicated little evidence of change. Even more remarkable, the measured heights are quite similar to heights estimated for skeletons recovered from archaeological sites dating between 1400 BC and 1500 AD.

Just as the continuing adverse environmental circumstance (poor nutritional status, high rates of disease) in Oaxaca has prevented positive secular trends, in some cases, very rapid secular changes have occurred in response to rapid environmental

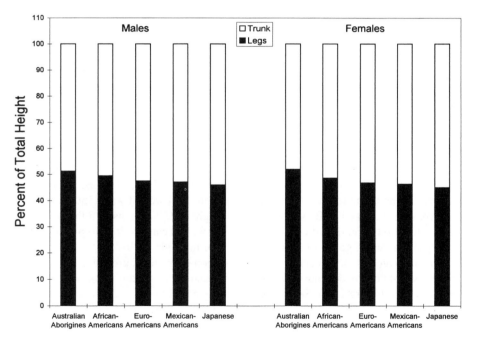

Figure 12.8   Body proportion differences among populations. Trunk length (open bars) is measured as height while sitting (sitting height), and leg length (closed bars) is determined by subtracting sitting height from total height. Australian Aborigines have relatively longer legs than other human populations. In fact, their trunks account for less of their height than do their legs. Populations of European and Asian ancestry have relatively longer trunks than African Americans. (*Source:* Data from Kimura 1984; Najjar and Kuczmarski 1989; Frisancho 1990; Norgan 1994)

change. The most noticeable cases of these rapid and large increases in body size involve migration to new environments. Recent Mayan refugees from Guatemala to the United States average 5.5 cm (>2 inches) taller and 4.7 kg (>10 lb) heavier than Mayan children living in Guatemala (Bogin and Loucky 1997). Just as positive environmental change can lead to positive secular trends, negative environmental change can lead to negative secular trends such as decreases in body size. British soldiers born between 1769 and 1774 were shorter than those born in the years immediately preceding or following, reflecting the diminished nutritional resources in their early childhood (Steegmann 1985). Negative secular trends also have occurred during the 20th century in several African populations (Tobias 1985).

EFFECTS OF NUTRITION AND DISEASE DURING THE LIFE CYCLE

In the previous section, I focused on how growth can differ in contrasting environmental circumstances (urban/rural, high SES/low SES, large family/small family, etc.). What are the actual factors that lead to growth differences between children living under different circumstances? Poor nutrition and recurrent disease are the main environmental factors that affect growth. In this section, you will see that nutrition and disease do not have equal effects on growth at all stages of the life cycle.

The rate of growth is fastest prenatally and in infancy, then slows down until it increases again at the time of puberty. We might expect that adverse environmental circumstances would be most disruptive at the times of fastest growth; however, this expectation is true only to a certain extent.

Prenatal Growth

Although methods such as ultrasound are now available for measuring growth in utero, birth weight is the measurement most frequently used to assess prenatal growth. In the United States in 1997, the median birth weight was 3350 grams (7.4 lb) (Ventura et al. 1999). Babies with birth weights <2500 grams (5.5 lb) are classified as **low birth weight** infants. Reduced birth weight can result from **prematurity**, slow growth during gestation, genetic causes, or multiple births.

One indicator of the effect of maternal nutrition on prenatal growth is the decrease in birth weight that occurs during acute famines and under conditions of seasonal food shortage (see Figure 12.9). Severe food scarcity affected previously well-nourished women in St. Petersburg, Russia and in the Netherlands during World War II. St. Petersburg was under siege by the Germans from 1941 to 1943, and in the Netherlands, a transportation embargo from 1944 to 1945 resulted in little food entering major cities (Antonov 1947; Smith 1947; Stein et al. 1975). In the Netherlands, the calories available to pregnant women fell to only ~800 kcal (~3350 kJ) a day during the worst of the "Hunger Winter." Birth weight decreased substantially in both areas during the time of food shortage, averaging almost 500 grams less (a little more than 1 lb) in normal gestation–length births in St. Petersburg and >300 grams less in the Netherlands. During the height of the siege of St. Petersburg, almost 50% of babies were low birth weight.

The effects of maternal nutrition on prenatal growth are also seen by examining changes in birth weight that occur when women are provided with additional food during pregnancy (Institute of Medicine 1990). Table 12.1 summarizes the results from studies in Guatemala, Colombia, Taiwan, and Gambia of mild to moderately undernourished women whose diets were supplemented during pregnancy. Although these studies varied considerably in method, all show some increase in birth weight associated with maternal supplementation during pregnancy. These increases were due to improved prenatal growth rather than longer gestation length. The changes in birth weight were relatively small, ranging from 95 to 226 grams (0.2 to 0.5 lb). Additionally, in three of the studies, the effect of supplementation was not evident in all infants but was limited to male infants or to a particular season. In Guatemala and Gambia, the percentage of low birth weight infants decreased quite dramatically, but this was not the case in the Colombian and Taiwanese studies.

Pregnancy supplementation of nutritionally at-risk women in affluent countries also has been studied. This supplementation sometimes produced increases in birth weight, sometimes not. Increases in birth weight were generally ~50 grams, less than those observed in the studies of women in developing countries.

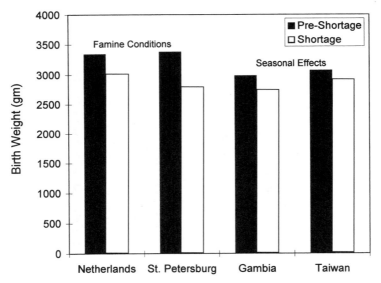

Figure 12.9 Changes in birth weight associated with food shortages (closed bars, before food shortage; hatched bars, during food shortage). Birth weight decreased under famine conditions during World War II in the Netherlands and Russia. Birth weight also decreases during seasonal food shortages. Food shortages that occur during the yearly "hungry season" in Gambia are intensified because they occur at a time of high agricultural work loads. A similar situation occurred in rural Taiwan in the late 1960s, when the season of greatest agricultural work did not coincide with the season of greatest food availability. (*Source:* Data from Antonov 1947; Stein et al. 1975; Prentice et al. 1981; Adair and Pollitt 1983)

Table 12.1   Effects of maternal supplementation during pregnancy

	Rural Guatemala	Bogota, Columbia	Rural Taiwan	Rural Gambia
Type of comparison	Infants of women with supplement intake of >20,000 kilocalories (kcal) during pregnancy were compared with infants of women with supplement intake of <20,000 kcal during pregnancy	Infants of women supplemented during the last trimester of pregnancy were compared with infants of unsupplemented women.	Women were unsupplemented during their first pregnancy and then supplemented from 1 week after the birth of their first infant through their second pregnancy. The first and second infants were compared.	The infants born during four years of supplementation were compared with the infants born to the same women during the four years before supplementation began.
Birth weight difference (grams)	111[a]	95[b]	162[c]	225[d]
Low birth weight (%)	High supplement: 9 Low supplement: 19	ND	ND	Supplemented: 7.5 Unsupplemented: 23.7

[a]High supplement, 3105 grams; low supplement, 2994 grams.
[b]Effect only in male infants: supplemented, 3061 grams; unsupplemented, 2966 grams.
[c]Effect only in male infants: supplemented, 3216 grams, unsupplemented, 3054 grams.
[d]Effect only in wet season (time of food shortages): supplemented, 3033 grams; unsupplemented, 2808 grams.
ND, no difference between supplemented and unsupplemented.

(*Source:* Lechtig et al. 1975; Mora et al. 1979; Adair and Pollitt 1985; and Prentice et al. 1987)

Studies of maternal supplementation during pregnancy illustrate that maternal nutrition affects prenatal growth, but in many cases, the effect of supplementation was less than expected. Several factors may explain the relatively small increases in birth weight that occur when additional food is provided to pregnant women. The effect of maternal nutritional supplementation depends on the nutritional status of the mother; supplementation causes greater changes in birth weight among poorly nourished women in developing countries than among women in industrialized countries. Another reason for the relatively small effect of supplementation, even in poorly nourished women, is that supplementation during one pregnancy cannot compensate for a lifetime of undernutrition. The small body size of undernourished women in developing countries may physically limit fetal growth, and undernutrition during the mother's own growth may have resulted in impaired reproductive function. Finally, the effects of supplementation during pregnancy could be smaller than expected because pregnant women are able to partially buffer the fetus against undernutrition. Some evidence indicates that the **basal metabolic rate** and, consequently, energy expenditure of women decrease during pregnancy (Prentice et al. 1989) (see also Chapter 10). A decrease in maternal energy requirements would allow the mother to direct more resources to the fetus and protect it from some of the effects of undernutrition.

Infant and Childhood Growth

The rate of growth is fastest before birth; during this period, environmental insult has a small effect on growth. Growth disruptions during infancy and childhood are the main cause of the small adult body size observed in populations that live under adverse conditions. In this section, I use the terms **infancy** and **childhood** as defined in Chapter 11: Infancy (infant) refers to birth through 3 years of age (the weaning age in traditional societies), and childhood (child, children) refers to the period from 3 to 7 years of age.

Infant Feeding Practices

The nutrient content of breast milk varies among mammals (see Table 12.2). Offspring growth rate and the interval between feedings are two factors associated with this variation. Milk high in nutrients is necessary for offspring to survive long intervals between feedings. Rabbits and northern fur seals—which may go 1 and 8 days, respectively, between nursings—have milk with a very high fat content. Small mammals have fast rates of relative growth, and their milk is characterized by high **protein** content. The low fat and protein content of human milk (and primate milk in general) reflects frequent nursing and slow rates of postnatal growth.

One dietary decision that has important consequences for infant health is whether an infant is bottle-fed or breast-fed. Infant feeding methods other than breast-feeding have many disadvantages: lack of contraceptive effects (see Chapter 15), lack of emotional bonding between mother and infant, possible long-term negative health consequences for mother and child (Cunningham 1995), and the nutrition- and disease-related consequences discussed in this chapter. Because of the potentially harmful consequences of other feeding methods, the World Health Organization (WHO) has recommended that infants should be fed nothing but breast milk from birth to 6 months of age, and that they should continue to breast-feed (with the addition of nutritionally adequate supplemental foods) until 2 years of age or older.

One of the major disadvantages of feeding methods other than breast-feeding is that they do not provide the various anti-infective agents found in breast milk. Young infants have a poorly developed immune system and thus are less able to resist disease. Their lack of resistance is partially offset by the passage of **antibodies** from the mother across the placenta before birth and then augmented by the anti-infective agents contained in breast milk. These agents include antibodies (against specific **bacteria** and **viruses** that affect the digestive and respiratory system); white blood cells; and several proteins, fats, and carbohydrates that convey antimi-

---

BOX 12.1   HOW DO WE COMPARE GROWTH IN DIFFERENT POPULATIONS?

Our task is to compare the heights of girls in rural villages in Sudan (Eveleth and Tanner 1990) and in the United States. The average heights in the two groups are shown here:

	Height	
Ages (years)	Sudan	United States
4	99.0 cm (39.0 inches)	101.6 cm (40.0 inches)
10	134.0 cm (52.8 inches)	138.3 cm (54.4 inches)

One way to compare the groups would be to calculate the height difference between them at ages 4 and 10. However, by calculating that 4-year-old Sudanese girls are on average 2.6 cm shorter and that 10-year-old Sudanese girls are on average 4.3 cm shorter than US girls, we do not determine whether Sudanese girls are much shorter or only slightly shorter than US girls. Several methods for comparing growth allow us to assess the relationship between populations more accurately.

One method is the use of **percentiles**. Percentiles indicate what percentage of the population has measurements below a particular value. The 50th percentile, or the median, is the midpoint of the population, because one-half of the individuals have measurements below the median (and, of course, one-half have values above). At the 25th percentile, 25% of the population falls below and 75% above, and so on. If we compare the 4-year-old Sudanese girls with percentiles calculated for the US population on the basis of a large national survey (WHO 1983), we find that the Sudanese girls fall between the US 20th and 30th percentiles. This means that 20–30% of 4-year-old US girls have heights below the Sudanese average. Or, put another way, more than 70% of 4-year-old US girls are taller than the average Sudanese girl. At age 10, the height of the Sudanese girls is also between the US 20th and 30th percentiles. Although the absolute difference in height between Sudanese and US girls is greater at age 10 than at age 4, a comparison based on percentiles indicates that the mean height of Sudanese girls is similar relative to US girls at both ages.

Another frequently used method for comparing populations is **z scores**. A z score indicates how many **standard deviations** above or below the mean or median a measurement is. The z scores to assess growth usually are calculated by using growth data on US children as the reference population of healthy, well-nourished children. Whether this practice is appropriate in all cases is discussed later. For growth comparisons, z scores usually are computed relative to the median by using the formula

$$z \text{ score} = \frac{\text{measurement for individual} - \text{median in reference population}}{\text{standard deviation in reference population}}$$

for example, a 6-year-old boy who weighs 22 kg (48.5 lb) would have a weight z score of

$$\frac{22 - 20.7 \text{ (the median weight for eight year old U.S. boys)}}{2.3 \text{ (the standard deviation for eight year old U.S. boys)}} = +0.57$$

The positive z score indicates that this boy's weight is above the median for his age. Negative z scores indicate measurements below the median. The average 4-year-old Sudanese girl would have a negative height z score because her height falls below the US median.

---

crobial action (Institute of Medicine 1991). Breast milk produced soon after birth is especially rich in antibodies.

Even in developed countries, bottle-fed infants have higher rates of illness and **mortality** than breast-fed infants. Certainly, breast-fed babies benefit from the anti-infective agents in breast milk, but they also are exposed to fewer **pathogens,** particularly where lack of clean water supplies means that formula is prepared using contaminated water and placed in bottles that have been washed with water from the same source.

The quantity and nutritional quality of breast milk produced by undernourished and well-nourished women is similar. One indicator of the ability of nursing women to produce adequate breast milk even when maternal nutritional status is compromised is that the milk production of undernourished women in Gambia did not increase when they were given nutritional supplements during lactation (Prentice et al. 1983).

In developing countries, breast-fed babies grow more than bottle-fed babies during the first year of life. Several differences between breast- and bottle-feeding could contribute to the better growth of breast-fed infants. For the reasons outlined earlier, formula-fed infants have higher rates of illness than breast-fed babies. Formula also can be under- or overdiluted, whereas breast milk cannot. In addition, families living in poverty may have difficulty purchasing enough formula to meet the infant's needs.

Comparisons of the growth of breast- and bottle-fed infants in the first year of life in industrialized countries have produced more variable results. In some cases, the growth of the two groups of infants is similar, but in many cases, breast-fed infants grow more slowly than bottle-fed infants, especially in weight, after the first several months of age. Both evolutionary and biocultural explanations have been proposed to explain why breast-fed babies in industrialized countries do not always grow as quickly as formula-fed babies.

From an evolutionary point of view, it could be argued that breast milk is the nutritionally ideal food for infants because it is the food that has evolved to meet their needs. Based on this reasoning, the growth of breast-fed babies should be considered "normal" infant growth. The faster growth of formula-fed babies would be an indicator that these infants are being overfed, perhaps because bottle-fed babies have less control over their food intake than do breast-fed babies.

Another argument is that breast milk may not be adequate after the first few months of life and should be supplemented early with other foods. If maximal growth rates are considered ideal (which is highly questionable, especially given the high rates of obesity in countries such as the United States), then breast milk is not the ideal food for infants because it does not always lead to the same rate of growth as formula feeding. Human breast milk that supports adequate but not maximal rates of infant growth could evolve because breast milk composition is a compromise between the needs of the infant and what the mother can afford to produce (Fomon 1993).

Yet a third explanation for the slower growth of breast-fed infants in industrialized countries is the style of breast-feeding. Rather than feeding whenever the infant indicates a desire to nurse, most women in industrialized countries use a more scheduled feeding style. Feedings are separated by many hours—especially at night, when infants sleep apart from the mother—and an attempt is made to get the infant to sleep through the night as early as possible. Because evidence shows that the fat content of breast milk increases when feedings are more frequent, Dettwyler and Fishman

Table 12.2 Breast milk composition in selected mammals

Species	Fat (%)	Protein (%)	Sugar (%)
Rabbit	15.2	10.3	1.8
Northern fur seal	49.4	10.2	0.1
Rat	8.8	8.1	3.8
Dog	9.5	7.5	3.8
Pig	8.3	5.6	5.0
Cow	3.7	3.2	4.6
Horse	1.3	1.9	6.9
Baboon	4.6	1.5	7.7
Human	4.1	0.8	6.8

(*Source:* Data from Oftedal 1984)

BOX 12.2   USING GROWTH DATA TO ASSESS NUTRITIONAL STATUS

The growth of a child is highly dependent on the child's environment and is particularly influenced by nutritional status and disease. For this reason, growth measurements are frequently used for assessing nutritional status.

Currently, the most common method is based on the calculation of z scores for **height-for-age** and **weight-for-height**. The z scores are used to compare children in the study population with a reference population of healthy, well-nourished children. Height-for-age z scores indicate how the heights of children in the study population compare with the heights of children of the same age in the reference population. Weight-for-height z scores indicate how the weights of children in the study population compare with the weights of children of the same height (but not necessarily the same age) in the reference population.

These two z scores provide different kinds of information about nutritional status. A low weight-for-height z score indicates a child who is thin, whereas a low height-for-age z score indicates a child who is short. Obviously, an extremely thin child was suffering from undernutrition at the time that the data were recorded. Extreme shortness also results from undernutrition, but a very short child was not necessarily undernourished at the time the data were recorded.

How do we determine whether a child is extremely short or extremely thin? The decision about what z score value indicates undernutrition is based on properties of the standard deviation. As explained in Box 12.1, a z score indicates how many standard deviations above or below the mean or median a measurement is. The standard deviation is a statistical measure of the amount of variation in a population. It has properties similar to percentiles, because if the measurements (height or weight) obey certain rules about how they are distributed (a **normal distribution**), then a fixed percentage of measurements will fall within a specific distance of the mean. As shown in Figure 12.3, , 34.1% of measurements fall between the mean and the value that is 1 standard deviation lower than the mean; 15.9% of measurements are below the value that is 1 standard deviation smaller than the mean; but only 2.2% of measurements are below the value that is 2 standard deviations smaller than the mean.

Because z scores are in standard deviation units, they are easy to interpret. In the reference population of healthy children, 15.9% will have z scores for height-for-age (or weight-for-height) lower than −1 (one standard deviation below the mean). If we decide that z scores less than −1 indicate undernutrition, then 15.9% of the reference population would be classified as undernourished. Because it is extremely unlikely that undernutrition is this common in our healthy reference population, a cut-off z score of −1 is too high. However, only a little more than 2% of the reference population will have z scores below −2. For this reason, the World Health Organization has suggested that z scores lower than −2 be considered to indicate undernutrition (WHO Working Group 1986). They further suggest that low height-for-age and low weight-for-height be distinguished. Children with height-for-age z scores below −2 are considered **stunted**, and those with weight-for-height z scores below −2 are considered **wasted**.

(1992) argue that long intervals between feedings would result in breast milk with a lower fat content. The lower nutrient content of this milk could explain why breast-fed infants grow more slowly than formula-fed infants in studies where most mothers were breast-feeding in a fairly scheduled way. According to this view, women can produce breast milk that supports the rates of growth seen in formula-fed infants, but cultural factors (ideas about frequency of breast-feeding and whether babies sleep with the mother) prevent many women from doing so.

Growth in Infancy and Childhood

For the first few months after birth, infants who live under adverse conditions in developing countries and infants in affluent industrialized countries have similar rates of growth. After the age of 4–6 months, the rate of growth declines among infants

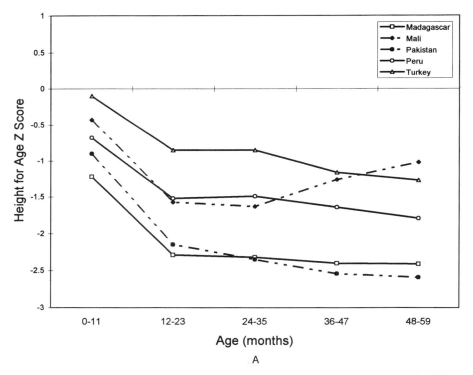

A

Figure 12.10 Height-for-age (A) and weight-for-height z scores (B) for infants and children under 5 years of age. The zero line is the median for US children. Heights of infants who live under disadvantaged conditions in developing countries are substantially lower than heights of infants who live under advantaged circumstances by the second year of life. In most cases, the weight-for-height of infants and children in developing countries is much more similar to that of infants and children in affluent countries. □, Madagascar; ■, Mali; ○, Peru; ●, Pakistan; △, Turkey. (*Source:* Data from Dettwyler 1991 and unpublished data from Macro International, Demographic and Health Surveys)

in disadvantaged circumstances in developing countries, and by the second year of life, they are substantially shorter than children in industrialized countries (Martorell and Habicht 1986). This slowing of growth is referred to as **growth faltering** and is illustrated in Figure 12.10A. By the age of 5 years, the mean height of children in the developing countries shown in Figure 12.10 is below the 30th percentile for US children and in many cases is below the 3rd percentile.

Infants and children who live under disadvantaged conditions in developing countries are more likely to be extremely short than they are to be extremely thin. In Figure 12.10B, the weight-for-height z scores are compared for infants and children in industrialized and developing countries. Although some infants and children in developing countries weigh less than those of the same height in industrialized countries, the differences in weight-for-height are less than the differences in height (the height-for-age z scores are lower than the weight-for-height z scores). In some cases, infants and children in developing countries weigh somewhat more than those of the same height in industrialized countries

Both poor nutrition and frequent disease contribute to growth faltering, but the exact causes of growth faltering are not totally clear (Waterlow and Schürch 1994). In terms of nutrition, growth faltering begins at the time when breast milk alone becomes insufficient to fully support growth. If foods other than breast milk are not given to the infant or if the supplementary foods given are low in nutritional value,

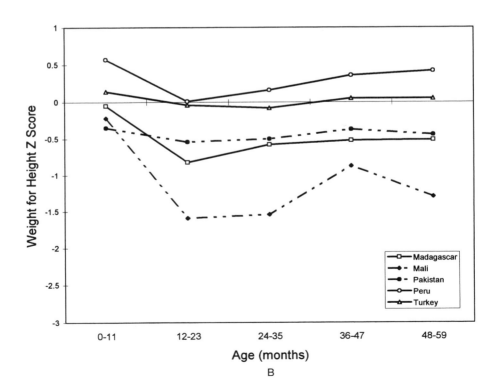

B

growth will be affected. When the main foods available are bulky starches, it is difficult for an infant or young child to consume the quantity of food necessary to meet nutritional needs, especially if he or she has only a few opportunities to eat during the day. The diets of infants and young children also are affected by cultural values about what is appropriate food for them and the extent to which they are allowed to make their own decisions about eating. Compared with most other mammals, humans are particularly vulnerable to these difficulties because we have a childhood stage during which children are no longer breast-fed but still are dependent on others for food.

One example of how decisions about feeding can affect the diets of infants and young children comes from rural southern Mali (Dettwyler 1991). In this area, infants do not receive supplementary foods until they are at least 7–8 months old; the widespread belief is that individuals should feed themselves and that younger infants cannot swallow food. The people of these rural villages are unaware of the nutritional value of different foods, and high-calorie and high-protein foods are reserved for adults. They believe that infants and young children won't appreciate the taste of these foods and that the young don't really deserve them because they haven't worked to produce them. Individuals are considered to be able to make their own decisions about eating. As a result, infants and young children are not offered much assistance with feeding and are not encouraged to eat if they do not feel like eating. The results in terms of infant and child growth are shown in Figure 12.10.

The infant feeding practices in rural Mali are not necessarily typical of populations in which growth faltering occurs. Growth disruption can occur even when children are preferentially given food. Leonard (1991) found that infants and children were more protected than adults from seasonal food shortages in rural highland Peru. During the preharvest season, when food supplies were limited, infants and children had energy intakes above their requirements. On the other hand, adult intakes during this same time period were substantially below their requirements. Even though women reduced the number of meals that they prepared each day during the preharvest season, they made sure that children received additional food as snacks. Even though infants and young children are favored in terms of feeding in this population, growth faltering still occurs.

The specific nutrients that are lacking in the diets of children who experience growth faltering certainly vary from place to place. In the case of rural Mali, infant feeding practices probably result in deficiencies of energy, protein, and **micronutrients** (vitamins and minerals). In other situations, the actual cause of growth faltering is much less clear (see Chapter 10). Because faltering occurs in some situations where intakes of energy and protein appear to be adequate, it has been suggested that micronutrient deficiency may play an important role.

Growth faltering is caused by not only nutritional insufficiencies but also disease. Disruptions in growth begin when infants' exposure to pathogens increases. The supplemental foods that become necessary after breast milk becomes insufficient to fully support growth are one source of pathogens, and food contamination is a potential danger to infants. Thus, whereas avoiding supplemental foods presents a nutritional risk, feeding supplemental foods increases the risk of infection. When infants become mobile (e.g., when they start to crawl), they are exposed to more pathogens in their physical environment in addition to those carried by people. One of the diseases thought to be a major cause of growth disruption is diarrhea. Diarrhea can have a substantial impact on nutritional status for two reasons: the nutrient losses that occur as a direct result of diarrhea and the reduced food intake that frequently coincides with disease (as a result of loss of appetite and cultural beliefs about food restriction during illness). Diarrhea is quite frequent in young children who live under disadvantaged circumstances. In rural Guatemala, infants 6–12 months of age (the same infants whose mothers were supplemented during pregnancy—see earlier in this chapter) were ill with diarrhea more than 10% of the time. Children under 7 years who were frequently sick with diarrhea grew an average of ~6% less in length and 10% less in weight than children who had fewer episodes of diarrhea (Martorell et al. 1975).

Juvenile and Adolescent Growth

In contrast to growth in infancy and childhood, **juvenile** and adolescent growth rates are much more similar between children who live in advantaged and disadvantaged circumstances.

The amount that height differs in childhood and in adulthood between US populations and populations living under adverse circumstances in developing countries is illustrated in Figure 12.11. In all cases shown in the figure, the height difference at 4–6 years old accounts for the majority of the height difference in adulthood. In one-half the cases shown (India, Peru, and Mexico), the difference in height is greater in adulthood than in childhood, suggesting that juvenile and adolescent growth are also somewhat slower in these populations than in the United States. In the other cases shown, the adult height difference is actually smaller than the height difference in childhood, suggesting that juvenile and/or adolescent growth has actually been greater in disadvantaged populations than in the United States. In all cases, the vast majority of the adult height difference is explained by growth differences that occurred before age 6. These data indicate that growth disruption occurs primarily during infancy and childhood and that juvenile and adolescent growth are much less affected by environmental factors than growth at previous stages.

Additional factors must be taken into account when interpreting Figure 12.11. Only the data from India and Gambia are based on **longitudinal** data; the other studies are based on **cross-sectional** data. In the cross-sectional studies, the same individuals were not measured as children and as adults. Thus, we cannot be certain how tall the children would have grown to be. Another potentially confounding factor is mortality. As discussed later, the smallest children are most likely to die. Part of the explanation for populations in Figure 12.11 in which the height difference is smaller in adulthood than in childhood may be that the smallest children do not live to adulthood.

Growth during the juvenile period probably is not greatly affected by environmental factors because it is characterized by slow growth rates. But why is adolescence, a period when growth is quite rapid, relatively unaffected by poor conditions? It appears that adolescent growth is under greater genetic control than is earlier growth.

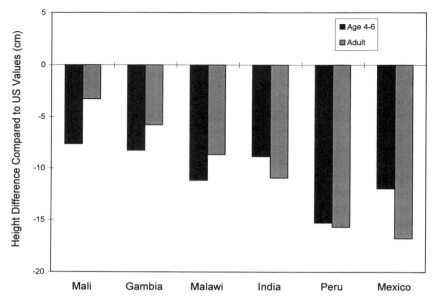

Figure 12.11   Heights of children (closed bars) and adults (hatched bars) who live under disadvantaged circumstances compared with heights of US children and adults. The line at zero is median US height. Bar length indicates how many centimeters shorter females living in developing countries are in childhood and in adulthood. Growth deficits accumulated during childhood are the major contributor to the small adult height observed in populations living under disadvantaged circumstances. (*Source:* Data from Eveleth and Tanner 1976; Hauspie et al. 1980; Malina et al. 1980, 1983; Billewicz and MacGregor 1982; WHO 1983; Dettwyler 1991, 1992; Pelletier et al. 1991)

A comparison of European American children (5–17 years old) with children attending an expensive private school in Guatemala (Johnston et al. 1976) illustrates this phenomenon. The children in Guatemala could be divided into two groups: those of Guatemalan ancestry (Spanish last names, Spanish spoken at home, and grandparents born in Guatemala) and those of European ancestry (home language other than Spanish, parents born in the United States or western Europe, excluding Spain). Based on frequencies of the ABO blood group, the children of Guatemalan ancestry had more Amerind admixture, so the two groups of Guatemalan children differed genetically but grew up in the same environment (see Table 12.3). The Guatemalan children of European ancestry and the European American children were genetically similar but grew up in different environments. Before adolescence, the two groups of children living in Guatemala grew similarly and differed from the European American children. But during adolescence, the European American children and Guatemalan children of European ancestry were most similar. Before adolescence, shared environment was more important than shared genes, whereas during adolescence, the opposite was true. The results of this study indicate that genetic effects on growth are more important during adolescence than during childhood.

Catch-Up Growth

Another indicator that the environment does not have a great effect on adolescent growth is that the amount of growth during adolescence does not always correlate with environmental circumstances.

In a comparison of British boys and rural boys in India, Satyanarayana et al. (1989) found that disadvantaged Indian boys actually grew more in height during adolescence than did British boys. As a result, the height difference between British and Indian boys was less at the end of the adolescent growth spurt than at the beginning. Like some of the groups shown in Figure 12.11, the disadvantaged children showed some **catch-up growth** relative to advantaged children.

The term catch-up growth was first used to refer to a period of faster growth that followed a period of growth disruption. Its meaning has been extended to include growth improvement that occurs as a result of a longer period of growth. The Turkana, a pastoralist group in Kenya, show catch-up growth that results from an extended growth period (see Figure 12.12A). The heights of young Turkana children are below the US 50th percentile and, at times, below the 5th percentile. However, Turkana adults reach heights similar to those of US adults. This discrepancy is easily explained: The Turkana grow for a longer time than Americans, and **skeletal maturation** occurs later in Turkana children than in US children. Although height growth in US females begins to level off at about age 16, Turkana females grow rapidly up to age 18. Increased mortality of the smallest individuals also may contribute to the appearance of catch-up growth.

Figure 12.12B illustrates catch-up growth in a young Peruvian girl that occurred because of faster rates of growth. Her growth improved as a result of changed circumstances, and by age 10, she had almost attained the average height for her age. Although complete catch-up growth can occur, in most cases it does not. (None of the examples in Figure 12.11 shows complete catch-up relative to US children). The main reason for lack of complete catch-up growth is that almost all children who experience growth faltering remain in the environment that produced the deficits. The potential for catch-up growth under these conditions is very limited.

Table 12.3 Comparison of preadolescent and adolescent growth in genetically different populations in the same environment and in genetically similar groups in different environments

	United States	Guatemala	
	European ancestry	European ancestry	Guatemalan ancestry
Environment	Different	More similar	More similar
Genetic makeup	More similar	More similar	Different
Preadolescent growth	Different	More similar	More similar
Adolescent growth	More similar	More similar	Different

(*Source:* Data from Johnston et al. 1976)

Biological factors also may limit the capacity for catch-up growth. Growth faltering begins in infancy, when growth rates are very high. Prolonged growth at the these rates may not be possible later in life. Another consideration is the relationship between skeletal maturation and linear growth. When the delay in skeletal maturation is equal to the delay in linear growth, the additional growth period is long enough to compensate for earlier slow growth. However, when the delay in skeletal maturation is less than the delay in linear growth, growth will never completely catch up. Although the period of growth may be lengthened somewhat under these conditions, skeletal maturation will be complete before the growth has made up for earlier faltering.

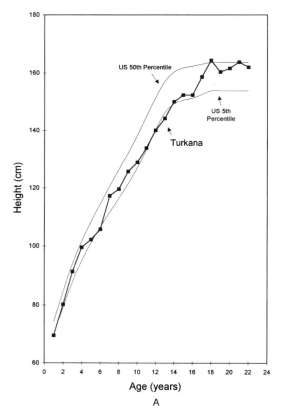

Figure 12.12 (A) Catch-up growth can occur through a longer period of growth, as is the case for Turkana females. (B) Catch-up growth also can occur through faster growth. The height of an undernourished Peruvian girl (heavy line) improved greatly during hospitalization. She showed little further catch-up growth during the next three years when she was in an orphanage and in foster care. A second episode of catch-up growth occurred after she was adopted, when she was a little over 8 years old. (*Source:* Data from Graham and Adrianzen 1972; Little et al. 1983; WHO 1983)

Sex Differences

Sex differences in body size and shape may be partly determined by genes. In addition, some evidence indicates that the sensitivity of growth to environmental factors may differ in males and females. In particular, prenatal growth appears to be more affected by the environment in males than in females. For example, some studies of maternal supplementation during pregnancy found that supplementation increased birth weights only in male infants.

It is more difficult to determine whether male postnatal growth is more sensitive to the environment because we cannot be certain that males and females are exposed to the exact same conditions after birth (Stinson 1985). However, in many cases, it appears that male growth is more affected by adverse postnatal circumstances. For example, the weights of 4- to 7-year-olds in rural Mali are farther below international reference data for boys than for girls (Dettwyler 1991). From an evolutionary point of view, it would be advantageous for females to be more buffered against environmental stress because they have to support pregnancy and lactation. The exact mechanisms that may lead to females being less sensitive to the environment are not known.

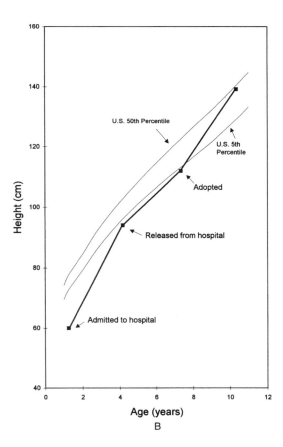

B

Is Small Body Size an Adaptation to Undernutrition?

Individuals in disadvantaged circumstances in developing countries grow up to be shorter adults, but their weight-to-height ratio is similar to that of individuals who grow up in affluent circumstances. It has been suggested that small body size (short stature [stunting], not accompanied by excessive thinness [wasting]) is advantageous under conditions of nutritional stress because small individuals are able to survive on fewer nutrients. The idea that stunting is a successful adaptation to undernutrition is controversial.

Many authors have pointed out functional disadvantages of stunting, especially the process that leads to stunting (e.g., Martorell 1989). Body size influences the ability to perform energy-demanding activities, so stunted individuals would be expected to be less productive, especially in activities requiring prolonged intense physical labor. The undernutrition that leads to stunting can hardly be considered advantageous. In addition to reducing immune function, undernutrition leads to a decrease in activity level that can negatively affect how young children interact with people and their surroundings. Although small body size does reduce nutrient needs, stunting can hardly be considered a no-cost response to undernutrition.

One way to test whether small body size is an adaptation to undernutrition is to determine whether small individuals have greater **fertility** under conditions of nutritional stress. The results of studies examining the relationship between maternal height and fertility in populations living under disadvantaged conditions in developing countries are variable (Stinson 1992). In some investigations, no relationship between height and number of surviving offspring was found. When an association between body size and fertility has been found, it has been variously found that taller women, shorter women, and women in the middle of the height distribution have the greatest number of surviving offspring. One problem in interpreting these results is that differences in adult height do not necessarily indicate genetic differences. Maternal body size also reflects the conditions experienced during growth and development. An association between fertility and maternal height therefore does not necessarily indicate natural selection based on genetic characteristics.

This problem is illustrated by another case in which body size is associated with survival. Under disadvantaged circumstances, the shortest, thinnest young children suffer the greatest mortality. Environmental conditions, rather than genes for small body size, are presumably the main cause of the children's size and weight. In this case, we do not have natural selection against those with certain genes but rather environmental factors that affect both body size and the probability of survival (Stinson 1992).

RELATIVE EFFECTS OF ENVIRONMENT AND GENETICS

In the previous sections, you have seen that many environmental factors can affect growth. Would the average body size be the same in all human populations if they all experienced the same environment? We can be fairly certain that the answer to this question is not invariably yes, because we already know that the small size of pygmy populations is probably the result of genetic factors affecting hormones. But what of the variation in body size among other human populations? Among populations, to what extent can growth differences be explained as the result of environmental differences, and to what extent are differences in average body size related to genetic makeup?

To determine the genetic component, we ideally compare genetically distinct populations who live in the same environment. By holding the environment constant, any growth differences between the populations should be the result of genetic factors. In practice, comparisons usually are made among affluent children. These healthy and well-nourished children should not have experienced major disruptions in their growth. However, because of the many environmental conditions that affect growth, it is difficult to be certain that even the circumstances of well-off populations are exactly the same.

African American, European, and European American children who live in affluent countries (e.g., the United States, Canada, western Europe, and Australia) generally show similar patterns of growth in height during childhood and adolescence (see Figure 12.13). Because the SES of African Americans is on average lower than the

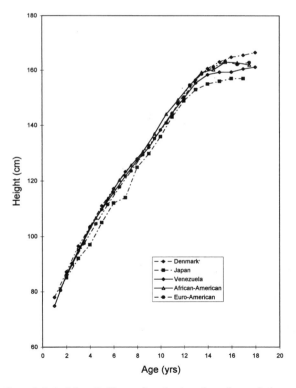

Figure 12.13 Growth in height of affluent females in selected populations. Affluent children in diverse populations generally have similar patterns of growth in childhood. Genetic effects on growth are more apparent during adolescence. (*Source:* Data from Eveleth and Tanner 1990)

SES of European Americans, the comparable growth in the two groups suggests that the average genetic growth potential is actually greater in African Americans. Although the growth of African American children is frequently used to represent the growth of African children, it should be remembered that African Americans have an appreciable amount of European admixture. In addition, because the African ancestry of African Americans is predominantly West African, African Americans certainly do not represent all of Africa.

Most studies of well-off children in developing countries (e.g., India and South and Central America) have found that growth in childhood does not differ greatly from the growth of children in affluent countries. In some cases, as shown for Venezuelan girls in Figure 12.13, differences increase during adolescence. It is difficult to determine the relative role of environmental and genetic factors as causes of these differences. Although elites in developing countries certainly do not suffer from the same undernutrition and disease as the poor, in most cases, they probably do not experience the same environmental conditions as children in affluent countries, either. They get different diseases, eat different diets, and are exposed to different environmental pollutants.

For many of the world's populations, we do not know what growth would be like under "ideal" conditions. Sometimes, information is lacking because the necessary research has not yet been done. In most populations, however, the simple fact is that few children are affluent. So, for example, we do not know how the native populations of Australia and New Guinea would grow under less adverse circumstances. In Latin America, high-SES individuals generally have a large amount of European admixture, so their growth pattern does not tell us about the growth potential of native South Americans.

One frequently cited exception to the finding that affluent children from genetically diverse populations show similar patterns of growth involves north Asian populations, such as the Japanese. Japanese children are shorter than children from other industrialized countries throughout the growth period, although the differences become more pronounced in adolescence (see Figure 12.13). By adulthood, the average Japanese stature is between the US 10th and 25th percentile. The larger divergence in adolescence and adulthood could partially reflect less favorable environmental conditions in childhood for the earlier-born, older Japanese **cohorts**. Thus, continuing secular trends may lessen the height difference. Another cause of this height difference is that Japanese children have an earlier adolescent growth spurt and thus grow for a shorter period of time than do children of European ancestry. Whether this difference is the result of genetic or environmental causes is not certain.

Both genetic makeup and the influence of the environment have important effects on growth. What conclusions can we reach about the relative importance of these two factors as explanations for variation in body size? Research indicates that usually, environmental rather than genetic causes account for the majority of variation among populations. Although evidence indicates that not all populations have the same average genetic growth potential, the difference between populations is small. The possibly genetic difference of almost 7 cm (2.75 inches) in average adult female height between Japanese and European Americans is only a fraction of that between European American women and poor women in rural Mexico (16.8 cm [6.6 inches]), which is mainly caused by the environment (see Figure 12.11). As discussed earlier, genetic factors play a greater role in growth during adolescence than during childhood. Therefore, most growth differences that take place during childhood are the result of environmental differences. At age 6, the average Japanese girl is only 2.4 cm (a little less than 1 inch) shorter than the average European American girl.

Because healthy, well-nourished children from genetically distinct populations have similar patterns of growth, especially before adolescence, it has been suggested that one set of reference data is appropriate to assess the nutritional status of children in diverse populations (see Box 12.2). In other words, the growth of children in the Netherlands, Mali, and any other country would be compared with the same set of growth data (i.e., the reference data). The reference data should be based on healthy, well-nourished children whose growth is what we expect under good conditions. The most frequently used reference data are based on US children (as in this chapter) because growth studies of US children are some of the largest, best-designed studies of presumably healthy, well-nourished children. A major advantage of using one set of reference data is that it makes it easy to compare data from different studies.

However, the use of one set of reference data has not been without controversy. First, evidence indicates that the average genetic growth potential is not the same in all populations. Although the genetic differences may not be extremely large, they can have a substantial effect in terms of the assessment of nutritional status using growth measures. Second, we simply do not know the average genetic growth potential for many populations. Finally, many researchers object to using data on US children as a growth reference because of the high (and increasing) rates of obesity in the United States.

CHAPTER SUMMARY

In this chapter, we reviewed the many factors that lead to variation in body size and shape within and among human populations. Growth is affected by genetic and environmental factors, which can be modified by cultural practices.

Many environmental factors are associated with growth. For example, high SES is correlated with greater stature. Growth differences also have been observed between residents of urban and rural areas and among individuals coming from different sized families, although the direction of these differences is not constant. Positive secular trends such as the increases in body size and rate of maturation that have taken place in industrialized countries over the past 150 years are good examples of environmental effects on growth. Secular trends and the effects of SES, residence (urban or rural), and family size are related to health and nutritional status. Toxins such as cigarette smoke, lead, and various pollutants are environmental factors whose effects on growth do not occur primarily through their influence on nutrition and disease.

Inadequate nutrition and high rates of disease are the two most important environmental factors that lead to poor growth. The effects of undernutrition on growth begin before birth. Infants born to women who have experienced food shortages during pregnancy are smaller than infants born to well-nourished women. Food supplements for undernourished pregnant women improve the prenatal growth of their infants, although the effects are not always as large as expected.

Infancy and childhood are the stages of life during which undernutrition and disease have the greatest effects on growth. Under disadvantaged conditions, growth faltering begins during the first 6 months of life. Infant feeding practices and diarrheal disease can have negative effects on early childhood growth, although the exact causes of growth faltering are not always clear. Some children who experience slowed growth early in life undergo catch-up growth later in life, but most never catch up completely.

Environmental effects on growth are smaller during the juvenile period and particularly during adolescence than during previous growth stages. For this reason, much of the smaller adult body size of individuals who live in adverse conditions in developing countries is the result of growth disruptions early in life. Short stature may be an adaptation to undernutrition, but no consistent evidence supports natural selection for small body size under conditions of nutritional stress.

Most studies of affluent children indicate the absence of major genetic differences in average growth potential among human populations, at least during childhood. However, information about growth potential is lacking for many populations. This lack of data has led to debate about whether it is appropriate to use one set of reference data for the growth assessment of children in diverse populations.

RECOMMENDED READINGS

Bogin B (1988) Rural-to-urban migration. In CG Mascie-Taylor and GW Lasker (eds.): Biological Aspects of Human Migration. Cambridge: Cambridge Univ. Press, pp. 90–129.

Bogin B (1999) Patterns of Human Growth, 2nd edition. Cambridge: Cambridge Univ. Press

Brown PJ and Konner M (1987) An anthropological perspective on obesity. Ann. N.Y. Acad. Sci. 499: 29–46.

Cassidy CM (1991) The good body: when big is better. Med. Anthropol. 13: 181–213.

Dettwyler KA (1994) Dancing Skeletons. Life and Death in West Africa. Prospect Heights, Ill.: Waveland Press.

Dettwyler KA and Fishman C (1992) Infant feeding practices and growth. Annu. Rev. Anthropol. 21: 171–204.

Eveleth PB and Tanner JM (1976) Worldwide Variation in Human Growth, 1st edition. Cambridge: Cambridge Univ. Press.

Eveleth PB and Tanner JM (1990) Worldwide Variation in Human Growth, 2nd edition. Cambridge: Cambridge Univ. Press.

Malina RM and Bouchard C (1991) Growth, Maturation, and Physical Activity. Champaign, Ill.: Human Kinetics Books.

Martorell R and Habicht J-P (1986) Growth in early childhood in developing countries. In F Falkner and JM Tanner (eds.): Human Growth. Volume 3. Methodology. Ecological, Genetic, and Nutritional Effects on Growth, 2nd edition. New York: Plenum, pp. 241–262.

Schell LM (1991) Effects of pollutants on human prenatal and postnatal growth: noise, lead, polychlorobiphenyl compounds, and toxic wastes. Yearbook Phys. Anthropol. 34: 157–188.

Stuart-Macadam P and Dettwyler KA (1995) Breastfeeding. Biocultural Perspectives. New York: Aldine De Gruyter.

Stinson, S (1992) Nutritional adaptation. Annu. Rev. Anthropol. 21: 143–170.

Tanner JM (1990) Fetus into Man, 2nd edition. Cambridge, Mass.: Harvard Univ. Press.

Ulijaszek SJ, Johnston FE, and Preece MA (eds.) (1998) The Cambridge Encyclopedia of Human Growth and Development. Cambridge: Cambridge Univ. Press.

van Wieringen JC (1986) Secular growth changes. In F Falkner and JM Tanner (eds.): Human Growth. Volume 3. Methodology. Ecological, Genetic, and Nutritional Effects on Growth. New York: Plenum, pp. 307–331.

REFERENCES CITED

Adair LS and Pollitt E (1983) Seasonal variation in pre- and postpartum maternal body weight measurements and infant birth weights. Am. J. Phys. Anthropol. 62: 325–331.

Adair LS and Pollitt E (1985) Outcome of maternal nutritional supplementation: a comprehensive review of the Bacon Chow study. Am. J. Clin. Nutr. 41: 948–978.

Antonov AN (1947) Children born during the siege of Leningrad in 1942. J. Pediatr. 30: 250–259.

Bailey RC (1991) The comparative growth of Efe pygmies and African farmers from birth to age 5 years. Ann. Hum. Biol. 18: 113–120.

Bielicki T and Waliszko H (1992) Stature, upward social mobility, and the nature of statural differences between social classes. Ann. Hum. Biol. 19: 589–593.

Billewicz WZ and McGregor IA (1982) A birth-to-maturity longitudinal study of heights and weights in two West African (Gambian) villages, 1951–1975. Ann. Hum. Biol. 9: 309–320.

Bogin B (1988) Rural-to-urban migration. In CG Mascie-Taylor and GW Lasker (eds.): Biological Aspects of Human Migration. Cambridge: Cambridge Univ. Press, pp. 90–129.

Bogin BA and MacVean RB (1978) Growth in height and weight of urban Guatemalan primary school children of low and high socioeconomic class. Hum. Biol. 50: 477–487.

Bogin B and MacVean RB (1981) Nutritional and biological determinants of body fat patterning in urban Guatemalan children. Hum. Biol. 53: 259–268.

Bogin B and Loucky J (1997) Plasticity, political economy, and physical growth status of Guatemalan Maya children living in the United States. Am. J. Phys. Anthropol. 102: 17–32.

Brown PJ and Konner M (1987) An anthropological perspective on obesity. Ann. N.Y. Acad. Sci. 499: 29–46.

Cameron N, Kgamphe JS, Leschner KF, and Farrant PJ (1992) Urban–rural differences in the growth of South African black children. Ann. Hum. Biol. 19: 23–33.

Cassidy CM (1991) The good body: when big is better. Med. Anthropol. 13: 181–213.

Chen LC, Huq E, and D'Souza S (1981) Sex bias in the family allocation of food and health care in rural Bangladesh. Pop. Dev. Rev. 7: 55–70.

Cunningham AS (1995) Breastfeeding: adaptive behavior for child health and longevity. In P Stuart-Macadam and KA Dettwyler (eds.): Breastfeeding. Biocultural Perspectives. New York: Aldine De Gruyter, pp. 243–303.

Dettwyler KA (1991) Growth status of children in rural Mali: implications for nutrition education programs. Am. J. Hum. Biol. 3: 447–462.

Dettwyler KA (1992) Nutritional status of adults in rural Mali. Am. J. Phys. Anthropol. 88: 309–321.

Dettwyler KA and Fishman C (1992) Infant feeding practices and growth. Annu. Rev. Anthropol. 21: 171–204.

Diamond JM (1991) Why are pygmies small? Nature 354: 111–112.

Dufour DL, Staten LK, Reina JC, and Spurr GB (1994) Anthropometry and secular changes in stature of urban Colombian women of differing socioeconomic status. Am. J. Hum. Biol. 6: 749–760.

Eveleth PB and Tanner JM (1976) Worldwide Variation in Human Growth, 1st edition. Cambridge: Cambridge Univ. Press.

Eveleth PB and Tanner JM (1990) Worldwide Variation in Human Growth, 2nd edition. Cambridge: Cambridge Univ. Press.

Flegal KM, Harlan WR, and Landis JR (1988) Secular trends in body mass index and skinfold thickness with socioeconomic factors in young adult women. Am. J. Clin. Nutr. 48: 535–543.

Fomon SJ (1993) Nutrition of Normal Infants. St. Louis: Mosby.

Frisancho AR (1990) Anthropometric Standards for the Assessment of Growth and Nutritional Status. Ann Arbor: Univ. of Michigan Press.

Garn SM (1985) Smoking and human biology. Hum. Biol. 57: 505–523.

Garn SM and Clark DC (1975) Nutrition, growth, development, and maturation: findings from the Ten-State Nutrition Survey of 1968–1970. Pediatrics 56: 306–319.

Geffner ME, Bersch N, Bailey RC, and Golde DW (1995) Insulin-like growth factor I resistance in immortalized T cell lines from African Efe pygmies. J. Clin. Endocrinol. Metab. 80: 3732–3738.

Georges E, Mueller WH, and Wear ML (1991) Body fat distribution: associations with socioeconomic status in the Hispanic Health and Nutrition Examination Survey. Am. J. Hum. Biol. 3: 489–501.

Graham GG and Adrianzen B (1972) Late "catch-up" growth after severe infantile malnutrition. Hopkins Med. J. 131: 204–211.

Hauspie RC, Das SR, Preece MA, and Tanner JM (1980) A longitudinal study of the growth in height of boys and girls of West Bengal (India) aged six months to 20 years. Ann. Hum. Biol. 7: 429–440.

Institute of Medicine (U.S.) Subcommittee on Nutritional Status and Weight Gain During Pregnancy (1990) Nutrition During Pregnancy. Washington, D.C.: National Academy Press.

Institute of Medicine (U.S.) Subcommittee on Nutrition During Lactation (1991) Nutrition During Lactation. Washington, D.C.: National Academy Press.

Instituto Nacional de Alimentação e Nutrição (1990) Pesquisa Nacional sobre Saúde e Nutrição. Brasilia, Brazil: Ministerio da Saúde.

Johnston FE, Borden M, and MacVean RB (1973) Height, weight, and their growth velocities in Guatemalan private school children of high socioeconomic class. Hum. Biol. 45: 627–641.

Johnston FE, Wainer H, Thissen D, and MacVean R (1976) Hereditary and environmental determinants of growth in height in a longitudinal sample of children and youth of Guatemalan and European ancestry. Am. J. Phys. Anthropol. 44: 469–476.

Kimura K (1984) Studies of growth and development in Japan. Yearbook Phys. Anthropol. 27: 179–213.

Kulin HE, Bwibo N, Mutie D, and Santner SJ (1982) The effect of chronic childhood malnutrition on pubertal growth and development. Am. J. Clin. Nutr. 36: 527–536.

Lechtig A, Habicht J-P, Delgado H, Klein RE, Yarbrough C, and Martorell R (1975) Effect of food supplementation during pregnancy on birth weight. Pediatrics 56: 508–520.

Leonard WR (1991) Household-level strategies for protecting children from seasonal food scarcity. Soc. Sci. Med. 33: 1127–1133.

Lin W-S, Zhu, F-C, Chen ACN, Xin W-H, Su Z, Li J-Y, and Ye G-S (1992) Physical growth of Chinese school children 7–18 years, in 1985. Ann. Hum. Biol. 19: 41–55.

Lindgren G (1976) Height, weight, and menarche in Swedish urban school children in relation to socio-economic and regional factors. Ann. Hum. Biol. 3: 501–528.

Lindgren GW and Hauspie RC (1989) Heights and weights of Swedish school children born in 1955 and 1967. Ann. Hum. Biol. 16: 397–406.

Lindgren GW and Cernerud L (1992) Physical growth and socioeconomic background of Stockholm schoolchildren born in 1933–63. Ann. Hum. Biol. 19: 1–16.

Little MA, Galvin K, and Mugambi M (1983) Cross-sectional growth of nomadic Turkana pastoralists. Hum. Biol. 55: 811–830.

Ljung B-O, Bergsten-Brucefors A, and Lindgren G (1974) The secular trend in physical growth in Sweden. Ann. Hum. Biol. 1: 245–256.

Malina RM, Selby HA, Buschang PH, and Aronson WL (1980) Growth status of schoolchildren in a rural Zapotec community in the Valley of Oaxaca, Mexico, in 1968 and 1978. Ann. Hum. Biol. 7: 367–374.

Malina RM, Selby HA, Buschang PH, Aronson WL, and Wilkinson RG (1983) Adult stature and age at menarche in Zapotec-speaking communities in the Valley of Oaxaca, Mexico, in a secular perspective. Am. J. Phys. Anthropol. 60: 437–449.

Martorell R (1989) Body size, adaptation and function. Hum. Organ. 48: 15–20.

Martorell R, Habicht J-P, Yarbrough C, Lechtig A, Klein RE, and Western KA (1975) Acute morbidity and physical growth in rural Guatemalan children. Am. J. Dis. Child. 129: 1296–1301.

Martorell R and Habicht J-P (1986) Growth in early childhood in developing countries. In F Falkner and JM Tanner (eds.): Human Growth. Volume 3. Methodology. Ecological, Genetic, and Nutrition Effects on Growth, 2nd edition. New York: Plenum, pp. 241–262.

Meredith HV (1976) Findings from Asia, Australia, Europe, and North America on secular change in mean height of children, youths, and young adults. Am. J. Phys. Anthropol. 44: 315–326.

Mora JO, de Paredes B, Wagner M, de Navarro L, Suescun J, Christiansen N, and Herrera MG (1979) Nutritional supplementation and the outcome of pregnancy. I. Birth weight. Am. J. Clin. Nutr. 32: 455–462.

Najjar MF and Rowland M (1987) Anthropometric Reference Data and Prevalence of Overweight, United States, 1976–1980. Vital and Health Statistics. Series 11, No. 238 Washington D.C.: US Government Printing Office.

Najjar MF and Kuczmarski RJ (1989) Anthropometric Data and Prevalence of Overweight for Hispanics: 1982–84. Vital and Health Statistics. Series 11, No. 239. Washington D.C.: US Government Printing Office.

Norgan NG (1994) Interpretation of low body mass indices: Australian Aborigines. Am. J. Phys. Anthropol. 94: 229–237.

Oftedal OT (1984) Milk composition, milk yield, and energy output at peak lactation: a comparative review. Symp. Zool. Soc. Lond. 51: 33–85.

Pawson IG (1986) The morphological characteristics of adult Samoans. In PT Baker, JM Hanna, and TS Baker (eds.): The Changing Samoans. New York: Oxford Univ. Press, pp. 254–274.

Pelletier DL, Low JW, and Msukwa LAH (1991) Malawi Maternal and Child Nutrition Study: study design and anthropometric characteristics of children and adults. Am. J. Hum. Biol. 3: 347–361.

Prader A (1984) Biomedical and endocrinological aspects of normal growth and development. In J Borms, R Hauspie, A Sand, C Susanne, and M. Hebbelinck (eds.): Human Growth and Development. New York: Plenum, pp. 1–22.

Prentice AM, Whitehead RG, Roberts SB, and Paul AA (1981) Long-term energy balance in childbearing Gambian women. Am. J. Clin. Nutr. 34: 2790–2799.

Prentice AM, Roberts SB, Prentice A, Paul AA, Watkinson M, Watkinson AA, and Whitehead RG (1983) Dietary supplementation of lactating Gambian women. I. Effect on breast-milk volume and quality. Hum. Nutr. Clin. Nutr. 37C: 53–64.

Prentice AM, Cole TJ, Foord FA, Lamb WH, and Whitehead RG (1987) Increased birthweight after prenatal dietary supplementation of rural African women. Am. J. Clin. Nutr. 46: 912–925.

Prentice AM, Goldberg GR, Davies HL, Murgatroyd PR, and Scott W (1989) Energy-sparing adaptations in human pregnancy assessed by whole-body calorimetry. Br. J. Nutr. 62: 5–22.

Roche AF (ed.) (1979) Secular Trends in Human Growth, Maturation, and Development. Monographs of the Society for Research in Child Development, Vol. 44, Nos. 3 and 4.

Russell M (1976) The relationship of family size and spacing to the growth of preschool Mayan children in Guatemala. Am. J. Public Health 66: 1165–1172.

Sanjeev, Indech GD, Jit I, and Johnston FE (1991) Skinfold thicknesses, body circumferences, and their relationship to age, sex, and socioeconomic status in adults from northwest India. Am. J. Hum. Biol. 3: 469–477.

Satyanarayana K, Radhaiah G, Murali Mohan KR, Thimmayamma BVS, Pralhad Rao N, Narasinga Rao BS, and Akella S (1989) The adolescent growth spurt of height among rural Indian boys in relation to childhood nutritional background: an 18 year longitudinal study. Ann. Hum. Biol. 16: 289–300.

Schell LM (1991) Effects of pollutants on human prenatal and postnatal growth: noise, lead, polychlorobiphenyl compounds, and toxic wastes. Yearbook Phys. Anthropol. 34: 157–188.

Schumacher A and Knussmann R (1979) Are the differences in stature between social classes a modification or an assortment effect? J. Hum. Evol. 8: 809–812.

Smith CA (1947) Effects of maternal undernutrition upon the newborn infant in Holland (1944–1945). J. Pediatr. 30: 229–243.

Steegmann AT (1985) 18th-century British military stature: growth cessation, selective recruiting, secular trends, nutrition at birth, cold and occupation. Hum. Biol. 57: 77–95.

Stein Z, Susser M, Saenger G, and Marolla F (1975) Famine and Human Development. New York: Oxford Univ. Press.

Stinson S (1980) Child growth and the economic value of children in rural Bolivia. Hum. Ecol. 8: 89–103.

Stinson S (1985) Sex differences in environmental sensitivity during growth and development. Yearbook Phys. Anthropol. 28: 123–147.

Stinson S (1989) Physical growth of Ecuadorian Chachi Amerindians. Am. J. Hum. Biol. 1: 697–707.

Stinson S (1992) Nutritional adaptation. Annu. Rev. Anthropol. 21: 143–170.

Tanner JM (1990) Fetus into Man, 2nd edition. Cambridge, Mass.: Harvard Univ. Press.

Tobias PV (1985) The negative secular trend. J. Hum. Evol. 14: 347–356.

Ventura SJ, Martin JA, Curtin SC, and Mathews TJ (1999) Births: Final Data for 1997. National Vital Statistics Report, Volume 47, Number 18. Hyattsville, MD: National Center for Health Statistics.

Waterlow JC and Schürch B (eds.) (1994) Causes and Mechanisms of Linear Growth Retardation. Eur. J. Clin. Nutr. 48 (Suppl. 1): S1-S216.

World Health Organization (WHO) (1983) Measuring Change in Nutritional Status. Geneva: WHO.

World Health Organization (WHO) Working Group (1986) Use and interpretation of anthropometric indicators of nutritional status. Bull. WHO 64: 929–941.

CHAPTER 13

Aging, Senescence, and Human Variation

GILLIAN J. HARPER AND DOUGLAS E. CREWS

INTRODUCTION

Today, humankind is a very long-lived species, a condition that likely did not characterize many of our ancestors. In the time of Aristotle (384–322 BC), boys became men in their teens. Alexander III (Alexander the Great, 356–323 BC) ascended the throne of Greece (Macedonia) at age 19, conquered the known world before his 25th birthday, and died at age 33. Joan of Arc (Jeanne d'Arc, the Maid of Orleans, 1412–1432 AD) led the French armies at age 15 and was put to death before her 20th birthday. As recently as 1850, US "white" males could expect to live 31 years after birth and US "white" females could expect to live 40 years (Erhardt and Berlin 1974). (Note: All attributions of race, ethnicity, and culture are used in this chapter as they appeared in original publications cited.)

As illustrated in Table 13.1, **life expectancy at birth** in the United States has increased rather steadily for both men and women since the beginning of the 20th century. In general, this phenomenon also has been observed for life expectancy at age 40. However, expectation of life at age 85 has neither increased as steadily nor as dramatically as that for younger persons. Until recently, increased life expectancy has been experienced most at young ages with little change at ages over 75 years. However, beginning in the 1980s, life expectancy at ages over 75 years has been steadily increasing, resulting in a large and increasing percentage of the US population aged 65 and over, today about 13%.

During **hominid** evolution, **life span**, **life histories**, estimated maximum life span, and life expectancy at all ages among humankind have changed and increased through the interplay of evolutionary and cultural forces. As late as 1959–61, only about 88% of persons lived to see their 50th birthday. At the beginning of the 21st century, it is an unlucky man or women who does not celebrate his or her 70th birthday, an age that at

Human Biology: An Evolutionary and Biocultural Perspective, Edited by Sara Stinson, Barry Bogin, Rebecca Huss-Ashmore, and Dennis O'Rourke
ISBN 0-471-13746-4   Copyright © 2000 Wiley-Liss, Inc.

least 53% of men and 70% of women could expect to reach in 1992 in more developed regions of the world (data for 1992 from National Center for Health Statistics 1996). Of great interest to human biologists is that **aging** and **senescence** are associated with greater biological, physiological, and sociocultural variation than is observed during any other period of the human life span. Set within the context of evolutionary biology and human variation, the interplay of biological and cultural processes by which human life span has evolved are explored in this chapter.

First, we briefly explore characteristics of the aged and some theories on aging. Then, we provide vocabulary and definitions to acquaint readers with some of the issues and jargon used in aging research, **gerontology**, and geriatrics. Next, we examine some possible models of age-related change in human physiology, including **hormonal** and immune changes and age-related **chronic degenerative diseases**, using **hypertension** and **dementia** as examples. In the final sections, we present current theories of senescence, mechanisms of aging, and human biological variation during aging and **longevity**.

Currently, the **elderly** (i.e., persons over 65) are the fastest growing segment of the populations in the United States and many other developed countries. This trend is partly attributable to the significant, worldwide increases in longevity at all ages during the past 100 years, and particularly the past 40 years. For example, from 1950 to 1980, the life expectancy of the US population aged 65 years or older increased an average of 2.5% a year (range 0.2% in 1951 to 3.6% in 1974; Guralnik et al. 1988; Verbrugge 1989; Brock et al. 1990). Although life expectancy is significantly lower in underdeveloped countries and among inner-city populations in developed countries (Susser et al. 1985; Grigsby 1991), the elderly population also is increasing in

Table 13.1   Male and female life expectancy at birth, age 40, and age 85 in the United States

	Life expectancy (years)					
	Men			*Women*		
Year	Birth	Age 40	Age 85	Birth	Age 40	Age 85
1900	46.6	68.0	88.8	48.7	69.1	89.1
1910	48.6	67.7	88.8	52.0	69.2	89.1
1920	54.4	69.1	89.0	55.6	69.9	89.1
1930	59.7	69.1	89.0	63.5	71.6	89.8
1940	62.1	69.9	89.0	66.6	73.0	89.3
1950	66.5	71.2	89.4	72.2	75.7	89.8
1960	67.4	71.6	89.3	74.1	77.1	89.7
1970	68.0	71.9	89.6	75.6	78.3	90.5
1980	70.7	74.0	90.0	78.1	80.1	91.3
1990	72.7	75.6	90.2	79.4	81.0	91.4
2000	74.3	76.9	90.4	80.9	82.0	91.7

(*Source*: Wright 1997)

these areas (Kinsella and Suzman 1992). Worldwide, the net increase of people aged 55-plus is 1.2 million per month. Of these, four-fifths reside in developing countries. The oldest old—those aged 85 or more years—are the fastest growing population subgroup in many countries (Rosenwaike 1985; Grigsby 1991; Kinsella and Suzman 1992). Although older cohorts are increasing partially due to a reduction in **mortality** at older ages, the primary factor producing increase in the population aged 55 and over has been decreased childhood mortality (Olshansky et al. 1990).

Increased numbers of elderly have led to increased funding of research on aging in many nations, particularly in the United States (Eveleth 1994). Current research interests range from the cellular and molecular basis of aging to the evolutionary biology of senescence, and from the **epidemiology** of aging to personal and social changes with aging. Numerous theories on the biology of aging have been proposed. Many of these fail to appreciate distinctions between proximate causes of aging in organs, systems, and individuals and the ultimate causes of senescence in sexually reproducing organisms. Many highly variable life history strategies are seen in the organic world. These variations have led to various theories on the evolution of finite life spans. Given that all sexually reproducing species show age-related alterations, research on the causes of aging has been hampered by specificity and a lack of theoretical consistency. Many theories of aging have failed to differentiate between two very different questions: why organisms age (i.e., ultimate cause) as opposed to how the aging process occurs in a particular organism (i.e., proximate cause; Austad 1992; Crews 1994).

One estimate is that there are 300 proximate explanations for aging, but only 2 theories as to the ultimate causes of aging (Austad 1992). One reason for this plethora of proposed mechanisms is that patterns, rates, and probably proximate causes of senescence differ across species, adding to the general confusion (Finch 1994).

DEFINITIONS

As with any area of scientific pursuit, the study of aging has its unique vocabulary. One basic division is between geriatrics (a branch of medicine that deals with problems and diseases of old age and aging individuals) and gerontology (a branch of knowledge that deals with aging and problems of the aged; Websters 1983). Another fundamental division is between aging (i.e., becoming old; showing the effects or characteristics of increasing age; Websters 1983) and senescence (i.e., the process of becoming old; the phase from full maturity to death, characterized by an accumulation of metabolic products and decreased probability of reproduction and survival; adapted from Websters 1977; see also Rose 1991); the two terms often are interchanged both colloquially and scientifically. All things, living or not, age. Bottles of wine improve with age, whereas rocks and socks weather and wear. However, only the living can senesce. Many physiological phenomena show age-related change, which may or may not be senescent.

Senescence refers to the biological processes by which organisms become less capable of maintaining **homeostasis** with increasing time. Aging is an elusive term that has multiple social and biological connotations, whereas senescence better serves current scientific discussion of mechanisms that preclude continued survival in sexually reproducing organisms (Finch 1994). Unfortunately, multiple definitions of senescence are used by different researchers, for example, gerontologists and population geneticists (Crews 1993). Comfort defined senescence as "a deteriorating process, with an increasing probability of death with increasing age" (1979, 8). Finch refined this definition as "age-related changes in an organism that adversely affects its vitality and function ... [associated with an] increase in mortality rate as a function of time" (1994, 5). Rose (1991) faulted these definitions for not including any aspect of reproduction, which is an essential component for any evolutionary model of senescence. Rose defined aging as "a persistent decline in age-specific **fitness** components of an organism due to internal physiological deterioration" (1991, 20). In a recent review of molecular aspects of aging, Kirkwood (1995) defined aging as "a progressive, generalized impairment of function resulting in a loss of adaptive response to stress and in a growing risk of age-related disease" that ultimately leads to an increased probability of death, whereas senescence was defined as "the process of growing old." In the same volume, Johnson et al. (1995) provide working definitions: "Aging is a naturally occurring, postdevelopmental process. Senescence is a progressive impairment of function resulting eventually in increased mortality, decreased function, or both." The view of Johnson et al. (1995) is that most, "but not all, degenerative diseases would thus be manifestations of senescence."

If these last two sets of definitions appear totally contradictory to you as a student reading this volume, you have some idea of the difficulties with terminology in this area. No current definitions of aging and senescence satisfy all gerontologists. For our purposes, it may clarify these issues to recall that all things can be said to age, whether living or nonliving. Thus, aging per se is simply the fact of existence through time, whereas senescence is a progressive degeneration following a period of development and attainment of **maximum reproductive potential** that leads to an increased probability of mortality, a process that only living organisms show.

Given this definition, scientifically speaking, senescence represents an evolutionary problem to be solved; medically speaking, it represents a process to be avoided, halted, or delayed. To do either, senescence must be understood within the context of **natural selection**. This understanding necessitates examinations of patterns of life history, that is, changes through which an organism passes in its development from its primary stage of life to its natural death, including in humans fertilization, embryogenesis, fetal development, birth, childhood, adolescence, reproductive adulthood, **menopause**, senescence, and extrinsic factors that affect individuals within populations (e.g., environment, diet, population density, culture, and social environment); (Finch 1994; Wood et al. 1994). Unfortunately, for most natural populations, these factors are difficult or impossible to measure, precluding the wide availability of accurate data (Finch 1994). Data that are available suggest that rates and patterns of senescence and perhaps even basic mechanisms of senescence differ within and between **phylogenetic** classes and across environmental contexts, even within the same species (Finch 1994; Johnson et al. 1995).

Inconsistency characterizes many other terms used in gerontological research and literature (Crews 1990; Finch 1994; Olshanksy et al. 1990). Concepts such as longevity, life span, expectation of life, average life span, life expectancy, maximum life span, mortality rate doubling time (MRDT), and **maximum life span potential** all have very specific meanings; however, they often are used either inappropriately or interchangeably in gerontological research. Expectation of life at birth or life expectancy at birth (e_0) is a demographic measure of average life span resulting from all-cause mortality of a **cohort** (i.e., a group of individuals born in the same year). Expectation of life (e_x) at any age (x) is a well-defined basic **life table** (an actuarial table based on mortality statistics that follows an entire cohort from birth to death) function (see Chapter 14 for a discussion of life tables). Despite being well defined, life expectancy data may be used misleadingly in aging research because they are based on both child and adult mortality rates (Olshansky et al. 1990). However, e_x may be calculated for any age, and ages other than birth may provide a less misleading comparative scale. (See Table 13.1, which presents life expectancies at birth and at 40 and 85 years old for the US population between 1900 and 2000.) Life expectancies vary considerably across populations. Current estimates of life expectancies in different countries are presented in Table 13.2.

Maximum life span potential (MLP) is a theoretical concept of the longest potential life span of a species based on the oldest known individual of that species (Weiss 1981; Hofman 1984). MLP typically is calculated on captive and domestic samples. It has been suggested that MLP represents the genetic capacity of a species for long-term survival (Cutler 1980; Hofman 1984; Susser et al. 1985). However, life expectancy is environmentally sensitive, highly variable both between and within species, and easily modulated under laboratory conditions. MLP may not be as labile; however, it has been observed to shift under certain laboratory regimens (e.g., dietary restriction, temperature variation) (Finch 1994). For example, the maximum life span of certain strains of rodents increases when their caloric intake is restricted. Thus, MLP represents a theoretical concept based on our current knowledge of the longest lived individual of a species.

MLP apparently has increased over evolutionary time and may change over the course of evolution of a species. Unfortunately, documentation of such change cannot be obtained directly from the fossil record. Instead, estimates of MLP for fossil specimens are based on **allometric** relationships (see Chapter 9) between life span (i.e., the duration of existence of an individual, the average length of life of a particular type of organism in a particular environment under specified circumstances) and either body size or brain size. One problem with such estimates is that allometric relationships are established by using modern, often domesticated, species. Another problem is that brain and body sizes are estimated by using several **regression** equations. Thus, estimates of life span and MLP for fossil specimens are error-prone and may differ significantly from investigator to investigator. Table 13.3 presents estimates of average and maximum life spans and ages at puberty in various modern species.

Longevity (i.e., a long duration of individual life) is an individual phenomena. The individual with the greatest longevity in any particular environment is an outlier,

Table 13.2   Life expectancy at birth in selected countries

Country or area	Life expectancy (years)	
	Male	Female
Afghanistan	47.82	46.82
Argentina	71.13	78.56
Australia	77.22	83.20
Bangladesh	60.73	60.46
Benin	67.23	71.26
Bolivia	58.51	64.51
Botswana	39.42	40.37
Brazil	59.35	69.01
Cambodia	46.81	49.75
Canada	76.12	82.79
Central African Republic	45.35	49.09
Chile	72.33	78.75
China	68.57	71.48
Croatia	70.69	77.52
Cuba	73.41	78.30
Ecuador	69.54	74.90
Egypt	60.39	64.49
Ethiopia	39.22	41.73
France	74.76	82.71
Hong Kong S.A.R	76.15	81.85
Hungary	66.85	75.74
Iceland	76.85	81.19
India	62.54	64.29
Indonesia	60.67	65.29
Iran	68.43	71.16
Iraq	65.54	67.56
Ireland	73.64	79.32
Israel	76.71	80.61
Laos	52.63	55.87
Mexico	68.98	75.17
Nepal	58.47	58.36
The Netherlands	75.28	81.17
North Korea	67.41	72.86
Peru	68.08	72.78
Philippines	63.79	69.5
Russia	58.83	71.72
Rwanda	40.84	41.80
Saudi Arabia	68.67	72.53
South Africa	52.68	56.90
South Korea	70.75	78.32
Sweden	76.61	82.11
Togo	56.93	61.64
United States	72.95	79.76

(*Source:* U.S. Census bureau: http://www.censu.gov/ipc/www/idbprint.html)

a truly unique individual. The maximum verified age among humans is 122 years (Jeanne Calment of France), 7 years above the maximum life span reported in Table 13.3. Use of comparisons, such as presented in Table 13.3, for studying senescence are fraught with problems. Data on maximum life spans from zoo specimens or capture–recapture studies in the wild, although often used for species comparisons and theory development about the role of senescence, are unlikely to provide the full story (Finch 1994). Paraphrasing Finch (1994, 12–13), little evidence about the role of senescence in limiting life span is garnered from these comparisons.

An additional term used to characterize population senescence is mortality rate doubling time (MRDT), or the amount of time that it takes a species' mortality rate to double, approximately every decade after age 30 in humans. MRDT is represented graphically as the natural log of the mortality rate against age; the slope of this line (G, the **Gompertz coefficient**; see Box 13.1) is the rate of acceleration of mortality with age. This rate is considered by many to be a simple estimate of the rate of senescence for a species (Finch 1994). However, using G as a measure of senescence has been critiqued. For one, MRDT is based on an exponential relationship using curve-fitting parameters; thus, deviations from expected values decrease as a function of increasing time (Hart and Tuturro 1987). As a result, the Gompertz coefficient becomes less sensitive with increasing age (Curtsinger et al. 1992). Second, accurate estimation of G assumes few survivors beyond the specific MLP (Hart and Tuturro 1987); therefore, possible values for G are restricted by the estimate of MLP. Although Gompertz coefficients of human populations in drastically different environments are remarkably similar (Finch 1994), no reports have examined the

Table 13.3   Estimated average and maximum life spans and ages at puberty for selected mammalian species

Species	Life span (months)		Age at puberty (months)
	Average	Maximum	
Human	849	1380	144
Gorilla	—	472	—
Chimpanzee	210	534	120
Rhesus	—	348	36
Cow	276	360	6
Swine	192	324	4
Horse	300	744	11
Elephant	480	840	21
Cat	180	336	2
Dog	180	408	2
Whale	—	960	12
Mouse	18	42	1.5
Rat	30	56	2
Guinea pig	24	90	2

(*Source:* Finch and Hayflick 1977, 9 [Table 2])

BOX 13.1 SURVIVORSHIP CURVES, ESTIMATES OF MORTALITY RATE DOUBLING TIME FOR VARIOUS SPECIES, AND THE GOMPERTZ EQUATION

Survivorship is a negative function of time—that is, mortality rates accelerate with age in most populations that live to show senescence (Figure 13.1A). The Gompertz model has been a mainstay in gerontological research and is commonly expressed as an exponential function. The log of the age-specific mortality rate (i.e., the fraction of survivors that die in the time interval t) increases linearly with age over most of the adult life of humans. When mortality rate is graphed by age on a semilog scale (in mortality), the curve fits a straight line from the age of sexual maturation through the average life expectancy. The observed slope of this line is the Gompertz coefficient (G), which represents the acceleration of mortality with increasing age. Conversely, survivorship (Figure 13.1B) is graphed as percent of a cohort still alive at age x.

In high mortality populations, the survivorship curve drops sharply during early life and then flattens out over the remaining life span (B1); in low-mortality populations, it becomes more rectangular (B4). In populations that do not show any age-related increase in the force of mortality, the curve appears concave (B2), whereas in populations where mortality is a constant number of deaths at each age, the **survivorship curve** is a straight line from 100% to 0 over the life span (B3). Type B1 applies to most traditional living populations with high infant and childhood mortality and relatively high mortality rates throughout the life span (e.g., Yanomami, !Kung, and Turkana). Type B4 applies to many modern-day populations with low infant, childhood, and young adult mortality, such as the United Kingdom, Japan, Sweden and the United States. In these populations, the majority of deaths occur after reproduction has ceased for women and most men. Pattern B2 is found among both the living and the nonliving; the best examples include some animal populations, hydra, songbirds, and water glasses in a restaurant. Pattern B3 is found among populations where mortality remains at a constant level, for example, toughened water glasses in a restaurant and sheep in a zoo (Arking 1998, 27–28; Wilson and Bossert 1971, 113).

Table 13.4 **Mortality rate doubling time (MRDT) in selected species**

Species	Initial mortality rate	MRDT (years)
Human female US 1980	0.0002	8.9
Dutch civilians 1945	0.0014	7.8
Rhesus	0.02	8.0
Mouse	0.01	0.3
Rat	0.002	0.3
Dog	0.02	3.0

(*Source:* Finch and Hayflick 1971)

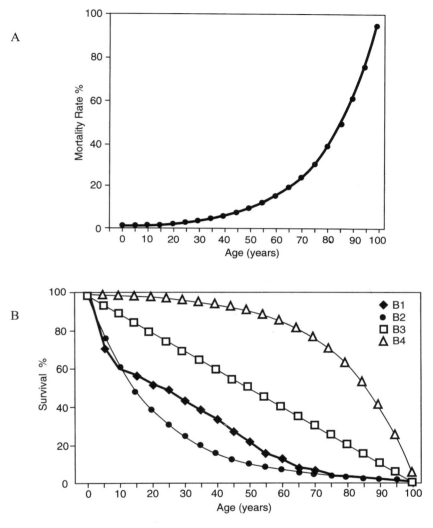

Figure 13.1   Mortality (A) and survivorship (B) curves.

magnitude of variation between populations or across different cohorts within populations (Hart and Tuturro 1983). In addition, the conformity of mortality parameters to **Gompertz equations** may be related to the intensity of selection acting on age-specific mortality rates, a phenomenon that changes as patterns of **pleiotropy** change within a population (Rose 1991). Finally, empirical data to complete such analyses do not exist for most living populations, nor are they available for fossil specimens (Finch 1994). Thus, any practical, rather than theoretical, applications of Gompertz coefficients are generally not possible.

HUMAN AGING, SENESCENCE, AND DISEASE

Not all aspects of physiology show decline with increasing age; some show stability, whereas others may show age-related enhancement (Finch 1994). To illustrate this phenomenon, a summary of findings for human immune system aging is reprinted as Table 13.5 (Mayer 1994); other systems show at least as much variability. One thing that is likely true for all physiological parameters is that they show greater variation in level and degree of change both within and between populations and between individuals within any population with increasing age.

As age increases, the variation—as measured by **standard deviations**—in almost all physiological parameters increases in concert, suggesting that senescence may be as much individualistic a characteristic as are fingerprints or **DNA**. Volumes have been written about the processes of senescence and age-related disease. In the remainder of this chapter, we present a general overview of age-associated changes in several areas of long-term interest to human biology: the hormonal, immune, cardiovascular, body composition, bone, and reproductive systems. Before reviewing these changes, it is necessary to understand what constitutes a physiological parameter that shows "senescent" change as opposed to one that simply shows "age-related" change.

Arking (1998), following Strehler (1982), enumerated five characteristics that must be met for an observed age-related change to be regarded as an aspect of senescence. The phenomena must be cumulative, universal, progressive, intrinsic, and deleterious (CUPID) within the population. Although the details may be argued for any specific parameter (e.g., some may suggest that senescent changes must be irreversible or degenerative in addition to these five criteria), Arking's CUPID criteria provide us with a baseline to begin evaluating age-related change. If any of these five criteria are not inherent aspects of an age-related change, that change may not be a strong candidate for being a senescent alteration. Unfortunately, almost nothing is universal among aging humans; there seems to be an exception to every condition and trend. However, these exceptions may represent outliers, like those extremely long-lived few who end up in *The Guinness Book of World Records*.

Given exceptions to every generalization, numerous well-documented age-related changes in humans appear to be senescent alterations (e.g., progressive dementia, lowered immune function, decreased reproductive capability, declines in hormone production, loss of reserve capacity in respiratory and circulatory systems, atrophy of muscle, and loss of bone). Others appear to be age-related disease processes (e.g., **Alzheimer's disease**, **non-insulin-dependent diabetes mellitus**, **atherosclerosis**, and **osteoporosis**) that may afflict a large proportion of elders in some societies, but few members of other societies. One factor that seems to differentiate these two kinds of changes is that the former are systems-level alterations in function, whereas the latter are specific degenerative processes that manifest as disease conditions. Separation of these two processes in any realistic fashion may be impossible at best and meaningless at worst. Rather, it has been suggested that it may be best, and greatly clarify research issues, if those studying cardiovascular disease, cancer, dementia, adult-onset diabetes, and other age-related chronic degenerative diseases work together with geriatricians and gerontologists as persons studying aging/senescence and consider their research as contributing to aging research in general (Miller 1995; Arking 1998).

Hormonal Aging

As discussed in Chapter 11, **hypothalamic** regulation of the **pituitary gland** and the actions of the endocrine hormones are the major central organizers for human growth and development. As such, the hypothalamic–pituitary axis has great potential to be a major modulator and promoter of longevity.

For example, the ovarian steroids **estradiol** and **progesterone** increase at sexual maturation and decline at menopause in human females. Furthermore, several tissue types and cells follow closely the life-long patterns of these hormones; vaginal and uterine tissues in particular show **hyperplasia** during puberty and hypoplasia following menopause (Arking 1998). Women become more susceptible to various circulatory conditions and cancers after menopause, and those who experience natural menopause early (before age 44) show higher mortality than those with natural menopause at 50–54 years (Snowden et al. 1989). Any senescent degeneration of function in this system could contribute to overall loss of homeostatic control and allow or promote additional senescent alterations in a cascading pattern.

Likewise, alterations of the **sympathetic–adrenal medullary system**, which is responsible for physiological arousal and stress response, could have a general effect on homeostasis in other systems. Both hormonal systems show age-related change that may or may not be senescence (Finch and Hayflick 1977). Many elders show hypercortisolemia (high **cortisol** levels) and respond more slowly to stressors than younger adults do; this response may be secondary to a reduced stress-sensing capability that accompanies age (Sapolsky 1990; Frolkis 1993). Among elders, both the hypothalamic–pituitary–adrenal axis and the sympathetic–adrenal medullary system appear to lose reserve capacity with age. When older individuals encounter stressors, their hormonal systems may be inadequate to maintain or return to physiological homeostasis (Lakatta 1990). This deficiency may result in long-term damage to these systems, which may then contribute to the onset and progression of disease and ultimately affect longevity.

Aging Changes in the Immune System

The human immune system shows numerous age-related alterations that appear to be senescent (see also Table 13.5). One change is involution of the thymus (a gland of largely lymphoid tissue that functions in the development of the body's immune system), largely due to atrophy of the cortex. The thymus is present in the young of most vertebrates in the upper anterior chest or base of the neck and tends to disappear or become rudimentary in adults. Despite thymic involution, total circulatory **lymphocytes** remain relatively stable through life, although proportions of various

Table 13.5   Summary of reported findings in human immune system aging

Molecular and biochemical

0	Complement function and levels
0	Interleukin-1 (IL-1) production
0	Prostaglandin production
0	B-cell growth factor production by T-cells
−	Interleukin-2 (IL-2) production
−	Number of chromosomes (especially X and Y) per cell
−	Induction of IL-2 and IL-2 receptor mRNA by PIIA activation
−	Production of immunoglobins by T-cell–independent stimuli
+	Autoanti-idiotypes (not necessarily associated with disease)
+	Synthesis of prostaglandin
0/−	Antibody production
0/−/+	γ-Interferon production
0/−/+	Serum levels of immunoglobulins A, G, and M

Cellular

0	Number of monocytes, macrophages, neutrophils
0	Number of pluripotent hematopoietic stem cells
0	Antigen processing by macrophages
0	Phagocytosis
0	B-cell function (0/− in very old >90 years)
+	Number of "uninducible" null (non-T, non-B) cells in bone marrow
+	Sensitivity to prostaglandin inhibition of lymphocyte stimulation (age >70 years)
+	Lymphocyte membrane rigidity (− membrane fluidity)
+	Percentage of resting T-cells in the circulation
+	Suppressor activity of monocytes/macrophages
+	Helper T-cell stimulation
−	Migration ability of neutrophils
−	Suppressor T-cell suppression of B-cell immunoglobulin production
−	Activity of cytotoxic T-cells
−	Number of oligopotent stems cells (precursors to T-cell subsets)
−	Proliferation and differentiation of B-cells to plasma cells
−	Oxidative burst in macrophage host defense response (neutrophils)
−	Protein synthesis following mitogenesis or activation
0/−	Proliferative response of T-cells to mitogens, antigens, lectins (e.g., phytohemagglutin, PHA)
−	Number of stimulatable cells
−	Rate of entry into DNA synthesis phases of cell cycle
−	Number of times stimulated cells divide
−	Total number of cell divisions in vitro (Hayflick limit)
0/+	Time needed for cells to enter DNA synthesis phase
0/+	Duration of cell cycle
0/−	Number and proportion of T-cells
0/−	Number of B-cells (−/+ in very old >90 years)
0/+	Number of natural killer (NK) cells (non-T, non-B cells)
0/+	Number of peripheral immature T-cells

(continued)

Table 13.5   *(Continued)*

−/+	Number of T-suppressor/cytotoxic cells (CD8+)
0/−/+	Number of T-helper/inducer cells (CD4+)
0/−/+	Ratio of CD4+ cell to CD8+ cells
0/−/+	Platelet function
0/−/+	Cytotoxicity of NK cells
Physiological	
+	Incidence and severity of infections
+	Autoimmunity (but not autoimmune disease)
+	Benign monoclonal gammapathics (homogeneous immunoglobulins)
0/−	Levels of thymic hormone
−	Hormone activity
−	Response to immunization (−response duration, −antibody levels)
−	Cell-mediated immunity (+reactivation of shingles, tuberculosis)
−	Primary antibody response (T-cell-dependent)
0	Secondary antibody response
0/−	Inflammatory response
0/−	Vigor of delayed-type hypersensitivity skin reaction

0, no change with age; −, decrease with age; +, increase with age.
(*Source:* Mayer 1992, 184–185 [Table 7-1] reprinted with permission.)

subpopulations of T-lymphocytes change (Arking 1998). Apparently, with age, the body tends to increase its production of autoreactive **antibodies** (i.e., antibodies that attack the self). Older adults also appear more susceptible to many infections and to suffer more severe consequences (Mayer 1994).

Another age-related alteration in the immune system is decreased production of **interleukin-2** with increasing age (Lakatta 1990; Mayer 1994). Although immunosenescence appears to be a fact of life in humans, numerous components of the immune system do not decline in a postmaturational age-related fashion, and some may even increase (see Table 13.5).

Cardiovascular Aging

Age-related increases in the incidence of cardiovascular diseases including hypertension (discussed more thoroughly later), **myocardial infarctions** (heart attacks), and **cerebrovascular accidents** (strokes) are well-documented in cosmopolitan societies of the 20th century and may reflect age-related changes in the cardiovascular system. However, similar increases have not been reported in numerous societies that still retain traditional lifestyles (Friedlaender 1987; James and Baker 1990).

Table 13.6 lists mortality rates from cardiovascular diseases for various populations. In cosmopolitan societies, arteries of individuals in their third decade of life (i.e., 20–29 years) already show atherosclerotic plagues, whereas after age 40, cardiovascular diseases lead cause-of-death statistics. Unfortunately, most people in less cosmopolitan societies succumb to **infectious diseases** and die at earlier ages. However, among those who survive, age-related disease contributes to mortality. Still, although some data suggest that hypertension, for example, may be an age-related and lifestyle-associated disease rather than an aspect of senescence, comparisons of age-specific prevalence rates for hypertension or mortality rates for cardiovascular disease between populations are biased by high infant and childhood mortality in less cosmopolitan societies (James and Baker 1990).

Atherosclerosis (a chronic disease characterized by yellowish plaques of cholesterol, lipids, and cellular debris on the walls of the arteries) has been reported in both captive and feral chimpanzees and other nonhuman primates; this condition may be characteristic of senescence in all long-lived primates. Atherosclerosis is so common that it is considered by some to be possibly the best example of an age-related senescent change (Arking 1998). The progression of atherosclerosis into atherosclerotic disease and related complications, however, appears to be an age-related disease that results from a combination of genetic predisposition and environmental exposures because it affects many members of cosmopolitan societies.

Body Composition and Aging

Human body composition changes continually through the life span. At birth, we are the "brainiest" we will ever be because ~20% of our body weight is brain tissue. At all other points in our life span, our brain will make up a decidedly smaller proportion of body weight. As we grow older, the general tendency is toward greater robusticity in body size and an increased percentage of body fat until the late 60s, after which the percentage of body fat decreases in long-lived individuals (Garn 1994).

Concurrent with increasing amounts of body fat, fat-free mass tends to decrease with increasing age. Thus, muscles tend to weaken and bones tend to demineralize in many elderly adults. Some portion of such declines may be offset by maintenance of high levels of physical activity and exercise in the latter decades of life. Still, loss of muscle and bone mass with age appears to be another CUPID aspect of senescent change. Current variations in body habitus (i.e., those aspects of body morphology that are associated with disease, such as **BMI**, weight, height, **skinfolds**, and waist-to-hip ratio) at ages 70–79 across various populations are listed in Table 13.7.

Bone Aging

Bone mass does not appear to decline in a linear fashion or to decline equally in men and women with age. Prior to menopause, bone loss of women is only slightly greater than that of men. However, after menopause, women apparently lose bone mass at a rate about two to three times that of same-age men (Stini 1990).

In young adults, bone mass may increase in response to diet and exercise; however, among elders, neither the overall amount nor the rate of bone loss appear to be greatly modifiable by lifestyle or dietary changes (e.g., ingestion of large doses of calcium, vitamins, or other nutritional supplements). **Estrogen** replacement therapy reduces rate of bone loss in postmenopausal women; however, previously lost bone

Table 13.6   Cross-country comparisons of cardiovascular mortality rates

| | Cardiovascular mortality rate (per 100,000 population) | | | | | |
| | Hypertensive disease | | Ischemic heart disease | | Cerebrovascular disease | |
Country	Female	Male	Female	Male	Female	Male
Albania	8.49	9.52	214.26	305.85	70.09	71.63
Argentina	10.02	10.58	50.09	65.26	62.12	67.83
Armenia	0.29	1.00	231.49	269.62	98.66	88.35
Barbados	8.96	5.55	32.27	42.6	96.08	79.6
Belarus	4.22	5.55	194.47	304.36	104.57	107.86
Belize	26.01	39.60	44.83	34.59	51.33	29.09
Brazil	6.56	7.13	20.67	34.52	30.35	35.52
Bulgaria	21.05	16.40	119.57	165.15	139.74	103.03
Canada	7.97	5.68	99.66	95.52	64.81	50.55
Chile	11.55	8.99	54.18	68.13	72.64	63.32
China	13.74	13.12	42.17	43.55	82.17	74.18
Columbia	25.42	18.72	71.66	85.50	56.90	46.73
Costa Rica	14.03	9.80	72.13	83.37	46.67	36.66
Cuba	8.43	7.79	129.16	135.60	61.30	51.92
Czech Republic	4.12	4.34	134.50	191.86	103.03	89.23
Dominica	147.03	51.92	24.25	32.65	59.59	66.5
Egypt	35.68	26.94	22.55	37.12	35.27	39.48
Estonia	8.77	8.32	222.68	321.38	133.68	119.96
Guatemala	4.24	3.32	129.16	135.60	37.13	27.51
Guyana	32.72	30.00	24.51	70.21	57.73	46.02
Hungary	28.97	22.55	126.82	208.69	98.98	108.49
Israel	7.76	4.99	83.49	105.25	54.07	39.91
Jamaica	38.04	NA	35.14	NA	114.98	NA
Kazakhstan	19.04	21.13	203.61	299.34	145.22	145.77
Kuwai	83.52	42.21	69.24	129.88	19.88	23.54
Kyrgyzstan	11.89	14.79	145.07	190.66	137.16	131
Latvia	1.27	2.41	222.12	389.99	158.16	159.47
Lithuania	4.02	4.23	214.26	305.85	83.28	78.97
Martinique	37.79	41.91	NA	NA	55.63	60.40
Mexico	17.67	11.22	57.67	67.37	45.96	36.30
Poland	13.53	12.71	45.53	104.45	50.80	53.06
Puerto Rico	22.39	23.03	55.97	75.57	27.15	25.91
Romania	62.41	46.05	115.03	161.95	116.05	103.99
Russia	7.71	9.44	198.66	360.10	176.96	182.78
Slovenia	12.18	7.09	222.12	389.99	73.65	76.75
Sri Lanka	14.82	17.33	22.17	51.81	22.48	27.11
St. Lucia	36.85	35.67	29.31	70.27	55.63	60.40
Tajikistan	31.63	29.13	139.56	154.14	97.43	80.69
The Bahamas	46.06	37.97	50.85	77.34	70.74	44.16
The Dominican Republic	18.21	13.88	40.61	46.99	47.91	37.86
Trinidad and Tobago	36.96	31.26	113.34	149.44	57.73	46.02
Ukraine	1.82	2.94	181.09	267.15	244.56	245.98
Uruguay	8.80	6.77	57.95	85.67	56.56	61.59
Uzbekistan	32.12	33.22	270.88	287.30	127.47	112.41
Venezuela	NA	NA	71.56	90.39	70.59	46.90
Yugoslavia	16.16	10.74	40.07	76.22	89.98	81.53

(*Source:* http://www.cvdinfobase.ic.gc.ca)

Table 13.7 Height, weight, and body mass index (BMI) of ethnically diverse 70- to 90-year-old men and women

Name of study	Sample size		Height (m)		Weight (kg)		BMI[1]	
	Men	Women	Men	Women	Men	Women	Men	Women
Guatemala	31	30	1.56(0.06)	1.40(0.07)	49.3(7.7)	41.6(9.0)	20.2(2.2)	21.4(4.4)
Beijing, China	44	54	1.61(0.04)	1.54(0.04)	63.1(9.2)	55.1(9.0)	24.2(3.5)	23.0(3.5)
China (national)	217	241	1.61(0.08)	1.48(0.06)	54.1(9.0)	45.5(8.6)	21.0(3.3)	20.7(3.6)
Hong Kong	246	479	1.62(0.06)	1.48(0.60)	55.7(10.1)	49.4(9.8)	24.2(3.4)	22.4(4.0)
Italy (17 sites)	187	206	1.62(0.07)	1.51(0.07)	67.3(12.5)	64.7(13.2)	25.5(4.3)	28.5(5.4)
Brazil (national)	634	698	1.63(0.10)	1.50(0.08)	61.3(15.1)	56.6(18.0)	22.9(5.0)	25.0(7.4)
Anglos, Australia	24	26	1.64(0.10)	1.66(0.08)	71.6(12.0)	70.9(8.7)	26.6(3.7)	25.6(3.0)
Padua, Italy	297	386	1.64(0.07)	1.51(0.06)	71.6(11.3)	65.1(13.0)	26.5(3.8)	28.4(5.3)
Sparta, Greece	26	20	1.66(0.06)	1.51(0.06)	75.7(13.8)	64.8(10.6)	27.5(4.4)	28.3(4.5)
Greek, Australia	64	59	1.65(0.06)	1.50(0.05)	76.4(10.9)	68.9(11.3)	28.0(3.6)	30.7(5.1)
Chinese, Australia	11	9	1.65(0.08)	ND	62.6(8.9)	ND	23.0(2.7)	ND
Tainjin, China	180	181	1.65(0.08)	1.53(0.06)	61.4(10.6)	52.2(10.9)	22.2(3.3)	22.2(4.1)
Barbados	31	30	1.66(0.06)	1.59(0.07)	69.9(15.4)	73.2(18.0)	25.3(5.6)	29.2(6.9)
Nigeria	21	13	1.66(0.06)	1.58(0.07)	57.2(9.7)	58.1(10.0)	20.6(2.9)	21.5(6.0)
Cameroon	16	9	1.67(0.06)	ND	63.2(7.9)	ND	22.6(2.7)	ND
Finland (national)	273	487	1.69(0.06)	1.55(0.06)	73.3(12.4)	64.3(11.7)	25.6(3.7)	26.8(4.5)
United States (national)	656	894	1.72(0.08)	1.59(0.05)	75.7(12.2)	64.9(12.7)	25.6(3.7)	25.7(4.9)
Rotterdam, the Netherlands	360	900	1.73(0.06)	1.59(0.07)	77.0(10.5)	68.7(11.0)	25.8(3.3)	27.1(4.3)
Johanneberg, Sweden	51	75	1.74(0.06)	1.61(0.05)	76.9(10.5)	62.5(12.0)	25.4(3.2)	24.1(4.5)

Values are means and standard deviations. BMI is weight in kg/height in m^2. ND, not determined (n <10 individuals).

(*Source:* Launer and Harris 1996, Table 11)

is not regenerated. Consequently, elderly women suffer from osteoporosis and hip fractures at three to four times the rate of same-age elderly men. These changes along with numerous others may lead to functional limitations in the elderly (Boult et al. 1994; Mulrow et al. 1994). For example, osteoporosis increases risk for hip fractures and subsequent muscular atrophy, bed sores, and secondary infections (Cummings and Nevitt 1989).

Osteoarthritis may occur in concert with osteoporosis and lead to even greater loss of locomotor abilities and ability to care for oneself and complete required ac-

---

BOX 13.2   ACTIVITIES OF DAILY LIVING AND INSTRUMENTAL ACTIVITIES OF DALY LIVING

Activities of daily living (ADLs) and instrumental activities of daily living (IADLs) include those activities that a person needs to complete in order to survive independently (see list below). ADLs are basic self-care activities, and IADLs are more complex functions. IADLs involve sequences of many activities; for example, shopping requires an individual to not only perform basic self-care activities but also being able to leave the house, arrange for transportation, and purchase needed items.

ADL impairments are typically due to physical impairments, whereas IADL impairments result more frequently from cognitive impairment. Elders with loss of or declining abilities to complete ADLs or IADLs are at increased risk for loss of social networks, community, and family activities; increased dependency; lower self-perceived and general health; increased infections; and higher mortality (Baltes et al. 1990; Guralnik et al. 1994; Mulrow et al. 1994).

Data from the 1984 National Health Survey indicate that difficulty completing one or more ADLs is experienced by 15%, 22%, and 40% of males and 18%, 31%, and 53% of females aged 65–74, 75–84, and 85 and older in the United States (National Center for Health Statistics 1987; Brock et al. 1990). Having two or more ADL restrictions is highly predictive of institutionalization after age 65 and tends to be more frequent in surviving females; however, by age 70, almost half of all men ever born have died, whereas 70% of women still survive. The sex difference may be a result of inherent frailty of women, differential survival of women or greater reporting by women.

Physical ADLs	Instrumental ADLs
Bathing	Using Telephone
Dressing	Transportation
Grooming	Laundry
Transferring (moving from chair to bed	Shopping
or chair to toilet and vice versa)	Taking Medicine
Feeding	Managing Money
Ambulating	

tivities of daily living (ADLs) and instrumental activities of daily living (IADLs) (Brock et al. 1990; Miles and Brody 1994) (see Box 13.2) Variability in anatomical assessments with age are illustrated in Table 13.8.

Dementia

The dementias include a heterogeneous assortment of disorders that affect the central nervous system (CNS). They progressively lead to several kinds of loss: orientation in time and space, muscular control, long- and short-term memory, and general neurological function. Included in this diverse group are **vascular dementia**, Alzheimer's disease, and **Parkinson's disease**.

One current suggestion is that the dementias represent a dysfunctional end point of age-related neurological alterations that occur in all human brains over the normal human life span (Strong and Garruto 1994; Broe and Creasy 1995; Arking 1998). Thus, individuals who develop dementia likely are those whose neurological systems undergo these alterations at a rapid pace or who have additional risk factors—genetic, environmental, lifestyle—that predispose them to earlier symptomology.

The human brain is the most complicated structure we know of, and its proper function is dependent upon the integration of a complex set of chemical and electrical messages. Any loss or damage may produce specific mental defects if that loss is in an area responsible for a specific task, such as speech, hearing, vision,

Table 13.8 Changes in anatomical measurements and indices over the life span

Parameter	Observed change
Weight	Increase to 50, decline from 60
Stature	Increase through 30–34, decline from 40
Span	Increase through 30–34, decline from 40
Thoracic index	Increase through 70–74
Biacromial diameter	Increase through 35–39, decline from 55
Relative shoulder breadth	Increase through 45–49
Chest breath	Increase through 50–54
Chest depth	Increase through 50–54
Sitting height	Increase through 35–39, decline thereafter
Relative sitting height	Slow decline after 40
Head circumference	Increase through 35–39
	Slow decline after 54
Cephalic index	Increase to 40 and slight decline thereafter
Cephalo-facial index	Rise through 75–79
Total face height	Increase 30–34; decline thereafter
Facial index	Increase through 25–29; decline thereafter
Upper face index	Increase through 30–34, decline from 55
Nose height	Increase through 55–59
Nose breath	Increase throughout age groups

(*Source*: Rossman 1977, © 1977 Litton Educational Publishing; reprinted with permission)

motor capabilities, or memory. However, damage or loss to an area with more general function, such a central role in the processing or distribution of information used by other areas of the brain, may give rise to more generalized functional defects.

Human brains lose both volume and weight with increasing age. A progressive decline in brain weight occurs between ages 55 and 80 years, averaging about 11% loss over the adult ages (Davison 1987). The results of **cross-sectional** studies indicate that human brain volume remains relatively stable, or enlarges slightly, from the third (ages 20–29) through the sixth decade of life (ages 50–59), and some individuals show volume expansion due to enlargement of both extracerebral and intracerebral spaces (Arking 1998). Neuronal numbers seem to decrease mostly after age 60. The most extensive losses, ranging from 25 to 45%, occur in regions that contain associative neurons of the cerebral cortex—areas where mental processes such as thought and memory reside (Duara et al. 1985).

Age-related alterations in brain function also may follow from a reduction in the active synthesis of neurotransmitters over the life span, a process that may lead to progressive changes in function with age and to the development of age-related behavioral alterations or disorders (Arking 1998). Age-related reductions in activity have been reported for acetylcholine, dopamine, norepinephrine, serotonin, and λ-amino butyric acid (GABA) in various areas of the brain (Whitbourne 1985). Variations in activity levels of these neurotransmitters also have been linked to mental health problems among young and middle-aged adults. Accumulation of lipofuscin (an inert yellowish pigmented waste product of cellular metabolism) also may contribute to malfunction and death of nondividing neurons.

The frequency of Alzheimer's disease (AD) increases with age and represents a leading cause of **morbidity** among the elderly. With a prevalence of 600 afflicted individuals per 100,000 population, AD affects ~5% of all persons aged 65 and over (Arking 1998). AD is characterized by a high incidence of neurofibrillary tangles and amyloid plaques in the anterior temporal lobe compared with "normal" individuals (Arking 1998). Neurofibrillary tangles are insoluble twisted fibers on neurons, the chief component of which are β-amyloid **proteins**. Plaques are dense deposits of τ-proteins that aggregate to form insoluble deposits.

At least two types of AD have been identified: familial (early onset) and sporatic (late onset). Familial AD (FAD) is a rare form of AD that affects <10% of individuals with AD. Onset of FAD is typically at age 60–65. **Genes** on chromosomes 1 (presinilin 2), 14 (presinilin 1), and 21 (amyloid precursor proteinn, APP) have been associated with FAD; the dominant forms of these **mutations** result in the development of the disease. The precursor protein of β-amyloid is coded for on chromosome 21; thus, individuals with trisomy 21 (Down's syndrome) develop symptoms indistinguishable from AD if they live long enough, because of increased gene dosage. Another gene on chromosome 21 is associated with some cases of AD (Tanzi et al. 1987). Together, presinilin 1, presinilin 2, and APP account for approximately 50% of all FAD cases.

The **apolipoprotein** (Apo) ε4 **allele** (chromosome 19) is associated with both sporadic and familial AD as well as vascular dementia (Marin et al. 1998; Mayeux and Ottman 1998). Risk estimates vary with studies, but carriers have a 25–40% risk of developing AD during their lifetime (Mayeux and Ottman 1998). Apo ε4 **homozygotes** have a significantly greater risk of developing AD than individuals with one or no copies of the apo ε4 allele. The frequency of the ε4 allele varies by population, and lower risks associated with the allele have been observed in some ethnic groups (e.g., Hispanics, Africans, African Americans) (Farrer et al. 1997).

Apo ε2 alleles are rare but may have a protective effect against AD. Numerous other **candidate genes** for AD, including human leukocyte antigen-A2 (**HLA**-A2), are being studied; however, additional testing is needed before these can be labeled susceptibility genes. Environmental factors have also been implicated in AD, including aluminum exposure (Savory and Garruto 1998). One cautionary note should be mentioned about the study of AD and its causes: Diagnosis of dementia type is difficult during a person's lifetime (Jorm 1990). AD is defined by the presence of neurofibrillary tangles, which cannot be observed in a living person. Thus, dementia is assessed by a battery of cognitive tests, and in the absence of any evidence for vascular dementia, AD is often assigned. Therefore, studies on AD prevalence and causes in the absence of confirmation of diagnosis with a postmortem exam should be viewed with caution.

Reproductive Aging

Attainment of maximum reproductive potential (MRP; the point in life at which an organism is not only sufficiently mature to bear or sire offspring but also best able to rear and **fledge** (i.e., to rear an offspring until it is prepared to live without **parental investment**) any offspring produced with maximum efficiency marks the onset of age-related decline and senescence.

Menopause (i.e., the cessation of monthly menstrual cycling and absence of menses in adult women) marks completion of the reproductive span in women. In men, reproductive senescence is not as obvious. Some men may sire children as late as their 80s and 90s, whereas others may decline in reproductive potential in their 50s and 60s. Whether all men experience an andropause (i.e., a cessation of **androgen** production) remains a debatable topic; however, all women who survive sufficiently long do experience menopause. Menopause seems to be a uniquely human characteristic; however, bonobos and macaques may experience something similar to human menopause when in captivity (Graham 1986; Pavalka and Fedigan 1991).

The basis of human menopause has been debated widely and extensively. Is human menopause an evolved characteristic that has resulted from natural selection to curtail reproduction, or has menopause developed due to the absence of selective pressures? It is extremely difficult to comprehend how total loss of reproductive potential could ever not be selected against. However, we can understand how evolutionary pressures over the majority of human existence shaped human **ovaries** so as to produce mature ovum only over the active human female reproductive life span—likely a maximum of about 40 and an average of less than 25 years—over most of the past 2 million years. One indication of this evolution is that at menopause, ovaries seem to no longer have sufficient primary ovarian **follicles** capable of producing a mature ovum to stimulate follicular maturation and expulsion of a fertile ovum (Wood et al. 1994).

Age-related **atresia** of oocytes is well-documented; numbers decline from about 3,500,000 in the 4-month-old fetus to 733,000 at birth, to 165,000 at ages 25–31, and finally to 11,000 at ages 39–45 (Finch and Hayflick 1977, 318–356). Thus, close to the point of their MRP (estimated at about 20 years), women have more that 20 times more primary oocytes in their ovaries than they do as they approach menopause at age 45. The average age at menopause is 50 years old but may vary by several years across populations. Numerous factors may contribute to population variation in age at menopause, including marital status, parity, heredity, altitude, nutrition, and **socioeconomic status** (Pavelka and Fedigan 1991).

Men show a very different pattern of reproductive senescence. Spermatogenesis continues in most men into the eighth and ninth decades of life. However, the number of motile and well-formed sperm declines with age, as does the number of sperm per unit volume of semen (from 100–200 million per milliliter to 30–60 million or less per milliliter). Among 20–39 year old men, 90% of their **seminiferous tubules** have **spermatids**; at ages 50–59, only 50% do; after age 70 years, this proportion declines by about 10% per decade (see Finch and Hayflick 1977 for a review of reproductive senescence). **Testosterone** production in men also appears to peak during the 20s and to decline gradually over the remainder of the life span. Taken together, these data suggest the existence of some degree of andropause in men.

CHRONIC DEGENERATIVE DISEASES

Numerous physiological systems decline in their functional and reserve capacity as humans age. These declines often contribute to the onset and progression of chronic degenerative diseases (CDDs; conditions that lead to progressive deterioration in one or more clinical, metabolic, or physiological traits that can be measured on a continuous scale) (Crews and James 1991; Crews and Gerber 1994; Gerber and Crews 1999).

CDDs often are debilitating to individuals who experience them. They also are among the leading causes of morbidity and mortality in cosmopolitan societies and represent an emerging concern for developing nations. As a result, the importance of CDDs as a major public health problem cannot be overemphasized. To understand, prevent, and treat CDDs, their distributions, etiologies, and underlying risk factors must be documented. Numerous biological and cultural factors have contributed to the current "epidemic" of CDDs, and untangling these factors is a major goal of epidemiology, public health, gerontology, medical and biological anthropology, and human population biology.

Although CDDs probably were rare through much of human prehistory, abundant evidence indicates that they did occur among members of earlier societies who lived to be elders or who were among the elite in preindustrial stratified societies. Examples include the Neandertal of La Chapelle aux Saints, France, who may have suffered from **rickets**, **osteomalacia**, and arthritis. Similarly, several Egyptian mummies from dynastic times reportedly show evidence of gallstones and atherosclerosis. In addition, ancient Egyptian papyrus scrolls and ancient Chinese and Greek writings include clinical descriptions of what appear to be cancer, heart disease, and diabetes. Socrates is widely regarded as having described both **insulin-dependent** and non-insulin-dependent diabetes (juvenile- and adult-onset, or Type I and Type II, respectively) and to have prescribed some of the first lifestyle interventions for disease: reduce fat intake and move to the countryside.

Because many CDDs commonly occur during the sixth and seventh decades of life, in the past, very few members of any population lived long enough to show such conditions. However, the alleles that predispose individuals to CDDs would have had little or no impact on **fertility**, so the individuals who carried the alleles could easily and unknowingly have passed them on to their modern-day descendents. Thus, alleles with late-acting detrimental effects could easily have accumulated in the **hominid** gene pool.

An example of one such allele may be that for **Huntington's disease**. This **autosomal dominant** condition commonly strikes individuals in their late 40s and early 50s, well after the age of MRP. In populations in which only a few individuals live past age 40, such a condition is easily passed on generation after generation and the disease will seldom, if ever, be observed.

Changes in lifestyle also have placed people at greater risk for developing CDDs (Crews and Gerber 1994; Gerber and Crews 1999). In many ways, cultural evolution has outpaced biological evolution among humans. Thus, we maintain a physiological system better suited to prehistoric diets of natural unrefined foods and mobile uncrowded lifestyles rather than modern highly refined, high-calorie foodstuffs; sedentary activities and entertainment; and population densities of thousands of persons per square kilometer.

Thrifty/Pleiotropic Genotypes

In this section, we examine briefly one evolutionary model for CDDs and two major CDDs: hypertension and dementia.

Several human populations—notably, the Pima and Dogrib Indians, Mexican Americans, African Americans, Native Americans, Asian Indians residing overseas, Australian Aborigines, Melanesians, and Polynesians—show high prevalence rates of non-insulin-dependent diabetes mellitus (NIDDM; see also Szathmáry 1985; Bindon and Crews 1990, 1993; Zimmet et al. 1990; Stinson 1992; Crews 1994). To aid in explaining the high prevalence of diabetes in certain Amerind populations, Neel (1962) proposed the **thrifty genotype** hypothesis. Since then, thrifty genes also have been proposed for other traits (e.g., salt-sensitive hypertension in African Americans and chronic diseases in general [Crews and Gerber 1994]). Recently, several candidate genes associated with salt-sensitive hypertension have been reported (Gerber and Crews 1999).

As discussed in Chapters 7 and 9, Neel (1962) originally suggested that populations with high NIDDM prevalence today likely were exposed to alternating periods of food scarcity and abundance in earlier times. In such populations, individuals who were capable of storing excess calories as fat during periods of plenty would have been better equipped to survive times of relative scarcity. Therefore, these "thrifty individuals" would have greater fitness than the "less thrifty" when food supplies fluctuated between abundance and scarcity. In today's settings, with a constant food supply, an overabundance of calories, and sedentary lifestyles, individuals with thrifty genes still store excess calories as fat, but times of scarcity do not occur. Thus, their continued quick trigger and storage of excess energy predisposes thrifty individuals in industrialized societies to obesity and diabetes. Natural selection continues to favor genes that are beneficial in early life because they improve survival and reproductive success, so natural selection cannot act to reduce the frequency of such alleles. However, due to pleiotropy, the basis for a model of aging proposed by Williams (1957), thrifty genes may be detrimental to the long-term survival of thrifty individuals.

Crews and Gerber (1994; Gerber and Crews 1999) elaborated a model for the evolution of some CDDs based on combining thrifty genotypes with **antagonistic pleiotropy**. In short, this model states that over evolutionary time, a multitude of thrifty genes and genotypes that show antagonistic pleiotropy have become incorporated into humanity's gene pool. Humankind may share some of these genes and genotypes with organisms as diverse as fruit flies, whereas others may be found exclusively in mammals or only in primates. Still others may be unique to humans (Crews and Gerber 1994; Gerber and Crews 1999).

This thrifty/pleiotropic model suggests that many of today's major CDDs and their associated risk factors arise, in part, secondary to mechanisms that evolved to extract nutrients that previously were scarce in early hominid/**hominoid** diets— energy, salts, lipids, **cholesterol**, **micronutrients**, and protein—with high efficiency (thrifty). In a nutritionally abundant setting those with thrifty genotypes show the pleiotropic effects of overextraction and retention leading to obesity, hypertension, atherosclerosis, and hyperlipidemia among adults.

Hypertension

High blood pressure (BP) is an important risk factor for cardiovascular disease in the Western world. Furthermore, epidemiological studies in the West demonstrate that BP increases with age (see James and Baker 1990 for a review of these issues).

This increase with age was, at one time, believed to be an inevitable consequence of the aging process; however, numerous studies of non-Western societies have shown that an increase in BP is not necessarily concomitant with increasing age (Crews 1993; Harper et al. 1994; Silva et al. 1995). However, common risk factors associated with hypertension (e.g., obesity, fat patterning, and diet) are related to BP even in low-BP societies (Harper et al. 1994; Silva et al. 1995). Such populations include the Yanomami of Brazil, the Gainji of Papua New Guinea, and the Cofan of Ecuador (see also James and Baker 1990).

The one common factor in low-BP populations is the retention of a traditional lifestyle, including diet and activity patterns. Variation in BP across multiple populations indicates a need for ecological studies that examine genes, culture, and environment and their joint and interactive influences on human variation. At present, the exact reason for low mean BP in these populations is not clear (James and Baker 1990). With more research, it likely will become apparent that many established risk factors and **marker loci** for high BP are not unique to each society; however, we also may anticipate that numerous specific alleles contributing to elevated BP are unique to certain populations and kindreds within populations (Crews and Williams1999).

Many studies of BP in traditional societies have been conducted in the Pacific Islands. Inhabitants of many Pacific Islands have low BP in general and the association of BP with age is either slightly negative or nonexistent. In a review, Beaglehole (1992) noted that among many of the Pacific populations who have begun to experience a more Western lifestyle and **epidemiological transition**, the average BP is increasing. (Epidemiological transition refers to a change in disease patterns in which the primary causes shift from infectious diseases to CDDs.) As average BP rises in many of these traditional populations, associations with age and other common risk factors also develop.

Although many populations in the Pacific Islands are undergoing an epidemiological transition, several populations in Papua New Guinea and Vanuatu continue to have high infant mortality rates, a great prevalence of infectious disease, little chronic disease, and relatively short life spans. These populations maintain a traditional lifestyle and for the most part have not shown any evidence of an epidemiological transition or elevated BP. However, even in very low BP populations, such as the Yanomami of Brazil and the Ganji of New Guinea, where no age-related increase in BP is found, clear associations of body habitus with BP are observed (Crews and Mancilha-Carvalho 1993; Harper et al. 1994).

The genetic basis of BP variation and cardiovascular disease has been extensively studied (for a recent review, see Crews and Williams 1999). Most candidate genes produce proteins that are involved in BP regulation. Elevated BP (or hypertension) can result from increased vascular resistance or increased **plasma volume**. Plasma volume is the amount of water relative to blood cells and proteins. Plasma volume is determined by the relative balance of sodium (Na^+) and potassium (K^+). Much of the research on the genetic basis of BP variation or the genetic basis of hypertension has focused on the renin–angiotensin–aldosterone system, which is a prime regulator of plasma volume through its influence on the Na^+ balance (Crews and Williams1999).

Figure 13.2 illustrates the main components of the renin–angiotensin–aldosterone system. Angiotensinogen, which is produced in the liver, is cleaved by renin (released from the kidney) to form angiotensin I. Angiotensin I is converted to angiotensin II by the action of angiotensin-converting enzyme (ACE), which is found in high concentrations in the lungs. Angiotensin II has several functions, including regulating the feedback of renin, stimulating aldosterone secretion (maintains Na^+ balance), causing **vasoconstriction** of the systemic and renal **arterioles** (results in an increase in BP), stimulating antidiuretic hormone (acts on kidneys to reduce Na^+ and water excretion), enhancing salt reabsorption in kidney, and stimulating thirst.

Genetic **polymorphisms** in renin, ACE, and angiotensinogen have been examined in relation to hypertension. Although many polymorphic sites of the renin locus have been identified that vary across populations, no consistent associations have been found between renin **RFLP**s and plasma renin activity, BP, or hypertension (Jeunemaitre et al. 1992). One variant of the angiotensinogen gene (*M235T*) appears to be associated with susceptibility to hypertension in some environments but not in others (Crews and Williams 1999). The ACE locus has two major alleles, one with a

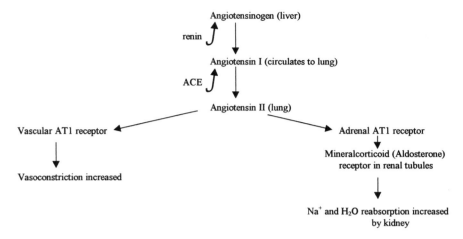

Figure 13.2 Blood pressure regulation. ACE, angiotensin converting enzyme; AT1, angiotensin I.

deletion and one with an insertion of 287 **base** pairs. Frequencies of the insertion and deletion alleles vary across populations but are not consistently associated with BP variation. However, people with a DD genotype are at increased risk for myocardial infarction and stroke. In addition, the haptoglobin allele *Hp1* has been associated with salt sensitivity, probably through **linkage** to an as-yet unidentified gene.

The search for genes that explain BP variation and hypertension is an active and constantly evolving field. Interested students should read Crews and Williams (1999), Luft and Weinberger (1997), and Lifton (1996) for more information.

THEORIES OF SENESCENCE

One of the seminal theories of senescence, developed in the early 20th century by Pearl (1928) and others, was that longevity was an epiphenomenon of other aspects of life history such as metabolic rate, body size, and reproduction (see also Comfort 1979; Finch 1994).

All else being equal, small animals have higher **basal metabolic rates** and shorter life spans than larger animals (Hofman 1983). It also is generally true that metabolic rate in mammals decreases with brain weight (Harvey and Bennet 1983; Hofman 1983) and that both body size and brain size appear allometrically related to life span (Cutler 1976, 1980; Sacher 1980; Hofman 1983; Finch 1994). Basic to this theory is the proposition that because smaller animals have more rapid metabolic rates, they expend their life's allotment of energy more rapidly and die more quickly than larger animals with slower metabolic rates.

A modern revival of this theory followed observations of cellular damage due to the byproducts of metabolic processes, particularly **free radicals** (i.e., highly reactive charged molecules with unpaired electrons; Harman 1984; Adelman et al. 1988; Cutler 1991). Although rate-of-living theories have been popular in gerontology, traditional theories have not withstood the close scrutiny of experimental data (Hart and Tuturro 1983; Rose 1991; Austad 1992; Finch 1994). Furthermore, research demonstrating that many age-related changes in physiology *do not* correlate with life span also has failed to support rate-of-living theories (Finch 1994).

An often-cited comparison of birds and mammals demonstrates some problems with rate-of-living theories (Austad 1992). According to rate-of-living theories, in both birds and mammals, mass-specific metabolic rates per unit of body weight should decrease with increasing size. However, at any given body size, birds show twice the metabolic rate of mammals. Therefore, one would expect that at a particular body size, mammals will live about twice as long. There are several cases where this is not true. Based on exhaustive review of many body systems, from whole body metabolism to cellular and biochemical levels, Finch (1994) demonstrated that there is neither a strong nor consistent relationship among body size, metabolism, and life span.

Primates, our nearest relatives, provide another example. **Folivorous** primates have lower metabolic rates than their body sizes would predict, possibly because they also have relatively smaller brains than **frugivorous** species, but perhaps also because leaves are not energetically dense and require greater energy for processing (Harvey and Bennet 1983). Thus, the relationship between body size and metabolic rate is affected by diet and ecology. Furthermore, there is no necessary reason to conclude that rate of senescence is related to length of life. Selection pressures may be quite variable for these two traits. For example, Pacific salmon show no senescent changes until after they swim upstream to spawn, and then they rapidly senesce and die. In this way, the length of time that they live is relative to their timing of reproduction, not the speed at which they senesce. Why we senesce and why we live a certain life span are two related but separate questions. In their current incarnation, rate-of-living theories address the question of how we senesce, not why, and therefore address proximate causes of senescence in specific organisms or organ systems, not the ultimate causes.

Evolutionary models of senescence were developed in concert with quantitative population genetics. Medawar (1946) developed the "wear and tear" theory based on observations that with age, organisms have an increasing likelihood of dying due to extrinsic factors (i.e., coming from outside the individual, such as being eaten by a lion). Medawar illustrated this point by using the example of test tubes in a lab (Austad 1992). Test tubes do not senesce. Still, over time, they tend to break due to laboratory mishaps (extrinsic factors). If most test tubes die (break) within two years, a change (mutation) that affects test tubes only after two years will have little effect on longevity in the overall population. However, a change that improves a 1-year-old test tube's survival will have a substantial effect on survival and longevity in that test tube population. Medawar also noted that selection pressure decreases with age as a result of decreasing fertility; thus, selection pressures against CDDs that occur after the point of MRP are likely to be minimal to nonexistent. One such consequence is that increases in human life expectancy have mainly resulted from improvements in childhood, not adult, survivorship.

Following Medawar's conceptualization of the decreasing force of natural selection with age, Williams (1957) elaborated the concept of senescence due to pleiotropy. In Williams' model, traits that increase survival and/or fertility at young ages are favored by natural selection, even when they carry additional detrimental pleiotropic effects that will be expressed in later years. Williams developed four "deductions" from evolutionary theory that any proximate model of aging must explain:

1. "Senescence should be found wherever the conditions specified are met (a soma) and should not be found where these conditions are absent" (Williams 1957, 403). In other words, clonal species, where the soma (i.e., the entire body of an organism with the exception of the germ cells) and germ plasm (i.e., reproductive cells; in sexually reproducing species, sperm and ovum) are the same, should not age; current data support this deduction (Finch 1994).

2. "Low adult death rates should be associated with low rates of senescence, and high death rates with high rates of senescence" (Williams 1957, 404). The continuum of senescence patterns outlined by Finch (1994) fits this prediction. Rapid senescence occurs in annual plants, marsupial mice, and Pacific salmon; these populations experience exponential increases in adult mortality after maturation. At the other extreme, negligible senescence occurs in populations with little adult mortality, such as anemones, clams, sharks, tortoises, and trees (Finch 1994).

3. "Senescence should be more rapid in those organisms that do not increase markedly in **fecundity** after maturity than those that do show such an increase" (Williams 1957, 405). Mice do not show increased fecundity (i.e., the quality or ability to produce offspring) with age and senesce rapidly, whereas sharks and tortoises show both increased fecundity and fertility (i.e., the production of offspring, process of having reproduced a new individual) with age and senesce slowly.

4. "Where there is a sex difference, the sex with the higher mortality rate and lower rate of increase in fecundity should undergo the more rapid senescence" (Williams 1957, 406). In most mammals, females outlive males (Williams 1957; Hazzard 1986). This difference may be explained partially by the variable reproductive strategies used by males and females. By nature, females invest more in producing and rearing offspring, and males invest more in maximizing access to mates (Dunbar 1988), including competition with other males. Thus, most mammalian males have a restricted window of reproductive opportunity during their prime reproductive years due to male–male competition, higher morbidity and mortality, and female choice; whereas most mammalian females reproduce early and throughout their life span. This is par-

ticularly true in group-living species such as lions, gorillas, baboons, and wild dogs. Therefore, selection pressures on females may have led them to maintain their soma longer than males, who not only show higher mortality rates (while obtaining mates) but also usually do not care for offspring, given that their parental investment (i.e., the cost of producing, caring for, and fledgling off-spring) often ends with insemination. In primate groups, some males actually adopt a strategy of "opportunistic mating" (Dunbar 1988). Such males may ex-perience less adult mortality due to reduced male–male competition and con-sequently experience longer reproductive periods and life spans than their conspecifics, a question in need of continued research.

Evolution of an extended postreproductive period is possible through natural se-lection (Hamilton 1966). If late-life curtailment of reproduction leads to an increase in **inclusive fitness** (e.g., through benefits to younger relatives), then such positive selection pressures could lead to increased postreproductive survival. Interestingly, in cosmopolitan settings men, who do not lose their ability to reproduce at older ages (although this ability may be greatly reduced) tend to die at younger ages than women. However, women, who after menopause are unable to reproduce, live longer than men. This discrepancy suggests that maintenance of reproductive potential into the latter decades of life has not been a successful reproductive strategy for women. However, the ability to sire offspring at even extreme old age—the seventh and later decades of life—likely has been a successful strategy for men.

Over evolutionary time, women who reproduced at older ages may have failed to fledge their children born to them at older ages. Consequently, they may have expe-rienced higher **completed fertility** (i.e., total offspring ever born) rates than their conspecifics who ceased reproduction at earlier ages but lower rates of fledgling these offspring. Thus, humankind's evolutionary stage was set for the evolution and development of grandmotherhood and the extended investment of women in their kindred's inclusive fitness (i.e., an individual's own reproductive success stripped of all those components that are due to the influence of its relatives plus that proportion of each relative's reproductive success that was due to the influence of the individual in question once it has been devalued by the **coefficient of relationship** between them) (Hamilton 1966).

Because grandmothers know not only who their children are but also who their ma-ternal grandchildren are, they could provide ongoing parental investment to their daughter's descendants, thereby increasing their adaptive value and producing positive selection pressures for increased postreproductive survival (Swedlund et al. 1976; Mayer 1991; Wood et al. 1994). This hypothesis is a challenging one to test; however, several investigators have looked at modern hunter-gatherers (Hawkes et al. 1989, 1997; Hill and Hurtado 1991), and Mayer (1982) examined data from New England women born between 1675 and 1875 in an attempt to test the "grandmother hypothe-sis." Mayer found that women who survived past menopause had greater reproductive success than those who died before menopause. In their analysis of fertility and mor-tality patterns, Hill and Hurtado (1996) found no increase in inclusive fitness benefits for women who helped older children and grandchildren compared with those who continued to bear more offspring. Hawkes et al. (1997) found that grandmothers contributed to the nutritional intake of weaned children who had difficulty providing for themselves when mothers were nursing and less able to provide for their young. The evolution of human menopause is an active area of research; interested readers should read Hawkes et al. (1989, 1997) for more on the topic.

Another general finding is that rates of senescence are higher in organisms with higher fertility rates and lower in those with lower fertility rates (Hamilton 1966; Rose 1991), illustrating that life history characteristics such as reproduction, longevity, and patterns of senescence are intimately related. Experimental results from roundworms, fruit flies, and mosquitoes bred for extended life spans have served to further solidify observed associations between reproductive output and somatic longevity (Rose 1991). In general, extensions of average and maximum life spans in these animal models are associated with reduced and/or delayed fertility. Furthermore, the best way to develop long-lived strains is to select only those individuals capable of reproducing at ages above the average age of reproduction in the baseline population (Rose 1991). Life span extension in rodents by dietary restriction also is associated with delayed reproductive maturation and late fertility, suggesting that caloric restriction may enhance longevity through its influence on growth, development, and maturation.

MECHANISMS OF SENESCENCE

Numerous mechanisms leading to senescence have been observed in different organs, organ systems, and species (Gavrilov and Gavrilova 1991; Rose 1991; Arking 1998). These observations have led to a proliferation of mechanistic theories of aging and senescence—wear and tear, damage due to reactive oxygen species including free radicals, error catastrophe, rate of living, loss of cell proliferation capacity, chromosomal **telomeric** shortening, and glycation. One, more than one, or all of these particular mechanisms may affect any particular organ or physiological system. Thus, the presence of one mechanism of senescence does not exclude the possibility that others may affect the same organ or system. In fact, it is likely that multiple mechanisms act to produce variation in patterns of senescence both within and between species and within and between different organs.

One of the earliest such theories was based on Hayflick's finding that fibroblasts have a finite doubling capability. Contrary to then-prevailing theory, Hayflick (1965) demonstrated, using very specific culturing techniques, that fibroblasts showed a limit of 50 population doublings when cultured in vitro (i.e., outside the body). This finding had a major impact on models of senescence developed thereafter (Finch 1994), but later, this doubling limit was found not to apply to all the body's cells (Weiss 1981). It is now clear that animals do not "run out" of cells due to limited cell proliferations (Hart and Tuturro 1983). In fact, this limit to cell doubling in vitro may be an artifact of the culturing techniques used (i.e., halving the fibroblast cell population during each subsequent generation) and may have little to do with mammalian aging (Kirkwood 1995). Others suggested that a genetic mechanism may code for programmed aging, such as loci associated with programmed cell death, and that such relatively parsimonious explanations may underlie organismal aging (Driscoll 1995; Miller 1995). However, little experimental evidence to date supports this claim (Rose 1991; Stadtman 1995).

Several other mechanistic theories have been categorized as "garbage can" models (Gibbons 1990). They include the accumulation of free radicals as a byproduct of metabolism (Adelman et al. 1988; Cutler 1991); accumulation of defective macromolecules as a byproduct of nonenzymatic glycation (Cerami 1985; Brownlee et al. 1988); an increase in faulty proteins and DNA as a result of DNA mutation, damage, or damage due to reactive oxygen species (Armstrong et al. 1984; Pryor 1984; Richter 1995; Sohal and Orr 1995; Swartz and Mäder 1995); and the accumulation of inert waste materials such as lipofuscin in nondividing cells (Gravilov and Gravilova 1991; Harrington and Wischik 1995). Current data do not contradict any of these mechanistic theories. For example, an enhanced ability to repair damage resulting from metabolic byproducts is associated with longer life spans in multiple species (Adelman et al. 1988; Cutler 1985; Rose 1991). However, each of these is but a proximate mechanism occurring in cells, organs, and systems without innate evolved abilities to control such degenerative processes.

LONGEVITY

Various models have been advanced to explain variation in longevity between individuals and between species. Some have suggested that longevity is an epiphenomenon of other biological processes, including metabolic rate (Sacher 1975 1980; Hofman 1983), brain and body size (Cutler 1980; Hofman 1984), degree of **encephalization** (Weiss 1981), growth and development patterns (Pfeiffer 1982), and reproductive strategies (Mayer 1982; Rose 1991;). Because these processes are part of most species' life history, likely under similar evolutionary pressures as longevity, such theories are easily mired in chicken-or-egg controversies. All are necessarily correlated with life span, but which is the prime mover and which is the outcome is far from clear.

Numerous papers have been published on the evolution of human longevity in general and on the evolution of the extended human postreproductive period in particular (Cutler 1975, 1980; Weiss 1981; Hofman 1984). Many researchers explain increased longevity in humans as an evolutionary strategy to increase inclusive fitness. For example, postmenopausal women may increase their fitness by ensuring the survival of their already born children and grandchildren. Other researchers suggest that the reproductive value of older individuals may have increased due to humankinds' reliance on learning, culture, and social reciprocity. However, it is not likely that until very recently (the past 20,000–50,000 years) that many individuals survived to be grandparents and great-grandparents. Thus, there likely was little pressure for extended postreproductive survival during the majority of human evolution. Other researchers explain current human longevity simply as a byproduct of modern cultural life, with no necessary need to evoke an evolutionary basis.

Limits to Human Longevity

Today, one of the most debated topics in gerontology is whether limits to human longevity exist. In the United States, life expectancy at birth has increased from 40 to 75 years since the mid-19th century (Olshansky et al. 1990). This increase has been due mostly to decreased infant and child mortality (Fries 1983; Guralnik et al. 1988; Brock et al. 1990; Olshansky et al. 1990). Less dramatic changes in older adult life expectancy have been observed. Any additional increase in longevity at older ages will necessitate decreased mortality from CDDs during middle and older ages (Fries 1983; Olshansky et al. 1990).

Some researchers see life span as a genetically fixed characteristic of species-specific senescence patterns (Fries 1980). These researchers predict that life expectancy will not increase beyond 85 years among humans. However, other researchers have demonstrated that maximum life span potential is modifiable with environmental change and artificial selection in laboratory models—rodents, insects, worms (Curtsinger 1992; Carey 1997). Still, others do not accept that any life span limit exists (Rose 1991). In fact, an increase in life expectancy after age 85 already has been reported for some Scandinavian populations (Vaupel et al. 1998).

Two opposing views on the limits to human life span currently are debated:

- life span is a genetically determined species-specific characteristic and human longevity will not increase beyond 120 years, the proposed upper limit, and
- life span is an environmentally and genetically labile characteristic, and there are no species-specific life spans.

According to the latter perspective, there is no reason to assume that humankind has achieved its limit, because no limit exists. These two contrary views lead to very different predictions of demographics and health care costs in the future; they also produce an interesting and revealing argument about evolution and the scientific method.

Fries (1980, 1983 and Fries and Grapo, 1981), a proponent of species-specific life spans, predicts that human life expectancy will not increase beyond 85 years and that the maximum life span is around 120 years. Even though the rate of mortality from infectious diseases has decreased close to 99% in the United States over the past century, Fries contends that only small decreases in chronic disease mortality are possible. This idea is essentially an extension of Hayflick's fibroblast work and follows from the notion of genetically determined senescence (Barinaga 1991). According to Fries, the species-specific average life span (life expectancy) is the average length of life that humans might expect to live. When his theory was originally published, average life span was 73 years (77 for women, 70 for men); today it is 75. Fries interpreted Hayflick's limit in fibroblast cultures as indicating a finite life span of organs. Thus, he concluded that there had to be a biologic limit to life, a point at which the soma can no longer function in the face of environmental insults and increased physiological failure. Additional support was generated using the Gompertz plot of linearly increasing mortality with age.

Fries (1980) further suggested that available mortality data demonstrate a point at which any additional reduction in chronic disease morbidity and mortality will have little effect on life expectancy, because senescence will lead to increasing mortality. Fries (1980, 1983) observed that between 1900 and 1980, survivorship curves (i.e., graphs illustrating what percentage of a cohort born in the same year survive to each subsequent year of life; see Box 13.1) appeared to become increasingly rectangular, providing additional support for this model. Over these decades, Fries suggested that survivorship curves have moved up and to the right as the proportion of deaths prematurely decreased. This process allegedly has yielded an increasingly flat survivorship curve with an increasingly sharp downward slope at latter ages in recent decades (see Figure 13.1B).

Fries (1980, 1983) contended that the increased flatness of survivorship curves through age 60, in association with convergence of cause-specific mortality curves at approximately 85 years, demonstrate a specific limit to human longevity. The point of downward inflection in the curve has remained approximately the same, about 60–70 years, since 1900. An additional component of this model is that human morbidity curves also will become more rectangular as the onset of disability from chronic disease is postponed to later years of life (Fries 1980). Because average and minimum ages at death are fixed in this model, chronic morbidity would be compressed into only a few years preceding death due to senescent processes (Fries 1980).

Limits to human life expectancy have been estimated from US life expectancy curves at birth and at age 65 over the past century (Fries 1980). Extrapolation of these lines produces a point of convergence at approximately 85 years, which Fries has proposed as the upper limit to human life expectancy. Fries presented an interesting argument; however, both the data for and the logic of his arguments have been seriously questioned. Although no evidence or theoretical support suggests that humans will one day be immortal, no clear and undeniable evidence indicates species-specific life expectancies, either. For example, we have no logical basis to assume that the intersection point of previous and current life expectancies represents a limit to life expectancy (Barinaga 1991). In humans, this intersection point is influenced by cultural and ecological factors, such as childhood mortality rates, which are unrelated to the biological limits of life span (Barinaga 1991).

Several gerontologists take an opposing view, that there is no theoretical reason to suggest limits to life span and demonstrate that life span is an environmentally and genetically liable characteristic. These scientists point to experimental evidence in invertebrates and current mortality data for the "oldest-old" people (people over 85). For example, experiments clearly demonstrated that the probability of mortality among Mediterranean fruit flies increases with age over the first one-third of their life, but thereafter, mortality rates level off (Barinaga 1991; Carey et al. 1992), similar results have been reported for common laboratory fruit flies (Curtsinger et al. 1992).

Furthermore, data from long-lived humans also indicate decreasing mortality rates at very old ages (80 years and over) (Perls 1995). Under different environmental conditions, different samples of the same species may have very different average and maximum life spans. These results have been observed in roundworms, fruit flies, mosquitoes, rodents, canines, primates, and humans (Resnick 1985; Arking 1991; Carey 1997). Such observations tell us little about the biology of senescence, except that species vary as much in longevity as they do in reproduction, growth and development, adult size, and other life history characteristics.

Curtsinger et al. (1992) demonstrated a shift to the right in the survivorship curves for fruit flies by manipulating only environmental factors. Similar results have been obtained with roundworms, mosquitoes, and rodents (Rose 1991). Currently available evidence suggests that similar environmental manipulation may be capable of shifting the human survivorship curve to the right (reviewed in Crews 1994). Finally, it is not clear that the human survivorship curve has become increasingly rectangular or has a sharp downward trend at about age 65, as contended by Fries. Rather, one observes wider variation in ages at death after age 65 than observed in earlier generations, life extension after age 65 for all five-year age groups, and a stretching out of the survivorship curve when observed as one-year increments rather than as five-year aggregate age classes.

It has been asserted that to increase life expectancy further, it is first necessary to eliminate CDDs, a prospect that many believe is unlikely (Olshansky 1990). If all circulatory diseases, diabetes, and cancer were eliminated, then life expectancy at birth in the United States would increase by only 15.82 years for females and 15.27 years for males (Olshansky 1990). If these estimates are accurate, then life expectancy for women would increase to 93 years and for men to 85 years. At least for women, 93 years is well beyond Fries' 85-year limit. Others argue that even if mortality from certain diseases were to decrease, then other diseases would take their place and life expectancy would increase little at older ages (Sacher 1977; Brody 1983). Even if death rates at 85 were reduced, currently too few people survive to over 85 to significantly change life expectancy statistics (Barinaga 1991). This argument may seem to support Fries' theory but if true, may represent only currently practical limits based on current culture and disease patterns, not biological limits (Barinaga 1991; Olshansky et al., 1990).

CHAPTER SUMMARY

Having read this chapter, you should have an idea of the evolutionary basis for longevity and senescence and how these represent a trade-off between reproductive success and somatic maintenance. Longevity represents a single outcome of several interacting factors—genes, culture, and environment—and is unlikely to be determined by any single switch. Rather, any individual's longevity is determined by a complex set of ongoing interactions that contribute to his or her specific life history.

It is clear that humans will never rival Methuselah in length of life, but we all are likely to live longer, more productive lives than lived by any previous generation or predicted for our generation. Predictions of life expectancy, MLP, and frailty always are based on past experience and thus must be interpreted with caution (Guralnik et al. 1988). In February 1997, the oldest known person, Jeanne Calment of France, turned 122 years old, and today's centenarians experience lower mortality and morbidity rates than were predicted just a decade ago. Will human life spans increase beyond the currently recorded 122 years? Although we have to wait and see, it seems like a good bet.

ACKNOWLEDGMENTS

We thank the editors for their commitments and efforts in developing this volume. Also, we gratefully acknowledge the efforts of Kristen Ossa for helping to find references and demographic data used in developing this chapter and of Susan Bean for typing multiple revisions of the manuscript.

RECOMMENDED READINGS

Borkan GA, Hults DE, and Mayer PJ (1982) Physical anthropological approaches to aging. Yearbook Phys. Anthropol. 25: 181–202.

Crews DE and Garruto RM (1994) Biological Anthropology and Aging: Perspectives on Human Variation over the Life Span. Oxford: Oxford Univ. Press.

Crews DE (1994) Biological anthropology and human aging: some current directions in aging research. Annu. Rev. Anthropol. 22: 395–423.

Finch CE (1994) Longevity, Senescence, and the Genome. Chicago, Ill.: Univ. of Chicago Press.

Perls TT (1995) The oldest old. Sci. Am. 272: 70–75.

Rose MR (1991) Evolutionary Biology of Aging. New York: Oxford Univ. Press.

Shock NW (1984) Normal Human Aging: The Baltimore Longitudinal Study of Aging (NIH Publication No. 84-2450). Washington, D.C.: US Department of Health and Human Services.

Williams GC (1957) Pleiotropy, natural selection, and the evolution of senescence. Evolution 11: 393–411.

REFERENCES CITED

Adelman R, Saukm RL, and Ames BN (1988) Oxidative damage to DNA: relation to species metabolic rate and life span. Proc. Natl. Acad. Sci. U.S.A. 85: 2706–2708.

Arking R (1998) Biology of Aging: Observations and Principles, 2nd edition. Englewood Cliffs, NJ: Prentice Hall.

Armstrong D (1984) Free radicals involvement in the formation of lipopigments. In D Armstrong, RS Sohal, RG Cutler, and TF Slater (eds.): Free Radicals in Molecular Biology, Aging, and Disease. New York: Raven Press, pp. 129–142.

Austad S (1992) On the nature of aging. Nat. Hist. 2: 25–57.

Baltes MM, Wahl HW, and Schmis-Furstoss U (1990) The daily life of elderly Germans: activity patters, personal control, and functional health. J. Gerontol. 45: P173–P179.

Barinaga GT (1991) How long is the human life span? Science 254: 936–938.

Beaglehole R (1992) Blood pressure in the South Pacific: The impact of social change. Ethnicity and Disease 2: 55–62.

Bindon JR and Crews DE (1993) Changes in some health status characteristics of American Samoan men: a 12-year follow-up study. Am. J. Hum. Biol. 5: 31–38.

Bindon JR and Crews DE (1990) Biocultural responses to modernization: lessons from Samoa. In: Human Ecology and Anthropology: Lessons for the 21st Century, Lectures. Zagreb, Yugoslavia: Institut za Medincinska Istrazivanja i Medicinu Rada, pp. 103–115.

Boult C, Kane RL, Louis TA, Boult L, and McCaggrey D (1994) Chronic conditions that lead to functional limitations in the elderly. J. Gerontol. 49: M28–M36.

Brock DB, Guralnik JM, and Brody JA (1990) Demography and epidemiology of aging in the United States. In EL Schneider and JW Rose (eds.): Handbook of the Biology of Aging. San Diego, Calif.: Academic Press, pp. 3–23.

Brody J (1983) Limited importance of cancer and competing risk theories of aging. J. Clin. Exp. Gerontol. 5: 141–154.

Broe GA and Creasey H (1995) Brain aging and neurodegenerative diseases: a major public health issue of the twenty-first century. Perspect. Hum. Biol. 1: 53–58.

Brownlee M, Cerami A, and Vlassara H (1988) Advanced glycosylation end products in tissue and the biochemical basis of diabetic complications. Semin. Med. (Beth Israel Hospital, Boston) 318(20): 1315–1321.

Carey J (1997) What demographers can learn from fruit fly actuarial models and biology. Demography 34: 17–30.

Carey JR, Liedo P, Orozco D, and Vaupel JW (1992) Slowing of mortality rates at older ages in large medfly cohorts. Science 258: 447–461.

Cerami A (1985) Hypothesis: glucose as a mediator of aging. J. Am. Geriatr. Soc. 33(9): 626–634.

Comfort A (1979) Aging: The Biology of Senescence, 3rd edition. New York: Elsevier.

Crews DE (1990) Anthropological issues in biological gerontology. In RL Rubinstein (ed.): Anthropology and Aging: Comprehensive Reviews. Boston: Kluwer Academic Publishers, pp. 11–38.

Crews DE (1993) Biological anthropology and aging: Current directions in aging research. Ann. Rev. Anthropol. 22: 395–423.

Crews DE and Gerber LM (1994) Chronic degenerative diseases and aging. In DE Crews and RM Garruto (eds.): Biological Anthropology and Aging: Perspectives on Human Variation over the Life Span. Oxford: Oxford Univ. Press, pp. 154–181.

Crews DE and James GD (1991) Human evolution and the genetic epidemiology of chronic degenerative diseases. In CGN Mascie-Taylor and G. Lasker (eds.): Applications of Biological Anthropology to Human Affairs. Cambridge Studies in Biological Anthropology. Cambridge: Cambridge Univ. Press, pp. 185–206.

Crews DE and Mancilha-Carvalho JJ (1993) Correlates of blood pressure in Yanomami Indians of northwestern Brazil. Ethnicity Dis. 3(4): 362–371.

Crews DE and Williams SR (1999) Molecular aspects of blood pressure regulation. Hum. Biol. 71: 475–503.

Cummings SR and Nevitt NC (1989) A hypothesis: the cause of hip fractures. J. Gerontol. 44: M107–M111.

Curtsinger JW, Fukui HH, Townsend DR, and Vaupel JW (1992) Demography of genotypes: failure of the limited life-span paradigm in *Drosophila melanogaster*. Science 258: 461–463.

Cutler RG (1976) Evolution of longevity in primates. J. Hum. Evol. 5: 169–202.

Cutler RG (1980) Evolution of human longevity. In C Borek, CM Fenoglis, and DW King (eds.): Aging, Cancer, and Cell Membranes. New York: Thieme-Straton Inc.

Cutler RG (1985) Antioxidants and longevity of mammalian species. Basic Life Sci. 35:15-73.

Cutler RG (1991) Antioxidants and aging. Am. J. Clin. Nutr. 53: 373S–9S.

CVD Infobase http://www.cvdinfobase.ic.gc.ca

Davison AN (1987) Pathophysiology of the aging brain. Gerontology 33: 129–135.

Driscoll M (1995) Genes controlling programmed cell death: relation to mechanisms of cell senescence and aging? In K Esser and GM Martin (eds.): Molecular Aspects of Aging. New York: John Wiley and Sons, pp. 45–60.

Duara R, London ED, and Rapoport SI (1985) Changes in structure and energy metabolisms of the aging brain. In EL Schnieder and CE Finch (eds.): Handbook of the Biology of Aging, 2nd edition. New York: VanNostrand Reinhold, pp. 595–616.

Dunbar RIM (1988) Primate Social Systems, Studies in Behavioral Adaptation. London: Croom Helm.

Eveleth PB (1994) Role of the National Institute on Aging. In DE Crews and RM Garruto (eds.): Biological Anthropology and Aging: Perspectives on Human Variation over the Life Span. Oxford: Oxford Univ. Press, pp. 426–436.

Erhardt CL and Berlin JE eds. (1974) Mortality and Morbidity in the United States; Cambridge, Harvard Univ. Press.

Farrer LA, Cupples LA, Haines JL, Hyman B, Kukull WA, Mayeux R, Myers RH, Pericak-Vance MA, Risch N, and vanDuijn CM (1997) Effects of age, sex, and ethnicity on the association between apolipoprotein E genotype and Alzheimer disease. A meta-analysis. JAMA. 278:1349–5.

Finch CE (1994) Longevity, Senescence, and the Genome. Chicago, Ill.: Chicago Univ. Press.

Finch CE and Hayflick L (eds.) (1977) Handbook of the Biology of Aging. New York: Van Nostrand Reinhold Company.

Friedlaender JS (1987) The Solomon Islands Project. New York: Oxford Univ. Press.

Fries JF (1980) Aging, natural death, and the compression of morbidity. N. Engl. J. Med. 303: 130–135.

Fries JF (1983) The compression of morbidity. Milbank Memorial Fund Q. 61: 397–419.

Fries JF and Crapo LM (1981) Vitality and Aging: Implications of the Rectangular Curve. San Francisco: WH Freeman and Company.

Frolkis VV (1993) Stress-age syndrome. Mech. Aging and Dev. 69: 93–107.

Garn S (1994) Fat, lipid, and blood pressure changes in adult years. In DE Crews and RM Garruto (eds.): Biological Anthropology and Aging: Perspectives on Human Variation over the Life Span. Oxford: Oxford Univ. Press, pp. 301–320.

Gerber LM and Crews DE (1999) Evolutionary perspectives on chronic degenerative diseases. In W Trevathan, J McKenna, and N Smith (eds.): Evolutionary Medicine. New York: Oxford Univ. Press, pp. 443–469.

Gibbons A (1990) Gerontology research comes of age. Science 250: 622–625.

Graham CE (1986) Endocrinology of reproductive senescence. In WR Dukelow and J Erwin (eds.) Comparative Primate Biology, Volume 3: Reproduction and Development. New York: Alan R Liss, Inc., pp. 93–99.

Gavrilov LA and Gavrilova NS (1991) The Biology of Life Span: A Quantitative Approach. New York: Harwood Academic Publishers (Revised and Updated English edition, V.P. Sku-lachev, ed. Translated from the Russian by John and Linda Payne).

Grigsby JS (1991) Paths for future population aging. Gerontologist 31(2): 195–203.

Guralnik JM, Yanagishita M, and Schneider EL (1988) Projecting the older population of the United States: lessons from the past and prospects for the future. Milbank Memorial Fund Q. 66: 622–625.

Guralnik JM, Simonsick EM, Ferrucci L, Glynn RJ, Berkman LF, Blazer DG, and Wallace RB (1994) A short physical performance battery assessing lower extremity function: association with self-reported disability and prediction of mortality and nursing home admission. J. Gerontol. 49: M85–M94.

Hamilton WD (1966) The moulding of senescence by natural selection. J. Theoret. Biol. 12:12–45.

Harman D (1984) Free radicals and the origination, evolution, and present status of the Free Radical Theory in aging. In D Armstrong, RS Sohal, RG Cutler, and TF Slater (eds.): Free Radicals in Molecular Biology, Aging, and Disease. New York: Raven Press, pp. 1–12.

Harper GJ, Crews DE, and Wood JW (1994) Lack of age-related blood pressure increase in the Gainj, Papua New Guinea: another low plood pressure population. Am. J. Hum. Biol. 6: 122–123.

Harrington CR and Wischik CM (1995) Pathogenic mechanisms in Alzheimer syndromes and related disorders. In K Esser and GM Martin (eds.): Molecular Aspects of Aging. New York: John Wiley and Sons, pp. 227–240.

Hart RW and Tuturro A (1983) Theories of aging. In Review of Biological Research in Aging, Volume 1. New York: Alan R. Liss Inc., pp. 5–18.

Hart RW and Tuturro A (1987) Part I: Evolution of life span in placental mammals. In HR Warner, RN Butler, RL Sprott, and EL Schneider (eds.) Modern Biological Theories of Aging. New York: Raven Press, pp. 5–18.

Harvey PH and Bennet PM (1983) Brain size, energentics, ecology, and life history patterns. Nature 306: 314–315.

Hawkes K, O'Connell JF and Blurton Jones NG (1989) Hardworking Hadza grandmothers. In V Standen and R Foley (eds.) Comparative Socioecology of Mammals and Man. London: Blackwell, pp. 341–366.

Hawkes K, O'Connell JF and Blurton Jones NG (1997) Hadza women's time allocation, offspring provisioning, and the evolution of long postmenopausal life spans. Current Anthropol. 38: 551–577.

Hayflick L (1965) The limited in vitro lifetime of human cell strains. Exp. Cell Res. 37: 614–636.

Hazzard WR (1986) Biological basis of the sex differential in longevity. J. Am. Geriatr. Soc. 34: 455–471.

Hill K and Hurtado AM (1991) The evolution of reproductive senescence and menopause in human females: an evaluation of the grandmother hypothesis. Human Nature 2:313–350.

Hill K and Hurtado AM (1996) Ache Life History: The Ecology and Demography of a Foraging People. Hawthorne, NY: Aldine de Gruyter.

Hofman WR (1984) Energy metabolism, brain size, and longevity in mammals. Q. Rev. Biol. 58: 495–512.

Hofman WR (1983) On the presumed coevolution of brain size and longevity. J. Hum. Evol. 13: 371–376.

James GD and Baker PT (1990) Human population biology and hypertension: evolutionary and ecological aspects of blood pressure. In JH Laragh and BM Brenner (eds.): Hypertension: Pathophysiology, Diagnosis, and Management. New York: Raven Press, pp. 137–145.

Jeunemaitre X, Lifton RP, Hunt SC, Williams RR, and Lalouel JM (1992) Absence of linkage between the angiotensin converting enzyme locus and human essential hypertension. Nat. Gen. 1: 72–75.

Johnson TE, Arkin R, Bertrand H, Driscoll M, Esser K, Griffiths AJF, Harley CB, Jazwinski SM, Kirkwood A, and Osiewacz HD (1995) Research on diverse model systems and the genetic basis of aging and longevity. In K Esser and GM Martin (eds.): Molecular Aspects of Aging. New York: John Wiley and Sons, pp. 5–15.

Jorm AF (1990) The Epidemiology of Alzheimer's Disease and Related Disorders. London: Chapman and Hall.

Kinsella, K and Suzman R (1992) Demographic dimensions of population aging in developing countries. Am. J. Hum. Biol. 4: 3–8.

Kirkwood TBL (1995) What can evolution theory tell us about the mechanisms of aging? In K Esser and GM Martin (eds.): Molecular Aspects of Aging. New York: John Wiley and Sons, pp. 5–15.

Lakatta EG (1990) Heart and circulation. In EL Schneider and JW Rowe (eds.): Handbook of the Biology of Aging. San Diego: Academic Press, Inc., pp. 181–218.

Launer LJ and Harris T (1996) Weight, height and body mass index distributions in geographically and ethnically diverse samples of older persons. Age and Aging 26:300–306.

Lifton RP (1996) Molecular genetics of human blood pressure variation. Science 272:676–80.

Luft FC and Weinberger MH (1997) Heterogeneous responses to changes in dietary salt intake: the salt-sensitivity paradigm. Am. J. Clin. Nutr. 65(2 Suppl):612S–617S.

Marin DB, Breuer B, Marin ML, Silverman J, Schmeidler J,Greenberg D, Flynn S, Mare M, Lantz M, Libow L, Neufeld R, Altstiel L, Davis KL, and Mohs RC (1998) The relationship between apolipoprotein E, dementia, and vascular illness. Atherosclerosis 140:173–80

Mayer PJ (1982) Evolutionary advantage of the menopause. Hum. Ecol.10: 477–494.

Mayer PJ (1991) Inheritance of longevity evinces no secular trend among members of six New England families born in 1650–1874. Am. J. Hum. Biol. 3: 49–58.

Mayer P (1994) Human immune system and aging: approaches, examples, and ideas. In DE Crews and RM Garruto (eds.): Biological Anthropology and Aging: Perspectives on Human Variation over the Life Span. Oxford: Oxford Univ. Press, pp. 182–213.

Mayeux R and Ottman R 1998. Alzheimer's disease genetics: Home runs and strikeouts. Ann. Neurol. 44: 716–719.

Medawar PB (1946) Old age and natural death. Mod. Q. 1: 30–56.

Miles TP and Brody JA (1994) Aging as a worldwide phenomenon. In DE Crews and RM Garruto (eds.): Biological Anthropology and Aging: Perspectives on Human Variation over the Life Span. Oxford: Oxford Univ. Press, pp. 3–15.

Miller RA (1995) Geroncology: The Study of Aging as the Study of Cancer. In K Esser and GM Martin (eds.): Molecular Aspects of Aging. New York: John Wiley and Sons, pp. 265–280.

Mulrow CD, Gerety MB, Cornell JE, Lawrence VA, and Kanten DN (1994) The relationship of disease and function and perceived health in very frail elders. J. Am. Geriatr. Soc. 42: 374–380.

National Center for Health Statistics (1987) Aging in the eighties, functional limitations of individuals age 65 years and over. In: Advance Data from Vital Health Statistics (No. 133. DHHS Publication No. [PHS] 87–1250). Hyattsville, Md.: Public Health Services.

National Center for Health Statistics (1996) Vital Statistics of the United States, 1992, Vol. II. Mortality, Part A. Washington, D.C.: Public Health Service.

Neel JV (1962) Diabetes mellitus: a "thrifty" genotype rendered detrimental by "progress." Am. J. Hum. Genet. 14: 353–362.

Olshansky SJ, Carnes BA, and Cassel C (1990) In search of Methuselah: estimating the upper limits to human longevity. Science 250: 634–640.

Pavalka MSM and Fedigan LM (1991) Menopause: a comparative life history perspective. Yearbook of Physical Anthropology 34: 13–38.

Pearl R (1928) The rate of living. London: University of London Press.

Perls TT (1995) The oldest old. Sci. Am. 272: 70–75.

Pfeiffer S (1982) The evolution of human longevity: distinctive mechanisms? Can. J. Aging 9: 95–103.

Pryor WA (1984) Free radicals in autoxidation and in aging. In D Armstrong, RS Sohal, RG Cutler, and TF Slater (eds.): Free Radicals in Molecular Biology, Aging, and Disease. New York: Raven Press, pp. 13–42.

Resnick D (1985) Costs of reproduction: an evaluation of the empirical evidence. Oikos 44: 257–267.

Richter C (1995) Oxidative damage to mitochondrial DNA and its relationship to aging. In K Esser and GM Martin (eds.): Molecular Aspects of Aging. New York: John Wiley and Sons, pp. 99–108.

Rose MR (1991) Evolutionary Biology of Aging. New York: Oxford Univ. Press.

Rosenwaike L (1985) A demographic portrait of the oldest old. Milbank Memorial Fund Q. 63: 187–205.

Rossman I (1977) Anatomical and body composition changes with aging. In CE Finch and L Hayflick (eds.): Handbook of the Biology of Aging. New York: Van Nostrand Reinhold Company, p. 203.

Sacher GA (1975) Maturation and longevity in relation to cranial capacity in hominid evolution. In R Tuttle (ed.): Primate Functional Morphology and Evolution. The Hague: Mouton, pp. 417–441.

Sacher GA (1977) Life table modification and life prolongation. In L Hayflick and CE Finch (eds.): Handbook of the Biology of Aging. New York: Academic Press, pp. 582–638.

Sacher GA (1980) Mammalian life histories: their evolution and molecular genetic mechanisms. In C Borek, CM Fenoglis, and DW King (eds.): Aging, Cancer, and Cell Membranes. New York: Thieme-Straton Inc.

Sapolsky RM (1990) The adrenocorticoid axis. In EL Schneider and JW Rowe (eds.): Handbook of the Biology of Aging, 3rd edition. San Diego: Academic Press, Inc., pp. 330–348.

Savory J and Garruto RM 1998. Aluminum, tau protein, and Alzheimer's disease: An important link? Nutr. 14: 313–314.

Silva HP, Crews DE, and Neves WA (1995) Subsistence patterns and blood pressure variation in two rural "Cabocolo" communities of Marajó, Pará, Brazil. Am. J. Hum. Biol. 7: 535–542.

Snowden DA, Dane RL, Beeson GL, Sprafka M, Bitter J, Iso H, Jacobs DR, and Philips RL (1989) Is early natural menopause a biologic marker of health and aging? Am. J. Public Health 79: 709–714.

Sohal RS and Orr WC (1995) Is oxidative stress a causal factor in aging? In K Esser and GM Martin (eds.): Molecular Aspects of Aging. New York: John Wiley and Sons, pp. 109–138.

Stadtman ER (1995) The Status of Oxidatively Modified Proteins as a Marker of Aging. In K Esser and GM Martin (eds.): Molecular Aspects of Aging. New York: John Wiley and Sons, pp. 129–144.

Stini WA (1990) Changing patterns of morbidity and mortality and the challenge to heath care delivery systems of the future. Coll. Anthropol. 14: 189–195.

Stinson S (1992) Nutritional adaptation. Annu. Rev. Anthropol. 21: 143–170.

Strehler B (1982) Time, Cells, and Aging. New York: Academic Press.

Strong MJ and Garruto RM (1994) Neuronal aging and age-related disorders. In DE Crews and RM Garruto (eds.): Biological Anthropology and Aging: Perspectives on Human Variation over the Life Span. Oxford: Oxford Univ. Press, pp. 214–231.

Susser M, Watson W, and Hopper K (1985) Sociology in Medicine. New York: Oxford Univ. Press.

Swartz K and Mäder K (1995) Free radicals in aging: theories, facts, and artifacts. In K Esser and GM Martin (eds.): Molecular Aspects of Aging. New York: John Wiley and Sons, pp. 77–98.

Swedlund AC, Tenkin H, and Meindl RS (1976) Population studies in the Connecticut Valley: a prospectus. J. Hum. Ecol. 5: 75–81.

Szathmáry EJE (1985) The search for genetic factors controlling plasma glucose levels in Dogrib Indians. In: R Chakroborty and EJE Szathmáry (eds.) Diseases of Complex Etiology in Small Populations: Ethnic Differences and Research Approaches, New York: Alan R. Liss, pp. 199–225.

Tanzi RE, Haines JL, Watkins PC, Stewart GD, Wallace MR, Hallewell R, Wong C, Wexler NS, Conneally PM and Gusella JF (1987) Genetic linkage map of human chromosome 21. Genomics 3:129–36.

Vaupel JW, Carey JR, Christensen K, Johnson TE, Yashin AI, Holm NV, Iachine IA, Kannisto V, Khazaeli AA, Liedo P, Longo VD, ZengY Manton KG, and Curtsinger JW (1998) Biodemographic trajectories of longevity. Science 280:855–60.

Verbrugge LM (1989) The dynamics of population aging and health. In SJ Lewis and P Baran (eds.): Aging and Health. Chelsea: Lewis Publishers, Inc, pp. 23–40.

Webster's New Universal Unabridged Dictionary: Deluxe 2nd edition (1983) New York: Simon and Schuster.

Weiss KM (1981) Evolutionary perspectives on aging. In PT Amoss and S Harrell (eds.): Other Ways of Growing Old. Anthropological Perspectives. Stanford: Stanford Univ. Press, pp. 25–58.

Whitbourne SK (1985) The Aging Body: Physiological Changes and Psychological Consequences. New York: Springer Verlag.

Williams GC (1957) Pleiotropy, natural selection and the evolution of senescence. Evolution 11: 393–411.

Wilson EO and Bossert WH (1971) A Primer of Population Biology. Stamford, Conn.: Sinauer Associates, Inc.

Wood JW, Weeks SC, Bentley GR, and Weiss KM (1994) Human population biology and the evolution of aging. In DE Crews and RM Garruto (eds.): Biological Anthropology and Aging: Perspectives on Human Variation over the Life Span. Oxford: Oxford Univ. Press, pp. 19–75.

Wright RO (1997) Life and Death in the United States. Jefferson, N.C.: McFarland and Company.

Zimmet P, Dowse G, Finch C, Serjeantson S, and King H (1990) The epidemiology and natural history of NIDDM: lessons from the South Pacific. Diabetes Metab. Rev. 6: 91–124.

CHAPTER 14

Demography

TIMOTHY B. GAGE

INTRODUCTION

Demography is best defined as the empirical study of **mortality, fertility,** and **migration** and their relationship to population growth, family formation, and human ecology. As a result, interest in demography is shared by researchers from many disciplines: economics, sociology, history, biology, geography, anthropology, and human biology, to list only a few. Each of these disciplines contributes slightly different theoretical perspectives to the field of demography. Economists, not surprisingly, are interested in the interaction between economies and population, whereas sociologists concentrate on the interaction of population and social institutions, such as the family. Anthropologists contribute to demography by studying the small, nonindustrial, often non-Western populations that have existed over the past few million years.

This chapter considers demography from the point of view of human biologists, biological anthropologists, and human ecologists. These evolutionary social sciences stress both the biological and cultural influences on the demography of past and present peoples. Historically, the development of demography within these disciplines derives primarily from the biological sciences (see Gage 1997 for a short history of demography in biological anthropology). Consequently, the perspective of human biologists differs from that of most researchers in the other social sciences—particularly sociology, economics, and even cultural anthropology, where biological explanations are often not considered. The approach presented here considers the temporal and interpopulation variation in human demographic rates and the genetic, developmental, physiological, environmental, and behavioral factors that may influence this variation.

This is the first of two chapters concerning demography. In this chapter I consider some of the problems in estimating demographic rates; review the literature on the history of human mortality patterns, its variation, and the causes of this variation; and briefly consider the demographic literature on human migration. Chapter 15 is

Human Biology: An Evolutionary and Biocultural Perspective, Edited by Sara Stinson, Barry Bogin, Rebecca Huss-Ashmore, and Dennis O'Rourke
ISBN 0-471-13746-4 Copyright © 2000 Wiley-Liss, Inc.

concerned with the variation in human fertility patterns as well as population growth, and regulation.

Several other chapters in this volume also consider demographic issues. Chapters 3 through 5 consider the genetic structure of human populations. Three of the four forces of evolution are demographic: mortality and fertility drive **natural selection**, migration is associated with **gene flow**, and population size and density influence the rate of **genetic drift**. Furthermore, Chapter 11 concerns the evolution of **life histories** and Chapter 13 the evolution of phenomena that influence demographic rates, such as **senescence**, **menarche**, and **menopause**. The theories underlying the evolution of these aspects of fertility and mortality all involve the interaction of genes and demographic rates.

DEMOGRAPHIC RATES AND THEIR ESTIMATION

The basic data used in demography are mortality, fertility, and migration rates (see Box 14.1). Rates by definition incorporate a time element and are typically presented as the number of events (births, deaths, and migrations) per year per 1000 individuals. Demographic rates are often broken down by age and sex; that is, rates are computed separately for each age and sex. Of course aggregate rates are also computed. Fertility and migration rates are similar, but both are more complicated than mortality. See Box 14.1 for additional details.

The **life table** is an array of mortality rates by age (and some derived measures of mortality) for a particular population. It has been an important methodological device for studying mortality for more than 200 years but can be applied to any kind of "duration," for example, length of life, a marital union, or time between births. As a result, modern life table analysis, survival analysis, or event history analysis and its extensions have become very important tools for the social sciences. The derivation of a life table is presented in Box 14.2. The interpretation of demographic rates is sometimes difficult, even when the data on which they are based are completely accurate. However, demographic data often are inaccurate, incomplete, and/or biased, which further complicates the problem. Demographic data may be inaccurate because people are missed by the census, births and/or deaths are not recorded, and ages and/or sex are either misestimated or misreported. The problem is compounded when these errors are unequally distributed among segments of the population. For example, inaccurate age estimates may be more frequent among the very old. Further, bias in demographic estimates may occur when demographic rates are estimated "indirectly." Indirect estimation typically uses a population theory to derive a rate or rates from incomplete data. However, the use of theories such as **stationary population theory** to estimate mortality or fertility results in bias if the assumptions underlying the theory are violated. (See Box 14.3 for further discussion.)

Finally, an important simplifying assumption of much demographic research is that the population is biologically and culturally homogenous. Where heterogeneity is known, the subpopulations usually are examined separately. For example, mortality is generally considered separately for males and females. When life tables are estimated without distinguishing between the sexes, the younger age categories of the life table are based on about equal proportions of males and females, whereas the older age categories of the life table are based predominately on females (because females tend to live longer than males, at least in most contemporary populations). Consequently, the estimated life table would not accurately represent the age patterns of mortality of either males or females, or even a consistent average of males and females. Some features of the age patterns of the resulting life table might be due to heterogeneity and not to mortality per se. However, the resulting life table still faithfully represents the deaths at each age to the population as a whole. To complicate matters further, in many cases, subpopulations are not readily identifiable or the sources of heterogeneity are not known or understood. For example, prior to the 1960s the impact of cigarettes on mortality was not known. It is highly likely that additional factors will be identified in the future. Until they are identified, the effects of heterogeneity cannot be completely controlled for or assessed. Even the best empirical demographic data must be interpreted with caution. (See Wood et al. 1992 for an extensive review of these problems with respect to anthropological demography.)

BOX 14.1   DEMOGRAPHIC RATES

Commonly reported demographic rates include measures of mortality, migration and fertility (Pollard et al. 1974). In general, the ideal demographic data consists of **longitudinal** observations on a population over the period in question with complete reporting of births, deaths, and changes in residence. Typically, however, demographic rates are estimated from **cross-sectional** data and may not be complete. Demographic rates are usually categorized into two types: crude rates and age-specific rates.

Crude rates refer to the population as a whole. As described in Chapter 7, the **crude mortality rate** is defined as

$$m = (\frac{d}{r})1000 \tag{14.1}$$

where m is the crude mortality rate, d is the number of deaths during a year, and r is the number of individuals exposed to the risk of dying. It is traditional to present crude rates based on 1000 individuals. If longitudinal data are available, the number of person-years exposed to risk can be calculated exactly (Lee 1980). However, exposure to risk is often estimated as the "midyear" population, that is, the average of the population at the beginning of the year and at the end of the year. The crude birth, migration, marriage, and divorce rates can be calculated simply by replacing d with the number of the appropriate vital event and r with the appropriate exposure to risk.

Most demographic rates change with age and/or sex. Hence, demographic rates often are computed specifically for each age and sex. **Age-specific mortality rates** are computed as

$$m_x = \frac{d_x}{N_x} \, 1000 \qquad (14.2)$$

where m_x is the death rate (a.k.a. the central death rate, or force of mortality), d_x is the number of deaths to individuals in the xth age category during a year, and N_x is the number of individuals in the xth age category exposed to the risk of dying during a year.

Exposure to risk can be computed exactly from longitudinal data (Lee 1980) but in demographic work usually is based on cross-sectional data; the midyear population of individuals aged x is used as an estimate. Typically, age-specific death rates are computed for each sex by 5-year age categories, although 1-year age categories are sometimes used. The value of x represents the beginning of the age interval. Age- and sex-specific rates usually are computed for deaths, births, migrations, and marriages.

Demographic rates for national populations generally are based on censuses and vital registration systems (e.g., of births, deaths, and marriages). International migration is also registered in most contemporary nations, but internal migration frequently is not registered, particularly among the democracies. Great Britain is an exception, because of the national medical system. As a result, the derivation of internal migration rates is particularly problematic. Often net migration rates (the sum of in-migration and out-migrations) are computed from censuses, births, and deaths.

In particular, the net number of migrants (M) is computed by rearranging the equation:

$$N_{t+1} = N_t - D + B + M \qquad (14.3)$$

where N_{t+1} is the population at time $t+1$, N_t is the population at an earlier time t, and D and B are deaths and births to the population between times t and $t+1$. Obviously, any errors in the census or in birth or death registrations will be transferred to migration. But even when migration is registered, as in Great Britain, the variation in medical district sizes and densities makes the resulting migration rates very difficult to interpret.

Estimating demographic rates for small populations is even more difficult because census materials and/or vital registration may be lacking. If everything is lacking, then there is no hope. But when some data exists, it may be possible to derive demographic rates indirectly. These processes are discussed in more detail in Boxes 14.2 and 14.3.

In general, demographic data on small nonindustrial, non-Western populations tend to be of poor quality whether ethnographically or archaeologically collected. The problems of census coverage, missing vital events, and misestimating age are generally more severe than in large industrial populations. Furthermore, most published life tables available for these populations have been estimated by using stationary population theory, without knowledge that the population was in fact stationary. However, these small populations may be biologically and culturally more homogeneous than large national populations, reducing but probably not completely eliminating the misleading effects of heterogeneity. In general, small-population demography is particularly challenging. It has been aptly described as an attempt to "infer the unobservable" (Weiss 1989).

Demographic data from national European populations (after 1850 AD) are generally considered most accurate, whereas the data from the rest of the world's nations run the gamut from very good in developed nations to completely lacking in some developing nations. Chad and Oman, for example, have never conducted a census (Haub 1987). As a consequence of the association of "good" demographic data with European populations, demography (even anthropological demography) has an intrinsic national, industrial-era European bias. This situation is improving as more data become available from other time periods and regions of the world. However, the tendency still is to use the recent European demographic experience as "demographic analogies" for non-European populations, both contemporary and prehistoric. The rationale is the "uniformitarian" assumption that we are all human and hence should respond in at least roughly similar ways to similar environments (Howell 1976). This is not an unreasonable assumption, because as yet, little convincing evidence indicates that much human variation in demographic rates is due to genetic differences among populations. Even so, it is unlikely that industrial-era Europe encompasses a complete sample of the biological and cultural environments that human populations have been exposed to around the world and over the past few million years.

In the following sections, I begin with the demography of historic European and contemporary developed nations (national populations) and then extend these patterns as carefully as possible to nonindustrialized nations and to the contemporary and prehistoric populations typically studied by anthropologists (small populations).

MORTALITY

Demographers divide mortality into two independent phenomena: the level of mortality, commonly measured as an expectation of life, and the changes in mortality with age, that is, the age patterns of mortality. Both of these phenomena are estimated by using life tables (Box 14.2). Expectation of life is the average number of years remaining to an individual and can be calculated for all exact ages reported in a life table. However, **life expectancy at birth** is the most commonly reported measure of the level of mortality. High expectations of life at birth denote low levels of mortality, and low expectations of life at birth denote high levels of mortality.

BOX 14.2   MORTALITY ANALYSIS: THE LIFE TABLE

The life table is the standard method used for analyzing mortality data but has very wide applications within the social and biomedical sciences. It is useful for measuring the duration of any phenomenon: lives, marriages, birth intervals, hospital stays, and more. As a result, life table analysis is now commonly referred to as event history analysis to emphasize its broad applicability.

There are two kinds of life tables: actuarial, and Kaplan–Mier. The major difference is that actuarial life tables categorize individuals into age categories (usually 5-year age categories in abridged human life tables), whereas Kaplan–Mier life tables account for each individual by exact age. The actuarial approach will be presented here because it is the best-known method in human biology. However, it is not necessarily the best procedure for use with small population data. (For a useful introduction to life table analysis, see Lee 1980 or Pollard et al. 1974.)

A typical actuarial life table for a small population is presented in Table 14.1. It represents the female population of the Trio of Surinam in 1-year age categories from birth to age 5 and 5-year age categories for ages 5–80. The last age category is open-ended and includes all individuals older than 80 years.

The five standard columns to a life table are as follows:

q_x = the probability of an individual that survives to age x dying within the interval, that is, between age x and $x + 1$ or $x + 5$, depending on the interval

l_x = the number (or portion) of individuals born who survive to exact age x, that is to the beginning of the interval

L_x = the person-years lived between age x and $x + 1$ or $x + 5$, depending on the interval

T_x = the total future person-years lived after exact age x

e_x = the expectation of life at age x

It is traditional to assume that the life table represents an imaginary cohort of individuals living through each of the age categories. An arbitrary size for the imaginary cohort, called the radix of the life table, is selected. In this case, 100,000 (see the value for l_0) was chosen. The q_x and l_x columns are related and can be estimated from each other; n is the length of the interval.

$$q_x = \frac{l_x - l_{x-n}}{l_x} \tag{14.4}$$

e.g., for ages 5–9, $0.01161 = (89,257 - 88,220)/89,257$

$$l_{x+n} = radix(\prod_{0}^{x}(1 - q_x)) \tag{14.5}$$

e.g., for ages 5–9, 89,257 = 100,000[(1 − 0.0749)(1 − 0.01728)(1 − 0.00823) (1 − 0.00557)(1 − 0.00444)]

$$L_x \cong \frac{n}{2}(l_x + l_{x+n})$$ (14.6)

e.g., for ages 5–9, 443,693.0 = (5/2)(89,257 + 88,220)

$$T_x = \sum_{\infty}^{x} L_x$$ (14.7)

e.g., for ages 5–9, 5,459,542 = 443,693.0 + 439,112.4 + 434,238.9 + 427,599.1 + 419,583.0 + 410,532.5 + 400,304.1 + 388,596.1 + 374,732.7 + 357,089.9 + 333,890.2 + 302,552.5 + 260,417.5 + 206,126.7 + 142,540.9 + 118,538.8

$$e_x = \frac{T_x}{l_x}$$ (14.8)

e.g., for ages 5–9, 61.7 = 5,459,542/443,693

There are several difficulties in estimating L_x exactly. Equation 14.6 works reasonably well at most ages and assumes that deaths are evenly distributed across the interval. However, during the first year of life, the majority of deaths occur within a few months of birth. Consequently, L_0 often is estimated as

$$L_0 \cong .3l_0 + .7l_1$$ (14.9)

This equation gives less weight to the number of individuals entering the interval, because many of them will die early in the interval and do not contribute heavily to person-years lived.

In Table 14.1, additional corrections suggested by Coale and others (1983 (assuming a "West" age pattern of mortality) were used, namely,

$$L_0 = .27491l_0 + .72509l_1$$ (14.10)

e.g., for age 0, 94,564.0 − [0.27491(100,000)] + [0.72509(92,503)]

Furthermore, the last entry in the L_x column (L_{80+}) is difficult to estimate. The best strategy is to carry the life table to an age at which all individuals are dead and then aggregate the results back to age 80. However, this information often is not available. In **paleodemography**, it is often assumed that

$$L_{x+} = \frac{n}{2} l_{x+}$$ (14.11)

e.g., for age 80+, 118,538.8 = (5/2)21,860
where n is chosen to be the length of the age category preceding the last open-ended age category. In general, this equation underestimates L_{80+} and the columns to the right of L_x to a greater or lesser extent depending on the proportion of the population still alive at the beginning of the last age category. More complex methods are available.

In Table 14.1, the open-ended category was estimated by extrapolating l_x to the end of life by using a mathematical hazard model such as the **Gompertz equation** (Coale et al. 1983; Gage 1989). This procedure is probably more accurate but is still arbitrary.

The first column of the life table maybe either l_x or q_x. National life tables generally approximate q_x from the age-specific death rate, central death rate, or force of mortality (see Box 14.1)

$$m_x = \frac{d_x}{N_x} \qquad (14.12)$$

without multiplying by 1000 (Pollard et al. 1974). The central death rate differs slightly from q_x, which is the proportion of individuals entering the interval that die in the interval. The q_x value is always slightly smaller than m_x because the denominator of q_x is the number entering the interval, whereas the denominator of m_x is the number still alive at the middle of the interval (hence the name, central death rate). Provided that deaths are linearly distributed across the age interval,

$$q_x \cong \frac{2m_x}{2+m_x} \qquad (14.13)$$

In small-population demography, the life table often begins with the l_x column. For example, in paleodemography, it is simplest to estimate l_x from an age at death distribution, assuming that the population is stationary (Weiss 1973) (see Box 14.3):

$$l_x = radix \left(\frac{\sum\limits_{a=\infty}^{x} d_a}{\sum\limits_{a=\infty}^{0} d_a} \right) \qquad (14.14)$$

The age pattern of mortality is usually measured as a list of age-specific death rates (Box 14.2). A typical human (or for that matter, mammalian) age pattern of mortality is presented in Figure 14.1. Mortality tends to be high in the few days and weeks after birth but declines rapidly, reaching a low in human populations, usually between 5 and 10 years of age. Mortality remains relatively low throughout the next 20 years of life and then begins to increase at an ever-increasing rate at least until 70 or 80 years of age. The result is a U-shaped curve. Other more controversial features of the typical human age pattern of mortality are a bump in the mortality distribution during childhood (not shown in Figure 14.1), a hump in the mortality distribution during early adulthood (shown in Figure 14.1), and a slowing of the rate of increase in mortality among the very oldest individuals (shown in Figure 14.1).

A bump in the mortality distribution during childhood is often attributed to increased mortality associated with weaning. However, this explanation is not well-supported empirically and consequently is not included in Figure 14.1. I am aware of only one study that provides convincing evidence of a bump that is attributable to weaning (Rosetta and O'Quigley 1990). There are no reports of this pattern among national populations with accurate demographic data. Consequently, most demographers believe that mortality declines monotonically throughout the first few years of life. When a weanling bump in mortality occurs, it is generally

Table 14.1   Life table: Trio females, 1963–78

Age (years)	q_x	l_x	L_x	T_x	e_x
0	0.07497	100,000	94,564.0	5,915,701.0	59.16
1	0.01728	92,503	91,703.8	5,821,137.0	62.93
2	0.00823	90,905	90,530.6	5,729,433.0	63.03
3	0.00557	90,157	89,905.6	5,638,902.0	62.55
4	0.00444	89,655	89,455.7	5,548,998.0	61.89
5–9	0.01161	89,257	443,693.0	5,459,542.0	61.17
10–14	0.00902	88,220	439,112.4	5,015,849.0	56.86
15–19	0.01319	87,425	434,238.9	4,576,737.0	52.35
20–24	0.01741	86,271	427,599.1	4,142,498.0	48.02
25–29	0.02010	84,769	419,583.0	3,714,899.0	43.82
30–34	0.02307	83,065	410,532.5	3,295,316.0	39.67
35–39	0.02681	81,148	400,304.1	2,884,784.0	35.55
40–44	0.03176	78,973	388,596.1	2,484,481.0	31.46
45–49	0.03972	76,465	374,732.7	2,095,885.0	27.41
50–54	0.05474	73,428	357,089.9	1,721,154.0	23.44
55–59	0.07579	69,408	333,890.2	1,364,064.0	19.65
60–64	0.11341	64,148	302,552.5	1,030,176.0	16.06
65–69	0.16843	56,873	260,417.5	727,624.0	12.79
70–74	0.25663	47,294	206,126.7	467,206.0	9.88
75–79	0.37822	35,157	142,540.9	261,080.0	7.43
80+	1.00000	21,860	118,538.8	118,538.8	5.42

q_x, probability of the individual who survives to age x dies within the interval, that is, between age x and $x + 1$ or $x + 5$, depending on the interval; l_x, number (or portion) of individuals born who survive to exact age x, that is, to the beginning of the interval; L_x, person-years lived between age x and $x + 1$ or $x + 5$, depending on the interval; T_x, total future person-years lived after exact age x; e_x, expectation of life at age x.
(*Source:* Adapted from Gage et al. 1984b)

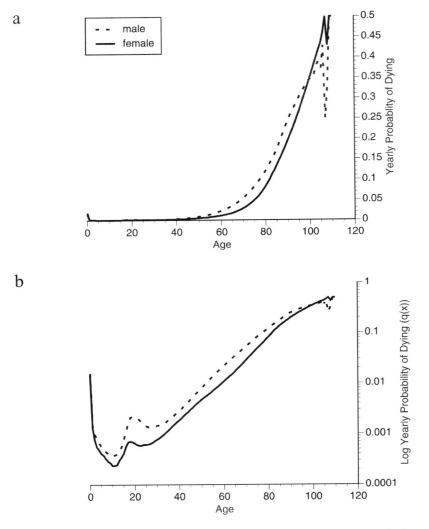

Figure 14.1   Australian mortality, 1970–72. This low-mortality population displays a particularly strong "accident hump" during early adulthood. (a) Plotted on a linear scale. (b) Plotted on a logarithmic scale. (*Source*: Data from Heligman and Pollard 1980)

thought to be due to inaccurate or incomplete data, probably underreporting of neonatal deaths.

The hump in the mortality distribution during early adulthood is frequently attributed to accidental deaths, particularly among young males who tend to engage in high-risk behaviors. This feature of the human mortality curve is deeply ingrained in the literature of national demography and occurs in some populations, primarily low-mortality western European populations (either living in Europe or elsewhere, such as in the United States and Australia). However, statistical analyses suggest that it is not a consistent or particularly important feature of human mortality (Gage and Mode 1993), even in those populations where it occurs. The importance attributed to this phenomenon by national demographers may be due to the common practice of graphing the age patterns of mortality on a log scale, which greatly accentuates this feature of the mortality distribution (see Figure 14.1b).

Finally, evidence is growing that the increase in mortality among the oldest old (90 years or older) does not continue to increase at an ever-increasing rate (as it does between the ages of 30 and 80) but increases more slowly, or even levels off, at some maximum level of mortality (Perls 1995). Once again, however, there are potential problems with data: Small numbers of individuals survive to these ages, and the data on the elderly are generally of low quality. Thus, the leveling off of the mortality curve at the oldest ages maybe due to to "bad data," (Coale and Kisker 1986; Elo and Preston 1994) since age is misreported more often in the elderly than in younger individuals. Alternatively the leveling off of mortality may be due to the fact that only physiologically exceptional individuals survive to the oldest ages (Vaupel et al. 1979; Economos 1982; Perls 1995), that is, due to heterogeneity in frailty within populations.

The human mortality curve appears to be characteristically U-shaped, at least prior to age 80 years. The height of the human mortality curve above the x-axis is a function of the level of mortality (see Figure 14.1a). Some populations may show additional features—a weanling bump or an accident hump—under special conditions. These phenomena remain controversial and require further research. The well-documented variation in the age patterns of mortality, which is presented in detail later, does not refer to the presence or absence of the various qualitative features mentioned earlier but to the quantitative variation in the levels of infant, childhood, and adult mortality among populations while statistically controlling for differences in expectation of life. For example, infant and childhood mortality may be relatively low and adult mortality high in one population and vice versa in another population, even though both have the same general level of mortality.

In the remainder of the section on mortality, I present several issues of human biological interest: empirical evidence for variation in the level of mortality among populations and between the sexes, empirical evidence for geographical variation in the age patterns of mortality among national populations as well as between national populations and small populations, and the causes of variation in the level and age patterns of mortality.

Variability in the Level of Mortality

The level of human mortality varies widely among human populations, from expectations of life at birth of less than 15 to more than 80 years (see Figure 14.2). The factors that account for the temporal and geographic variation in the level of mortality are not yet well understood. It is clear, however, that the industrial revolution and the process of industrial development are closely associated with dramatic changes in the level of mortality.

The distributions of expectation of life for prehistoric, preindustrial, contemporary developing, and contemporary developed populations are presented in Figure 14.2. Prior to the industrial revolution, expectation of life at birth was short, less than 40 years. However, few contemporary countries, regardless of the state of development, have expectations of life at birth less than 40 years. In general, expectation of life at birth among developing nations ranges from 40 to 70 years, whereas expectations of life greater than 60 years are characteristic of developed populations (Weiss 1973; Behn and Vallin 1982). Clearly, the range of variation in mortality is greater among contemporary populations than among prehistoric populations.

The variation in expectation of life at birth and at age 15 years among 29 archaeologically studied populations by level of socioeconomic development, is presented in Table 14.2. These estimates are all based on paleodemographic methods—that is, skeletal samples and stationary population theory—and must be interpreted with caution. The two lowest expectations of life in this sample represent New World populations around the time of European contact. The very high mortality rates in these populations may represent the **epidemics** associated with contact or defects in the data (e.g., the population may not be closed to migration due to the disruption of contact—including the epidemic itself). The two populations

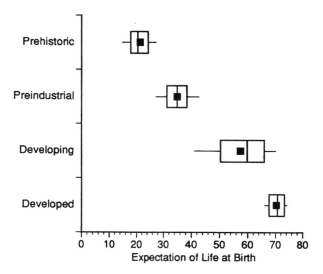

Figure 14.2 Approximate range of expectations of life at birth, at various levels of sociocultural complexity. (*Source*: Data from Weiss 1973 and Behm and Vallin 1982; adapted from Gage et al. 1989)

Table 14.2   Expectations of life among prehistoric small populations

Population	e_0 (years)	e_{15} (years)	Economy type
Mobridge I	12.48	22.69	P
Larson	13.14	19.19	P
Kulubnarti I	14.42	23.53	A
Grasshopper	15.59	19.02	H
Lerna	16.56	19.36	A
Averbuch	16.76	15.26	H
Arroyo Hondo	16.87	20.50	H
Mobridge II	18.19	20.52	P
Dickson Mounds Mid. Mississipian	18.52	18.63	H
Oneota	18.64	23.21	H
Indian Knoll	18.96	14.60	G
Mesa Verde (Late)	19.27	13.14	H
Tipu	19.48	15.32	P
Libben	19.88	18.72	H
Tlajinga 33	20.52	22.84	H
Taforalt	21.38	26.95	G
Meinarti	21.80	15.51	A
Mesa Verde (early)	22.19	14.82	H
Carlston Annis	22.38	19.45	G
Kulubnarti II	22.92	23.51	A
Carrier Mills	23.85	20.01	G
Kane	24.07	19.04	H
Black Mesa	25.26	25.87	H
Copan	25.53	21.15	H
Dickson Mounds (transitional)	25.85	23.16	H
Sopronkohida	26.65	31.38	A
Gibson/Klunk	27.77	25.11	H
Keszthely-Dobogo	35.28	32.33	A
Tiszapolgar-Basatanya	36.36	28.43	A

Economies: P, cultural present; G, hunter-gatherer; H, horticulturalist; A, agriculturalist. e_0 and e_{15}, expectations of life at birth and at age 15 years, respectively.
(*Source:* Adapted from O'Connor 1995, Tables 41 and 42)

with the greatest expectation of life are Mediterranean populations reported by Acsadi and Nemeskeri (1970). It may be that Mediterranean populations had lower mortality than other populations of this type, or that the aging methods used in constructing these life tables inflated the estimation of adult ages, at least compared to the methods used to age other paleodemographic populations. In any event, most estimates of prehistoric life expectancy at birth vary within a surprisingly narrow range, from perhaps 18 to 25 years.

Given the association between level of mortality and the industrial revolution, it is commonly assumed that the agricultural revolution may also be associated with changes in mortality. Some maintain that mortality increased (see papers in Cohen and Armelagos 1984 and Cohen 1989); others argue that it could just as well have decreased given the problems associated with estimating mortality in paleodemographic samples (Wood et al. 1992). On the basis of the data presented in Table 14.2, there are few significant differences among these populations. The mean expectations of life for hunter-gatherers is 21.6 years (**standard deviation** [SD] = 2.1 years), for horticulturalists is 21.2 years (SD = 3.9 years), and for agriculturalists is 24.9 years (SD = 8.5 years). The mean for hunter-gatherers is between horticulturalists and agriculturalists. None of these differences, however, are statistically significant. What is clear is that mortality during these times was high and apparently did not vary a great deal from one population to another.

Among contemporary nations, expectation of life ranges from about 40 to above 80 years (see Figure 14.2). Japan currently has the longest life expectancy, 75.86 years for males and 81.90 years for females (Keyfitz and Flieger 1990). The current estimate of expectation of life at birth for the world's population as a whole slightly exceeds 60 years (World Factbook 1994). Industrial development has been associated with dramatic declines in the level of mortality and increases in the variation of the level of mortality worldwide. In the 1970s and 1980s, the developing nations typically had expectations of life between 30 and 60 years at birth, whereas the developed nations had expectations of life above 60 years (Weiss 1973; Behn and Vallin 1982). As indicated in Figure 14.2, which includes more recent data, the range of expectations of life among the developing nations is now from 40 to 65 years, with the "break" between the developing and developed countries occurring around 65 years.

Additionally, it has been observed in recent years that a few developing nations have surprisingly low mortality given their low gross domestic products, indicating that the decline in mortality is not strictly a function of economic development. Caldwell (1986) described three of these populations—Costa Rican, Sri Lankan, and Kerala—in detail. Among developed nations, the decline in mortality appears to have begun during the 19th century. In developing countries, the shift occurred later, around 1920 (see Figure 14.3). Mortality may have been ameliorated world-wide in the developing countries beginning in the 1920s (Baker 1987; Gage et al. 1989), that is, independent of the level of industrial or economic development. The cause of this decline is not clear.

Expectation of life at birth among a sample of contemporary small populations, studied ethnographically, suggests that the level of mortality is lower among contemporary small populations (Table 14.3) than among archaeologically studied small populations (Table 14.2). The mortality of contemporary small populations appears more similar to the mortality observed in the contemporary developing nations than to the mortality of socioeconomically similar archaeological populations presented in Table 14.2. With the notable exceptions of the contemporary Yanomama and perhaps of the rural Chinese in the 1920s, expectation of life in contemporary small populations is greater than 30 years. Either contemporary small populations (after 1920) have considerably lower mortality than the archaeological populations presented in Table 14.2, or the data have large and consistent defects for the contemporary populations, the archaeological populations, or both. If these differences are real, then contemporary small populations may not provide a useful ethnographic analogy of prehistoric populations, at least with respect to the level of mortality. Contrasts among contemporary small populations by economic system are not possible because of the small number of populations for which expectation of life is available. Comparisons based on infant and childhood mortality, however, indicate broad variation in mortality among these populations, but no statistically significant variation in mortality by economic system (Hewlett 1991). Perhaps these contemporary small populations also participated in a worldwide decline in the level of mortality that began in about 1920.

It is generally considered biologically "normal" for women to live longer than males of the same population (Lopez and Ruzicka 1983; Waldron 1983). However, the sex differentials in mortality may not always have favored women, at least to the extent observed among contemporary nations. For example, in England and Wales between 1861 and 1985, the sex differential in mortality increased from 2.6 years to 5.7 years (see Figure 14.3). Similarly, in Chile the sex differential increased in favor of women from 3.1 to 7.0 years between 1909 and 1985. In a few contemporary nations, males have greater **longevity** than females (e.g., Nepal, Bangladesh, and Pakistan) (Keyfitz and Flieger 1990). This pattern, however, is uncommon, although it may have been more common in the past. In any event, the secular decline in mortality that has occurred over the past few hundred years appears to have benefited females more than males.

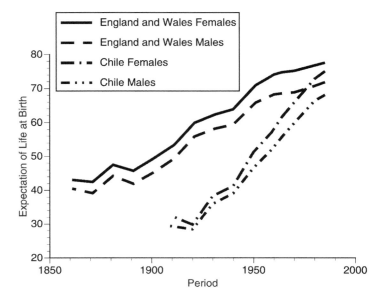

Figure 14.3   Historical trends in level of mortality in England and Wales and in Chile, by sex. *(Source*: Data from Preston et al. 1972 and Keyfitz and Flieger 1990)

Among contemporary small populations, higher mortality in females than in males appears to be considerably more common. Among the seven populations in Table 14.3 for which comparisons can be made, only two indicate higher mortality in males than in females. The prehistoric data on sex differentials in mortality measured as an expectation of life at birth are not available because of the difficulties of sexing the skeletal remains of children. As a result, published paleodemographic life tables typically pool the sexes prior to the adult ages. However, both Angel (1972) and Acsadi and Nemeskeri (1970) examined the mean age of death in adults and found that adult female mortality is generally higher than adult male mortality prior to 1900 among the Mediterranean populations. Whether expectation of life at birth was lower for females is not clear from these analyses, because the sex differential in mortality during childhood, which generally favors females, is not incorporated in these analyses. What is clear is that the large sex differentials in favor of females are characteristic of contemporary national populations—not universal—and probably a relatively recent development.

Variation in the Age Patterns of Mortality

The age patterns of mortality are thought to vary geographically among national populations and may vary between national populations and small populations. The age patterns are considered to vary independently of the level of mortality because the geographical distribution of age patterns of mortality has been remarkably consistent throughout the historical decline in mortality, at least in Europe (Coale et al. 1983).

Table 14.3   Expectations of life among contemporary small populations

Population	e_0 Males	e_0 Females	e_{15} Males	e_{15} Females	Economy Type
Yanomama[a]	17	15	22	24	H
Yanomama[b]	21	20	25	28	H
Rural China, 1920 (North)	29	20	37	28	A
Rural China, 1920 (South)	22	21	29	29	A
!Kung sexes combined	35		42		G
Trio	42	51	43	44	H
New Guinea[c]	53	51	45	44	Mixed
Cocos Islands[d]	50	47			Mixed
Semai Senoi	31	32	33	30	H

Economies: G, hunter-gatherer; H, horticulturalist; A, agriculturalist
[a] Estimated by Gage(1988)
[b] Estimated by Neel and Weiss (1975)
[c] Indigenous population as a whole
[d] Malay peoples, migrated to Cocos in the 1820s
(*Source*: Adapted from Gage et al. 1989)

In their seminal work, Coale and others (1983) identified four patterns of mortality that they named North, East, South, and West, because they appeared to be characteristic of the age patterns of mortality of northern, eastern, southern, and western Europe, respectively. The sample they studied consisted almost entirely of European populations and populations of European ancestry (e.g., in the United States, Canada, and Australia). Only a few truly non-European populations were included (several life tables from Japan and Taiwan). Other non-European populations were excluded because of the low quality of data available. Nevertheless, Coale and Demeny considered West the "average" human pattern of mortality and the remaining patterns deviations from that pattern. More recent studies have attempted to extend these analyses to include the age patterns of mortality of non-European populations (United Nations 1982; Gage 1990).

Gage (1990) reports that seven age patterns of mortality (clusters in Figure 14.4) are required to encompass the range of variation found in a worldwide sample of contemporary national populations. Four of these are closely associated with the Coale et al. (1983) model life tables (see Figure 14.4). West is equivalent to the average of Gage's (1990) sample and similar to Cluster 2, whereas East is similar to Cluster 1, South to Cluster 5, and North to Cluster 4. Gage's study identified three new patterns that deviate from the Coale et al. patterns and that are not prevalent in Europe (Table 14.4).

From a statistical viewpoint, the most significant aspect of the variation is due to adult mortality. This difference is contrasted in Figure 14.5. Three patterns have adult mortality that increase at the older ages relative to the overall average pattern of mortality (see Figure 14.5a), whereas four patterns have adult mortality that declines (see Figure 14.5b). The three patterns with increasing mortality at the oldest ages are European, whereas Cluster 2 (West), the only European pattern with mortality that declines at the oldest ages, is the least extreme of the four patterns shown in Figure

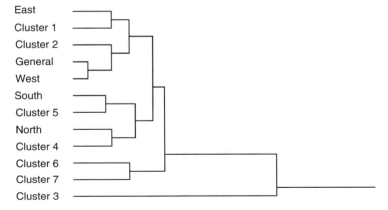

Figure 14.4   Hierarchical cluster analysis indicating the relative differentiation among the seven mortality patterns identified by Gage (1990) and the Coale and others (1983) patterns of mortality. The farther to the right that the clusters "join," the greater the differentiation among the life tables. (*Source*: Adapted from Gage 1990)

14.5b. The age patterns tend to vary geographically, although a tight correlation of age pattern with region can be questioned. Cluster 7 consists predominately of North American, southern African, and southern South American populations. Cluster 6 is predominately an Asian pattern of mortality. Interestingly, some Asian life tables from South Africa (where racial groups are reported separately) are included in Cluster 6 rather than in Cluster 7, with the life tables for the other races of southern Africa. Cluster 3 consists largely of central African populations.

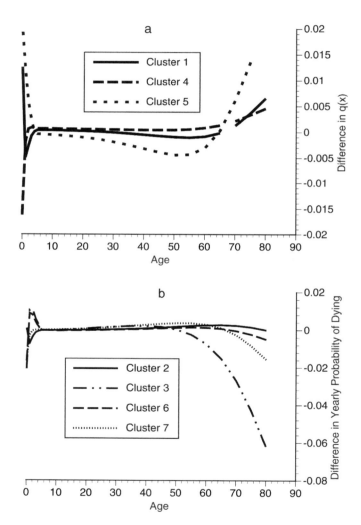

Figure 14.5   Difference in q_x (by year) between the age pattern of mortality of each cluster compared with an overall average pattern. (a) Clusters that display increasing mortality with respect to the general pattern at 80 years or age. (b) Clusters that display decreasing mortality relative to the overall average pattern at age 80. (*Source*: Adapted from Gage 1990)

Cluster 3 is the most extreme of the age patterns, with mortality at the oldest ages declining very rapidly compared with the overall average age pattern of mortality (see Figures 14.4 and 14.5b). It also has the lowest infant and highest childhood and adult mortality prior to the oldest ages. This age pattern is generally less U-shaped than other age patterns of mortality. Because of the poor quality of demographic data from Africa, this pattern may simply represent errors in the underlying life tables (e.g., underenumeration of population and/or deaths, misstatement of age). However, the African pattern may represent mortality in Africa more closely than the European patterns of mortality, which are based on "better" data but from a radically different physical, biological, and cultural environment. Interestingly, comparisons of differences in mortality between African American (Black) and European American (White) populations in the United States show similar variation in the age patterns of mortality. The Black pattern is less U-shaped and shows lower mortality among older individuals than the White pattern. This phenomenon is often referred to as the Black–White mortality crossover (Gage 1989). Its causes are currently attributed to "bad" data and/or heterogeneity and remain subject to heated debate.

Very little empirical evidence—ethnographic or archaeological—exists concerning the age patterns of mortality characteristic of small populations. This paucity of data is due to the difficulties in collecting sufficient data and the potential for error in the data collected. The most progress has been made with respect to archaeological investigations (Gage 1988; O'Connor 1995). Less is known about the age patterns of mortality in contemporary small populations, largely because studies often depend on **model life tables**, usually the West pattern of mortality, to estimate a life table (e.g., Howell 1979; Gage et al. 1984b). Consequently, the West pattern has been imposed on the data, and independent evidence of the age pattern of mortality is not available. Hewlett (1991) collected some evidence concerning the relative rates of infant and childhood mortality in contemporary small populations.

O'Connor (1995) developed a model life table for prehistoric populations by using archeological data and the methods developed by Gage and Dyke (Gage 1988;

Table 14.4   The worldwide distribution of age patterns of mortality

Region	Cluster						
	1	2	3	4	5	6	7
Africa	1.9	15.0	85.7	9.8	47.4	25.0	42.9
N. America	30.8	15.0	—	5.9	21.1	21.9	21.4
S. America	24.0	27.5	—	3.9	—	6.3	14.3
Asia	2.9	7.5	14.3	29.4	5.3	37.5	10.7
Europe	40.4	30.0	—	39.2	26.3	6.3	10.7
Pacific	—	5.0	—	11.7	—	3.1	—
Total	100.0	100.0	100.0	100.0	100.0	100.0	100.0

(*Source*: Adapted from Gage 1990)

Gage and Dyke 1988; Dyke et al. 1993). O'Connor's analysis includes 12 life tables based on age at death distributions. Each of these life tables was estimated assuming stationary conditions (see Box 14.3). As a result, any deviation from the stationary state will incorporate error into the individual life tables, either over- or underestimating the level of mortality for that particular population, depending on whether the population is growing or declining, respectively. However, by combining a series of these life tables, the average growth rate should tend toward zero, the stationary condition. This tendency is based on the assumption that prior to the industrial revolution, mean population growth rates were very small. In any event, the resulting paleodemographic model life table has a life expectancy of about 20 years at birth.

This age pattern of mortality is compared with Gage's (1990) average national pattern at a life expectancy of 20 years in Figure 14.6. The paleodemographic age pattern indicates very low infant mortality but extraordinarily high adult mortality. These results could be a result of underrepresentation of infant deaths, misestimation of age at death, or both. The underrepresentation of infant deaths alone probably is not the cause, because it would imply an expectation of life considerably lower than 20 years. As is discussed later, very high **total fertility rates** (greater than 6.5 [Coale and Watkins 1986]) would be necessary to maintain a population at a life expectancy lower than 20 years at birth.

Hewlett (1991) assembled the mortality data available for various hunter-gatherer, horticulturalist, and pastoralist populations. The data consist of infant mortality (the proportion of the population who dies in the first year of life) and childhood mortality (the proportion of the population who dies in the first 15 years of life). These data suggest that, on average, hunter-gatherer populations have a mean infant mortality rate of about 20.3% (SD = 8.6%), whereas horticulturalists have a mean infant mortality rate of 21.0% (SD = 4.3%). Mean childhood mortality rates for hunter-gatherers is 43.4% (SD = 11.1%), whereas for horticulturalists and pastoralists, the mean childhood mortality is 38.1% (SD = 10.4%).

These results for infant and child mortality are comparable to the model life table derived by O'Connor (1995). The archaeological data provide an estimate of infant mortality of 18.6% and child mortality of 46.6%. These estimates are well within the range of variation reported by Hewlett for contemporary small populations. Although none of these results are statistically significant, there is a trend of increasing infant mortality and decreasing childhood mortality from the archeological data to the contemporary hunter-gatherer data, to the contemporary horticultural data. The age pattern of mortality has become more U-shaped over time, at least at the younger ages. Of course, the very high adult mortality rates of the archaeological model life table at the older ages may still be spurious. Given the similarities during the infant and childhood ages, the differences in expectation of life estimated for contemporary and paleodemographic small populations must be due to higher adult mortality (estimates) in the paleodemographic populations. In this regard, Howell (1982) questioned the accuracy of paleodemographic life tables because of the unusually high adult mortality. As she pointed out, the very high adult mortality rates imply high rates of orphanhood and marital disruption. She postulated that either the social organization of these societies was different than expected in contemporary national populations or the paleodemographic mortality rates are incorrect. However, some evidence indicates that orphanhood and marital disruption are common in contemporary small populations (Hewlett 1991).

Model life tables are particularly useful and important as an aid for estimating the level of mortality in situations where the demographic data are incomplete or defective. Thus, a model life table can be used to estimate the overall level of mortality from estimates of childhood mortality, which may be easier to obtain. This task could be as simple as selecting the Coale et al. West model life table with the same childhood mortality as the observed population. Variations of this procedure have been used to estimate expectations of life for several small populations (e.g., Howell 1979; Gage et al. 1984b). The assumption is that an appropriate model life table is both available and selected for this purpose. For example, is the West age pattern an appropriate model life table to use in Central Africa or among small populations? Probably not, unless the extreme age patterns of mortality of Central African and small populations are simply and completely a result of poor data.

To resolve these issues, a greater understanding of the causes of variation in human age patterns of mortality is necessary. This area of research has largely been ignored by national demographers. Consequently, little is known about the causes—genetic, environmental, cultural, and behavioral—that influence the age patterns of mortality. But it is likely to be of particular interest to human biologists, given their focus on understanding human biocultural variation.

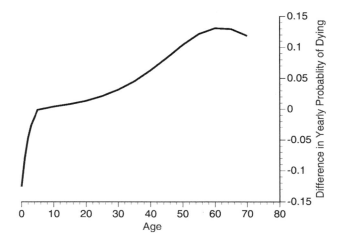

Figure 14.6   Difference in q_x (by year) between the age pattern of mortality of the paleodemographic model life table and the overall average human mortality pattern from Gage (1990) at the same level of mortality. Note that the scale is different then that used in Figure 14.5. (*Source*: Data from Gage 1990 and O'Connor 1995)

Proximate and Ultimate Causes of the Variation in Mortality

In this section, I consider the factors that may account for variation in the level and age patterns of mortality. The section is divided into three parts. In the first, I consider the variation in the distribution of causes of death from a theoretical and an empirical perspective. Variation in causes of death is clearly a proximate determinant of the human variation in mortality. Proximate causes, however, only raise questions concerning the determinants of the proximate determinants. In the second section, I present some potential ultimate determinants of the variation in level of mortality. In the third section, I briefly consider the proximate and ultimate determinants of the variation in age patterns of mortality.

Fenner (1970) described hypothetical changes in human disease ecology throughout human prehistory and history on the basis of several simple **epidemiological** principles:

- the characteristics of **host–pathogen** interaction (development of **immunity**),
- the frequency of contact between hosts and pathogens (population size density and movement), and
- the lifestyle of the host population (nomadic, sedentary, urban).

Given the rise in population associated with the agricultural and the industrial revolutions, the increasingly sedentary lifestyle as a result of the adoption of agriculture, and the acceleration of urbanism with industrial development, Fenner (1970) argues that the types and prevalence of diseases and hence causes of death characteristic of human populations must have changed (Table 14.5).

Population density is an important epidemiological variable, because the kinds of **infectious disease** continuously maintained within a population (**endemic**) are a function of the number of individuals in contact with each other. Many of the childhood infectious diseases common during the historic period result in life-long immunity among survivors. These diseases cannot be continuously maintained in small- or low-density populations because of insufficient contact between infectious individuals and **susceptible** individuals. When introduced into a small- or low-density population, these diseases sweep through the population, infecting the susceptible individuals. However, after all the susceptible individuals are infected, the disease cannot find new hosts and goes extinct. Consequently, many of the common childhood diseases of the historic period may not have been important causes of death in low-density populations such as hunter-gatherers (Table 14.5).

Some of these diseases may have evolved as human pathogens only relatively recently. Measles is perhaps the best example, because it is not endemic in contemporary island populations of less then half a million inhabitants. Consequently, it could not have originated before the origins of intensive agriculture (Fenner 1970). An increasingly sedentary lifestyle also may have changed the disease environment, because it brought humans in more direct contact with human waste products and **parasites**. Thus, the infectious diseases spread by **fecal–oral transmission** should have increased with the development of sedentary lifestyles. The emergence of

Table 14.5 Emergence and growth of the major infectious diseases as a function of human population size

Years BP/ Economy type	Size of human communities	Major infectious causes of death			
		Zoonose	Helminths	Bacteria	Viruses
1,000,000/ Hunter-gatherer	Scattered nomadic bands <100, mean size ~25	scrub typhus, plague, encephalitis, Rocky Mountain spotted fever	trichinosis, trematodes, filariasis	tuberculosis, yaws, malaria, staphylococcus	chickenpox, herpes
10,000/ Horticulturalist	Settled villages <300	—	trichinosis,[a] trematodes,[a] etc.	tuberculosis,[a] malaria,[b] dysentery, diarrhea	—
5,500/ Intensive agriculture	A few cities >100,000, mostly villages <300	—	schistosomiasis[a]	tuberculosis,[a] dysentery, diarrhea	smallpox, measles, rubella, influenza
250/ Steam power	Cities >500,000 villages 1,000	—	—	tuberculosis,[b] dysentery,[b] diarrhea[b]	smallpox,[b] measles,[b] rubella,[b] influenza,[a] and colds
130/ Sanitation	Decline of most infectious diseases as a cause of death, except polio[a], influenza[a], and colds[a] Rise of pandemics				

[a] Prevalence of this cause of death increased over previous periods.

[b] Prevalence of this cause of death increased a relatively large amount over previous periods.

(*Source:* Adapted from Fenner 1970)

urbanism as a result of agriculture and its acceleration with industrialization brought many people in face-to-face contact, hence providing the large, dense populations necessary to support the infectious diseases that characterize human populations today (e.g., respiratory tuberculosis, measles, mumps). "Emerging" pathogens are not a recent development in human affairs; the appearance of human immunodeficiency virus (HIV) and Ebola in modern times is not unprecedented. Many—even most—of these diseases appear to have been originally introduced into human populations by contact with other animals (e.g., measles and canine distemper [Fenner 1970]).

Fenner's (1970) theory suggests two hypotheses concerning the variation in mortality: that infectious disease mortality should increase and that infant infectious disease mortality should become relatively more important as populations become denser, increasingly sedentary, and more urbanized. The first hypothesis is based on the assumption that the increase in the number of infectious diseases should cause infectious disease mortality to increase. The second hypothesis is based on the assumption that individuals are likely to be infected at an earlier age if the disease is endemic. The empirical data, however, do not support the first hypothesis. The prehistoric data is contested, but as far as can be determined, mortality changed little over the course of prehistory and with the introduction of agriculture (Table 14.2). Furthermore, the recent expansion of populations during the historic period has clearly been associated with a *decline* in mortality, not an *increase*. Finally, those causes of death that declined most during the historic period are precisely those predicted to increase: the infectious diseases.

Empirical studies have demonstrated that during the historic period, mortality declined and the cause of death structure changed. In particular, the structure of causes of death shifted from predominately infectious diseases (e.g., measles, respiratory tuberculosis) to predominately **chronic degenerative diseases** (e.g., cardiovascular disease, cancers), a process that has been called the **epidemiological transition** (Omran 1977). An accurate empirical understanding of the epidemiological transition is complicated by competition between causes of death and by secular changes in the reporting and reliability of diagnosis of the various causes of death. Clearly, the proportion of deaths shifted from predominately infectious causes to predominately degenerative causes. It is often assumed that this shift necessarily indicates an increase in the risk of the degenerative diseases as a result of maladaptation to modern lifestyles (largely characterized by overnutrition, reduced exercise, and increased psychosocial stress) (Trowell and Burkitt 1981; Rose 1982; McKeown 1983; Frisancho 1993). However, the epidemiological transition could simply be due to a decline in the infectious causes of death. Everyone must die; consequently, if the risk due to the infectious causes of death declines, then the proportion of deaths due to degenerative causes must increase, even if the risk of the degenerative causes remains constant. Furthermore, observed "increases" in the risk of the degenerative diseases may simply result from secular improvements in diagnosis; that is, deaths due to degenerative diseases, which in the past may have been misclassified as "ill-defined" causes of death (e.g., old age) are increasingly classified into the appropriate degenerative category (e.g., cardiovascular disease, cancer). As a result of these problems, the exact nature of the epidemiological transition is still a matter of debate.

Studies of the epidemiological transition that attempt to account for these problems tend to conclude that the risks of both infectious and degenerative causes of death declined (Preston 1976; Gage 1994). These results are consistent with trends in these same causes of death that have been observed over the past several decades in developed countries, during which the reporting of degenerative causes of death is considered to be accurate (Greenberg 1983; US Bureau of the Census 1988). Thus, the transition from predominantly infectious deaths to predominantly degenerative deaths appears to be due to the faster decline of the infectious than the degenerative causes of death. Furthermore, evidence indicates that the decline in infectious disease mortality was greater for males than for females, whereas the decline in degenerative disease mortality was greater for females than for males. The shifts in the sex differentials in mortality that occurred during the decline in mortality are apparently due to greater sex differentials in the decline of the degenerative causes compared with the infectious causes (Lopez and Ruzicka 1983; Gage 1994).

Although the overall combined risk of degenerative causes of death has declined, one exception is worthy of note. The risk of cancer has clearly increased, and the increase has been greater in men than in women. The sex differential in this cause is largely responsible for the slower decline in degenerative mortality in males than in females. This difference probably represents cancer of the lung and is largely a function of the secular increases and gender differences in the prevalence of smoking. However, the declines in the other degenerative causes of death have more than compensated for the increased incidence of cancers (Gage 1994).

These trends in the sex differentials in the national populations might explain the unusual sex differences in mortality in favor of males in the contemporary small populations shown in Table 14.3 and noted earlier. If infectious disease mortality declined in small populations as a result of global reductions beginning in about 1920 (Baker 1987; Gage et al. 1989), but degenerative diseases remained at prehistoric levels or at least did not decline to the extent they have among contemporary national populations, then the sex differentials might be expected to favor males. However, this notion is contrary to the anthropological dogma that the degenerative diseases are absent in traditional small populations.

In any event, infectious diseases clearly declined with the modern rise of population, at least until recently. This trend is clearly counter to epidemiological predictions suggested by Fenner (1970) and requires additional examination of the more distal causes of the decline in mortality. For example, what factors did Fenner leave out that might explain the decline in infectious disease mortality?

Ultimate Determinants of the Variation in Mortality

A classic study of the cause of the historical decline in mortality (McKeown 1976) examined four potential causes of the decline in mortality: host–parasite interactions (**coevolution**), the development of modern medicine, the introduction of effective sanitary systems, and improvements in nutrition. These factors are not considered in Fenner's theory. McKeown based his identification of the cause or causes of the decline on the changes in the empirical structure of cause of death associated with the decline in mortality in England and Wales from 1860 to 1971 (Table 14.6).

Host–parasite interactions refer to the coevolutionary process by which natural selection tends to reduce the **virulence** of the pathogen and increase the resistance of the host. McKeown (1976) eliminated this cause of the decline on the basis that there was insufficient time, less than 200 years, for significant evolution of the human populations and because of the lack of any evidence for the amelioration of virulence in the pathogens. Scarlet fever is an exception because evidence shows cyclical changes in virulence of this pathogen; however, the decline in mortality requires the general decline in virulence of most pathogens for which there is no compelling evidence.

McKeown (1976) also argued that sanitation was not responsible for most of the historic decline in mortality. He defines sanitation as the introduction of effective measures to eliminate contact between host and pathogen populations. Only 21% of the overall decline in mortality is due to causes of death that can easily be controlled using sanitary measures, that is, the water- and food-borne diseases (Table 14.6). McKeown maintains that sanitary measures are unlikely to have affected the air-borne diseases because of the lack of effective methods of isolation.

Similarly, McKeown (1976) claimed that modern medicine was not responsible for the decline in mortality. He defines modern medicine as the development of "effective treatments," usually drug therapies. However, the decline in mortality and the epidemiological transition generally precede the development of what are considered effective treatments today. For example, the initial decline of respiratory tuberculosis (the largest single component of the decline in mortality in England and Wales) preceded the development of streptomycin, the first effective treatment of respiratory tuberculosis, by at least 100 years (see Figure 14.7).

McKeown (1976) concluded that the decline in mortality was due to improvements in nutrition. He based this conclusion on Sherlock Holmes' principle that if all other possible explanations can be eliminated, the one remaining explanation must be cor-

Table 14.6 Declines in cause of death in England and Wales, 1850–1971

Cause of death	Death rate per million population		Reduction (%)[a]	Reduction prior to 1901 (%)[b]
	1850 AD	1971 AD		
Infectious causes				
Air-borne diseases	7,259	619	40	32
Water- and food-borne diseases	3,562	35	21	46
Other conditions	2,144	60	13	35
Total infectious causes	12,965	714	74	37
Noninfectious causes	8,891	4,070	26	10
Grand total	21,856	5,384	100	30

[a] Percent of the total reduction in mortality explained by reduction in this disease.
[b] Percent reduction in this disease prior to 1901.

(*Source*: Adapted from McKeown 1976)

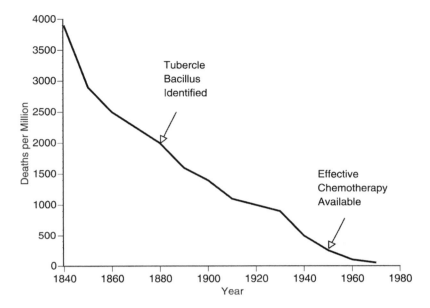

Figure 14.7   Decline in respiratory tuberculosis in England and Wales and the development of effective treatment. (*Source*: Adapted from McKeown 1976)

rect. Unfortunately, little direct empirical evidence indicates that nutrition improved over the period in question. Certainly, a historical (secular) increase in stature (see Chapter 12) has accompanied the decline in mortality (Floud et al. 1990). Additionally, nutritional availability based on national food balance sheet estimates is closely correlated with the level of mortality (Gage and O'Connor 1994). It is not clear, however, whether nutrition is a cause of the decline in mortality or simply a correlate.

Furthermore, it is not clear that McKeown's (1976) list of causes is complete. For example, advances in medical knowledge (e.g., the **germ theory of disease** [Chapter 7]), coupled with education, may have been responsible for improvements in primary care and a decline in associated infectious disease mortality (Preston and Haines 1991). Thus, no clear explanation has been established for the decline in mortality during the industrial era. Fenner (1970) may be correct in the long run. If it is not known why mortality declined, it also is not known how to keep infectious diseases from reemerging.

Cause of Death and the Age Patterns of Mortality

The factors that affect human variation in the age patterns of mortality have received less attention than studies of the level of mortality. Preston (1976) showed that variation in the age patterns of mortality are closely correlated with the geographical distribution of cause of death. For example, the deviations of the North pattern of mortality from the West pattern are thought to be due to the relatively high rates of

BOX 14.3   THEORIES OF POPULATION DYNAMICS

Theories of population growth are used for two purposes: to predict future population growth and characteristics, and to estimate demographic rates indirectly. Predicting future population levels is an important practical application of demography because this information is used by governments to plan for the future. This kind of application is most commonly used among national populations.

The second application of population theories, indirect estimation, is critical to small-population demography because the data on these populations are often incomplete and/or defective, so demographic rates cannot be estimated directly. The theories of population dynamics indicate the functional relationship among mortality, fertility, migration, and the age structure of a population. In principle, if three parameters are known, then the fourth can be estimated. Frequently, to simplify the theory, the population is assumed to be closed to migration. Because little is known about migration in small populations, this convention will be followed here.

The three theories are stable population theory, stationary population theory, and the so-called variable r (nonstable) population theory. Historically, the most influential theory has been stable population theory. It states that if age-specific mortality and fertility rates are constant through time, then the "stable" age structure will be defined as

$$N_x = B \exp(-rx)l_x \qquad (14.15)$$

where N_x is the number of individuals of age x, B is the number of births, exp is the mathematical constant e, r is the **intrinsic rate of increase**, and l_x is the proportion of births that survive to age x (see Box 14.2).

A population perturbed from its stable age structure—perhaps by war—will return to a stable age distribution over time, provided that mortality and fertility remain constant. A population's age structure will effectively achieve its stable configuration in about 70 years after a change in demographic rates. Consequently, most human populations appear to have stable age structures. One well-known exception to the rule is the United States that has not yet achieved a stable age structure after the baby boom following World War II. Nevertheless, it is this "ergodic" or stable feature of population growth that makes stable population theory so useful. In any event, after a population has achieved its theoretical stable age distribution, it will grow or decline at a constant rate defined by r, the intrinsic rate of increase. Furthermore, each age category will grow or decline at a constant rate defined by r.

The importance of Eq. 14.15 is that a life table (l_x) can be estimated from a census and the growth rate of the population, that is, an enumeration of population size at two points in time. If the population age structure is stable, then a census provides an estimate of N_x and the number of births B within the past year, the observed growth rate of the population provides an estimate of r, and an accurate life table can be estimated. An estimate of the distribution of ages at death from an archaeologically recovered skeletal collection and an estimate of the growth rate of the population also may be used to compute a life table (Moore et al. 1975). Similar procedures can be applied to the estimation of fertility (Weiss 1975). These methods are particularly useful when births and deaths are not registered or are incomplete and a life table cannot be computed directly from the data. Suitable populations include those in many less developed nations and most of the small populations studied by anthropologists. The necessary assumption is that the age structure of the population is theoretically stable, but human age distributions naturally tend to be close to their stable states.

Often, only a census is available for a population, and the growth rate of the population (an estimate of r) is not known. In such cases, investigators often assume that the population is stationary (i.e., not growing or declining). Stationary population theory requires two assumptions:) that the age structure is stable and that $r = 0.0$. A life table then can be computed from Eq. 14.15 and a census (N_x and B). The problem is that whereas the assumption of stability is generally reasonable, because populations naturally tend toward stable age distributions, the assumption of no population growth or decline is much more unlikely. If the population is growing and stationary assumptions are imposed, then mortality will be overestimated; if the population is declining, then mortality will be underestimated. Consequently, mortality estimates based on stationary population theory must be interpreted with great caution. It has been argued that stationary estimates of mortality are better estimates of the variation in fertility than mortality (Sattenspiel and Harpending 1983). This argument has even been used to estimate variation in prehistoric levels of fertility (Buikstra et al. 1986). However, the application of stationary population theory to fertility estimation is as prone to error as its application to mortality.

Stable population theory has been extended to nonstable population dynamics (Preston and Coale 1982). The extension consists of simply replacing r with age-specific rates of growth r_x. Equation 14.15 thus becomes

$$N_x = B\exp\left(-\int_0^x r_x\right)l_x \qquad (14.16)$$

This equation provides a series of indirect methods that can be applied to any population—nonstable, stable, or stationary—provided that estimates of N_x and r_x are available. These values can be estimated from two censuses on the same population at two different points in time (Gage et al. 1984a, 1984b; Gage 1985). If the data are available, this method provides assumption-free estimates of the level of mortality and fertility and should be preferred over other estimates.

respiratory tuberculosis found in northern European countries. Gage and O'Connor (1994) showed that variation in nutritional availability could be responsible for the observed variation in the age patterns of mortality (Gage 1990). In particular, abnormally low-calorie and low-protein diets are associated with the African pattern of mortality. Whether nutrition is a causal or simply an associated phenomenon is not yet known. However, the geographical distribution of causes of death and hence the age patterns of mortality within Europe, for example, might be partially due to differences in national cuisines.

Changes in population density and endemicity of infectious diseases might be responsible for some changes in the age patterns of mortality, that is, the second prediction suggested by Fenner's (1970) theory of disease and society. In particular, infant mortality appears to have increased and childhood mortality decreased among small populations when comparing the archeological data with the contemporary hunter-gatherer data and the contemporary horticultural data, as noted earlier. A similar trend occurs with the historical decline in mortality and across contemporary national populations when ranked by level of mortality (Gage 1993; Gage 1994). This trend could be the result of contracting these diseases at earlier ages, as population density increases and more diseases become endemic. If this view is correct, then model life tables based on historical European mortality patterns may not be appropriate for use in low-density prehistoric hunter-gatherer populations.

MIGRATION AND THE SPATIAL DISTRIBUTION OF HUMAN POPULATIONS

Compared with mortality and fertility, the patterns of migration that influence the geographic distribution of populations have not been studied as demographic processes by many human biologists. In this section, I briefly review

- the characteristic spatial distributions of human populations past and present,
- the definition of migration and its measurement,
- the variation in levels and age patterns of migration, and
- the determinants of migration.

It is an attempt to bring together the literature on aspects of migration as a demographic process that might be of interest to human biologists.

Spatial Distribution of Human Populations

People are not uniformly distributed across the landscape. Like other social animals, we live in groups. Among human populations, the detail of the spatial organization is closely related to the mode of subsistence or economic production and to the distribution of resources within the environment. A simple classification scheme presented here follows the same general categories presented in Table 14.5.

Hunter-gatherers are thought to have a bipartite social structure that consists of **ex-ogamous** bands of 25 or so members loosely affiliated with other bands who speak the same basic dialect and number perhaps 500 persons (Table 14.5). In general, the bands are nomadic and hence move around, following their subsistence resources. Bands may fission or fuse depending on the spatial distribution of subsistence resources at any given time. When subsistence resources are abundant and/or concentrated in one area, bands will come together around these resources. If subsistence resources are low and/or evenly distributed across the landscape, then the population may break up into smaller bands. Furthermore, if a band increases in size due to population growth to the point where dissension occurs among the members or the band can no longer efficiently obtain subsistence resources, then the band may break apart. However, a band whose population has been depleted for some reason may join with another. The result is a spatial distribution of the population that is exquisitely tuned to the current availability of subsistence resources (see Lee and Devore 1968).

Horticulturalists tend to live in small nucleated villages of perhaps a few hundred people (Table 14.5). These populations typically practice a nonintensive form of agriculture, such as slash and burn (i.e., the forest is cleared, often burned over, and cultivated for several seasons before clearing new fields and allowing the old fields to return to forest). As with hunter-gatherers, the villages are loosely affiliated with larger populations (other villages) who speak the same basic dialect. In many cases, "clan" memberships cut across the entire population, that is, a tribal level of social organization. Horticultural villages may move from one location to another to be as close as possible to the fields currently in use. The rate of movement, however, is slower than among hunter-gatherers; horticultural villages remain in the same locations for years. As with hunting bands, horticultural villages fission or fuse as a result of internal growth or decline in numbers (Steward 1955; Service 1962; Vayda 1969; Chagnon 1974).

The development of intensive agricultural and the industrial revolution are associated with the rise of cities and state levels of social organization (Table 14.5). Intensive agriculture makes possible the production of a surplus, so that portions of the population are engaged in nonagricultural modes of production. These people tend to collect in cities. Thus, the distribution of the population among intensive agriculturalists is to superimpose cities and other "urban" areas on the "rural" agricultural villages. The development of "state" levels of social organization may have reduced warfare among villages that make up the state.

The result often is a three-tiered population distribution. Individual households may be located on their land holdings, several households may occur together in small nucleated villages, and the remaining population lives in a few larger (urban) cities. This spatial distribution began with the advent of intensive agriculture but intensified during the industrial revolution (Steward 1955).

Migration

As with mortality, demographers study migration in two ways: the level of migration and the age patterns of migration (Rogers and Castro 1986). The level of migration is particularly difficult to estimate because migration is difficult to define, particularly across cultures (see Box 14.1).In fact, even the Migration and Settlement Study, conducted by the International Institute for Applied Systems Analysis (a cooperative effort to study migration in several European countries), was unable to produce data sufficiently consistent across countries to conduct meaningful comparisons (Rogers and Castro 1986). For our purposes, migration is defined as a permanent or semipermanent change in location to a completely new geographical region. The intention is to separate "mobility" (or circulation or partial displacement migration) due to nomadism (i.e., the yearly movements of hunter-gatherers and pastoralists within their territories), tourism, commuting (in modern societies), attending professional meetings, or changing residences within a metropolitan area from true migration or total displacement migration (Clark 1986; Cavalli-Sforza 1962). Mobility means that the individual continues many of his or her activities within the old geographic region, whereas the true migrant transfers most of his or her activities to a new geographical region. In this regard, migration implies a new hunting territory, a new farm, or a new city. The movement of peoples within an old hunting territory, the movement of slash-and-burn agricultural fields around the same village, and the relocation of a family within an urban area are classified as mobility, not migration.

There is no standard demographic measure of the level of migration. Many studies simply resort to estimating the raw number of people that migrate, or the crude outmigration rate (Davis 1974). However, migration is highly age-dependent (see later), so some control for the age structure of the population is desirable. The measure used by the Migration and Settlement Study was the sum of the individual age-specific migration rates (Rogers and Castro 1986). This calculation has been called the "gross migraproduction rate" due to its similarity to the **gross reproduction rate**.

Like mortality, the age pattern of migration—at least in contemporary national populations—has a characteristic shape (see Figure 14.8). The risk of migration tends to be high in infants and children, and in young adults 20–30 years of age. The high migration rate in children is thought to be an artifact of families moving, because families with parents in their mid-20s are likely to have young children. In contemporary national populations, high migration rates also occur at 60–70 years. This peak presumably represents retirees moving to new locations (Clark 1986; Rogers and Castro 1986).

Migration rates do not appear to vary between the sexes. In Sweden, for example, the rate of migration among females is only slightly higher than that of males (see Figure 14.8). These principles appear to apply to the United States, are probably true for Western Europe, and may apply to the developing countries as well. Whether these principles hold for humans in general, and for small populations, in particular is not known.

Variability in Level of Migration

Difficulties in defining migration consistently across populations make it impossible to conduct meaningful cross-cultural comparisons of the variability in level of migration. Nevertheless, there are considered to be some general trends in migration

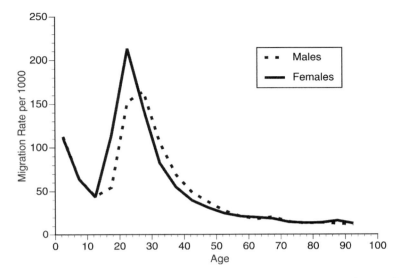

Figure 14.8 Typical age-specific migration rates by sex. These represent the population of Sweden in 1966. (*Source*: Data from Shryock and Siegel, 1975, 666)

patterns with development. In this section, I consider the trends in internal migration, invasion (international) migration, and forced migration.

Characteristic changes in the level of internal migration accompany the process of industrial development (Figure 14.9) (Zelinsky 1971). This mobility transition accompanies the historical decline in mortality (discussed earlier) and the decline in fertility (the **demographic transition**; see Chapter 15) as well as the epidemiological transition (discussed earlier). With industrial development, rural-to-urban migration increases during the early developing phases, contributing to the growth of cities, followed by a decline in the rural-to-urban flow in the later phases (i.e., in developed nations). Urban-to-urban migration also increases with development, as more of the population is located in cities, and migration between cities predominates. Furthermore, circulation for business and vacation increases, apparently due to a decline in the relative costs of transportation. Finally, Zelinsky (1971) predicted that circulation may decline in the future, at least for business purposes, as result of improved communication.

The developed nations are considered to be in the latest phases of this transition. Although not a part of Zelinsky's (1971) original description, evidence now shows a reversal of the rural-to-urban trend, that is, an urban-to-rural flow that began in the developed countries in the early 1960s. Urban-to-rural migration appears to have exceeded rural-to-urban migration in the early 1970s (Oosterbaan 1980; Jones 1990). This trend is probably due to improvements in transportation and communication (e.g., the Internet) that reduce the advantages of cities as major economic hubs.

Zelinsky (1971) assumed that migration in small populations and in prehistoric times was generally low. This is not the case. Circulation migration is high among nomadic hunter-gatherer populations and declined prehistorically only with the development of horticulture and particularly intensive agriculture (i.e., the situation as Zelinsky saw it at the advent of the industrial revolution) (Lee and Devore 1968; Fenner 1970). Of course, pastoral populations (who are generally associated with agriculturalists) have maintained high levels of circulation, as have those hunter-gatherers that survived the expansion of agricultural. Additionally, rural-to-rural migration has always occurred, at least with respect to finding mates. Furthermore, there must have been a prehistoric "mobility transition" with the introduction of intensive agricultural and the original emergence of cities (Table 14.5). It has been argued that the growth of these early cities depended on migration from rural areas because early urban populations did not have low enough mortality and high enough fertility to replace themselves (Davis 1974).

Prehistory is replete with migratory expansions of human populations (Table 14.7) (Cavalli-Sforza et al. 1993). These expansions have been called invasion migration and **demic expansion** and might include modern international migration. The demic expansion of European populations into the rest of the world observed historically is only the latest of these phenomenon. Clearly, the data in Table 14.7 indicate that humans—even prehistoric humans—are highly migratory. Of course,

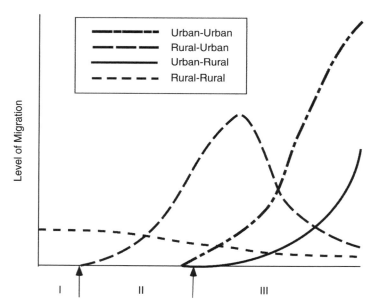

Figure 14.9   A graphical depiction of the likely relative changes in level of migration by type of migration. Phase I represents hunter-gatherers as well as horticulturalists, phase II represents intensive agriculture and the emergence of cities, and phase III represents trends in migration during industrial development. (*Source*: Adapted from Zelinsky 1971)

Table 14.7 Some invasion migration events

Origin of migrants	Destination of migrants	Period (years before present × 1000)	Contested or uncontested	Advantage
Africa (*Homo erectus*)	Old World	>1,000	Uncontested	NA
Africa (*Homo sapiens sapiens*)	Old and New World	100–30	Contested (Old World) Uncontested (New World, and Austriala)	Advances in tool production
Middle East	Europe, North Africa, and Southwest Asia	10–5	Contested (by hunter-gatherers)	Horticulture and intensive agriculture
Central America and Northern Andes	New World	9–2	Contested (by hunter-gatherers)	Horticulture and intensive agriculture
West Africa, including Bantu expansion	Sub-Saharan Africa	4–0.3	Contested (by hunter-gatherers)	Horticulture
Eurasian Steppes	Eurasia	5–0.3	Contested (by agri-culturalists)	Pastoral nomadism (horses and warfare)
Europe	Much of the world	0.4–0.04	Contested (by agri-culturalists and hunter-gatherers)	Advances in transportation and, later, industrial development (the demo-graphic transition)
Developing countries	Developed countries	0.04–present	Contested (by industrialists)	Industrial development (the demo-graphic transition)

NA, not applicable.
(*Source*: Adapted from Cavalli-Sforza et al. 1993)

prehistoric migrations are underreported, because the evidence for these expansions is mainly inferred from the distribution of genes. Any migrations that did not influence the contemporary distribution of genes have been overlooked (see Chapter 4).

Invasion migration takes two forms: uncontested migration, when a population moves into a region previously unoccupied by human populations, or contested migration, when a population moves into a region occupied by an other population. Invasion migration is generally thought to be the result of social, biological, or technical innovations that increase the rate of population growth in a particular population. The resulting population pressure is relieved by outmigration. The rate of a population's geographical expansion from a central locus is simply a result of the population growth rate and the rate of migration (Davis 1974; Cavalli-Sforza et al. 1993). There are no direct ethnographic analogies for this process among hunter-gatherers, because most contemporary hunter-gatherer populations are not growing rapidly. In fact, they have been pushed into marginal environmental regions by the more recent expansion of agriculturalists. However, expansions of hunter-gatherer populations out of Africa (first as *Homo erectus* and later as *Homo sapiens sapiens*) to the rest of the world must have occurred prehistorically (Table 14.7). It is assumed bands continually grow in size and then split apart, increasing the number of bands with simultaneous displacement and expansion of hunting territories of the band that split and the bands around them. New linguistic "tribes" might form as expansion continued and exchange of mates and communication between widely separated bands declined.

Horticultural populations, have been observed during periods of expansion (Vayda 1969; Chagnon 1974; Early and Peters 1990). As for hunter-gatherers, the basic model is one of local population growth followed by the breaking apart of villages (in the case of horticulturalists) and the concomitant displacement of the surrounding local groups. Vayda (1969) proposed two models on the basis of ethnographic observations. The first is characterized by intratribal warfare, that is, warfare among the villages, including and perhaps particularly among recently split villages (Chagnon 1974). In this case, hostility among the villages provides the motivating force for the dispersal of the population. Examples include the Tiv, the Tupinamba, the Maori (Vayda 1969), and the Yanomama (Chagnon 1974; Early and Peters 1990). The second model is characterized by intratribal peace, relatively higher rates of mobility of peoples among villages, and the availability of "unclaimed" cultivable land. If the expansion is contested, then the availability of unclaimed cultivable land is maintained by extratribal warfare and conquest. The Iban of Sarawak have been proposed as an example of this type of expansion. In this case, new villages in virgin territory were often colonized by the younger families. The initial colonization may have been carried out by younger males in advance of the "main body of settlers" (Vayda 1969).

Invasions carried out by societies at the "state" level of socioeconomic integration probably resemble the Iban model, that is, relative internal peace (maintained in this case by the state) and warfare against other groups. Some invasions probably took place as explicit acts of conquest. How much migration actually took place when the Mongols under Genghis Kan overran much of Asia and Eastern Europe is not known (Davis 1974); perhaps very little. Other invasions may have largely replaced the original populations, such as the European colonization of North America (Weiss 1988). The European invasion of North America and Australia is not dissimilar to the process of Iban expansion described above except that "state" governments may actively promote the process of expansion. It is interesting to point out that the European invasion of underdeveloped areas of the rest of the world during the historic period has recently been replaced by a counter-flow of migrants from the developing countries to the developed countries (e.g., the migration of Hispanic populations into the United States) (Table 14.7).

One question with respect to contested migration is why the invading population replaces the indigenous population in some cases and in other cases does not (Weiss 1988). Modern analogies, such as the European invasion of the New World, are varied. In North America and in Costa Rica, for example, the indigenous populations were largely replaced, whereas in Mexico and much of South America, the indigenous populations remained an important segment of society. Three possible explanations of these differences are the size and density of the indigenous population, the relative differences in technology, and/or the relative differences in the level of sociopolitical integration. Clearly, where indigenous populations were low, replacement occurred. However, the relative differences in technology and level of sociocultural integration also were large in these locations.

We generally think of migration as voluntary and based on the decisions of individuals or families. However, migration may be forced—that is, not voluntary from the point of view of the migrant. This category would include the forced transport of Africans to the New World, the slave trade within Africa and Europe, deportation of convicts to Australia, and even the capture of wives by Yanomama (Chagnon 1974; Early and Peters 1990). The movement of refugees during periods of war and civil strife is also considered forced migration. This form of migration appears to be increasing with development. Davis (1974) estimated that over the 55-year period from 1913 to 1968, 71.1 million individuals were forced to migrate, whereas over the 90-year period 1840 to 1930 (the height of the European expansion), an estimated 52 million left Europe voluntarily. More recently, the breakup of the Soviet Union and civil strife in Africa, Asia, and Central America have contributed to the volume of people forced to migrate. Forced migration is clearly an important and perhaps increasing component of total migratory flows.

Variability in the Age Patterns of Migration

Almost no quantitative data on the age patterns of migration exist, outside of a few developed nations. However, some circumstantial evidence suggests that the age patterns of migration in small populations are qualitatively similar to those plotted in Figure 14.8. With respect to contemporary hunter-gatherers, Fix (1977) reported the highest rates of migration during "childbearing ages" among the Semai Senoi. Similar findings have been reported for horticulturalists such as the Yanomama (Early and Peters 1990) and the Iban (Vayda 1969).

It appears likely that most migration occurs just before or shortly after the beginning of the reproductive career. However, adjustment might need to be made for lower ages of marriage that characterize many small populations compared with contemporary industrialized societies. However, forced migration and the pattern of village fissioning along family lines in hunter-gatherer and horticultural populations may not be well represented by Figure 14.8. There simply are no good quantitative data on the age patterns of migration in these situations. Additional research is necessary to determine the extent of this variation.

Proximate Causes of Migration

The proximate causes of migration are sometimes divided into causes that push people out of their current area of residence and causes that pull people to new locations. In voluntary migration, the proximate cause of migration clearly is the difference between the conditions in the original location and those at the potential new location. Proximate causes may stem from a new marriage, economic reasons, or be a form of forced migration.

Marital migration occurs in all populations, as individuals move in search of mates or couples marry and set up new households in a new region. In hunter-gatherers, bands are typically exogamous and exchange mates among bands within the same linguistic "tribe" (Service 1962; Lee and Devore 1968). Horticulturalists may or may not practice village exogamy but tend to practice "clan" exogamy. Again, mates are exchanged within subunits of the linguistic "tribe"(Service 1962). Among agriculturalists, marital migration occurs among households (or farms) within the agricultural proportion of the population, between the agricultural and urban portions of the population, or within the urban portions of the population. In any event, because this process is conceived of as an *exchange of mates*, the results do not greatly change the demographic structure of the populations involved. However, this kind of migration may have profound effects on the genetic structure of the population (see Chapter 4).

In general, people migrate voluntarily because they believe that moving will enhance their economic well-being, that is, when they believe that the benefits accrued in the future exceed the costs of migration (Sjaastad 1962). In developed countries, individuals and/or families clearly benefit from migrating (DaVanzo 1976; DaVanzo 1978; Grant and Vanderkamp, 1980). Whether the same is true of migrants in the developing countries, particularly migrants to developing cities, is less clear. Todaro (1976) has argued that migrants in these areas may not be better off economically because of the high rates of unemployment and/or underemployment that characterize cities in developing countries. Todaro argues that migration may be driven by the *perception* of being better off. Unfortunately, sufficient economic data are not available to formally test these models on a large scale in the developing countries. However, studies of the biological indicators of well-being, such as faster growth and maturation and lower mortality and fertility, suggest that long-term migrants from rural to urban areas of the developing countries are generally better off than nonmigrants (Bogin 1988) (see Chapter 12). Similar economic and/or ecological considerations are thought to have driven the invasion migrations (Cavalli-Sforza et al. 1993; Davis 1974).

The reasons for forced migration are more idiosyncratic. Some forced migration may be economic. Certainly, the slave trade was based on the need (by the enslavers) for a labor force (Davis 1974). Other forced migration may not be as much economic as political. It includes the growing ranks of refugees resulting from international and civil strife. As Davis (1974) pointed out, much of this forced, political, migration stems ironically from the principle of "national self-determination" originally proposed by President Wilson, but carried to its extreme, "ethnic purity."

A question that has concerned human biologists is whether migrants (voluntary migrants; the question is moot for forced migrants) are truly better off. This question is difficult to answer because it requires a comparison with how well off the migrant would have been had he or she not migrated. Comparisons with nonmigrants are frequently used as a surrogate measure of the effects of migration on the individual. However, migrants are not a random sample of the individuals who choose to stay; furthermore, the simple fact that a migrant leaves changes the environment that the nonmigrant is exposed to. The evolutionary hypothesis suggests that migrants should be worse off because they migrated from an environment to which they are presumably well-adapted (genetically, developmentally, physiologically, and/or socially) to a different environment, to which they may not be well adapted (Little and Baker 1988). Blood pressure, for example, a risk factor for cardiovascular disease, is generally higher in rural-to-urban migrants than in nonmigrants (Little and Baker 1988). These phenomenon, in addition to moving and setting up a new household and finding employment (Sjaastad 1962) are the costs of migrating. As mentioned earlier, however, overall the biological and economic data currently available suggest that the benefits of moving outweigh the costs at least for voluntarily migrants (DaVanzo 1976; DaVanzo 1978; Grant and Vanderkamp. 1980; Bogin 1988; Little and Baker 1988). Still it is not clear whether all those who choose to migrate are truly better off. Comparisons of migrants and nonmigrants are based on studies of the migrants that actually succeed at migrating. Some migrants die during the migration process or fail to adapt to new surroundings and either move on or more back to where they came from. Thus, the apparent improved living conditions of migrants may be due to heterogeneity.

CHAPTER SUMMARY

In this chapter, we considered some of the problems in estimating demographic rates, the history of variation in human mortality patterns, and human migration.

One basic tool for demographic analysis is the life table, an array of mortality rates by age for a particular population. The interpretation of demographic rates is sometimes difficult, even when the data on which they are based are completely accurate. However, demographic data often are inaccurate, incomplete, and/or biased. In general, demographic data on small nonindustrial, non-Western populations tend to be of poor quality whether collected ethnographically or archaeologically.

All human populations show a similar U-shaped age pattern of mortality. Mortality declines from birth to about age 5 to age 10, remains relatively low throughout the next 20 years of life, and then begins to increase at an ever-increasing rate at least until 70 or 80 years of age. Some populations may show additional features—such as a weanling bump or accident hump—under special conditions.

Age patterns of mortality vary among national populations and may vary between national populations and small populations. For example, there is a trend of increasing infant mortality and decreasing childhood mortality from archeological data to contemporary hunter-gatherer data and to contemporary horticultural data. That is, the age pattern of mortality has become more U-shaped over time, at least at the younger ages.

The level of mortality varies widely among past and living human populations, from expectations of life at birth of less than 15 to more than 80 years. Most estimates of prehistoric life expectancy at birth vary within a surprisingly narrow range, from perhaps 18 to 25 years. The means are similar for hunter-gatherers, horticulturalists, and agriculturalists, indicating that mortality during these times was high and apparently did not vary a great deal from one subsistence type to another. In historic populations prior to the industrial revolution, expectation of life at birth was short—less than 40 years. Among contemporary nations, expectation of life ranges from about 40 to more than 80 years. Expectation of life at birth among contemporary small populations, studied ethnographically, suggests that the level of mortality is lower among contemporary small populations than among archaeologically studied small populations. The mortality of contemporary small populations appears more similar to the mortality observed in the contemporary developing nations than to the mortality of socioeconomically similar archaeological populations.

Several hypotheses have been suggested to explain changes in mortality level over time, but none are entirely satisfactory. Another apparent change in mortality over time is that the large sex differentials in favor of females characteristic of contemporary national populations are not universal and probably are a relatively recent development.

During the historic period, the major causes of death changed from predominately infectious diseases to predominately chronic degenerative diseases, a process that has been called the "epidemiological transition." The risks of both infectious and degenerative causes of death declined, and the transition from a predominance of infectious deaths to degenerative deaths appears to be due to the faster decline of the infectious than degenerative causes of death.

Several kinds of migration were considered: internal migration, invasion migration, and forced migration. Migration rates have changed over time and vary with age. Migration rates tend to be particularly high in young adults between the years of 20 and 30 years of age.

ACKNOWLEDGMENTS

I would like to thank Drs. O'Connor, Wilkinson, Relethford, and Swedlund for comments on an original manuscript and suggestions concerning the treatment of migration.

RECOMMENDED READINGS

Caldwell JC (1986) Routes to low mortality in poor countries. Popul. Dev. Rev. 12: 171–220.

Cavalli-Sforza LL, Menozzi P, and Piazza A (1993) Demic expansions and human evolution. Science 259: 639–646.

Cohen MN (1989) Health and the Rise of Civilization. New Haven, Conn.: Yale Univ. Press.

Davis K (1974) The migrations of human populations. Sci. Am. 231: 92–107.

Fenner F (1970) The effects of changing social organization on the infectious diseases of man. In SV Boyden (ed.): The Impact of Civilization on the Biology of Man. Canberra: Australian National Univ. Press, pp. 48–76.

Floud R, Wachter K, and Gregory A (1990) Height, Health, and History. Cambridge: Cambridge Univ. Press.

Gage TB (1989) The bio-mathematical approaches to the study of mortality. Yearbook Phys. Anthropol. 32: 185–214.

Gage TB, McCullough JM, Weitz CA, Dutt JS, and Abelson A (1989) Demographic studies in human population biology. In M Little and J Haas (eds.): Human Population Biology: A Transdisciplinary Science. Oxford: Oxford Univ. Press, pp. 45–68.

Hewlett BS (1991) Demography and childcare in preindustrial societies. J. Anthropol. Res. 47: 1–39.

Howell N (1979) Demography of the Dobe !Kung. New York: Academic Press.

Howell N (1982) Village composition implied by a paleodemographic life table: the Libben Site. Am. J. Phys. Anthropol. 59: 263–269.

Lee ET (1980) Statistical Methods for Survival Data Analysis. Belmont: Life Time Learning Publications.

Little MA and Baker PT (1988) Migration and adaptation. In CGN Mascie-Taylor and GW Lasker (eds.): Biological Aspects of Human Migration. Cambridge: Cambridge Univ. Press, pp. 167–215.

McKeown T (1976) The Modern Rise of Population. New York: Academic Press.

Moore JA, Swedlund AC, and Armelagos GJ (1975) The use of life tables in paleodemography. In AC Swedlund (ed.): Population Studies in Archaeology and Biological Anthropology: A Symposium. Memoirs of the Society for American Archaeology, American Antiquity 40(2):57–70, Memoir 30.

Pollard AH, Yusuf F, and Pollard GN (1974) Demographic Techniques, 2nd edition. Sydney, Australia: Pergamon Press.

Sjaastad LA (1962) The costs and returns of human migration. J. Polit. Econ. 70: 80–93.

Todaro MP (1976) Internal Migration in Developing Countries. Geneva: International Labor Office.

Vayda AP (1969) Expansion and warfare among swidden agriculturalists. In AP Vayda (ed.): Environment and Cultural Behavior. Garden City, N.Y.: The Natural History Press, pp. 202–220.

Weiss KM (1988) In search of times past: gene flow and invasion in the generation of human diversity. In CGN Mascie-Taylor and GW Lasker (eds.): Biological Aspects of Human Migration. Cambridge: Cambridge Univ. Press, pp. 167–215.

Zelinsky W (1971) The hypothesis of the mobility transition. Geogr. Rev. 61: 219–249.

REFERENCES CITED

Acsadi G and Nemeskeri J (1970) History of Human Life Span and Mortality. Budapest: Akademiai Kiado, p. 346.

Angel JL (1972) Ecology and population in the eastern Mediterranean. World Archeol. 4: 88–105.

Baker PT (1987) Modernization and human biological responses. In GA Harrison, JM Tanner, DR Pilbeam, and PT Baker (eds.): Human Biology. New York: Oxford Univ. Press, pp. 529–543.

Behn H and Vallin J (1982) Mortality differentials among human populations. In SH Preston (ed.): Biological and Social Aspects of Mortality and Length of Life. Liège, Belgium: Ordina Editions, pp. 11–38.

Bogin B (1988) Rural-to-urban migration. In CGN Mascie-Taylor and GW Lasker (eds.): Biological Aspects of Human Migration. Cambridge: Cambridge Univ. Press, pp. 90–129.

Buikstra JE, Konigsberg LW, and Bullington J (1986) Fertility and the development of agriculture in the prehistoric midwest. Am. Antiquity 51: 528–546.

Cavalli-Sforza L (1962) The distribution of migration distances: models and the applications to genetics. In J Sutter (ed.): Human Displacement. Monaco: Hachette.

Caldwell JC (1986) Routes to low mortality in poor countries. Popul. Dev. Rev. 12: 171–220.

Cavalli-Sforza LL, Menozzi P, and Piazza A (1993) Demic expansions and human evolution. Science 259: 639–646.

Chagnon NA (1974) Studying the Yanomamo. New York: Holt, Rinehart, and Winston, Inc., p. 270.

Clark WAV (1986) Human Migration. Beverly Hills, Calif.: Sage, p. 95.

Coale AJ and Kisker EE (1986) Mortality crossovers: reality or bad data? Popul. Stud. 40: 380–401.

Coale AJ and Watkins SC (1986) The Decline of Fertility in Europe. Princeton, N.J.: Princeton Univ. Press.

Coale AJ, and Demeny P, with Vaughan B (1983) Regional Model Life Tables and Stable Populations. New York, N.Y.: Academic Press.

Cohen MN (1989) Health and the Rise of Civilization. New Haven, Conn.: Yale Univ. Press.

Cohen MN and Armelagos GJ (1984) Paleo-pathology at the Origins of Agriculture. Orlando, Fla.: Academic Press.

DaVanzo J (1976) Why families move: a model of the geographic mobility of married couples (R-1972-DOL). Santa Monica, Calif.: Rand Corporation.

DaVanzo J (1978) Does unemployment affect migration? Evidence from micro data. Rev. Econ. Stat. 60: 504–514.

Davis K (1974) The migrations of human populations. Sci. Am. 231: 92–107.

Dyke B, Gage TB, Ballou JD, Petto AJ, Tardif SD, and Williams LJ (1993) Model life tables for the smaller New World monkeys. Am. J. Primatol. 37: 25–37, 1995.

Early JD and Peters JF (1990) The Population Dynamics of the Mucajai Yanomama. San Diego: Academic Press.

Economos AC (1982) Rate of aging, rate of dying, and the mechanisms of mortality. Arch. Gerontol. Geriatr. 1: 3–27.

Elo IT and Preston SH (1994) Estimating African-American mortality from inaccurate data. Demography 31: 427–458.

Fenner F (1970) The effects of changing social organization on the infectious diseases of man. In SV Boyden (ed.): The Impact of Civilization on the Biology of Man. Canberra: Australian National Univ. Press, pp. 48–76.

Fix AG (1977) The Demography of the Semai Senoi. Ann Arbor: Museum of Anthropology, Univ. of Michigan.

Floud R, Wachter K, and Gregory A (1990) Height, Health, and History. Cambridge: Cambridge Univ. Press.

Frisancho AR (1993) Human Adaptation: A Functional Interpretation. Ann Arbor: Univ. of Michigan Press.

Gage TB (1985) Demographic estimation from anthropological data: a review of the non-stable methods. Curr. Anthropol. 26: 644–647.

Gage TB (1988) Mathematical hazard models of mortality: an alternative to model life tables. Am. J. Phys. Anthropol. 76: 429–441.

Gage TB (1989) The bio-mathematical approaches to the study of mortality. Yearbook Phys. Anthropol. 32: 185–214.

Gage TB (1990) Variation and classification of human age patterns of mortality: analysis using competing hazards models. Hum. Biol. 62: 589–617.

Gage TB (1993) The decline of mortality in England and Wales 1861 to 1964: an analysis using competing hazards models. Popul. Stud. 47: 47–66.

Gage TB (1994) Population variation in cause of death: level, gender, and period effects. Demography 31: 271–296.

Gage TB (1997) Demography. In F. Spencer History of Physical Anthropology, an Encyclopedia. New York: Garland Press, Vol. 1, pp. 323–330.

Gage TB and Dyke B (1988) Model life tables for the larger old world monkeys. Am. J. Primatol. 16: 305–320.

Gage TB and Mode CJ (1993) Some laws of mortality: how well do they fit? Hum. Biol. 65: 445–461.

Gage TB and O'Connor K (1994) Nutrition and variation in level and age pattern of mortality. Hum. Biol. 66: 77–103.

Gage TB, Dyke B, and Riviere PG (1984a) Estimating fertility and population dynamics from two censuses: an application to the Trio of Surinam. Hum. Biol. 56: 691–701.

Gage TB, Dyke B, and Riviere PG (1984b) Estimating mortality from two censuses: an application to the Trio of Surinam. Hum. Biol. 56: 489– 502.

Gage TB, McCullough JM, Weitz CA, Dutt JS, and Abelson A (1989) Demographic studies in human population biology. In M Little and J Haas (eds.): Human Population Biology: A Transdisciplinary Science. Oxford: Oxford Univ. Press, pp. 45–68.

Grant KE and Vanderkamp J (1980) The effects of migration on income: a macro study with Canadian data. Can. J. Econ. 13: 381–406.

Greenberg MR (1983) Urbanization and Cancer Mortality: The United States Experience, 1950–1975. New York: Oxford Univ. Press.

Haub C (1987) Understanding population projections. Popul. Bull. 42: 1–43.

Heligman L and Pollard JH (1980) The age pattern of mortality. J. Inst. Actuaries 107: 49–80.

Hewlett BS (1991) Demography and child care in preindustrial societies. J. Anthropol. Res. 47: 1–37.

Howell N (1976) Toward a uniformitarian theory of human paleo-demography. In RH Ward and KM Weiss (eds.): The Demographic Evolution of Human Populations. London: Academic Press, pp. 25–40.

Howell N (1979) Demography of the Dobe !Kung. New York: Academic Press.

Howell N (1982) Village composition implied by a paleodemographic life table: the Libben Site. Am. J. Phys. Anthropol. 59: 263–269.

Jones H (1990) Population Geography. London: Paul Chapmen.

Keyfitz N and Flieger W (1990) World Population Growth and Aging. Chicago, Ill.: The Univ. of Chicago Press, p. 608.

Lee ET (1980) Statistical Methods for Survival Data Analysis. Belmont: Life Time Learning Publications, p. 557.

Lee RB and Devore I (1968) Man the Hunter. Chicago, Ill.: Aldine Atherton, p. 415.

Little MA and Baker PT (1988) Migration and adaptation. In CGN Mascie-Taylor and GW Lasker (eds.): Biological Aspects of Human Migration. Cambridge: Cambridge Univ. Press, pp. 167–215.

Lopez AD and Ruzicka LT (1983) Sex Differentials in Mortality. Canberra: Australian National Univ. Press, p. 498.

McKeown T. (1983) A basis for health strategies: a classification of disease. British Medical Journal. 287: 594–596.

McKeown T (1976) The Modern Rise of Population. New York: Academic Press.

Moore JA, Swedlund AC, and Armelagos GJ (1975) The use of life tables in paleodemography. In AC Swedlund (ed.): Population Studies in Archaeology and Biological Anthropology: A Symposium. Memoirs of the Society for American Archaeology, American Antiquity 40(2):57–70.

Neel J. V. and Weiss, K. (1975) The genetic structure of a tribal population, the Yanomama indians. Am J. Phys. Anthrop. 42:25–52.

O'Connor KA (1995) The age pattern of mortality: a micro-analysis of Tipu and a meta-analysis of twenty-nine paleodemographic samples. Dissertation, University at Albany, State Univ. of New York.

Omran AR (1977) Epidemiological transition in the U.S. Popul. Bull. 32: 1–41.

Oosterbaan J (1980) Population Dispersal. Lexington, Mass.: D.C. Heath and Company.

Perls TT (1995) The oldest old. Sci. Am. January 272(1): 70–75.

Pollard AH, Yusuf F, and Pollard GN (1974) Demographic Techniques, 2nd edition. Sydney, Australia: Pergamon Press.

Preston SH (1976) Mortality Patterns in National Populations. New York: Academic Press.

Preston SH and Coale AJ (1982) Age structure, growth, attrition, and accession: a new synthesis. Population Index 48: 217–259.

Preston SH and Haines MR (1991) Fatal Years: Child Mortality in Late Nineteenth-Century America. Princeton, N.J.: Princeton Univ. Press.

Preston SH, Keyfitz N, and Schoen R (1972) Causes of Death Life Tables for National Populations. New York, N.Y.: Seminar Press.

Rogers A and Castro LJ (1986) Migration and Settlement. In A Rogers and FJ Willekens (eds.): Migration and Settlement. Dordrecht, the Netherlands: D. Reidel Publishing Company, pp. 157–210.

Rose J (1982) Preface. In J Rose (ed.): Nutrition and Killer Diseases: The Effects of Dietary Factors on Fatal Chronic Diseases. Park Ridge, N.J.: Noyes Publications.

Rosetta L and O'Quigley J (1990) Mortality among Serere children in Senegal. Am. J. Hum. Biol. 2: 719–726.

Sattenspiel L and Harpending H (1983) Stable populations and skeletal age. Am. Antiquity 48: 489–498.

Service ER (1962) Primitive Social Organization. New York: Random House.

Shryock HS and Siegel JS (1975) The Methods and Materials of Demography. Washington, D.C.: U.S. Government Printing Office.

Sjaastad LA (1962) The costs and returns of human migration. J. Polit. Econ. 70: 80–93.

Steward JH (1955) Theory of Culture Change. Urbana: Univ. of Illinois Press.

Todaro MP (1976) Internal migration in developing countries. Geneva: International Labor Office.

Trowell HC and Burkitt DP (1981) Western Diseases: Their Emergence and Prevention. London: Edward Arnold.

United Nations (1982) Model Life Tables for Developing Countries. New York: United Nations.

U.S. Bureau of the Census (1988) Statistical Abstract of the United States. Washington, D.C.: U.S. Government Printing Office.

Vaupel JW, Manton KG, and Stallard E (1979) The impact of heterogeneity in individual frailty on the dynamics of mortality. Demography 16: 439–454.

Vayda AP (1969) Expansion and warfare among swidden agriculturalists. In AP Vayda (ed.): Environment and Cultural Behavior. Garden City, N.Y.: The Natural History Press, pp. 202–220.

Waldron I (1983) The role of genetic and biological factors in sex differences in mortality. In AD Lopez and LT Ruzicka (eds.): Sex Differentials in Mortality: Trends, Determinants, and Consequences. Canberra: Australian National Univ. Press, pp. 53–120.

Weiss KM (1973) Demographic models for anthropology. Memoirs of the Society for American Archaeology, American Antiquity 38(2)Part 2: 1-186. Memoir 27.

Weiss KM (1975) The application of demographic models to anthropological data. Hum. Ecol. 3: 87–103.

Weiss KM (1988) In search of times past: gene flow and invasion in the generation of human diversity. In CGN Mascie-Taylor and GW Lasker (eds.): Biological Aspects of Human Migration. Cambridge: Cambridge Univ. Press, pp. 167–215.

Weiss KM (1989) A survey of human biodemography. J. Quant. Anthropol. 1: 79–151.

Wood JW, Milner GR, Harpending HC, and Weiss KM (1992) The osteological paradox. Curr. Anthropol. 33: 343–370.

World Factbook (1994) The World Factbook. Washington D.C.: Central Intelligence Agency.

Zelinsky W (1971) The hypothesis of the mobility transition. Geogr. Rev. 61: 219–249.

CHAPTER 15

Population Growth and Fertility Regulation

PETER T. ELLISON AND MARY T. O'ROURKE

INTRODUCTION

The history of human population growth is closely interwoven with the history of human ecology. Ever since Malthus directed attention toward the general tendency of human populations to grow exponentially, there has been considerable interest in the mechanisms that regulate such growth. After hundreds of thousands of years of barely perceptible growth, the human population has expanded at an accelerating rate over the past thousand years and at an explosive rate over the past century. Understanding the mechanisms of human growth regulation thus seems crucial to our future as well as important to the fuller understanding of our past.

Human population growth results from the net effect of two dynamic processes: birth and death. (A third process, **migration**, can be important to the dynamics of local populations but is largely ignored in this chapter.) Both of these processes are modulated by cultural and biological factors. In this chapter, we consider the way that birth and death rates interact to produce population growth and the way that **fertility**, in particular, is regulated. Particular attention is devoted to the recent changes in human fertility patterns that are best documented and that have shaped the recent surge in human population numbers. The dramatic changes in **mortality** and fertility rates that have been occurring globally over the past century provide substantial insights into the interaction of biological and social factors in regulating population growth. They also provide an empirical basis for extrapolation into the past and the future.

Human biologists and biological anthropologists bring a particular set of perspectives to bear on these issues. They can use an understanding of the fundamental processes that shape human population growth as a window into the past to illuminate some of the processes that underlie human evolution. They also appreciate the varia-

Human Biology: An Evolutionary and Biocultural Perspective, Edited by Sara Stinson, Barry Bogin, Rebecca Huss-Ashmore, and Dennis O'Rourke
ISBN 0-471-13746-4   Copyright © 2000 Wiley-Liss, Inc.

tion that exists among people today, not only in genetic and cultural endowments but also in the range of ecological conditions in which humans survive and perhaps prosper. Although clinical studies have contributed greatly to our understanding of human biology, clinicians by definition deal with pathology or deviation from a standard definition of "normal function." The focus of human biologists, by contrast, is on human variation itself and its potential functional significance. This chapter closes with an effort to generalize from these perspectives in considering present conditions and future prospects.

POPULATION GROWTH

Exponential Growth

Ecologists are fond of telling a story about a human settlement on one side of a pond. A small floating duckweed establishes itself on the far side of the pond, where it divides in two every 24 hours. It takes years for the people in the settlement even to notice the duckweed colony, and then only when they paddle their canoes to the far end of the lake. Some years later, they realize that the duckweed mat is growing, but it still covers only a few percent of the pond surface. They decide not to worry about it until the mat covers half the pond and starts to encroach on their side. When that day comes, they establish a committee to deal with the situation. How long does the committee have to carry out its mission before the pond is completely covered in duckweed?

Populations of organisms, including humans, tend to grow like the duckweed mat—not doubling in size every day, necessarily, but doubling over some constant unit of time. If we disregard the effects of migration for the moment (as would be appropriate if we were considering the global human population, for example), the growth rate of a population (r) is equal to the birth rate (b) minus the death rate (d). If more people are born than die per unit time, then r will be positive and the population will grow. If more people die than are born, then r will be negative and the population will decline. A stationary population—neither growing nor declining—is a very special case in which births and deaths just balance each other, and r is equal to zero.

Birth rates and death rates, however, are functions not only of time but also of the starting number of people. More people will have more babies in a given stretch of time, even if each individual is having babies at the same rate. More people will die in a larger population as well. Thus, birth rates and death rates are expressed per unit of time, per unit of population. The birth rate in Ghana in 1994, for example, was 42 per 1000 people per year, while the death rate was 12 per 1000 people per year. If population growth rates are simply the difference between birth and death rates, then they must have the same units. The birth and death rates for Ghana, for example, imply a population growth rate of 30 per 1000 people per year. All three rates are often expressed in decimal annual rates (0.042, 0.012, and 0.030, respectively) or as percentages (4.2%, 1.2%, and 3.0%, respectively).

This compounding of population with time, similar to the compounding of interest on a bank account, is referred to as **exponential growth** (Figure 15.1). Just as money placed in the bank at an annual interest rate i will double in $0.693/i$ years[1], so a population growing at an annual exponential rate r will double in $0.693/r$ years. Conversely, a population doubling in x years implies a growth rate of $0.693/x$ averaged over that time. The population of the world at the time of Caesar Augustus, in 0 AD, is estimated to have been about 250 million. It is thought to have reached 500 million at about 1650, implying an average annual growth rate of 0.0004. It doubled again by 1900, implying a growth rate of 0.0028, again by 1950, implying a growth rate of 0.014, and again by 1985, implying a growth rate of nearly 0.02. In comparison, if its 1995 growth rate were to continue, the population of Ghana would double from 17.5 million to 35 million in a little more than 23 years.

Malthus (1798, 1830) called the tendency of human populations to grow exponentially the "principle of population." He also recognized, however, that exponential growth can never persist in the long term. Continuing to double every 30 years would spread the human population over the entire surface of the globe, dry land and sea bed, to the density of Manhattan before we are halfway through the next millennium. Human population growth must eventually be checked, whether voluntarily or involuntarily. Because we cannot yet leave the planet, that regulation must be effected either through increased mortality or through decreased fertility. Malthus proposed two classes of checks on population. Positive checks (e.g., war, pestilence, and famine) are those that increase mortality, whereas preventive checks—essentially sexual restraint, achieved mainly through delaying age of marriage—decrease fertility. Malthus suggested that we had better learn how to regulate our fertility voluntarily, or nature would regulate our mortality for us, an argument that many continue to find compelling.

The history of the human population can be told in terms of these two great engines of population growth and regulation, mortality and fertility. Changes in mortality have had the most dramatic impact on historical populations: The Black Death (bubonic plague) reduced the population of Western Europe by more than a quarter in the 14th century; smallpox and other **communicable diseases** that were new to the Americas may have reduced the native population by 90% or more in the century after Columbus; recent changes in ecology, medicine, and public health in many populations have reduced infant mortality rates by 95% and extended life expectancy by more than 30 years in the 20th century. Other chapters in this book consider the processes of disease and aging that largely determine human mortality patterns, so they will not be dealt with in detail here (see Chapters 7, 8, 13, and 14).

Variation in fertility, between different populations and within populations over time, is also a potent force in the dynamics of human populations, though less dramatic in its short-term effects than mortality. The biological limits on the rate at which women can bear children make catastrophic swings in fertility unlikely. But, as

---

[1]Under continuous compounding, a bank account grows by a factor of e^{it} in t years, where e is the base of the natural logarithms and i is the annual interest rate. Setting this factor equal to 2 and solving for t allows us to determine how quickly money invested at interest rate i will double, a formula that many investors carry in their heads.

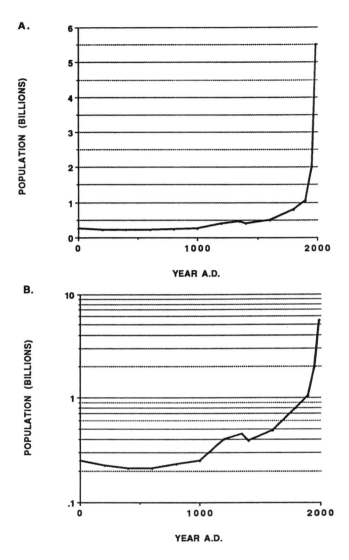

Figure 15.1 World population growth over the past 2000 years on absolute (A) and logarithmic (B) scales. Reductions in world population following the decline of the Roman Empire and the spread of bubonic plague in Europe are the only notable interruptions in an otherwise accelerating trajectory of population growth.

Malthus pointed out, fertility provides the positive force behind population growth. Mortality can provide only a negative constraint on such growth. Malthus assumed that human fertility was not regulated physiologically but could only be restrained behaviorally. We now know that natural variation in fertility can be substantial.

Carrying Capacity and Logistic Growth

Inherent in Malthus' notion that an unrestrained population would eventually outgrow its food supply is a concept that we now call the *carrying capacity of the environment*. As a concept, carrying capacity implies a feedback between the size of a population in a given environment and its birth and death rates (Wilson and Bossert 1971). Malthus assumed that the only feedback occurred by way of mortality rates and that it would tend to occur precipitously in some catastrophe such as a plague or famine.

We can now imagine a more continuous sort of feedback operating through both mortality and fertility. For example, as a population grows in a certain environment, its ability to extract food from that environment may not keep pace with the number of mouths to feed. Lack of food may then lead to increased infant mortality and decreased resistance to disease. At the same time, increasing nutritional stress might prolong the interval between births for women, as we explain later. Consequently, as the population grows to the point of diminishing food returns, mortality rates would increase and fertility rates decrease, thus slowing population growth. If mortality rates continue to increase with increasing population size and fertility rates continue to decrease, then the population growth rate will eventually fall to zero (or even below zero, if the population grows too large). The population size at which the growth rate equals zero is referred to by ecologists as the carrying capacity of the environment as if it were a discrete, measurable quantity, although its exact measurement for any population and environment is difficult at best. Note that although the notion of carrying capacity as a feedback relationship between a population and its resources is implicit in Malthus' thinking, he did not imagine the feedback as smooth, nor did he imagine that fertility would naturally decrease.

Feedback between population size and birth and death rates can be formalized into a model of population growth known as **logistic growth**. In many simple biological systems—yeast cells in a culture medium, for example—logistic growth provides a good fit to observed patterns of population growth. For a while in the early part of this century, logistic growth enjoyed a certain vogue among demographers, and confident predictions of a stable US population below 200 million were made (Cohen 1995; Figure 15.2). The problem is that logistic growth models assume that the relationship between population size and birth and death rates is fixed for a given environment, an assumption that would be true only if there never were any technological change altering the ability of the population to extract food from the environment, avoid disease, or otherwise alter the feedback relationship. Such technological innovation seems to be a hallmark of the human species, so the assumptions of the logistic growth model may never be met, or never for very long (Spooner 1972; Boserup 1981; Hassan 1981). Human populations seem to be constantly pushing the carrying capacities of their environments to higher levels. This process itself has been the object of intense speculation by anthropologists, economists, and demographers interested in whether population growth precipitates technological innovation or whether technological innovation precipitates population growth (Boserup 1965, 1981; Cohen 1977).

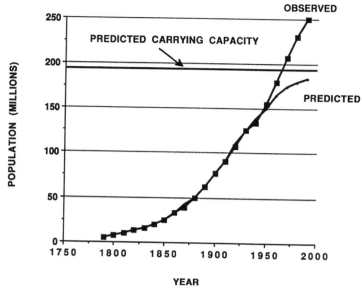

Figure 15.2   Observed population growth in the United States since 1790 compared with the logistic trajectory predicted by Pearl and Reed (1920). They projected a stable population of approximately 190 million, a figure that was surpassed in 1965 with little evidence of any deceleration in population growth.

The links among population, environment, and subsistence are central to human ecology, demography, and anthropology. These links feature prominently in most accounts of human cultural and population history. But the links can be imagined in different ways. Two schools of thought in particular have adopted contrasting interpretations of these links in understanding the way population growth and technological innovation are related. The transition to agriculture is a prime example.

An important transformation of human population dynamics appears to have occurred between 12,000 and 8000 years ago. In a few different places around the world, a new mode of human subsistence emerged based on cultivation of crops, domestication of animals, and settled habitation of towns. At the same time, local populations began to grow at increasing rates. Subsequently, the new mode of subsistence and the higher population growth rates that went with it spread beyond the original core areas to adjacent and outlying regions. Archaeologists in the middle of the 20th century such as Childe (1942, 1950) assumed that the human populations living in the core areas "discovered" agriculture, perhaps as a result of some environmental change that caused them to intensify their exploitation of wild cereal resources. Once this technological innovation had been adopted, however, the increased rate of caloric return for human labor raised the carrying capacity of the environment, leading to increased population growth rates.

However, Danish economist Boserup (1965) turned this story on its head. She argued that most technological innovations in subsistence, particularly those associated with the adoption and subsequent intensification of agriculture, increase the number of calories that can be extracted from an acre of land but decrease the yield per hour of person-effort. Farmers work harder than foragers for their daily bread, but they can produce more of it from a given environment. According to Boserup, human populations don't "discover" agriculture; they are forced to adopt it when their numbers become too large to be supported by foraging. Rather than population growth being the result of technological innovation, innovation is seen as caused by population growth.

To support her thesis, Boserup (1965) pointed out that during periods of population decline, such as followed the collapse of the Roman Empire in Central and Western Europe, subsistence practices often revert to less intensive forms. Furthermore, many extant populations of foragers know about agriculture but are resistant to adopting it, despite government programs to the contrary. "Why should we plant," asked one !Kung San forager from Botswana, "when there are so many Mongongo nut trees in the world?" (Lee 1979).

It may never be possible to fully determine whether population growth is, like necessity, the mother of innovation or her daughter. Indeed, demographer Lee (1994) has argued that both relationships may exist, producing an upward spiral of population growth that we are still trying to reign in.

BIOLOGICAL REGULATION OF HUMAN FERTILITY

Female Reproductive Physiology

At the population level, human fertility essentially means *female* fertility, because the pace of female childbearing is an ultimate constraint. A male's reproductive capacity is limited primarily by the number of females he can inseminate, not by the number of sperm he can produce.

Female **fecundity** (i.e., the biological potential for childbearing) is ultimately a function of **oocyte** production, the maintenance of an appropriate environment for **fertilization** of the oocyte and the subsequent **implantation** and gestation of an embryo. These activities are carried out primarily by the **ovary** in interaction with other elements of the female reproductive system. (See Box 15.1; cf. Johnson and Everitt 1988 or Jones 1991 for comprehensive treatment of human reproductive physiology.)

BOX 15.1   HORMONES, FECUNDITY, AND THE MENSTRUAL CYCLE

Five primary **hormones** are involved in the regulation of female fecundity. **Gonadotropin-releasing hormone** (GnRH) is produced by the **hypothalamus**. **Follicle-stimulating hormone** (FSH), and **luteinizing hormone** (LH) are produced by the anterior **pituitary**. FSH and LH collectively are known as gonadotropins, because they stimulate the gonad (ovary) to produce hormones in its turn. **Estradiol** is the predominant hormone of the first portion of the **menstrual cycle** (i.e., before **ovulation**), and **progesterone** is the predominant hormone of the second (postovulatory) portion.

Figure 15.3 illustrates this "hypothalamic–pituitary–ovarian–uterine axis," an interactively connected set of tissues, cells, and their products that regulate the menstrual cycle and thus regulate female fecundity. The hormone products can act in either stimulatory (+) or inhibitory (–) fashion. Generally, GnRH stimulates the production of FSH and LH, which stimulate the growth of egg-bearing ovarian **follicles** and the production of the ovarian hormones. LH is also proximately responsible for causing ovulation to occur at midcycle. Estradiol and progesterone act on the uterus; estradiol causes the uterine lining to thicken before ovulation, and progesterone increases the blood supply after ovulation. These hormones regulate their own production by "feeding back" on the hypothalamus and pituitary. Progesterone is inhibitory at these organs; at low circulating levels, estradiol is also inhibitory at the hypothalamus and pituitary but acts positively on the ovary to further stimulate its own production. At high levels, estradiol has a stimulatory effect on the hypothalamus; the **LH surge** that leads to ovulation is thus induced by the estradiol peak that precedes it.

The characteristic pattern of ovarian hormones in a fully functional cycle is schematically illustrated in the top panel of Figure 15.4. It shows the exponential rise and sharp fall of estradiol in the first portion of the cycle (the **follicular phase**). The parabolic rise and fall of progesterone in the second portion (the **luteal phase**) reflects the presence of a functional **corpus luteum**, a hormone-producing tissue that develops anew each cycle on the ovary from the remains of the cells that surrounded the oocyte before ovulation. If pregnancy does occur, an adequate supply of progesterone from the corpus luteum is absolutely necessary to maintain it. Suppression of either the follicular or the luteal phase is reflected in lower profiles of the respective hormones. Ovulatory failure is reflected in the absence of a luteal progesterone rise and a clear midcycle drop in estradiol. These first three stages of ovarian suppression can occur without any recognizable disruption of menstrual regularity.

The interactive nature of this axis means that control of fecundity can be, and undoubtedly is, exerted at any of its levels. The "starvation **amenorrhea**" of anorexia nervosa, for example, is caused ultimately by disrupted hypothalamic function. Current evidence suggests that some other aspects of energy balance that affect fecundity in well-nourished breast-feeding or exercising women may be mediated directly at the ovary. In any case, ovarian function serves as the "final common pathway" by which many factors—ecological, constitutional, and energetic—affect reproduction.

It is relatively easy to obtain repeated measures of estradiol and progesterone in women (see Box 15.2) and thus to plot ovarian function over time for an individual woman or collectively, for populations. Such measures allow for much more detailed analyses of fecundity than simply tracking menstrual cycles or their absence.

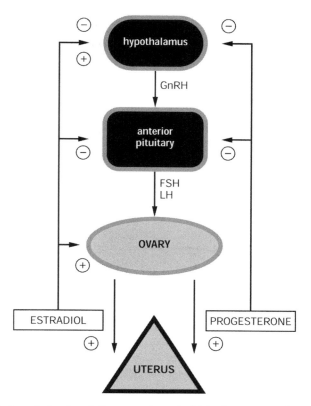

Figure 15.3   The hypothalamic–pituitary–ovarian axis. Major hormone products of each unit are noted, along with their major sites of action, where these hormones may act in a stimulatory or positive fashion (+) or in a suppressive or negative fashion (–). GnRH, gonadotropin-releasing hormone; FSH, follicle-stimulating hormone; LH, luteinizing hormone.

In a fecund nonpregnant woman, ovarian function follows a quasimonthly cycle of oocyte production, preparation of the uterus for implantation, and shedding of the uterine lining in menstruation if conception does not occur. This monthly cycle is produced by the hormonal feedback between the ovary and the hypothalamus and pituitary gland. The absence of menstruation for an extended period (amenorrhea) indicates the absence of ovarian activity and zero fecundity. An important advance in our understanding of the regulation of female fecundity has been the realization that there is a continuum of fecundity between the extremes of amenorrhea and "textbook" menstrual cycles reflected in a continuum of ovarian function (Ellison 1990). Oocytes can be produced that vary in their maturation or fertilizability, and the uterus can vary in its ability to receive an implanting embryo, depending on the ovarian production of estrogen in the follicular phase of the cycle. The fate of an implanting embryo can be further supported or compromised

THE CONTINUUM OF OVARIAN FUNCTION

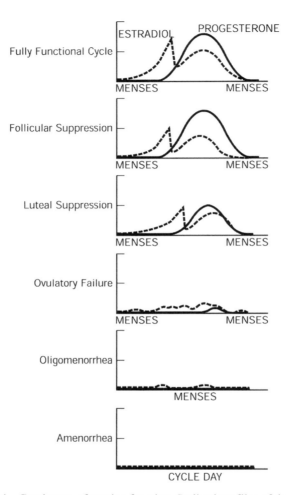

Figure 15.4 Continuum of ovarian function. Stylized profiles of the two principal ovarian hormones, estradiol and progesterone, represent the various stages along the continuum. (See text for discussion.)

depending on ovarian production of progesterone in the luteal phase of the cycle. Ovulation does not necessarily occur in every cycle, and the interval between successive ovulations can be longer or shorter, again depending on levels of ovarian estrogen production.

All of this variation can occur in women who are menstruating regularly, but it requires rather sophisticated methods such as hormonal measurements (Box 15.2) or ultrasound imaging to detect. Recent advances in our ability to make such observa-

BOX 15.2   MEASURING HORMONES

Estradiol and progesterone are particularly interesting hormones to examine because of their position in the regulatory pathway (see Box 15.1) as the most proximal hormonal influences on uterine and ovarian function. Conveniently, they have some chemical properties that render them relatively easy to measure. Both hormones belong to a class of chemicals known as steroids, which are soluble in both lipids and water. They are carried throughout the body by the blood and pass readily through cell membranes, distributing themselves among all the body's cells and fluids, including saliva, urine, and feces. They are highly stable molecules as well, very resistant to degradation by heat, light, time, and other factors. Steroids have even been identified in preserved fecal samples that are several thousand years old.

Saliva is a biological fluid of choice in many field situations. Unlike blood or urine, which usually must be frozen unless it is to be analyzed immediately, saliva samples last for months or longer at room temperature with only the addition of a bacteriocide. Only small samples are needed (a few milliliters); they are easy and painless to collect, so people may be willing to give samples daily or even more often. All of these considerations make saliva an especially useful medium in tropical and/or remote areas, such as central Zaire and the Himalayas of Nepal, two places where this procedure has been used.

Steroids can be measured by a technique known as radioimmunoassay (RIA), one of a class of so-called "competitive binding" assays. RIA uses a radioactively labeled steroid (the "competitor") and specially made **antibodies** (the "binder") to the steroid in question. The principles are as follows.

Imagine a row of test tubes. Into each one, put a small known amount of an antibody to the steroid in question—let's say progesterone. (This antibody has been generously provided by the immune system of a lab animal, often a rabbit or a rat. Blood is taken from the animal and centrifuged to separate the cells; the diluted, antibody-rich serum is used in the assay. Antibodies do not ordinarily react with steroids, so some tricks, not discussed here, have to be used to fool the animal's immune system into producing the antibody you want.) Also add to each tube a known amount of progesterone that has a radioactive tag ("hot" progesterone), allowing it to be quantified with a beta or gamma counter, depending on the radionuclide used. All the tubes contain the same quantity of antibody (less than would be needed to bind all the steroid in the tube) and the same quantity of hot progesterone.

To each tube, add some saliva from one of your samples. In this case, unlike the antibody and hot progesterone, you know only the volume you are adding, but not how much progesterone ("cold" progesterone, because it has no radioactive tag) it contains. Let the molecules in the tubes react, then remove the progesterone (both hot and cold) that has not bound to antibody by adding some charcoal and centrifuging. Then, use your radioactivity counter to measure how much hot progesterone is left in each tube. At this point, all the progesterone in the tube is bound to antibody.

By simply doing this much, you would be able to estimate how the samples differed in the amount of cold progesterone. Why? The hot and cold progesterone had to compete for binding sites on the antibody, so the more hot progesterone there is remaining at the end of the process, the less cold progesterone there was to start with. Quantifying the actual amount of cold progesterone in each tube requires only a small additional step. Instead of adding samples of unknown concentration to all your tubes of antibody and hot progesterone, add known concentrations of cold progesterone to some of them. Then use these tubes to construct a "standard curve," in which a known concentration of cold progesterone is plotted against the amount of hot that was able to bind to the antibody. Because you know the amount of hot progesterone remaining in each tube of sample, it's easy to convert this amount to the concentration of progesterone that was in the sample initially by reading it from the standard curve (Figure 15.5). The hot steroid is usually plotted as "% bound"—if there is no cold steroid at all, then the antibody is said to be 100% bound. Increasing amounts of cold competitor reduce the amount of hot steroid that can bind, so that lower percent bindings are associated with higher starting concentrations of cold steroid.

There are a number of variations on this technique, and some newer versions allow the use of nonradioactive (e.g., fluorescent or luminescent) tags. The basic principles, however, remain the same.

tions, especially under nonclinical conditions, have allowed us to gain a greater understanding of the biological regulation of female fecundity by factors such as age, energetics, and lactation.

Age and Female Fecundity

The French demographer Henry (1961) first pointed out that the age pattern of fertility among married women in natural fertility populations (i.e., populations without conscious family size limitations) is remarkably constant. As Figure 15.6 shows, despite wide variation in the level of fertility between populations, the proportional increase in fertility rates in the teens and early 20s and the proportional decrease after the 30s differs very little. Although Henry suggested that this pattern was likely the manifestation of physiological regulation, the suggestion was largely ignored until recently. Declining fertility after age 30 usually was attributed to declining frequency of intercourse. Moreover, studies of menstrual patterns suggested that female ovarian function was relatively constant between the few years after menarche and the few years preceding **menopause** (Treloar et al. 1967). But menstrual regularity alone, we now know, is a very insensitive index of ovarian function and female fecundity. More sensitive indices such as levels of ovarian hormones, the frequency of ovulation, the size of the follicle before ovulation, and the thickness of the uterine lining show variation with age parallel to age variation in natural fertility (Ellison 1996).

Clinical studies also have provided compelling evidence of age variation in female fecundity. With increasing age over 30 years, women are less likely to conceive, even through artificial insemination by donor semen or **in vitro fertilization** (Fédération CECOS et al. 1982; FIVNAT 1993). Fetal loss rates increase with increasing female age over 30 as well. Aging of ovarian function appears to reflect aging both of the follicles that produce ovarian hormones and of the oocytes that they contain. In vitro experiments show that follicles from older women produce lower levels of hormones than those from younger women under the same conditions. Ovum donation programs, which allow women incapable of producing their own oocytes to conceive through in vitro fertilization and subsequent implantation of someone else's ova, have higher success rates with the oocytes of younger than older donors (cf. review in Ellison 1996).

The increase in fecundity through the teen years in women is less often studied. However, an important corollary fact is that the rate at which fecundity increases

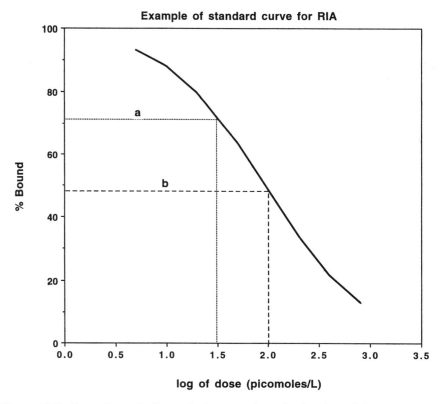

Figure 15.5   Example standard curve for hormone determinations by radioimmunoassay. For each sample, the percent bound (y-axis) is already known; the point on the x-axis at which that y-axis value intersects the curve represents the hormone concentration in the sample. (a) At 72% binding, the value is 31.6 pmol/L (antilog of 1.5). (b) At 48%, the value is 100 pmol/L.

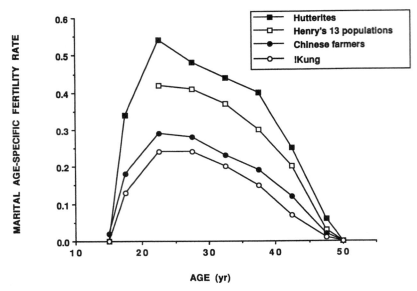

Figure 15.6 Age-specific rates of fertility among married women in natural fertility populations, that is, populations that do not consciously limit family size. Notice that whereas the levels of fertility vary substantially between populations, the age pattern is quite consistent. (*Source*: Redrawn from Howell 1979)

with age in teenage women appears to be related to the overall tempo of maturation (Apter and Vihko 1983; Ellison 1996). Fecundity rises more rapidly and to higher absolute levels in populations with early maturation than in populations with late maturation. The fact that our society seems more concerned with teenage pregnancy than with teenage subfecundity may be a consequence of the fact that we are among the earliest maturing populations on the planet. The same concern may soon emerge in populations that are now in the process of a **secular trend** toward earlier maturation.

As for age patterns of natural fertility, the age pattern of female fecundity (at least as it is reflected in ovarian hormone production and rates of fetal loss) also appears to be constant across widely differing populations. Populations may vary in the typical levels of ovarian hormones circulating at a given age, as do the three populations illustrated in Figure 15.7, but the pattern of steadily rising ovarian function through the early 20s and steady decline after the early 30s seems to be universal and not dependent on environmental circumstances or reproductive history. It seems to be an endogenous feature of human reproductive biology that all populations share and probably contributes to the relatively constant age pattern of natural fertility that Henry first observed.

Energetics and Female Fecundity

Female fecundity varies with several factors related to female energetic state, such as nutritional status (e.g., fatness or relative weight), energy balance (e.g., energy intake greater or less than expenditure), and energy flux (e.g., energy intake and expenditure both high or both low) (Ellison et al. 1993). Early attempts to relate body fat to menstrual patterns met with equivocal results. Women who are under severe nutritional or energetic stress are very likely to stop menstruating (Figure 15.8), but the idea of a simple threshold of body fatness sufficient to establish or arrest menstruation has been clearly rejected (cf. Ellison 1990). Recent attention has focused on more sensitive measures of female fecundity, such as ovarian hormone levels and ovulatory frequency, and more dynamic measures of energetic stress, such as energy balance and energy flux.

In American women, weight loss as a consequence of caloric restriction dieting and high energy expenditure as a consequence of voluntary **aerobic** exercise are associated with increases in menstrual irregularity and lower hormonal indices of ovarian function. Both are also associated with diminished fecundity in epidemiological studies. In a study in Washington State, for example, it was found that women who exercise more than 1 hour a day or who are less than 85% of conventional **weight-for-height** standards are five to six times more likely to be unable to conceive within a year of trying than nonexercising, normal-weight women (Green et al. 1988) The degree of weight loss or exercise necessary to produce a measurable effect on ovarian function does not need to be severe. Although highly trained endurance athletes may show profound suppression of ovarian function, often to the point of amenorrhea (Bullen et al. 1985; Cumming 1989), moderate en-

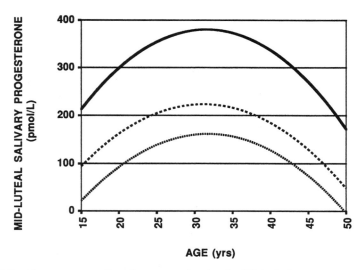

Figure 15.7   Curves representing the average levels of midluteal phase progesterone measured in the saliva of women from three different populations: middle-class women in Boston (solid line), Lese horticulturalists from the Ituri Forest of Zaire (dashed line), and Tamang agropastoralists from the highlands of central Nepal (dotted line). As with natural fertility, the levels of ovarian function represented by this hormonal index vary substantially between populations, but the age pattern is very consistent.

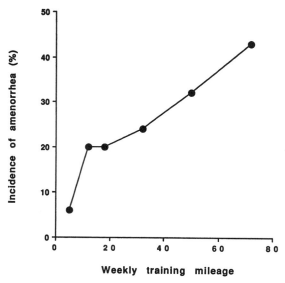

Figure 15.8   Relationship between incidence of amenorrhea and weekly training mileage run in a group of college athletes. (*Source*: Redrawn from Feicht et al. 1978)

ergetic stress appears to produce moderate suppression of ovarian hormone levels in a dose–response manner, often without any change in menstrual regularity (Ellison and Lager 1985). The effects also are readily reversible on weight gain or cessation of exercise regimes.

Some carefully controlled studies have demonstrated that the effects of weight loss and exercise are independent; that is, exercise suppresses ovarian function even when stable weight is maintained (Bullen et al. 1985). Figure 15.9 shows data from a study of women in their 20s, selected to be in good health and physically fit and to have normal menstrual function, who spent 2 months following a highly structured and well-supervised exercise regime. Although the subgroup that lost weight during the 2 months of exercise showed the greatest suppression of ovarian function, the subgroup without weight loss showed significant suppression as well. Both weight loss and exercise have also been associated with ovarian suppression in women of normal nutritional status, suggesting that ovarian function is sensitive to dynamic energetic state, not simply to fatness or body composition.

Similar effects can be observed in other populations in which variations in energetic state are not matters of individual choice and where energy expenditure occurs in the context of traditional subsistence work, rather than recreational exercise (Ellison et al. 1993). In Zaire (Ellison et al. 1989) and Nepal (Panter-Brick et al. 1993), for example, weight loss has been associated with reduced ovarian hormone levels, whether occasioned by low energy intake or high energy expenditure. The Lese women of Zaire (Figure 15.10) are subsistence farmers. In many years, if conditions have not been good, they experience "hunger seasons" preceding the next year's harvest, which begins in November. In populations studied in Nepal, workload is a primary factor; these women live in the Himalayas and travel with heavy loads among fields situated at different altitudes. Polish farm women also show variation in ovarian function during seasons of heavy manual labor; the degree of ovarian suppression correlates with average workload (Jasienska 1996).

The sensitivity of female fecundity to female energetics is likely a product of **natural selection**. The energetic investment that a woman must make in reproduction is both substantial and relatively inelastic. Studies of the energetics of pregnancy and lactation have shown that women must often divert energy from their own metabolic needs to meet the demands of reproduction (Prentice and Whitehead 1987; Lunn 1994, 1996) (see also Chapters 10 and 12). Thus, the ability to meet these demands and the cost that must be paid in meeting them is at least partly a function of energetic state. A woman hard-pressed to meet her own metabolic needs will have a hard time reproducing successfully without paying too high a price in terms of her own survival probabilities.

Among the physiological mechanisms that link female fecundity and energetic state is the sensitivity of ovarian function to circulating levels of hormones, such as

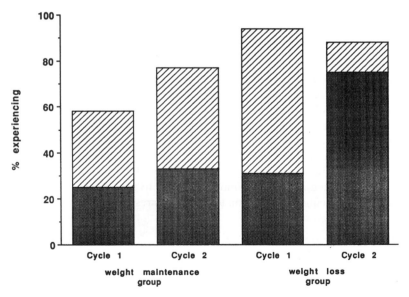

Figure 15.9   Independent and additive effects of exercise and weight loss on ovarian function. Women in the weight-maintenance group engaged in an intensive exercise regime for 2 months but maintained a stable body weight. Women in the weight loss group had the same exercise training but also lost modest amounts of weight over the 2 months. In both groups, the frequency and/or severity (anovulation [closed bars] vs. ovulation with abnormal luteal function [hatched bars]) of effects increased from the first to the second cycle, but the weight loss group experienced more—and more severe—impairment overall. (*Source*: Data from Bullen et al. 1985)

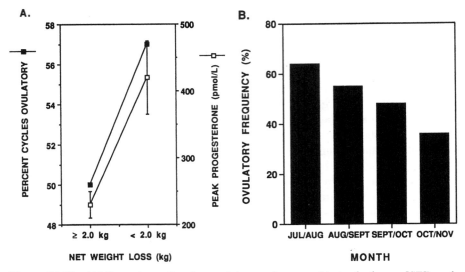

Figure 15.10 (A) Percentage of cycles ovulatory and average (± standard error [SE]) peak salivary progesterone levels in Lese women from the Ituri forest of Zaire. Women were classified on the basis of weight lost at the end of an extended hunger season following a failed harvest (July–November 1984). (B) Average ovulatory frequency of the same Lese women over the same period. (*Source*: Ellison et al. 1989)

insulin and **cortisol**, that regulate energy storage and mobilization in the body. High insulin levels favor energy storage—its transport into cells and conversion into fats for later use. Cortisol is concerned with the mobilization of energy for use in stressful situations. When a woman is in positive energy balance, high insulin levels promote fat storage and ovarian steroid production at the same time (Nahum et al. 1995). When a woman is in negative energy balance, high cortisol levels promote energy mobilization and inhibit ovarian hormone production (Figure 15.11). These linkages appear to be designed elements of human reproductive physiology that function to modulate female fecundity in relation to energetic stress.

Lactation and Female Fecundity

As recently as the mid-20th century, breast-feeding was not thought by physiologists and physicians to have significant suppressive effect on female fecundity. **Postpartum amenorrhea** was considered to reflect a refractory period on the part of the ovaries and uterus after the stresses of pregnancy. In the United States, most women who chose to breast-feed their offspring resumed menstruating after an average of 3 or 4 months, scarcely longer than those who chose not to breast-feed. After World War II, however, studies of postpartum amenorrhea in developing populations revealed more striking differences between lactating and nonlactating women (although in these cases, the absence of lactation was more often due to infant death than to maternal choice). However, considerable variation remained in the length of amenorrhea among lactating women and between populations. It was suggested that this residual variation might be due to differences in the "intensity" of lactation, but what that consisted of or how it should be measured was far from clear (cf. review in Ellison 1995).

In the 1970s, physiologists began to link amenorrhea to high levels of **prolactin**, the hormone that promotes milk production in lactating women. Prolactin, in turn, was shown to be released in response to suckling. On this basis, the hypothesis was advanced that the frequency of suckling might be the sought-for measure of intensity of lactation: Women who nursed their infants frequently might sustain higher prolactin levels than women who nursed less frequently but for the same total amount of time. Higher prolactin levels might convey a longer period of amenorrhea.

The hypothesis soon was put to the test in the field. Data, which have accumulated rapidly, can only be briefly summarized here (for more detailed discussions see Ellison 1995 and Vitzthum 1995). Women in Zaire who reported nursing their children more than six times a day had prolactin levels higher than women who reported nursing children of the same age one to three or four to six times a day and were more likely to be amenorrheic (Delvoye et al. 1977; Figure 15.12). !Kung hunter-gatherer women in Botswana were observed to nurse their children every 15 minutes, on average, throughout the day (Konner and Worthman 1980). Among the Gainj of highland Papua New Guinea, the observed frequency of nursing was correlated with prolactin levels (Wood et al. 1985). Longitudinal studies of nursing women in Edinburgh showed that resumption of ovarian activity was closely correlated in time with the introduction of supplementary foods in the infant's diet and associated changes in both nursing patterns and prolactin levels (Howie and McNeilly 1982). Demographic surveys showed that the average duration of lactation and the average duration of amenorrhea were highly correlated across populations; each additional month of breast-feeding was associated with an additional 0.8 months of amenorrhea, on average (Goldman et al. 1987).

Whereas some of the variation in the contraceptive effect of lactation seems to be associated with the frequency of nursing, some seems to be associated with maternal

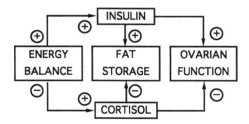

Figure 15.11   Some of the pathways thought to connect energy balance to ovarian function. Negative energy balance is associated with increased cortisol production (and decreased insulin production), which inhibits fat storage and suppresses ovarian hormone production. Positive energy balance is associated with increased insulin production (and decreased cortisol production), which promotes fat storage and stimulates ovarian hormone production.

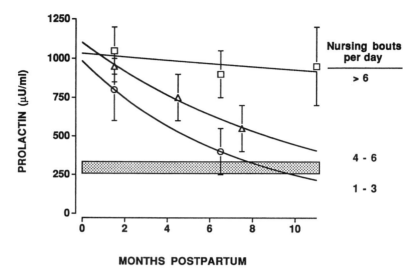

MONTHS POSTPARTUM

Figure 15.12   Prolactin levels in nursing mothers in Zaire versus the age of the infant, classified according to the reported number of nursing bouts per day. The shaded area represents the prolactin levels in non-nursing women. Notice that prolactin levels remain high for up to a year among women who nurse their infants at least six times a day. (*Source*: Redrawn from Delvoye et al. 1977)

condition. Well-nourished Amele women from lowland Papua New Guinea had low levels of prolactin and short periods of amenorrhea despite nursing as frequently as highland Gainj women (Worthman et al. 1993). Similarly, Australian aborigine women who nurse as frequently as the !Kung but are much heavier, had shorter periods of amenorrhea (Rich 1984). In Bangladesh, nursing frequency had no relationship to the duration of postpartum amenorrhea, whereas postpartum maternal weight did have a significant negative relationship (Huffman et al. 1987; Ford and Huffman 1988). In Chile, estrogen levels in the first 3 months postpartum were found to predict the duration of amenorrhea in women who were indistinguishable in their nursing patterns or prolactin levels (Diaz et al. 1995). In the Gambia, providing lactating women with a caloric supplement reduced prolactin levels, shortened the duration of amenorrhea, and led to shorter birth intervals but did not increase milk yields, the nutritional value of the milk, or alter nursing patterns (Lunn et al. 1984; Figure 15.13).

The relationship of lactation to female fecundity reflects the substantial energetic investment that milk production represents. Natural selection appears to have shaped female reproductive physiology to lower the possibility of conception when the metabolic investment in the most recent infant is still high. Reduction in nursing frequency is often a reliable indicator that the metabolic load of lactation is decreasing. But the mother's ability to meet a given lactational load also will depend on her own condition and energetic state. Prolactin is no longer thought to directly inhibit ovarian function, but it may be a good indicator of how hard a woman's metabolism must be pushed to produce the milk the baby needs. An undernourished woman may have to sacrifice a greater portion of her own metabolic needs to meet a given milk demand than a well-nourished woman and may experience more profound suppression of fecundity as a result (Lunn 1994, 1996).

Because lactation is such a powerful factor in the suppression of ovarian function it has been termed "nature's contraceptive" by some. In populations where lactational amenorrhea is the major mechanism of birthspacing, however, high rates of infant mortality can lead to high rates of fertility as a mother returns rapidly to a fecund state in the absence of a nursing child. By depleting maternal reserves, successive cycles of pregnancy and infant mortality can undermine a mother's health (Tracer 1991). Thus, in natural fertility populations, high fertility and high mortality often are concomitant.

Disease and Female Fecundity

Many diseases can have negative effects on female fecundity (McFalls and McFalls 1984; Mascie-Taylor 1996). These effects are more often pathological than adaptive. Sexually transmitted diseases such as gonorrhea and chlamydia can lead to blockage of the fallopian tubes. Infections contracted during childbirth can permanently damage the reproductive tract. Malaria can impair placental function and raise the rate of embryonic and fetal loss. Endocrine diseases, such as hypothyroidism resulting from iodine deficiency in the diet, can disrupt the normal regulation of ovarian function. There are populations in which each of these diseases is prevalent enough to have a substantial impact on fertility.

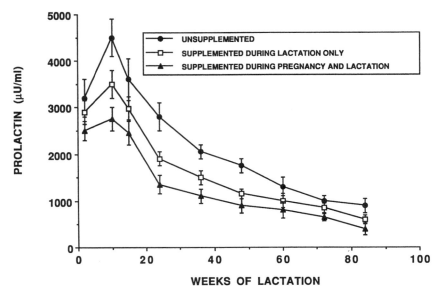

Figure 15.13   Prolactin levels among Gambian women versus the age of the infant, classified according to whether they received a substantial caloric supplement to their diets during pregnancy and lactation, during lactation only, or not at all. (*Source*: Redrawn from Lunn et al. 1984)

An intriguing but as-yet unstudied question concerns the nonspecific effect of disease on fecundity, mediated not by pathological effects but by altering the energetic state of the organism. Febrile infections raise basal metabolism. **Lymphocyte** proliferation and immunoglobulin production are energetically expensive processes. It may be that a high incidence of disease in a population represents an increased energetic burden on affected women and might depress female fecundity as a result.

Male Fecundity

Male fecundity usually is not a constraint on human population dynamics, but it can, of course, vary widely between individuals and can dramatically affect the fertility of individual couples. Most variation in male fecundity that we know about is idiosyncratic or pathological, with little systematic ecological variation. Male fecundity variation also is difficult to document. Although levels of **testosterone**, the primary testicular hormone, vary in males with age and energetic stress, they bear little relationship to sperm counts, sperm viability, or other physiological indices of male reproductive function. Nor do sperm counts have a direct relationship to the probability of fertilization. Although extreme caloric deprivation or high levels of aerobic exercise stress can completely suppress sperm production in males, moderate energetic stress does not appear to affect male reproductive physiology as sensitively as it does female reproductive capacity (Ellison and Panter-Brick 1996). Male fecundity may decline with age, at least **cross-sectionally**, but does not drop precipitously to zero at any age, as female fecundity does at menopause (cf. review in Campbell and Leslie 1994).

Natural selection probably has acted on male reproductive physiology quite differently from the way it has acted on female physiology. Time and energy are primary constraints on the reproductive success of most female mammals, including humans. Human females must minimally invest up to a couple of years and an enormous number of calories during pregnancy and lactation for each episode of reproduction. In contrast, male reproductive success is more constrained by mating opportunity than by the direct energy or time investment in conception or gestation. The energy investment in sperm production is trivial compared with the female investment in gestation and lactation, so there presumably has been little selection pressure to modulate male fecundity in response to environmental conditions short of extreme deprivation.

Testosterone levels do vary with ecological conditions, however, especially with long-term fluctuations in energy availability. Rather than serving to modulate male fecundity, these variations probably modulate the maintenance of metabolically active muscle tissue (Bribiescas 1996). On average, males and females of the same height in the same population carry about the same absolute amount of fat on their bodies; however, males carry a much greater amount of muscle. This difference in body composition is presumably a result of sexual selection enhancing the ability of males to compete effectively with other males for reproductive opportunities. But muscle mass can be a handicap when energy availability is low, because it requires additional energy to maintain. Lowering testosterone, hence the physiological support for muscle building and maintenance, under conditions of chronic energy shortage may be an adaptive response.

Testosterone also may affect aspects of male behavior related to male–male competition and social dominance. When two male rhesus monkeys have an **agonistic** encounter to determine their relative dominance status, the winner emerges with higher testosterone levels than before the encounter and the loser with lower levels (Rose et al. 1971). It is postulated that these short-term hormonal changes, by interacting with receptors in the brain, reinforce the display of dominant behavior in the winner and submissive behavior in the loser. Interestingly, exactly the same pattern of testosterone changes has been observed in human males competing in situations ranging from wrestling bouts to chess matches (Elias 1981; Mazur 1992). Whether this is just a vestige of our evolutionary past or a functional aspect of our modern-day physiology has yet to be determined.

SOCIAL REGULATION OF FERTILITY

The social regulation of human fertility is exceedingly complex (Davis and Blake 1956; Bongaarts and Potter 1983). Most contemporary analyses identify two major routes by which the social regulation of fertility is effected: the exposure to intercourse and the probability of conception and birth given intercourse.

The first of these regulations includes social prescriptions regarding the formation of reproductive unions (the term "marriage" is often used for convenience, even though it is recognized that not all reproductive unions are marriages in the legal sense) between individual couples as well as social factors that might influence the pattern of intercourse within such unions. The second route includes all forms of deliberate contraception and abortion. Infanticide, though it may be practiced to limit effective family size, is not subsumed under fertility regulation in the strict sense, because it occurs postnatally.

Regulation of the Formation of Reproductive Unions

Human societies differ markedly in the degree to which they elaborate rules for the formation of reproductive unions and in the effect of those rules on the fertility of the population.

Among the most variable and the most potent aspects of such regulation are the average age at which unions are formed and the percentage of women who never enter a union. Early and nearly universal marriage clearly promotes the fertility of populations that practice it, whereas conventions that lead to late age at marriage and a high percentage of women never married have the opposite effect. Medieval Europe exemplifies the latter case. Reliance on subsistence agriculture plus constraints on arable land led to the practice of delaying marriage until a would-be husband had inherited or otherwise acquired sufficient land on which to support a family. Average ages at marriage in the early to late 30s were not uncommon. Nor was it uncommon for substantial proportions of people never to achieve the material conditions necessary for marriage. Even though illegitimate births occurred, the overall effect was to dramatically limit the average number of births per woman.

Social regulation of the formation of reproductive unions can occur through less direct means as well. Economic and educational opportunities in more industrialized societies often lead to delays in marriage and higher proportions of women and men remaining single through choice, rather than through constraint. Yet the career and economic advancement options available and the degree to which they are or are not compatible with marriage and family are subtle matters of social organization and practice as well.

Pattern of Intercourse within Reproductive Unions

Social factors can influence the frequency and timing of intercourse within reproductive unions in more and less direct ways. Spousal separation, due to the requirements of migrant labor, for example, has been found to affect the pattern of conceptions and births in numerous populations. Cultural practices may either mandate or allow variation in sexual activity at different times such as religious festivals or observances or during the infancy of the most recently born child. The traditional August vacation in France has been associated with a birth peak nine months later, while the practice of postpartum sexual abstinence contributes to the length of the interbirth interval in many African societies.

The frequency of intercourse in married couples declines significantly with the increasing age of both spouses *and* with the duration of the marriage. Figure 15.14 illustrates that women achieve their highest fertility early in marriage, whether that marriage occurs at age 20 or age 30. Furthermore, newly married 30-year-olds have much higher fertility than 30-year-olds who have been married for 10 years. These data strongly suggest that a high frequency of intercourse is associated with being newly married—the so-called "honeymoon effect." Yet even with the honeymoon effect, the fertility of newlywed couples declines with age. This may reflect physiological changes in female fecundity, male potency, and both male and female libido with age as well as social factors that influence behavior.

Behavioral Regulation of the Probability of Conception and Live Birth

Contraceptive usage has a demonstrable and dramatic effect on birth rates in many populations. Recent studies have shown clearly that the effective practice of contraception depends on having both the means and the motivation.

Demographers and others who study population issues agree that the prevalence of contraceptive practice in a population is a good predictor of the fertility of that population. They disagree, however, on what motivates people to use contraception and the extent to which there is a gap between the number of children people want and the (larger) number they actually have (Caldwell et al. 1992; Robey et al. 1993; Dasgupta 1995).

One position is that "development is the best contraceptive." This argument says that economic development itself—and with it economic, educational, and em-

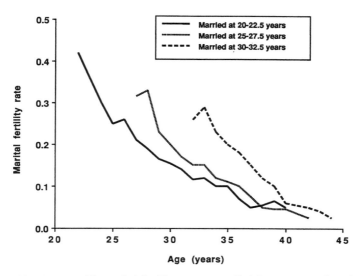

Figure 15.14   Age-specific marital fertility rates controlled for age at marriage (and thus also for duration of marriage) in a 20th-century British population. For any given age up to the late 30s, women married more recently have higher fertility than women married longer, but fertility decreases with marriage duration regardless of the age at which the marriage was contracted. For any given duration of marriage, women who married at older ages have lower fertility than women who married young. (*Source*: Redrawn from Henry 1976)

ployment opportunities, particularly for women—leads to less need for large families and a desire for smaller ones. In many agrarian societies, children are net producers of family income, as animal tenders, water fetchers, and such, rather than the net consumers they tend to be in industrialized countries. A woman's economic value in agrarian societies is primarily as a reproducer (childbearer) and as an agricultural laborer. Development generally expands a woman's economic role into production of goods and services outside the household, and also may give her a stronger voice in family decision making. Concurrently, the cost of children rises. Schooling, for example, even if free, removes children from productive agricultural roles. Supporters of this position point to strong negative associations, in many areas of the world, between women's levels of education or outside employment and their fertility. They suggest that the causal arrow runs from increased female economic opportunity, which in the long run benefits the whole family, to reduced fertility.

Others argue that whereas economic development may set the stage for a transition to lower fertility, it is by itself neither necessary nor sufficient and that "contraceptives are the best contraceptive." As evidence, they look to fertility reductions that have been achieved by intensive government campaigns, including public education through broadcast and other media and ready availability of contraceptives, even among relatively uneducated rural families. In particular, they point to areas of sub-Saharan Africa (e.g., Kenya, Botswana), where fertility has begun to decline after local governments adopted strong family-planning policies even in the face of a social system based on lineage membership that places high value on fertility.

Proponents of these positions tend to differ, not surprisingly, in their views of whether an "unmet need" for contraception exists and how large it is. An unmet need can be defined as the extent to which fertility would be reduced if everybody had only as many children as they wanted. If people do, in general, act rationally, having enough children to keep the farm going or to be assured of support in old age, then they don't truly have an unmet need under their particular circumstances—however, changing the circumstances may change their actions. In contrast, women may tell survey takers that they want more children than they do, especially if they already have a large family and they know that this is expected of them by their husbands or by society in general. In such instances, a local government's population policy can provide a welcome rationale and justification for smaller families.

Social factors that influence the motivation for contraceptive use can also affect the way in which it is used. In Southeast Asia, for example, contraception has been increasingly adopted as a means of family limitation, stopping childbearing when a couple has "enough" children. In Africa, on the other hand, contraception has primarily been used to space births in a traditional manner, taking the place of lactation.

The availability and suitability of various forms of artificial contraception depends on the infrastructure of the health delivery system in a given population and on the way in which individuals interact with that system. Making oral contraceptives available in clinics will not be effective if women can't get to the clinics regularly. Distance and cost may be barriers to regular attendance, but so may be a cultural system that does not sanction visits to clinics by women who are not ill or who are not accompanied by their husbands.

Abortion also can be practiced to limit the number of births in a population, but only at greater risk to female health than the risks associated with contraception. Nor is abortion likely to be a very effective means of fertility control, because it only adds a few months to the typical interbirth interval. Abortion as a form of birth control seems to be practiced primarily when the motivation to limit births exists but the means to achieve effective contraception are unavailable. Efforts to increase contraceptive availability and effectiveness should have a limiting effect on the practice of abortion.

LOOKING BACK AND LOOKING FORWARD

Population Regulation in the Past

Understanding the regulation of mortality and fertility rates allows us to hazard some informed guesses about human population history. Most human biologists now hypothesize that human populations in the Pleistocene were regulated by a combination of moderate mortality balanced by moderate fertility. Low population densities and a mobile foraging lifestyle would have limited the impact of **infectious diseases** as causes of mortality. At the same time, low availability of appropriate supplementary foods would have led to prolonged lactation, which, together with relatively high levels of energy expenditure, would have had a suppressive effect on female fecundity. Small populations and varying local conditions would have produced a great deal of **stochastic** variation in local population dynamics, and migration probably served to keep the human population distributed across the landscape in some relation to local resources.

A shift in subsistence and residence patterns associated with the adoption of plant and animal domestication would have changed the nature of human population regulation. Higher population densities and more sedentary residence patterns would have provided an environment for infectious disease mortality to increase (see Chapters 8 and 14). Not only would longer chains of disease transmission be possible—an initial case could ultimately be passed to far more people over a longer period of time than before—but **pathogen** exposure would have increased both for pathogens in human waste and for **zoonotic** pathogens in domestic animals. Cereals and dairy products would provide supplementary foods for infants, which could lead to earlier weaning but also may have led to higher rates of weanling diarrhea and early childhood mortality. Whether a shift to agricultural subsistence would have increased nutritional status is debatable. It probably did introduce an increased seasonality in food availability, which may have been reflected in increasing seasonality of births and deaths. Overall, this ecological shift is likely to have shifted both mortality and fertility rates higher. Because an acceleration in global population growth seems to date from this transition, it is assumed that the increase in fertility was greater than the increase in mortality.

The high mortality/high fertility regime seems to have persisted, with considerable local variation, well into the historical period. Social regulation of fertility through marriage is important in historical Europe in situations where populations began to outstrip local resources. Then, in the 18th century, the transition toward low mortality and low fertility began in Europe and North America, soon to spread throughout the world. Mortality dropped precipitously with the dramatic reduction in infectious disease mortality achieved primarily through changes in general nutrition and public health, such as the introduction of clean water and sewage systems. Fertility dropped as a result of conscious limitation of family size and the spread of knowledge and use of modern contraceptive methods. A key to the fertility transition was the motivation to limit family size, closely linked to economic development, educational and occupational opportunities for women, and the perceived costs of raising children. The transition from high fertility and mortality rates to low fertility and mortality rates is called the **demographic transition**.

The Present and the Future

Understanding the processes by which fertility is regulated is clearly critical to any informed discussion of future prospects for human population regulation. There are, however, additional implications for a number of health-related issues of current concern.

In Europe, North America, and industrialized countries worldwide, the idea of controlling reproduction has come to mean not only limiting fertility but limiting infertility. Many thousands of couples each year, for example, seek in vitro fertilization or related techniques, in which an egg from the woman herself or from a donor is fertilized (by her partner's sperm or by donor sperm) and returned to her uterus for gestation (FIVNAT 1993). Many more seek other help such as artificial insemination or ovulation induction by so-called "fertility drugs." Ironically perhaps, a good bit of what we know about the physiological and biochemical mechanisms underlying observed variation in natural fertility comes from studies of assisted reproductive techniques—physicians in these cases must try to mimic what occurs when reproduction proceeds without assistance. The flow of information is not unidirectional, however. As discussed earlier, studies of women in natural settings have elucidated a number of energetic, ecological, and constitutional variables that contribute to infertility (or infecundity) in the first place. Social developments that have led many women in Western societies to postpone childbearing to their 30s and 40s probably contribute a great deal to the demand for clinical infertility services in developed countries.

Similarly, there is a great deal of current interest in the ways that lifestyle variables, diet and exercise for example, influence a person's likelihood of developing cancer. It is becoming increasingly clear that, however characteristic of women in the industrialized world, early maturation and repeated menstrual cycles uninterrupted by pregnancy or lactational amenorrhea, combined with good nutrition and a sedentary lifestyle, are not the norm evolutionarily. For example, recent evidence has shown that permanent weight gain in adulthood is associated with an increased risk of breast cancer in women (Huang et al. 1997). Another study (Bernstein et al. 1994) has shown the beneficial effects of even moderate exercise during adolescence on the likelihood of developing breast cancer later. These and similar findings raise the exciting possibility that modest, essentially risk-free interventions, such as moderate caloric intake, weight control, and regular, moderate exercise, could have highly positive effects on women's health. These benefits would be achieved, paradoxically, not by engaging technological intervention but by drawing on what we know of our evolutionary history (cf. review in Ellison 1999).

The dramatic changes in mortality over the past few hundred years also mean that more women are surviving well past menopause into old age, when they are no longer producing appreciable amounts of estrogen. Although estrogen exposure is a risk factor for certain cancers, it has preventive effects on some other diseases, such as cardiovascular disease, **osteoporosis**, and possibly **Alzheimer's disease**. Increasing numbers of women now must make the decision—not a clear-cut one—of whether to undergo hormone replacement therapy after menopause.

Projections of the human population into the future vary, but most demographers predict a stabilization of world population by the end of the 21st century at something between 8 and 11 billion people, allowing for perhaps one last doubling of the world population. Even if fertility rates dropped to replacement levels worldwide today, the human population would continue to grow for some time as the human beings already born live out their reproductive lives.

We should remember, however, that a population growth rate of zero is a very special case of exactly balanced birth and death rates. The projections that foresee such a happy future for humanity are based on assumptions that we cannot test. They assume that the motivation for family limitation throughout the world will follow the pattern observed in Europe and North America as global economic development proceeds. They assume that global economic development will proceed in a way that will raise standards of living and social and economic opportunities everywhere to comparable levels. They assume that replacement level fertility will be a natural result of the interaction of the two former assumptions. All of these assumptions can be questioned. Already, evidence from parts of Africa suggests that the motivation for family limitation and the resulting pattern of use of contraception may be different in different populations, not as a function of level of economic development but as a function of deeply ingrained cultural ideologies. Global homogeneity of economic and social systems within a hundred years seems highly unlikely. And evidence from the populations with the highest standards of living, including Japan, Western Europe, and the United States, suggest that there is nothing to keep fertility levels from falling well below replacement when the disincentives for family building and the incentives of childlessness become very great.

There is reason to question the assumption that global mortality rates will approach the current level of the developed countries as well. The increase in global population densities, the ease, speed, and volume of international travel, the spread of drug-resistant pathogens—all set the stage for a potential resurgence of infectious disease mortality.

Malthus was right, of course, that the human population cannot continue to grow indefinitely. He is also right about the options: increased mortality or decreased fertility. Either we do it ourselves, or nature will do it for us.

CHAPTER SUMMARY

The history of human population growth is closely interwoven with the history of human ecology. Population growth is the product of two processes: birth and death, or fertility and mortality. If births exceed deaths, then populations will grow, and if unchecked, they tend to grow exponentially. Malthus (1798, 1830) pointed out, two centuries ago, that such unrestrained growth soon would lead to the population outgrowing its food supply and that mortality in the form of war, disease, and starvation then would act as a check on growth. Thus, population size is a dynamic process, driven by the engines of fertility and mortality in interaction with the environment. Malthus' notion has since been incorporated in the concept of carrying capacity, and many people continue to be concerned about the extent to which carrying capacity can be expanded or population constrained. Fertility control appears to provide the most promising solution to unfettered population growth.

Malthus (1798, 1830) thought that fertility could probably only be constrained behaviorally, but we have since learned that there is considerable natural variation in human fertility. Compared with males, females make the far larger commitment of time and energy to reproduction; hence, female reproductive rates tend to be limiting for a population. Female fecundity varies with age and also is constrained by energy availability. Teenaged women and those in their late 30s and 40s have lower fecundity and fertility than women between those ages. High energy expenditure and low energy intake, whether voluntary (as in exercise and dieting) or involuntary (e.g., as a consequence of subsistence ecology), are associated with lowered female fecundity. Breast-feeding women, especially if they are undernourished to begin with, also have suppressed fecundity. Disease can have an impact as well.

Fertility is also regulated socially, by cultural rules regarding the formation of reproductive unions and the patterns of intercourse within unions. Contraceptives can have a dramatic effect, but only if they are socially available to women who are motivated to use them.

Although it is still uncertain whether population increase led to the adoption of agriculture or agriculture permitted population increase, both events occurred about 8,000–12,000 years ago, and global population has increased rapidly since. Predictions for the future draw on knowledge of the past but also rest on a number of untestable assumptions. Nonetheless, most demographers predict stabilization at about 8 billion to 11 billion people by the end of the 21st century. The path to such stabilization, however, remains unclear.

RECOMMENDED READINGS

Bongaarts J and Potter RG (1983) Fertility, Biology, and Behavior: An Analysis of the Proximate Determinants. New York: Academic Press.

Boserup E (1965) The Conditions of Agricultural Growth. Chicago: Aldine.

Boserup E (1981) Population and Technological Change: A Study of Long-Term Trends. Chicago: Univ. of Chicago Press.

Caldwell, JC and Caldwell P (1990) High fertility in sub-Saharan Africa. Sci. Am. May: 118–125.

Caldwell JC, Orubuloye IO, and Caldwell P (1992) Fertility decline in Africa: a new type of transition? Popul. Dev. Rev. 18: 211–242.

Cohen, JE. 1995. How Many People Can the Earth Support? New York: Norton.

Cohen MN. 1977. The Food Crisis in Prehistory: Overpopulation and the Origin of Agriculture. New Haven, Conn.: Yale Univ. Press.

Dasgupta, PS (1995) Population, poverty, and the local environment. Sci. Am. February: 40–45.

Ellison PT (1995) Breastfeeding, fertility, and maternal condition. In P Stuart-Macadam and KA Dettwyler (eds.): Breastfeeding: Biocultural Perspectives. New York: Aldine de Gruyter, pp. 305–346.

Ellison PT (1991) Reproductive ecology and human fertility. In CGN Mascie-Taylor and GW Lasker (eds.): Applications of Biological Anthropology to Human Affairs. Cambridge: Cambridge Univ. Press, pp. 14–54.

Ellison PT, Panter-Brick C, Lipson SF, and O'Rourke MT (1993) The ecological context of human ovarian function. Hum. Reprod. 8: 2248–2258.

Howell N (1979) The Demography of the Dobe !Kung. New York: Academic Press.

Meadows DH, Meadows DL, Randers J, and Behrens WW 3rd (1971) The Limits to Growth. New York: Signet.

Meadows DH, Meadows DL, and Randers J (1992) Beyond the Limits: Global Collapse or a Sustainable Future. London: Earthscan.

Prentice AM and Whitehead RG (1987) The energetics of human reproduction. Symp. Zool. Soc. London 57: 275–304.

Robey AR, Rutstein SO, and Morris L (1993) The fertility decline in developing countries. Sci. Am. December: 60–67.

Wood JW (1994) Dynamics of Human Reproduction: Biology, Biometry, Demography. New York: Aldine de Gruyter.

Wrigley EA (1969) Population and History. New York: McGraw-Hill.

REFERENCES CITED

Apter D and Vihko R (1983) Early menarche, a risk factor for breast cancer, indicates early onset of ovulatory cycles. J. Clin. Endocrinol. Metab. 57: 82–88.

Bernstein L, Henderson BE, Hanisch R, Sullivan-Halley J, and Ross RK (1994) Physical exercise and reduced risk of breast cancer in young women. J. Nat. Cancer Inst. 86:1403-1408.

Bongaarts J and Potter RG (1983) Fertility, Biology, and Behavior: An Analysis of the Proximate Determinants. New York: Academic Press.

Boserup E (1965) The Conditions of Agricultural Growth. Chicago: Aldine.

Boserup E (1981) Population and Technological Change: A Study of Long-Term Trends. Chicago: Univ. of Chicago Press.

Bribiescas RG (1996) Testosterone levels among Aché hunter/gatherer men: a functional interpretation of population variation among adult males. Hum. Nat. 7: 163–188.

Bullen BA, Skrinar GS, Beitins IZ, von Mering G, Turnbull BA, and McArthur JW (1985) Induction of menstrual disorders by strenuous exercise in untrained women. N. Engl. J. Med. 312: 1349–1353.

Caldwell JC, Orubuloye IO, and Caldwell P (1992) Fertility decline in Africa: a new type of transition? Popul. Dev. Rev. 18: 211–242.

Campbell BC and Leslie PW (1995) Reproductive ecology of human males. Yearbook Phys. Anthropol. 38:1–26.

Childe VG (1942) What Happened in History. London: Penguin.

Childe VG (1950) The urban revolution. Town Planning Rev. 21: 3–17.

Cohen MN (1977) The Food Crisis in Prehistory: Overpopulation and the Origin of Agriculture. New Haven, Conn.: Yale Univ. Press.

Cohen, JE (1995) How Many People Can the Earth Support? New York: Norton.

Cumming DC (1989) Menstrual disturbances caused by exercise. In KM Pirke, W Wuttke, and U Schweiger (eds.): The Menstrual Cycle and Its Disorders. Berlin: Springer-Verlag, pp. 150–160.

Dasgupta, PS (1995) Population, poverty, and the local environment. Sci. Am. February: 40–45.

Davis K and Blake J (1956) Social structure and fertility: an analytical framework. Econ. Dev. Cult. Change 4: 211–235.

Delvoye P, Demaegd M, Delogne-Desnoeck J, and Robyn C (1977) The influence of the frequency of nursing and of previous lactation experience on serum prolactin in lactating mothers. J. Biosoc. Sci. 9: 447–451.

Diaz S, Cardenas H, Zepeda A, Brandeis A, Schiappacasse V, Miranda P, Seron-Ferre M, and Corxatto HB (1995) Luteinizing hormone pulsatile release and the length of lactational amenorrhoea. Hum. Reprod. 10: 1957–1961.

Elias M (1981) Serum cortisol, testosterone, and testosterone-binding globulin responses to competitive fighting in human males. Aggressive Behav. 7: 215–224.

Ellison PT (1990) Human ovarian function and reproductive ecology: new hypotheses. Am. Anthropol. 92: 933–952.

Ellison PT (1995) Breastfeeding, fertility, and maternal condition. In P Stuart-Macadam and KA Dettwyler (eds.): Breastfeeding: Biocultural Perspectives. New York: Aldine de Gruyter, pp. 305–346.

Ellison PT (1996) Age and developmental effects on adult ovarian function. In L Rosetta and NCG. Mascie-Taylor (eds.): Variability in Human Fertility: A Biological Anthropological Approach. Cambridge: Cambridge Univ. Press, pp. 69–90.

Ellison PT (1999) Reproductive ecology and reproductive cancers. In C Panter-Brick and C Worthman (eds.): Hormones and Human Health. Cambridge: Cambridge Univ. Press, pp. 184–209.

Ellison, PT and Lager C (1985) Exercise-induced menstrual disorders. N. Engl. J. Med. 313: 825–826.

Ellison PT and Panter-Brick C (1996) Salivary testosterone levels among Tamang and Kami males of central Nepal. Hum. Biol. 68: 955–965.

Ellison PT, Peacock NR, and Lager C (1989) Ecology and ovarian function among Lese women of the Ituri Forest, Zaire. Am. J. Phys. Anthropol. 78: 519–526.

Ellison PT, Panter-Brick C, Lipson SF, and O'Rourke MT (1993) The ecological context of human ovarian function. Hum. Reprod. 8: 2248–2258.

Fédération CECOS, Schwarz D, and Mayaux MJ (1982) Female fecundity as a function of age. N. Engl. J. Med. 306: 404–406.

Feicht CB, Johnson TS, Martin BJ, Sparkes KE, and Wagner WW Jr (1978) Secondary amenorrhea in athletes. Lancet 2: 1145.

FIVNAT (1993) French national IVF registry: analysis of 1986 to 1990 data. Fertil. Steril. 59: 587–595.

Ford K and Huffman S (1988) Nutrition, infant feeding, and post-partum amenorrhea in rural Bangladesh. J. Biosoc. Sci. 20: 461–469.

Goldman N, Westhoff CF, and Paul LE (1987) Variations in natural fertility: the effect of lactation and other determinants. Popul. Stud. 41: 127–146.

Green BB, Weiss NS, and Daling JR (1988) Risk of ovulatory infertility in relation to body weight. Fertil. Steril. 50: 721–726.

Hassan FA (1981) Demographic Archaeology. New York: Academic Press.

Henry L (1961) Some data on natural fertility. Eugen. Q. 8: 81–91.

Henry L (1976) Population: Analysis and Models. London: Edward Arnold.

Howell N (1979) The Demography of the Dobe !Kung. New York: Academic Press.

Howie PW and McNeilly AS (1982) Effect of breast feeding patterns on human birth intervals. J. Reprod. Fertil. 65: 545–557.

Huang Z, Hankinson SE, Colditz GA, Stampfer MJ, Hunter DJ, Manson JE, Hennekens CH, Rosner B, Speizer FE, and Willett WC (1997) Dual effects of weight and weight gain on breast cancer risk. J. Amer. Med. Assoc. 278:1407-1411.

Huffman SL, Chowdhury A, Allen H, and Nahar L (1987) Suckling patterns and post-partum amenorrhea in Bangladesh. J. Biosoc. Sci. 19: 171–179.

Jasienska G (1996) Energy and ovarian function in rural women from Poland (Ph.D. dissertation). Cambridge, Mass.: Department of Anthropology, Harvard University.

Johnson M and Everitt B (1988) Essential Reproduction. Oxford, UK: Blackwell.

Jones, RE (1991) Human Reproductive Biology. New York: Academic Press.

Konner M and Worthman C (1980) Nursing frequency, gonadal function, and birth spacing among !Kung hunter-gatherers. Science 207: 788–791.

Lee RB (1979) The !Kung San: Men , Women, and Work in a Foraging Society. Cambridge: Cambridge Univ. Press.

Lee R (1994) Human fertility and population equilibrium. Ann. N.Y. Acad. Sci. 709: 396–407.

Lunn PG (1994) Lactation, metabolic loads, and reproduction. Ann. N.Y. Acad. Sci..709: 77–85.

Lunn PG (1996) Breastfeeding practices and other metabolic loads affecting human reproduction. In L Rosetta and CGN Mascie-Taylor (eds.): Variability in Human Fertility: A Biological Anthropological Approach. Cambridge: Cambridge Univ. Press, pp. 195–216.

Lunn PG, Austin S, Prentice AM, and Whitehead RG (1984) The effect of improved nutrition on plasma prolactin concentrations and postpartum infertility in lactating Gambian women. Am. J. Clin. Nutr. 39: 227–235.

Malthus TR (1798) An Essay on the Principle of Population (Reprinted 1986). New York: Penguin Classics.

Malthus TR (1830) A Summary View of the Principle of Population (Reprinted 1986). New York: Penguin Classics.

Mascie-Taylor CGN (1996) The relationship between disease and subfecundity. In L Rosetta and CGN Mascie-Taylor (eds.): Variability in Human Fertility: A Biological Anthropological Approach. Cambridge: Cambridge Univ. Press, pp. 106–122.

Mazur A (1992) Testosterone and chess competition. Soc. Psychol. Q. 55: 70–77.

McFalls JA and McFalls MH (1984) Disease and Fertility. New York: Academic Press.

Nahum R, Thong KJ, and Hillier SG (1995) Metabolic regulation of androgen production by human thecal cells in vitro. Hum. Reprod. 10: 75–81.

Panter-Brick C, Lotstein D, and Ellison PT (1993) Seasonality of reproductive function and weight loss in rural Nepali women. Hum. Reprod. 8: 684–690.

Pearl R and Reed LJ (1920) On the rate of growth of the population of the United States since 1790 and its mathematical representation. Proc. Natl. Acad. Sci. USA 6: 275–288.

Prentice AM and Whitehead RG (1987) The energetics of human reproduction. Symp. Zool. Soc. London 57: 275–304.

Rich JW (1984) Patterns of breast feeding and lactational infertility: a comparison of the Yolngu with the !Kung San and La Leche League (Honors Thesis). Cambridge, Mass.: Department of Anthropology, Harvard University.

Robey AR, Rutstein SO, and Morris L (1993) The fertility decline in developing countries. Sci. Am. December: 60–67.

Rose RM, Holaday JW, and Bernsterin IS (1971) Plasma testosterone, dominance rank, and aggressive behavior. Nature 231: 366–368.

Spooner B (ed.) (1972) Population Growth: Anthropological Implications. Cambridge, Mass.: Massachusetts Institute of Technology.

Tracer DP (1991) Fertility-related changes in maternal body composition among the Au of Papua New Guinea. Am. J. Phys. Anthropol. 85: 393–405.

Treloar AE, Boynton RE, Behn BG, and Brown BW (1967) Variation of the human menstrual cycle through reproductive life. Int. J. Fertil. 12: 77–126.

Vitzthum VJ (1995) Comparative study of breastfeeding structure and its relation to human reproductive ecology. Yearbook Phys. Anthropol. 37: 307–349.

Wilson EO and Bosert WH. (1971) A Primer of Population Biology. Stamford, Conn.: Sinauer Associates.

Wood JW, Lai D, Johnson PL, Campbell KL, and Maslar IA (1985) Lactation and birth spacing in highland New Guinea. J. Biosoc. Sci. 9 (Suppl.): 159–173.

Worthman CM, Jenkins CL, Stallings JF, and Lai D (1993) Attenuation of nursing-related ovarian suppression and high fertility in well-nourished, intensively breastfeeding Amele women of lowland Papua New Guinea. J. Biosoc. Sci. 25: 425–443.

Glossary

Absolute risk: the incidence of a disease within a population.

Acclimatizations: physiological alterations during the lifetime of an organism that act to maintain **homeostasis** when exposed to environmental stress such as heat, cold, high altitude, nutrient imbalance, or disease.

Additive genetic variance: the **variance** of the sum of the individual genetic values across all **loci** that contribute to the **quantitative trait**.

Adenosine triphosphate (ATP): a chemical compound that stores energy for use by the cell.

Adolescence: the stage of development that lasts for 5-10 years after the onset of **puberty**; characterized by the **adolescent growth spurt** in height and weight, permanent tooth eruption almost complete, development of **secondary sexual characteristics**, sociosexual maturation, and intensification of interest in and practice of adult social, economic and sexual activities.

Adolescent growth spurt: the rapid increase in growth velocity that occurs during **adolescence**.

Adrenarche: the progressive increase in the secretion of adrenal **androgen hormones** usually occurring at about 6-8 years of age in most children.

Adrenocorticotropic hormone (ACTH): a **hormone** produced by the **pituitary** that stimulates the adrenal gland to produce **cortisol**.

Adulthood: the stage in the life cycle that begins when adult stature is attained; characterized by the achievement of full reproductive maturity.

Aerobic: using oxygen.

Aerobic power: the amount of oxygen the body can extract during exercise; a measure of the capacity for **aerobic** exercise (also called **maximum oxygen uptake, physical work capacity**, or **VO_2max**).

Age distribution: the proportion of people at each age within a population.

Agent: a living organism or nonliving material that can be the cause or part of the cause of a disease.

Age-specific mortality rate: the number of deaths among people of a certain age divided by the number of people of that age at risk of dying.

Age standardization: a method that allows direct comparison of health statistics in populations that differ in their **age distributions**.

Aging: to become old; to show the effects or characteristics of increasing age.

Agonistic: involving aggression.

Alleles: variant forms of a **gene**.

Allen's Rule: a biogeographic rule stating that cold climate populations tend to have short arms and legs relative to their height.

Allometry: scaling; the change in size of one biological measure with respect to another (usually body size). Such scaling relationships are described by the general equation: $Y = aX^b$, where X and Y are the biological measures, a is a constant and b is the scaling exponent.

Allopathic medicine: the common medical system of most industrialized countries; a system of medicine in which treatments are designed to produce physiological effects different from or opposite to those of the disease.

Alveoli: small sacs in the lung that contain air and are the site of gas exchange with the blood.

Alzheimer's disease: a progressively deteriorating form of **dementia** characterized by the presence of amyloid plaques and neurofibrillary tangles in the brain.

Ambient temperature: the temperature of the surrounding environment.

Amenorrhea: the absence of menses for an extended time.

Amino acids: the building blocks of **polypeptide chains**, which in turn compose **proteins**.

Amino acid score: a score reflecting the percent of the limiting **amino acid** in the dietary **protein** relative to the requirement level.

Androgen: a general term for an important class of **hormones** secreted by the **testes** of male mammals, of which **testosterone** is the most important; androgens are also secreted by the adrenal glands in both males and females.

Anorexia: the lack or loss of appetite for food.

Anovulatory: menstrual cycles in which **ovulation** does not occur.

Antagonistic pleiotropy: a theory that suggests that aging is due to **genes** which have **pleiotropic** effects such that the genes are beneficial early in life and detrimental later in life.

Anthropometry: the measurement of the human body using predefined anatomical points.

Antibody: a **protein** that can combine with a specific **antigen**.

Anticodon: a sequence of 3 bases in **tRNA** capable of bonding with the 3 bases of a **codon** in the **mRNA**.

Antigen: a substance that can produce an immune response.

APOB locus: a **gene** on **chromosome** 2 coding for **apolipoprotein** B, a primary constituent of low density lipoprotein.

APOE*4 allele:** one of the **alleles** for **apolipoprotein** E.

Apolipoprotein: the **protein** molecules in lipoproteins; lipoproteins carry lipids in the blood.

Apoptosis: the programmed cell death of a normally functioning cell.

Arterioles: small arteries.

Atherosclerosis: a chronic disease characterized by yellowish plaques of **cholesterol**, lipids and cellular debris on the walls of the arteries.

Atresia: degeneration, especially of female germ cells.

Attributable risk: the proportion of the **incidence** of a disease that can be attributed to a particular risk factor or cause.

Autoimmune disorders: disorders in which the body's own **antibodies** attack specific tissues or structures; e.g. multiple sclerosis, systemic lupus.

Autosomes: the **chromosomes** not involved in sex determination; chromosome pairs 1-22 in humans.

B cell: a white blood cell that produces **antibodies**.

β-cells: the cells in the pancreas that produce **insulin**.

Background risk: the risk for development of a disease in the absence of exposure to a risk factor.

Bacteria: microorganisms that lack a true **nucleus** and that usually have their **DNA** in a single molecule.

Basal metabolic rate (BMR): the minimum energy expenditure for maintenance of respiration, circulation, body temperature, and other vegetative functions.

Bases: the building blocks of **nucleic acids**; either **purines** or **pyrimidines**.

Bergmann's Rule: a biogeographic rule stating that in a widespread, warm-blooded species, populations in colder climates are generally larger than those in temperate and warm climates.

Bioelectrical impedance: a technique for determining the fat and lean composition of the body that involves measuring the passage of low energy electrical signals through the body.

Biomass: the total weight of living matter within an area.

BMI: body mass index.

Body mass index (BMI): a measure of relative weight calculated as weight (in kilograms) divided by height (in meters) squared.

Candidate gene method: a method for finding genes that relies on selecting **markers** that are at or near **loci** (candidate genes) whose biochemical effects might be expected to affect the quantitative **phenotype**.

Capillaries: the smallest blood vessels.

Cardiac ischemia: a decrease in blood flow to the heart muscle, usually as a result of deposits in the blood vessels.

Case: a person who has a disease being studied.

Case-control study: an **epidemiological** study design that defines two groups on the basis of disease status; the case group consists of people in the study population who have the disease of interest, the control group consists of people who do not have the disease.

Catch-up growth: either a period of faster growth following a period of growth disruption or growth improvement that occurs as a result of a longer period of growth.

CD 4 T cells: a subset of **lymphocytes** that have undergone a period of development in the thymus and contribute to the regulation of cell-mediated immunity.

Centers of ossification: points at which bone formation occurs.

Cerebrovascular accidents: strokes.

Chemokine receptor: a receptor on the surface of cells for **proteins** (chemokines) that attract immune system cells to injured or diseased tissue; these receptors also allow the HIV **virus** to enter the cell.

Childhood: stage in the life cycle from about ages 3-7 years; characterized by a moderate growth rate, dependency on older people for feeding, **midgrowth spurt**, eruption of first permanent molar and incisor, and cessation of brain growth by end of the stage.

Cholesterol: a fat that is important for the synthesis of many **hormones**.

Chromophores: chemical compounds that absorb light.

Chromosome: a large **DNA** molecule, wrapped around **protein**, which has **genes** arranged along its length.

Chronic degenerative diseases (CDDs): conditions that lead to progressive deterioration in one or more clinical, metabolic, or physiological traits that can be measured on a continuous scale.

Classical markers: non-molecular genetic markers, e.g., red cell **antigens** (blood groups), serum **proteins**, and **enzymes**.

Cline: a regular pattern of change in the mean value of a biological trait over geographic distance.

Coalescent: the time, in generations, to the most recent common ancestor for two or more **DNA** sequences.

Codominant: when neither allele is **dominant** or **recessive** to the other, both are expressed in the **phenotype** of the **heterozygote**.

Codon: a sequence of three **bases** that is "read" as part of the genetic code; most insert an **amino acid** into a **polypeptide chain**, some mark the end of the chain.

Coefficient of relationship: the expected proportion of **genes** that two individuals have in common, ranging from zero for unrelated individuals to one for monozygotic twins.

Coevolution: when evolutionary change in one species influences the evolution of other species.

Cohort: any designated group of people (or other organisms) who share characteristics of interest to a researcher.

Cohort study: an **epidemiological** study design that divides a study population into two or more **cohorts**; most commonly exposure to a risk factor is the basis for group formation.

Cold-induced vasodilation (CIVD): an increase in the diameter of blood vessels in response to cold; results in local tissue warming.

Communicable disease: see **infectious disease.**

Complement: a family of **proteins** that is activated by factors such as the combination of **antigen** and **antibody** and results in a variety of consequences, including cell destruction.

Complementary: bases in **nucleic acids** that pair together.

Completed fertility: total offspring ever born.

Complex segregation analysis: a statistical method for identifying Mendelian inheritance of **quantitative traits** from pedigree data, while allowing for polygenic inheritance.

Conduction: heat transfer by contact between matter of different temperatures.

Congenital: hereditary or inborn.

Consanguineous unions: matings between blood relatives.

Contagion: transmission of infection by direct contact between individuals, **droplet spread**, or contact with inanimate objects.

Control: a person who does not have a disease being studied.

Control region (of **mitochondrial DNA**): a non-coding region that evolves quite rapidly and thus is useful for population comparisons; also known as the D-loop.

Core temperature: internal temperature.

Corpus luteum: a **hormone** producing tissue that develops anew each cycle on the **ovary** from the remains of the cells that surrounded the **oocyte** before **ovulation.**

Correlation: a standardized measurement of association between two variables that ranges between negative one (a complete inverse relationship) to positive one (a "perfect" relationship); a zero correlation indicates no relationship between the variables.

Cortisol: a **hormone** produced by the adrenal gland that regulates energy balance; cortisol favors the mobilization of energy in a form that muscle and brain cells can use rapidly.

Covariance: the expected product of deviations around respective means for two variables.

Craniometric: involving the measurement of the skull.

Creatine phosphate: a high energy compound that can supply energy for **adenosine triphosphate**.

Cross-sectional: a research design in which each individual is examined at one point in time (see **longitudinal**).

Crude mortality rate: the total number of deaths divided by the total number of people at risk of dying (estimated by mid-year population).

Culture area: a geographical area in which people share similar ways of life, including cultural features such as food procurement strategies, sociopolitical organization, religion, etc.; groups within the same culture area do not necessarily speak the same language.

Cytogenetics: the study of **chromosomes**.

Cytokine: a **protein** that mediates cellular interactions and regulates cell growth and secretion.

Cytoplasm: the non-nuclear material within a cell.

Cytoskeleton proteins: proteins that support the cell.

Cytotoxic: destroying cells.

Dementia: an organic brain disorder, characterized by a decline in previously attained intellectual capacity, which interferes with social or occupational functioning. **Alzheimer's disease**, **Parkinson's disease**, and **vascular dementia** are examples.

Demic expansion: population growth characterized by an excess of births over deaths. By implication, expanding demes also occupy ever larger geographic areas.

Demographic transition: the historical transition from high **fertility** and **mortality** rates to low fertility and mortality rates.

Demography: the study of population statistics such as **mortality, fertility, migration**, and their relationship to population growth, family formation, and human ecology.

Dermis: the outermost vascular layer of the skin just below the **epidermis**.

Development: a progression of changes, either quantitative or qualitative, that lead from an undifferentiated or immature state to a highly organized, specialized, and mature state.

Developmental adaptation or developmental acclimatization: acclimatizations that occur during the period of growth.

Diaphysis: the shaft of the bone.

Diastolic blood pressure: the second number in a blood pressure measurement giving the pressure when the heart is between beats.

Dietary thermogenesis: see **thermic effect of food.**

Differential fertility: differences in **fertility** between mature organisms.

Differential mortality: differences in **mortality** between individual organisms prior to reproductive maturation.

Diploid: cells with two full sets (one maternal, one paternal) of **chromosomes**.

Directional selection: natural selection favoring one extreme of the distribution of trait values, resulting in a consistent increase (or decrease) in the mean value of the trait.

Disaccharide: a carbohydrate molecule formed by a combination of two simple sugars (**monosaccharides**), includes **lactose** (**glucose** + **galactose**).

Discrete characters: traits such as blood groups that have **phenotypes** in distinct categories.

Distance curve: a growth curve showing the amount of growth achieved from year to year.

DNA: deoxyribonucleic acid; the genetic material for most species.

Dominance: the ability of one form of a gene (an **allele**) to mask the presence of a variant form (see **recessive**).

Droplet transmission (also respiratory transmission): a mode of **infectious disease** transmission whereby infectious organisms are spread through droplets in the air when a person coughs, sneezes, or breathes.

Duffy blood group: one of the many blood groups that are determined by **antigens** on the surface of the red blood cells.

DYS: a notation for **markers** on the Y chromosome; numbers identify specific markers.

Effective population size: an estimate of the breeding size of an abstract population.

Elderly: persons over 65 years of age.

Encephalization: brain size in relation to body size; in general, primates are more encephalized (have larger brain-body weight ratios) than other mammals.

Endemic: a disease that is present in a population at a relatively constant (usually low) level at all times.

Endogamous: mating within the group.

Environment: all that which is external to an individual human **host** and a particular **agent** that causes disease.

Environmental Genome Project: a research initiative of the National Institutes for Environmental Health Science that aims to investigate the interrelationship of specific human genetic sequences and particular environmental factors.

Environmental variance: the **variance** in **quantitative traits** that is the result of the effects of the environment.

Enzymes: proteins that speed up chemical reactions.

Eosinophil: a type of white blood cell thought to be chiefly important in defense against **parasitic** infections.

Epidemic: a sudden and short-term increase above the expected or normal number of cases of a disease in a fairly localized area.

Epidemiological transition: the historical change in disease patterns such that the primary causes shift from **infectious diseases** to **chronic degenerative diseases**.

Epidemiologists: scientists who study **epidemiology**.

Epidemiology: a set of methods for determining the causes of disease from looking at who in the population is affected, where diseases occur in space and time, and the social, environmental, dietary, and lifestyle correlates of disease occurrence.

Epidermis: the outermost nonvascular layer of the skin.

Epiphysis: the growing end of the bone.

Estradiol: the main **estrogen** produced in females of reproductive age.

Estrogen: a general term for an important class of hormones of female mammals, produced primarily by the ovarian **follicle**. In humans, estrogen affects growth of the follicle and **oocyte** and development of the uterine lining during the first half of the **menstrual cycle**.

Eukaryote: organisms that have cells containing a **nucleus**.

Exogamous: mating outside the group.

Exon: a segment of a **gene** that is expressed, that codes for **protein**.

Expanding tissue: tissue such as the liver, kidney, and the endocrine glands, in which the cells retain their potential to undergo **mitosis** even in the differentiated state.

Exponential growth: growth that is compound, or greater than linear, like interest on a bank account.

F13A locus: a gene regulating subunit A of Factor XIII, the last **enzyme** in the cascade leading to blood clotting.

Fatty acid: a straight chain hydrocarbon with a carboxyl group at one end; with glycerol, makes up triglycerides.

Fecal-oral transmission: a mode of **infectious disease** transmission whereby infectious organisms are spread when **pathogens** are shed from the body in fecal material and then introduced into another person's mouth through contact with the infected fecal material.

Fecundity: the capacity to have offspring; contrasts with **fertility**, which refers to an individual's actual number of offspring.

Fertility: the production of offspring.

Fertilization: the process whereby the sperm penetrates the egg cell (**oocyte**) and the male's and female's genetic material combine.

Fitness: the reproductive success of a **genotype**.

Fixation: when an **allele** has a frequency of 1.0.

Fledge: to rear an offspring until it is prepared to live without **parental investment**.

Folivory: consumption of leaves and structural plant parts.

Follicle: the cells that enclose the female germ cell; as a particular germ cell proceeds toward **ovulation**, follicular cells multiply, support the maturation of the **oocyte** they enclose, and produce **hormones**, notably **estrogen**.

Follicle stimulating hormone (FSH): a **hormone** produced by the **pituitary** that travels to the **ovaries** or **testes**, where it stimulates the production and release of **estrogen** or **androgen** hormones.

Follicular phase: the first portion, approximately half, of the **menstrual cycle**. So named because emphasis at this time is on **follicular** development leading to **ovulation**.

Food-borne illness: an illness caused by eating contaminated food.

Free radicals: highly reactive molecules with unpaired electrons.

Frostbite: freezing of body tissue.

Frugivory: consumption of fruit and reproductive plant parts.

Galactose: a simple sugar (**monosaccharide**) that is one of the constituents of **lactose**.

Gametes: sex cells: sperm or eggs.

Gene: a sequence of **DNA** that either codes for a **protein** or for an **RNA** molecule that functions directly (e.g. **tRNA** genes).

Gene flow: genetic exchange between populations.

Gene pool: all of the **genes** carried by members of a population.

Genetic distance: a statistical measure of the degree of similarity or dissimilarity between the **allele** frequencies in two populations.

Genetic drift: random fluctuations in **allele** frequencies between generations, most pronounced in small populations.

Genetic kinship: the probability that two **alleles** sampled from the same **locus** in two individuals will be **identical by descent**.

Genome: the totality of **genes** found within a cell.

Genotype: the combination of alleles that one possesses for a particular trait (see **phenotype**).

Germ theory of disease: a theory proposed by Pasteur that disease was a consequence of the introduction of microorganisms into the body; the predominant theory in modern Western medicine about causes of **infectious disease**.

Gerontology: a branch of knowledge dealing with aging and problems of the aged.

Glucose: a simple sugar (**monosaccaride**) that is the body's primary metabolic fuel.

Glycogen: the **polysaccharide** used for storing carbohydrates in animal tissues.

Glycolysis: the chemical pathway for the breakdown of **glucose** to provide energy for the cell.

Gompertz coefficient: the slope of the line of the natural log of **mortality** rate against age; indicates the rate of acceleration of mortality with age.

Gompertz equation: a model of age-related **mortality** rates that assumes an exponential increase in mortality rates during adult ages.

Gonadarche: the reactivation of production and secretion of gonadal **hormones** at the start of **puberty**.

Gonadotrophin-releasing hormone: a **hormone** produced by the **hypothalamus** that causes the **pituitary** to release **follicle stimulating hormone** and **luteinizing hormone.**

Gross reproduction rate: an estimate of the average number of daughters born to a woman surviving to the end of her reproductive career (i.e. about 50 years).

Growth: a quantitative increase in size or mass.

Growth faltering: a reduction in the growth rate of a child relative to expected growth potential.

Growth plate region: the region between the **epiphysis** and **diaphysis** where elongation in bone length occurs.

Habituation: the gradual reduction of response to repeated stimulation or the perception of stimulation.

Haplogroup: a collection of related **haplotypes**.

Haploid: cells with one set of **chromosomes**; half the **diploid** state.

Haplotype: a set of **alleles** for two or more closely **linked loci** on a **chromosome**.

Hardy-Weinberg equilibrium: the relationship between **allele** and **genotype** frequencies in an equilibrium population.

Heat load: the amount of heat in the body.

Heat resistance: the ability to keep deep body temperature below dangerous levels when exposed to heat.

Heat tolerance: a greater capacity to function normally even when body **core temperatures** are nearing the upper limits.

Height for age: height in comparison to the heights of other individuals of the same age.

Helminths: worms.

Hematocrit: the percentage volume of blood occupied by red blood cells.

Heritability: in the broad sense (H^2), the proportion of **phenotypic variance** due to genetic (as opposed to environmental) effects; in the narrow sense (h^2), the proportion of phenotypic variance due to **additive genetic variance**.

Heterozygous: the condition of possessing two different **alleles** for some trait (see **homozygous**).

HLA system: human leukocyte antigen system; a **hypervariable** genetic system of closely **linked loci** on **chromosome** 6 that determines **antigens** on the surface of many body cells and is involved in immune response.

Homeostasis: the tendency of an organism to maintain stability of the internal environment in the normal physiological range.

Hominids: living humans and our fossil ancestors that lived after the last common ancestor between humans and apes; includes species of the genera *Australopithecus* and *Homo*.

Hominoids: living humans and apes and our fossil ancestors.

Homologous: chromosomes that have **alleles** for the same traits.

Homology: similarity that results from the inheritance of a feature from a common ancestor.

Homozygous: the condition of possessing two identical **alleles** for some trait (see **heterozygous**).

Hormone: a chemical produced by one type of cell that affects other cells.

Host: a person or other living organism who provides shelter, sustenance, etc. for a **parasitic** organism.

Huntington's disease: a degenerative neurological disease caused by an **autosomal dominant gene**; symptoms of Huntington's disease do not generally occur until about age 40.

Hyperendemic: a disease that is constantly present in high numbers in a population.

Hyperplasia: growth by an increase in cell number by **mitosis**.

Hypertension: systolic blood pressures of at least 140 mm of mercury or **diastolic pressures** of at least 90 mm.

Hypertrophy: growth by an enlargement in size of already existing cells.

Hypervariable: a **locus** that has a number of **alleles** at significant frequencies; no single allele is ubiquitous, thus there is a high degree of **polymorphism**.

Hypervariable regions (of **mitochondrial DNA**): segments of the mitochondrial **control region** that have exceptionally high substitution rates.

Hypothalamus: a region of the brain that has important roles in regulating the nervous system, **hormone** production, and body temperature.

Hypothermia: low body temperature.

Hypoxia: reduced oxygen in ambient air and reduced physiologically available oxygen, compared to sea level.

Hypoxic ventilatory response: an increase in depth and rate of breathing in response to **hypoxia**.

Identical by descent: two alleles that are the same because they were inherited from a common ancestor.

IFN-γ: interferon gamma; a **protein** that is important in the body's response to **viral** and **parasitic** infections.

IgE: a class of serum immunoglobulins that mediates allergies and is important in the immune response to **helminths**.

IgG: the major serum immunoglobulin.

Immunity: incapable of being infected by a particular **pathogen**; can be temporary or permanent.

Immunology: the study of the immune system.

Implantation: the process whereby the fertilized egg cell becomes embedded in the uterine lining, allowing it to be nourished by materials carried in the mother's bloodstream.

***In vitro* fertilization:** term generally used to describe a number of related techniques whereby an **oocyte** is fertilized by a sperm outside the body and then returned to the uterus for gestation. *In vitro* means literally "in glass."

Incidence: the number of *new* cases of a disease during a particular time period.

Inclusive fitness: an individual's own reproductive success stripped of all those components that are due to the influence of its relatives plus that proportion of each relative's reproductive success that was due to the influence of the individual in question once it has been devalued by the **coefficient of relationship** between them.

Independent assortment: Mendel's Second Law; states that which member of a pair of **chromosomes** (maternal or paternal) enters a **gamete** does not affect which member of another pair enters the same gamete.

Infancy: the stage of the life cycle from the second month of life to end of lactation, usually by 36 months of age; characterized by rapid growth velocity, deciduous tooth eruption, and many developmental milestones.

Infectious diseases: all diseases that are caused by specific infectious agents or their toxic products; these agents or toxins are transmitted from one person to another, either directly or indirectly.

Infinite allele model: a **mutation** model that assumes each mutation creates a new **allele**, or an allele that does not currently exist in the population.

Infrared radiation: light with wavelengths over 750 nanometers.

Insulin: a **hormone** produced by the pancreas that regulates energy balance; insulin favors the transport of energy into cells where it is converted to fat and stored for later use.

Insulin dependent diabetes: a form of diabetes that usually develops in childhood and requires injections of **insulin**.

Intergenic DNA: the **DNA** separating known **genes.**

Interleukin: a **protein** that acts as a growth and differentiation factor for the cells of the immune system.

Intrinsic rate of increase: the annual rate of increase of a stable population.

Intron: an untranslated section within a **gene** (see **exon**).

Isolation by distance: a regular decrease in genetic similarity between populations as the geographic distance between them increases.

Isometry: an **allometric** or scaling relationship between two biological measures, Y and X, such that Y increases as a constant multiple or fraction of X.

Jumping PCR: an amplification error that occurs when the extending primer "jumps" to another DNA sequence template during the **polymerase chain reaction** and polymerization continues, creating the equivalent of a **recombination** product.

Juvenile: the stage of the life cycle from about ages 7-10 years for girls, 7-12 years for boys; characterized by the slowest postnatal growth rate, capable of self-feeding, and cognitive transition leading to learning of economic and social skills.

Keratinocytes: epidermal skin cells.

Kilocalorie: a measure of energy; the amount of energy it takes to raise the temperature of 1 kilogram of water 1°C.

Kleiber's law (Kleiber relationship): the **allometric** relationship between **basal metabolic rate** (BMR) and body weight across mammalian species of different size. BMR scales to the ¾ power of body weight and is estimated by the equation: $BMR(kcal/d) = 70 \times Wt(kg)^{0.750}$.

Lactase: the intestinal **enzyme** that breaks **lactose** into its component sugars, **glucose** and **galactose**.

Lactase persistence: the maintenance of high levels of **lactase** throughout the life cycle.

Lactose: the sugar (**disaccaride**) found in milk, composed of the simple sugars **glucose** and **galactose**.

Lactose intolerance: symptoms such as abdominal bloating, pain and flatulence and diarrhea that occur in some **lactase** deficient individuals when they consume **lactose**.

Language family: a group of languages that descend from the same ancestral language.

Latent period: the length of time between infection and the ability to infect someone else.

Late-onset lactase deficiency: the normal human condition in which levels of **lactase** are not maintained after about five years of age.

LH surge: the increase in **luteinizing hormone** that leads to **ovulation.**

Liability: a continuous risk factor, that may have both genetic and environmental determinants; in the **threshold model** this unobserved liability is discretized by one or more thresholds.

Life expectancy at birth: a measure of the average **life span** of a **cohort**.

Life history: changes through which an organism passes in its development from its primary stage of life to its natural death; especially the strategy an organism uses to allocate its energy toward growth, maintenance, reproduction, raising offspring to independence, and avoiding death.

Life span: the duration of existence of an individual; the average length of life of a particular type of organism in a particular environment under specified circumstances.

Life table: an actuarial table based on **age-specific mortality** statistics that follows an entire **cohort** from birth to death.

Likelihood: a number proportional to the probability of obtaining the observed data conditional on **parameter** values.

Linkage: the state in which two **loci** are located so close together on a **chromosome** that they are inherited as a unit.

Linkage disequilibrium: the association of specific **alleles** for two different traits at frequencies other than random association or **segregation** would produce.

Linked: loci located on the same **chromosome**.

Locus: a location on a **chromosome** that is occupied by a particular **gene** (pl. loci).

LOD score: the base 10 logarithm of the odds of obtaining the observed data at an estimated recombination value with a **marker** locus versus at a recombination value of 0.5 (i.e., **unlinked**).

Logistic growth: a pattern of population growth that is produced when birth rates decline and/or **mortality** rates increase with increasing population size.

Longevity: a long duration of individual life.

Longitudinal: a research design in which the same individuals are examined over a period of time (see **cross-sectional**).

Low birth weight: a weight of less than 2500 grams for a full term birth.

Luteal phase: the second portion, approximately half, of the **menstrual cycle.** So named because emphasis at this time is on **progesterone** production by the **corpus luteum.**

Luteinizing hormone (LH): a **hormone** produced by the **pituitary** that affects sperm and egg production and that triggers **ovulation** in females.

Lymphocyte: a type of white blood cells that is part of the immune system.

Lysosomes: organelles in the cell that contain **enzymes** that can digest unwanted substances, including **bacteria**.

Lysozyme: an **enzyme** in tears, saliva, and **neutrophils** that attacks **bacteria**.

Macroevolution: evolution over long periods of time to produce new species, genera, families, etc.

Macroparasites: multi-celled disease-causing organisms.

Macrophage: a large **phagocytic** cell that is important in immune response.

Marker: a variable genetic trait that can be traced across generations and populations.

Maturity: the state of reaching functional capacity in biological, behavioral, and cognitive abilities.

Maximum life span potential (MLP): the longest potential life span of any one member of a species.

Maximum likelihood: the point in the **parameter** space at which the data have the highest probability.

Maximum oxygen uptake: the amount of oxygen the body can extract during exercise; a measure of the capacity for **aerobic** exercise (also called **aerobic power, physical work capacity,** or **VO₂max**).

Maximum reproductive potential (MRP): the point in life at which an organism is sufficiently mature to not only bare /sire offspring, but also best able to rear and **fledge** any offspring produced with maximum efficiency.

Mean pairwise distance: a measure of dissimilarity based on the average number of **nucleotide** differences in a collection of **DNA** sequences examined pairwise.

Megajoule: a measure of energy; one megajoule $\approx$ 240 **kilocalories**.

Meiosis: the process by which **haploid gametes** are produced.

Melanin: the major pigment responsible for human skin color.

Melanocytes: cells in the basal layer of the **epidermis** that synthesize **melanin**.

Melanosomes: small organelles inside **melanocytes** where **melanin** is synthesized.

Menarche: the first menstrual period.

Menopause: the cessation of the monthly **menstrual cycle** subsequent to the loss of **ovarian** function.

Menstrual cycle: regular cycle in human females (and some other primates) whereby, under the influence of **hormones**, the uterine lining develops, and an **oocyte** matures and is released; if pregnancy does not occur, the uterine lining is shed in menstruation and the cycle repeats.

Microevolution: changes in **allele** frequencies within a species.

Micronutrients: vitamins and minerals.

Microparasites: single-celled disease-causing organisms, including pathogenic species of **bacteria, viruses,** rickettsiae, prions, **protozoa**, and fungi.

Microsatellite DNA: a class of **satellite DNA** in which a short (~2-6 **bases**) DNA sequence is repeated a number of times in tandem.

Midgrowth spurt: a modest acceleration in growth velocity at about 6–8 years.

Migration: a permanent or semi-permanent change in location to a completely new geographical region.

Minimal erythemal dose: the smallest dose of **ultraviolet radiation** necessary to produce sunburn.

Minisatellite DNA: a class of **satellite DNA** in which a sequence of DNA, usually 10-100 **bases** long, is repeated many times in tandem.

Mismatch distribution: the distribution, generally displayed as a histogram, that describes the variation in the number of **base** pair differences in a stretch of **DNA** for a collection of pairs of individuals.

Mitochondria: organelles within the cell where energy-producing reactions take place.

Mitochondrial DNA (mtDNA)**:** the small, circular, maternally inherited **DNA** found in the **mitochondria**.

Mitosis: the process of cell division producing cells with the same **chromosome** number as the parental cell.

Model life tables: synthetic **life tables** characterizing the general human patterns of **age-specific mortality** that are often used to estimate **mortality** in populations with biased or missing mortality statistics.

Monosaccharide: a simple sugar; includes the 6 carbon sugars **glucose**, **galactose**, and fructose.

Morbidity: illness.

Mortality: death.

mRNA: messenger **RNA**; the RNA molecule that is transcribed from the **DNA** and in **eukaryotes** carries genetic information from the **nucleus** to the **ribosomes**.

Mutation: an alteration in the structure of a **DNA** sequence, can be as small as the addition, loss or substitution of one **base** for another.

Myocardial infarction: heart attack.

Natural selection: differential reproduction and survival of different **genotypes**.

Neonate: an infant from birth to 28 days of life.

Neuron: nerve cell.

Neutrophil: a type of white blood cell.

NK (natural killer) cells: lymphocytes that are important in immunity to **viruses** and are capable of killing certain tumor cells.

Noninfectious disease: diseases with an environmental cause, most kinds of cancer, genetic diseases, nutritional diseases, allergies, etc.

Non-insulin dependent diabetes mellitus (NIDDM)**:** adult-onset or Type 2 diabetes, a form of diabetes characterized by higher than normal levels of blood **glucose** and excess **insulin**; insulin injections are not required for treatment.

Norepinephrine: a **hormone** secreted by the adrenal gland that is important in the **sympathetic nervous system**.

Normal distribution: a symmetric bell shaped curve which is characterized by its location (mean) and spread (**variance**).

Nucleic acid: genetic materials, **RNA** and **DNA**, composed of repeating units of sugar and phosphate with **bases** attached to the sugar-phosphate backbone.

Nucleotide: a unit of **nucleic acid** composed of a sugar, phosphate and one nitrogen **base**.

Nucleus: a structure, bounded by a membrane and found in **eukaryotes**, that contains a cell's **chromosomes**.

Odds ratio: the ratio of the odds that a **case** is exposed to the risk factor to the odds that a **control** is exposed to the risk factor; used to assess the strength of an association between a particular risk factor and the development of a disease.

Oocyte: the female germ cell when it is at the stage of maturation at which **ovulation** and **fertilization** occur.

Osteoarthritis: a common form of arthritis due to breakdown of joint tissue resulting in pain, inflammation, and stiffness in joints.

Osteomalacia: a decalcification and softening of existing bone as a result of vitamin D deficiency in adulthood.

Osteoporosis: reduction in the amount of bone.

Ovary: the female gonad where **oocytes** are produced.

Overdominant selection: balancing selection; loss of both **alleles** in a two allele system, due to lowered **fitness** of both **homozygotes**, is balanced by the higher fitness of the **heterozygote**.

Ovulation: the release of the female germ cell (**oocyte**) from the **follicle**, on the **ovary**, into the oviduct; in humans, this occurs at approximately the midpoint of the **menstrual cycle**.

Paleodemography: the **demography** of past populations, usually based on skeletal samples.

Panmictic: random mating.

Parameter: a value that is fixed in the real world, but not necessarily known to a researcher; parameters are the controllable elements of a mathematical model; their values are estimated from data and these values are then used to predict the unknowns, or variables, in the model.

Parasite: an organism that depends upon the **host** for its own nourishment and survival to the detriment of the host.

Parental investment: the cost of producing, caring for, and **fledgling** offspring.

Parkinson's disease: a neurodegenerative disease resulting from a deficiency of dopamine-producing neurons in the basal ganglia.

Pathogen: an organism capable of causing disease.

Pathogenesis: the origin and development of a disease.

Pelvic inlet: the bony opening of the birth canal.

Percent skin reflectance: the percent of incident light reflected back from the skin; a measure of the amount of melanin in the epidermis.

Percentile: a statistical measure indicating the percentage of a population that has measurements below a particular value.

Performance effort: the proportion of maximal **aerobic power** expended in the performance of any given task.

Phagocytosis: the ingestion of material by cells.

Phenotype: the visible or measurable appearance of an organism for a trait (see **genotype**).

Photochemical reaction: a reaction initiated by electromagnetic light waves or particles.

Photons: particles of radiant energy.

Phylogenetic: dealing with the evolutionary relationships among organisms.

Physical activity level (PAL): the **total daily energy expenditure** as a multiple of the **basal metabolic rate**; reflects the proportion of energy *above* basal requirements that an individual spends over the course of a normal day.

Physical work capacity: the amount of oxygen the body can extract during exercise; a measure of the capacity for **aerobic** exercise (also called **aerobic power, maximum oxygen uptake**, or VO_2max).

Pituitary gland: a tissue at the base of the brain that produces a number of **hormones**, including **prolactin, follicle stimulating hormone** and **luteinizing hormone**.

Plasma volume: the volume of the fluid portion of the blood.

Plasticity: the ability of an organism to modify its biology or behavior to respond to changes in the environment, particularly when these are stressful.

Pleiotropy: the effect of one **gene** on different **phenotypes**.

Polymerase: a series of **enzymes** that catalyze the bonding of **nucleotides**.

Polymerase chain reaction (PCR): a laboratory procedure in which specific sequences of **DNA** can be amplified into many identical copies.

Polymorphism: a condition where at least two **alleles** exist at a **locus** and both are present at a frequency of greater than 1%.

Polypeptide chain: a string of **amino acids** bonded together with peptide bonds, the basic unit of **protein**.

Polysaccharide: a carbohydrate molecule composed of three or more simple sugars (**monosaccarides**).

Postpartum amenorrhea: absence of menses for a variable period of time following birth of a child.

Prematurity: birth prior to 37 weeks gestation.

Prevalence: the total number of cases of a disease in a given population during a particular time interval.

Principal component analysis: a statistical technique that reduces N variables to a maximum of N-1 components by extracting linear combinations of the original variables. Each component, therefore, accounts for some proportion of the variation inherent in the original suite of variables. Principal components are uncorrelated with each other.

Progesterone: a **hormone** produced by the **corpus luteum** that is particularly important during the second half of the **menstrual cycle** in humans, when it increases blood supply to, and secretory activity of, the uterus and sustains pregnancy if it occurs.

Prokaryotes: organisms that lack a **nucleus**.

Prolactin: a **hormone** produced by the **pituitary gland** that is concerned with milk production by the mammary glands.

Proportionate mortality ratio: the proportion of total deaths that are due to a specific cause.

Protein: a functional molecule typically composed of one or more **polypeptide chains** folded into a specific three dimensional structure. May contain materials other than just the **amino acids**.

Protein digestibility: the proportion of nitrogen from dietary **protein** that is available for use by the body.

Protozoa: single-celled organisms that lack a true cell wall; more animal-like than plant-like.

Pseudogene: a non-functional **DNA** sequence that bears a resemblance to a functional **gene**.

Puberty: the event of development that marks the onset of sexual maturation.

Purine: a class of **bases** with a double carbon-nitrogen ring structure, adenine and guanine.

Pyrimidine: a class of **bases** with a single carbon-nitrogen ring structure, cytosine, thymine and uracil.

Quantitative trait: traits such as height and weight that show continuous variation.

Quantitative trait loci: genetic **loci** that contribute to a **quantitative trait,** and may be identified by **complex segregation analysis** and **linkage.**

Radiation: heat transfer by **infrared** waves.

Rate: the proportion of events that occur during a specified time interval.

Ratio: one number divided by another; expresses the frequency of some characteristic relative to some other characteristic.

Recessive: a form of a trait, or the underlying **allele**, that can be masked by a dominant alternative (see **dominance**).

Recombination: the process of forming new associations of **genes** at different **loci** following **chromosomal** crossing-over.

Recovered: a person who has recovered from an **infectious disease**, cannot transmit the disease to another person and is not at risk for infection; one of the basic disease stages in epidemic models.

Recumbent length: the length of the body when lying down.

Refractory: unable to become infected with a disease.

Regression analysis: a statistical method used to examine and describe a linear relationship between two variables; how well the individual data points fit the regression line is determined by the coefficient of **correlation**.

Relative risk: the ratio of the calculated risk in the exposed group to the calculated risk in the unexposed group; used to assess the strength of an association between a particular risk factor and the development of a disease.

Renewing tissue: tissues including blood cells, **gametes,** and the **epidermis** in which mature cells are incapable of **mitosis**; new cells are produced from a reserve of undifferentiated cells.

Residuals: values indicating the degree to which observed measurements differ from the values predicted from a **regression analysis**.

Residual volume (RV): the volume of air remaining in the lungs after a complete expiration.

Resting metabolic rate: see **basal metabolic rate.**

Restriction enzyme: enzyme that cuts **DNA** at a specific, short sequence.

Restriction mapping: determining the location in the **genome** where individual **restriction enzymes** cut **DNA**. Determination of these sites for large numbers of restriction enzymes is referred to as high density restriction mapping.

RFLP: restriction fragment length **polymorphisms**; fragments of different lengths produced when **DNA** is cut by **restriction enzymes.**

Rh blood group: one of the many blood groups determined by **antigens** on the surface of the red blood cells.

Ribosome: the organelle in the cell where **mRNA** is **translated** into a **polypeptide chain**.

Rickets: inadequate calcification of bone as a result of vitamin D deficiency during growth and development.

RNA: ribonucleic acid; a **nucleic acid** very similar in composition to **DNA** except for the sugar in the backbone and the substitution of uracil for thymine (see also **mRNA** and **tRNA**).

RSP: restriction site polymorphism; **DNA** sequence variation identified by **restriction enzymes**.

SaO_2: oxygen saturation; the percent of arterial hemoglobin that is saturated with (carrying) oxygen.

Satellite DNA: a class of **eukaryotic DNA** that consists of many copies of short, repeated sequences.

Secondary sexual characteristics: characteristics such as the external genitalia, and sexual dimorphism in body size and composition.

Secular trend: changes over time, especially used to refer to changes in physical growth over the last several centuries.

Segregation: Mendel's first law; states that members of a pair of **chromosome** will enter different **gametes** during **meiosis**.

Segregation analysis: a method of genetic epidemiology in which family data on the prevalence of a particular **phenotypic** trait are analyzed statistically to determine whether the trait is inherited genetically, and if so, to determine its likely mode of inheritance.

Semi-conservative replication: the process in which one double strand of **DNA** serves as the basis for making two double strands; each of the new double strands contains one old and one newly synthesized strand.

Seminiferous tubules: the tissue within the **testes** where sperm production takes place.

Senescence: the process of becoming old; the phase from full maturity to death characterized by an accumulation of metabolic products, decline in function, and decreased probability of reproduction and survival.

Set point: the temperature threshold at which control cells are stimulated to act.

Sex chromosomes: the X and Y chromosomes that determine sex.

Sexual dimorphism: a difference in appearance or behavior between males and females.

Shivering: involuntary muscle contraction for the purpose of producing heat rather than work.

Skeletal age and **skeletal maturation:** the degree to which the skeleton has attained its adult form, measured by criteria including the degree of fusion of **epiphyses** and **diaphyses**.

Skinfold: the thickness of a double fold of skin and **subcutaneous fat**.

Socioeconomic status (SES): a measure of aspects of education, occupation, and social prestige of a person or social group.

Spatial autocorrelation: the pattern of correlation between **allele** frequencies across geographic distances.

Speciation: the evolution of new species.

Spermatids: sperm cells that are not completely mature.

Standard deviation: a measure of the spread around the mean calculated as the square root of the **variance**.

Static tissue: tissues such as nerve cells and striated muscle that lose the ability to divide by **mitosis** early in their lives.

Stationary population theory: a theory of population structure that assumes that the age structure is stable and that the population is not growing or declining.

Stochastic: random, determined by probability.

STR: short tandem repeat, see **microsatellite**.

Stunted: small in terms of **height for age**, but not necessarily excessively thin.

Subcutaneous fat: fat just below the skin.

Subscapular skinfold: the **skinfold** measured below the shoulder blade.

Surface area: the total area of the skin.

Surveillance: the regular and ongoing collection of data on the occurrence and spread of a disease, especially as needed for developing and applying effective control measures.

Survey: an investigation in which information is systematically collected.

Survivorship curve: a graph illustrating what percent of a **cohort** born in the same year survives to each subsequent year of life.

Susceptibility: the degree to which a person's physical state places them at risk for infection by a **pathogen**.

Susceptible: a person at risk for a disease; an uninfected person; one of the basic disease stages in epidemic models.

Sympathetic (adrenal medullary) nervous system: the part of the nervous system responsible for physiological arousal and stress response.

Systolic blood pressure: the first number in a blood pressure measurement giving the blood pressure at the point when the heart is contracting.

T cell: a white blood cell that is important in immune response either by directly killing cells or by secreting substances that regulate other cells in the immune system.

Telomere: the end of a **chromosome,** contains sequences that stabilize the ends.

Template: a **nucleic acid** sequence that serves as the basis for construction of a **complementary** strand.

Testes: the male gonads where sperm and **androgen hormones** are produced.

Testosterone: an important **hormone** in male mammals; testosterone is produced by the **testes** and is necessary for sperm production, libido, and the maintenance of muscle mass, and may have a role in modulating some aspects of aggressive or competitive behavior.

Thalassemia: a genetic disease in which hemoglobin chains are not produced or are produced in a reduced amount.

Thermic effect of food: the elevation of energy metabolism associated with the energy cost of digestion, absorption, transport, metabolism and storage of ingested food.

Thermoregulation: the regulation of body temperature.

Threshold model: a model for qualitative traits in which there is a continuously distributed liability that is cut into categories by one or more thresholds.

Thrifty genotype hypothesis: a model proposed by James Neel in 1962 to explain high rates of adult-onset diabetes; he suggested that diabetes was the result of a genetic adaptation being "rendered detrimental" by dietary and lifestyle changes.

Total daily energy expenditure (TDEE): the total amount of energy used by an individual in a day.

Total fertility rate: an estimate of the average number of children born to a woman surviving to the end of her reproductive career (i.e. about 50 years).

Transcription: the process of converting a **DNA** sequence into an **RNA** sequence.

Transition: a **nucleotide** substitution of one **purine** (A or G) for another, or one **pyrimidine** (C or T) for another.

Translation: the process of converting a **mRNA** sequence into an **amino acid** sequence.

Tribe: a group of nominally independent communities that occupy a specific region and share a common language and culture.

Tricarboxylic cycle: a metabolic pathway that completes the energy extraction from glucose that began with **glycolysis**.

Triceps skinfold: the **skinfold** measured on the back of the upper arm.

Trimesters: the three periods into which pregnancy is divided.

tRNA: transfer **RNA**; form of RNA that transports **amino acids** to their position in a **polypeptide chain**.

Typology: the description of clusters of measurements that presumably may be identified in an individual, e.g., the "Nordic" type.

Ultraviolet (UV) **radiation:** light with wavelengths of less than 400 nanometers.

Ultraviolet A (UV-A) **radiation:** light in the 315-400 nanometers range.

Ultraviolet B (UV-B) **radiation:** light in the 280-315 nanometers range.

Underwater (hydrostatic) **weighing:** a technique for determining the fat and lean composition of the body that involves determining body density based on the difference between weight measured in air versus that measured while submerged under water.

Unlinked: loci located on different **chromosomes.**

Variance: a statistic calculated as the average of squared deviations around the average value.

Vascular dementia: dementia of cardiovascular origin; severe **hypertension** is one of the most frequent causes.

Vasoconstriction: narrowing of blood vessels.

Vasodilation: widening of blood vessels.

Vector: a living organism, usually an arthropod, that carries **pathogens** from one person to another; the pathogen may complete part of its life cycle within the vector; common vectors include mosquitoes, lice, ticks, fleas, and flies.

Velocity curve: a growth curve representing the rate of growth.

Virulent: highly infectious.

Virus: a type of non-living obligate intracellular **parasite** with a simple structure consisting of a **protein** coat surrounding a molecule of either **DNA** or **RNA**, but not both, with little other internal structure; not capable of independent metabolic activities.

Visible radiation: light in wavelengths of 400-750 nanometers.

Vital capacity (VC): the maximum amount of air that can be expired after a maximal inspiration.

Vital statistics: systematically tabulated information on the number of vital events in a population, including births, deaths, marriages, divorces and separations; based on official registration of these events.

Vitamin D winter: time of the year when there is not enough **ultraviolet radiation** for the skin to produce vitamin D.

VO$_2$max: the amount of oxygen the body can extract during exercise; a measure of the capacity for **aerobic** exercise (also called **aerobic power**, **physical work capacity**, or **maximal oxygen uptake**).

Wasted: excessively thin.

Weight for height: weight in comparison to the weights of other individuals of the same height.

Xenobiotic: foreign; in the case of organ transplants, from another species.

Z score: a score indicating how many **standard deviations** a measurement is from the value in the reference population.

Zoonosis: an animal disease that can be contracted by humans.

Index